Hyperspectral Imaging for Food Quality Analysis and Control

Hyperspectral Imaging for Food Quality Analysis and Control

Edited by

Professor Da-Wen Sun

Director, Food Refrigeration and Computerized Food Technology,
National University of Ireland, Dublin (University College Dublin),
Agriculture & Food Science Centre

ELSEVIER

AMSTERDAM • BOSTON • HEIDELBERG • LONDON • NEW YORK • OXFORD
PARIS • SAN DIEGO • SAN FRANCISCO • SINGAPORE • SYDNEY • TOKYO

Academic press is an imprint of Elsevier

Academic Press is an imprint of Elsevier
32 Jamestown Road, London NW1 7BY, UK
30 Corporate Drive, Suite 400, Burlington, MA 01803, USA
525 B Street, Suite 1900, San Diego, CA 92101-4495, USA

First edition 2010

British Library Cataloguing-in-Publication Data
A catalogue record for this book is available from the British Library

Library of Congress Cataloging-in-Publication Data
A catalog record for this book is available from the Library of Congress

ISBN : 978-0-12-374753-2

For information on all Academic Press publications visit our website at elsevierdirect.com

Typeset by TNQ Books and Journals Pvt Ltd.
www.tnq.co.in

Printed and bound in United States of America

10 11 12 13 14 15 10 9 8 7 6 5 4 3 2 1

Contents

About the Editor

Born in Southern China, Professor Da-Wen Sun is a world authority in food engineering research and education, he is a Member of Royal Irish Academy which is the highest academic honour in Ireland. His main research activities include cooling, drying, and refrigeration processes and systems, quality and safety of food products, bioprocess simulation and optimisation, and computer vision technology. Especially, his innovative studies on vacuum cooling of cooked meats, pizza quality inspection by computer vision, and edible films for shelf-life extension of fruit and vegetables have been widely reported in national and international media. Results of his work have been published in more than 200 peer-reviewed journal papers and over 200 conference papers.

He received a first class BSc Honours and MSc in Mechanical Engineering, and a PhD in Chemical Engineering in China before working in various universities in Europe. He became the first Chinese national to be permanently employed in an Irish University when he was appointed College Lecturer at National University of Ireland, Dublin (University College Dublin) in 1995, and was then continuously promoted in the shortest possible time to Senior Lecturer, Associate Professor and Full Professor. Dr Sun is now Professor of Food and Biosystems Engineering and Director of the Food Refrigeration and Computerised Food Technology Research Group at University College Dublin.

As a leading educator in food engineering, Professor Sun has significantly contributed to the field of food engineering. He has trained many PhD students, who have made their own contributions to the industry and academia. He has also given lectures on advances in food engineering on a regular basis in academic institutions internationally and delivered keynote speeches at international conferences. As a recognized authority in food engineering, he has been conferred adjunct/visiting/consulting professorships from ten top universities in China, including Zhejiang University, Shanghai Jiaotong University, Harbin Institute of Technology, China Agricultural University, South China University of Technology, and Jiangnan University. In recognition of his significant contribution to food engineering worldwide and for his outstanding leadership in the field, the International Commission of Agricultural Engineering (CIGR) awarded him the CIGR Merit Award in 2000 and again in 2006, the Institution of Mechanical Engineers (IMechE) based in the UK named him "Food Engineer of the Year 2004", and in 2008 he was awarded CIGR Recognition Award in honour of his distinguished achievements as one of the top one percent of agricultural engineering scientists in the world.

He is a Fellow of the Institution of Agricultural Engineers and a Fellow of Engineers Ireland (the Institution of Engineers of Ireland). He has also received numerous awards for teaching and research excellence, including the President's Research Fellowship, and has twice received the President's Research Award of University College Dublin. He is a Member of CIGR Executive Board and Honorary Vice-President of CIGR, Editor-in-Chief of *Food and Bioprocess Technology – an International Journal* (Springer), Series Editor of the "Contemporary Food Engineering" book series (CRC Press/Taylor & Francis), former Editor of *Journal of Food Engineering* (Elsevier), and Editorial Board Member for *Journal of Food Engineering* (Elsevier), *Journal of Food Process Engineering* (Blackwell), *Sensing and Instrumentation for Food Quality and Safety* (Springer), and *Czech Journal of Food Sciences*. He is also a Chartered Engineer.

Contributors

Jacopo Aguzzi
Institut de Ciències del Mar (ICM-CSIC), Barcelona, Spain

Diwan P. Ariana
Michigan State University, Department of Biosystems and Agricultural Engineering, East Lansing, Michigan, USA

José Blasco
Instituto Valenciano de Investigaciones Agrarias (IVIA), Centro de Agroingeniería, Moncada (Valencia), Spain

Kuanglin Chao
US Department of Agriculture, Agricultural Research Service, Henry A. Wallace Beltsville Agricultural Research Center, Environmental Microbial and Food Safety Laboratory, Beltsville, Maryland, USA

Corrado Costa
CRA-ING Agricultural Engineering Research Unit of the Agriculture Research Council, Monterotondo (Rome), Italy

Gamal ElMasry
University College Dublin, Agriculture and Food Science Centre, Belfield, Dublin, Ireland; Agricultural Engineering Department, Suez Canal University, Ismailia, Egypt

Juan Gómez-Sanchis
Intelligent Data Analysis Laboratory (IDAL), Electronic Engineering Department, Universidad de Valencia, Burjassot (Valencia), Spain

Aoife A. Gowen
Biosystems Engineering, School of Agriculture, Food Science and Veterinary Medicine, University College Dublin, Belfield, Dublin, Ireland

Digvir S. Jayas
Biosystems Engineering, University of Manitoba, Winnipeg, Manitoba, Canada

Lu Jiang
Bio-imaging and Machine Vision Lab, The Fischell Department of Bioengineering, University of Maryland, USA

David Lewis
Radiance Technologies, Stennis Space Center, Mississippi, USA

Li Liu
Department of Bioresource Engineering, McGill University, Macdonald Campus, Quebec, Canada

Renfu Lu
USDA ARS Sugarbeet and Bean Research Unit, Michigan State University, East Lansing, Michigan, USA

Paolo Menesatti
CRA-ING Agricultural Engineering Research Unit of the Agriculture Research Council, Monterotondo (Rome), Italy

Enrique Moltó
Instituto Valenciano de Investigaciones Agrarias (IVIA), Centro de Agroingeniería, Moncada (Valencia), Spain

Michael O. Ngadi
Department of Bioresource Engineering, McGill University, Macdonald Campus, Quebec, Canada

Colm P. O'Donnell
Biosystems Engineering, School of Agriculture, Food Science and Veterinary Medicine, University College Dublin, Belfield, Dublin, Ireland

Jitendra Paliwal
Biosystems Engineering, University of Manitoba, Winnipeg, Manitoba, Canada

Gerrit Polder
Wageningen UR, Biometris, Wageningen, The Netherlands

Jianwei Qin
US Department of Agriculture, Agricultural Research Service, Henry A. Wallace Beltsville Agricultural Research Center, Beltsville, Maryland, USA

Chandra B. Singh
Biosystems Engineering, University of Manitoba, Winnipeg, Manitoba, Canada

Junichi Sugiyama
National Food Research Institute, Tsukuba, Ibaraki, Japan

Da-Wen Sun
University College Dublin, Agriculture and Food Science Centre, Belfield, Dublin, Ireland

Masoud Taghizadeh
Biosystems Engineering, School of Agriculture, Food Science and Veterinary Medicine, University College Dublin, Belfield, Dublin, Ireland

Yang Tao
Bio-imaging and Machine Vision Lab, The Fischell Department of Bioengineering, University of Maryland, USA

Mizuki Tsuta
National Food Research Institute, Tsukuba, Ibaraki, Japan

Gerie van der Heijden
Wageningen UR, Biometris, Wageningen, The Netherlands

Ning Wang
Department of Biosystems and Agricultural Engineering, Oklahoma State University, Stilwater, Oklahoma, USA

Haibo Yao
Mississippi State University, Stennis Space Center, Mississippi, USA

Bin Zhu
Bio-imaging and Machine Vision Lab, The Fischell Department of Bioengineering, University of Maryland, USA

Preface

Based on the integration of image processing and spectroscopy techniques, hyperspectral imaging is a novel technology for obtaining both spatial and spectral information from an object. In recent years, hyperspectral imaging has rapidly emerged as and matured into one of the most powerful and fastest-growing non-destructive tools for food quality analysis and control. Using the hyperspectral imaging technique, the spectrum associated with each pixel in a food image can be used as a fingerprint to characterize the biochemical composition of the pixel, thus enabling the visualization of the constituents of the food sample at pixel level. As a result, hyperspectral imagery provides the potential for more accurate and detailed information extraction than is possible with any other type of technology for the food industry.

In order to reflect the rapid developing trend of the technology, it is timely to publish *Hyperspectral Imaging for Food Quality Analysis and Control.* The book is divided into two parts. Part 1 deals with principles and instruments, including theory, image data treatment techniques, and hyperspectral imaging instruments. Part 2 covers its applications in quality analysis and control for various foods and agricultural products.

As the first book in the subject area, *Hyperspectral Imaging for Food Quality Analysis and Control* is written by the most active peers in this field, with both academic and professional credentials, highlighting the truly international nature of the work. The book is intended to provide the engineer and technologist working in research, development, and operations in the food industry with critical and readily accessible information on the art and science of the hyperspectral imaging technology. The book should also serve as an essential reference source to undergraduate and postgraduate students and researchers in universities and research institutions.

PART 1

Fundamentals

CHAPTER 1

Principles of Hyperspectral Imaging Technology

Gamal ElMasry [1,2], Da-Wen Sun [1]

[1] *University College Dublin, Agriculture and Food Science Centre, Belfield, Dublin, Ireland*

[2] *Agricultural Engineering Department, Suez Canal University, Ismailia, Egypt*

CONTENTS

1.1. INTRODUCTION

During the past few decades a number of different techniques have been explored as possible instrumental methods for quality evaluation of food products. In recent years, *hyperspectral imaging* technique has been regarded as a smart and promising analytical tool for analyses conducted in research, control, and industries. Hyperspectral imaging is a technique that generates a spatial map of spectral variation, making it a useful tool in many applications. The use of hyperspectral imaging for both automatic target detection and recognizing its analytical composition is relatively new and is an amazing area of research. The main impetus for developing a hyperspectral imaging system was to integrate spectroscopic and imaging techniques to enable direct identification of different components and their spatial distribution in the tested sample. A hyperspectral imaging system produces a two-dimensional spatial array of vectors which represents the spectrum at each pixel location. The resulting three-dimensional dataset containing the two spatial dimensions and one spectral dimension is known as the *datacube* or hypercube (Chen *et al.*, 2002; Kim *et al.*, 2002; Mehl *et al.*, 2004; Schweizer & Moura, 2001). The advantages of hyperspectral imaging over the traditional methods include minimal sample preparation, nondestructive nature, fast acquisition times, and visualizing spatial distribution of numerous chemical compositions simultaneously. The hyperspectral imaging technique is currently tackling many challenges to be accepted as the most preferable analytical tool in identifying compositional fingerprints of food products and their authentication. The need for fast and reliable methods of authenticity and object identification has increased the interest in the application of hyperspectral imaging for quality control in the agricultural, pharmaceutical, and food industries. Moreover, enhancement in instrumental developments, the availability of high-speed computers, and the development of appropriate chemometric procedures will allow this technique to be dominant in the future. This chapter presents the fundamentals, characteristics, configuration, terminologies, merits and demerits, limits and potential of hyperspectral imaging. Basics and theoretical aspects relating to this technique, the information that can be supplied, and the main features of the instrumentation are presented and briefly discussed. The final part of the chapter concerns a general overview of the main steps involved in analyzing hyperspectral images. The potential applications of hyperspectral imaging in food analysis will be explained in more detail in the relevant chapters of this book

1.1.1. The Necessity for Automating Quality Assessment

With increased expectations for food products with high quality and safety, the need for accurate, fast, and objective quality determination of these characteristics continues to grow. Quality assurance is one of the most important goals of any industry. The ability to manufacture high-quality products consistently is the basis for success in the highly competitive food industry. It encourages loyalty in customers and results in an expanding market share. The quality assurance methods used in the food industry have traditionally involved human visual inspection. Such methods are tedious, laborious, time-consuming, and inconsistent. As plant throughput increased and quality tolerance tightened, it became necessary to employ automatic methods for quality assurance and quality control (Gunasekaran, 1996). Also, the increased awareness and sophistication of consumers have created the expectation for improved quality food products. Consumers are always demanding superior quality of food products, i.e., higher quality for an individual food item, consistency of products in a batch, and enhanced food safety as a whole (Nagata *et al.*, 2005). This in turn has increased the need for enhanced quality monitoring. In general, automation of a quality assessment operation not only optimizes quality assurance but more importantly it also helps in removing human subjectivity and inconsistency. Moreover, automation usually increases the productivity and changes the character of the work, making it less arduous and more attractive. Considering the fact that the productivity of a person working in a mechanized and automated environments is approximately ten times that of a manual worker, this has stimulated progress in the development of many novel sensors and instruments for the food industry, often by technology transfer from other industrial sectors, including medical, electronic, and nonclinical sectors (Abdullah *et al.*, 2004). If quality evaluation is achieved automatically, production speed and efficiency can be improved drastically in addition to increased evaluation accuracy, with an accompanying reduction in production costs.

1.2. RELATIONSHIP BETWEEN SPECTROSCOPY, IMAGING, AND HYPERSPECTRAL IMAGING

In the past two decades, considerable progress has been accomplished in the development of new sensing technologies for quality and safety inspection of agricultural and food products. These new sensing technologies have provided us with unprecedented capabilities to measure, inspect, sort, and grade food products effectively and efficiently. Consequently, some smart methods to evaluate quality and quality-related attributes have been developed using advanced techniques and instrumentation. Most recently, the emphasis has been on developing sensors for real-time, nondestructive systems. As a result, automated visual inspection by computer-based systems has been developed in the food industry to replace the traditional inspection by human inspectors because of its cost-effectiveness, consistency, superior speed, and accuracy. *Computer vision* technology utilizing image processing routines is one alternative which became an integral part of the industry's move towards automation. Combined with an illumination system, a computer vision system is typically based on a personal computer in connection with electrical and mechanical devices to replace human manipulative effort in the performance of a given process (Du & Sun, 2006). Image processing and image analysis are the core of computer vision, involving mathematics, computer science and software programming. This system has a great ability in evaluation cycle to apply the principle: several objects per second instead of several seconds per object.

Unfortunately, the computer vision system has some drawbacks that make it unsuitable for certain industrial applications. It is inefficient in the case of objects of similar colours, inefficient in the case of complex classifications, unable to predict quality attributes (e.g. chemical composition), and it is inefficient for detecting invisible defects. Since machine vision is operated at visible wavelengths, it can only produce an image registering the external view of the object and not its internal view. Situations exist whereby food technologists need to look inside the object in a noninvasive and nondestructive manner. For instance, food technologists need to measure and map the water content of food in order to assess its microbiological stability and to implement risk analysis as defined by the hazard analysis critical control point (HACCP) (Abdullah *et al.*, 2004). Therefore, external attributes such as size, shape, colour, surface texture, and external defects can easily be evaluated by ordinary means (e.g. RGB colour camera). However, internal structures are difficult to detect with relatively simple and traditional imaging means, which cannot provide enough information for detecting internal attributes (Du & Sun, 2004).

Since quality is not a single attribute but comprises many properties or characteristics (Abbott, 1999; Noh & Lu, 2005), measurement of the *optical properties* of food products has been one of the most successful nondestructive techniques for quality assessment to provide several quality details simultaneously. Optical properties are based on reflectance, transmittance, absorbance, or scatter of polychromatic or monochromatic radiation in the ultraviolet (UV), visible (VIS), and near-infrared (NIR) regions of the electromagnetic spectrum which can be measured by spectral instruments. A quality index for the product can be based on the correlation between the spectral response and a specific quality attribute of the product, usually a chemical constituent (Park *et al.*, 2002). Diffusely reflected light contains information about the absorbers near the surface of a material. Recently, optical techniques using near-infrared spectroscopy (NIRS) have received considerable attention as a means for nondestructive sensing of food quality. NIRS is rapid, nondestructive, and relatively easy to implement for on-line and off-line applications. More importantly, NIRS has the potential for simultaneously measuring multiple quality attributes. In these spectroscopic techniques, it is possible to obtain information about the sample components based on the light absorption of the sample, but it is not easy to know the position/location information. On the other hand, it is easy to know the position of certain features by naked eye or computer vision systems, but it is not easy to conduct the quantitative analysis of a component. The combination of the strong and weak points of visible/near-infrared spectroscopic techniques and vision techniques is the *hyperspectral imaging* technique, which is also called *imaging spectroscopy* or *imaging spectrometry*, even though the meaning is different (spectrometry—"measuring", spectroscopy—"seeing", hyperspectral—"many bands"). Because hyperspectral imaging techniques overcome the limits of spectroscopic techniques and vision techniques, they have emerged as a powerful technique in agricultural and food systems. Based on hyperspectral imaging techniques, multispectral imaging system can be built for real-time implementations (Lee *et al.*, 2005).

While a grayscale image typically reflects the light intensity over the electromagnetic spectrum in a single band, a colour image reflects the intensity over the red, green, and blue bands of the spectrum. Increasing the number of bands can greatly increase the amount of information from an image. Hyperspectral images commonly contain information from several bands with different resolution values. Hyperspectral imaging has been invented to integrate spectroscopic and spatial (imaging) information which otherwise cannot be achieved with either conventional imaging or spectroscopic techniques. It involves measuring the intensity of diffusely reflected light from a surface at one or more wavelengths with relatively narrow band-passes. Hyperspectral imaging goes beyond conventional imaging and spectroscopy to acquire both spectral and spatial information from an object simultaneously. *Imaging* technique is essentially the science of acquiring spatial and temporal data information from objects using a digital camera, whereas *spectroscopy* is the science of acquiring and explaining the spectral characteristics of an object to describe light intensities emerging from its molecules at different wavelengths and thus provide a precise fingerprint of that object. Since image data are considered two-dimensional, by adding a new dimension of "spectrum" information, the hyperspectral image data can be perceived as a three-dimensional datacube (Chao *et al.*, 2001). Hyperspectral imaging, like other spectroscopy techniques, can be carried out in reflectance, transmission or fluorescence modes. While the majority of published research on hyperspectral imaging has been performed in reflectance mode, transmission and emission modes have also been investigated. In brief, the main differences and advantages of hyperspectral imaging over conventional imaging and spectroscopic techniques are outlined in Table 1.1.

Table 1.1 Main differences among imaging, spectroscopy, and hyperspectral imaging techniques

Features	Imaging	Spectroscopy	Hyperspectral imaging
Spatial information	√	×	√
Spectral information	×	√	√
Multi-constituent information	×	√	√
Building chemical images	×	×	√
Flexibility of spectral information extraction	×	×	√

1.2.1. Advantages of Hyperspectral Imaging

The rich information content and outstanding feature identification capabilities of hyperspectral imaging make it highly suitable for numerous applications. However, the technology also has some demerits that need to be considered before its implementation in food quality assessment regimes. These are covered in the following section. The foremost advantages of using hyperspectral imaging technology in food analysis can be summarized in the following points:

- No sample preparation is required.
- It is a chemical-free assessment method, which enables safety and environmental protection by thoroughly eliminating pollutant solvents, chemicals and/or potentially dangerous reagents during analyses.
- Once the calibration model is built and validated, it becomes an extremely simple and expeditious analysis method.
- It is a noninvasive, and nondestructive method, so that the same sample could be used for other purposes and analyses.
- It is eventually economic compared with traditional methods, owing to the savings in labor, time, and reagent cost in addition to the large saving in the cost of waste treatments.
- Rather than collecting a single spectrum at one spot on a sample, as in spectroscopy, hyperspectral imaging records a spectral volume that contains a complete spectrum for every spot (pixels) in the sample.
- It has the flexibility in choosing any region of interest (ROI) in the image even after image acquisition. Also, when an object or a ROI in the object presents very obvious spectral characteristics, that region could be selected and its spectrum is saved in a spectral library.
- Due to its high spectral resolution, hyperspectral imaging provides both qualitative and quantitative measurements.
- It is able to determine several constituents simultaneously in the same sample.
- One of the strategic advantages of hyperspectral imaging is that it allows for the visualization of different biochemical constituents presented in a sample based on their spectral signatures because regions of similar spectral properties should have similar chemical composition. This

process is called building *chemical images*, or chemical mapping, for constructing detailed maps of the surface composition of foods which traditionally requires use of intense laboratory methods. This approach will be explained in more detail in Chapter 6.

The greater spectral information residing in the spectral images allows many different objects to be detected and distinguished even if they have similar colors, morphological features or overlapping spectra.

The spatial distribution and concentration of the chemical composition in the product can be obtained, not just the bulk composition.

Its ability to build chemical images permits labeling of different entities in a sample simultaneously and quantitative analysis of each entity. Therefore, it enables documentation of the chemical composition of the product. Such documentation allows different pricing and labeling to be used in sorting food products with different chemical compositions according to market requirements, consumer preference, and/or product specifications.

If the high dimensionality of hyperspectral imaging were reduced to form multispectral imaging by choosing some optimal wavelengths for certain classifications, the technology would be incomparable for process monitoring and real-time inspection.

1.2.2. Disadvantages and Constraints of Hyperspectral Imaging

In spite of the aforementioned advantages, hyperspectral imaging does have some disadvantages, which can be summarized as follows:

Hyperspectral images contain a substantial amount of data, including much redundant information, and pose considerable computational challenges.

It takes a long time for image acquisition and analysis, therefore hyperspectral imaging technology has to a very limited extent been directly implemented in on-line systems for automated quality evaluation purposes.

From an analyst's point of view, one of the main analytical drawbacks of hyperspectral imaging technique is that it is an indirect method, which means that it needs standardized calibration and model transfer procedures.

Similar to all spectroscopic techniques, spectral data extracted from any location of the image contain a series of successive overlapping bands, which are difficult to assign to specific chemical groups.

One major factor that limits its industrial applications for food inspection is the hardware speed needed for rapid image acquisition and analysis of the huge amount of data collected.

Hyperspectral data suffer from the well-known problem of multicollinearity; although some multivariate analysis techniques like principal component regression (PCR) and partial least square (PLS) are often employed to overcome this problem. However, the effects of multicollinearity in data can only be reduced but not completely removed by PCR and PLS. In this aspect, variable selection is advantageous in the sense that not only can it improve the predictive power of the calibration model, but also it can simplify the model by avoiding repetition of information or redundancies and irrelevant variables.

Hyperspectral imaging is not suitable in some cases, such as liquids or homogenous samples, because the value of imaging lies in the ability to resolve spatial heterogeneities in samples. Imaging a liquid or even a suspension has limited use as constant sample motion serves to average spatial information, unless ultra-fast recording techniques are employed as in fluorescence correlation microspectroscopy or fluorescence lifetime imaging microscopy (FLIM) observations where a single molecule may be monitored at extremely high detection speed. Similarly, there is no benefit in imaging a truly homogeneous sample, as a single point spectrometer will generate the same spectral information. Of course the definition of homogeneity is dependent on the spatial resolution of the imaging system employed.

To identify and detect different objects unambiguously in the same image, these objects must exhibit characteristic absorption features. Furthermore, if an object has diagnostic absorption features, it must be present at a minimum concentration or converge in a pixel to be detected.

Depending on the spatial resolution and the structure of the sample investigated, spectra from individual image pixels may not represent a pure spectrum of one singular material, but a mixed spectrum consisting of spectral responses of the various materials that cover the region of interest (ROI) selected from the sample.

In a hyperspectral imaging system it is time-consuming to acquire the spectral and spatial information of the entire sample, and therefore it is not practical to implement such a system on-line as it is. However, by means of analyzing the hyperspectral imaging data, it is possible to select a few effective and suitable wavebands for building a multispectral imaging system to meet the speed requirement of production lines (Xing *et al.*, 2006). The problem caused by the huge amount of data generated in hyperspectral imaging can be overcome by using data reduction schemes in such a way that only those wavelengths and spatial locations of special interest are selected. In this way, the amount of data can be effectively reduced, which will benefit later data processing. Therefore the hyperspectral imaging experiment is usually conducted off-line in the laboratory to select some optimal wavelengths for later multispectral imaging measurements suitable for on-line applications (Chao *et al.*, 2002; Mehl *et al.*, 2004). Once the optimal inspection bands are identified, an automatic inspection system using only these bands can be designed and then industrially implemented. Such a method has been increasingly used with computers becoming faster and more powerful, and it has now entered a new era of industrial applications for on-line evaluation of food and agricultural products. Nowadays, a significant number of scientific articles are published annually on hyperspectral and multispectral imaging for various applications. Moreover, several manufacturers specialized in spectral systems have emerged in the market to sell not only the spectral components but also the whole hyperspectral imaging units.

1.3. FUNDAMENTALS OF HYPERSPECTRAL IMAGING

In order to use the hyperspectral imaging technology, a good understanding of the theory behind the technique is required. Therefore, some basic information about spectroscopy will be provided in this section. The electromagnetic spectrum and the nature of light and its properties are also described to allow the reader to gain knowledge about the importance of light in hyperspectral imaging. Furthermore, definitions of basic terms, such as wavelength, waveband, frequency, spectral signature, and spectrum, are briefly given. Detailed descriptions can be found in many optics and physics textbooks (e.g. Hecht, 2002).

1.3.1. Basics of Spectroscopy

The root of *spectrometric technique* dates back to 1665, when Sir Isaac Newton described the concept of dispersion of light and the optomechanical hardware of a spectrometer after he passed light through a *prism* and observed the splitting of light into colors. In particular, visible and near-infrared *spectroscopy* is an established technique for determining chemical constituents in food products. These instruments use gratings to separate the individual frequencies of the radiation leaving the sample. The development of an NIR spectrometric technique for assessing quality traits in food products relies on the collection of spectra of the produce and developing a calibration equation to relate this spectral data to the quality trait ascertained using a standard laboratory method. In NIR quantitative analysis, this is typically called a calibration equation. The difference between failing and succeeding in this task is greatly dependent on the quality of the reference values associated with the samples in the calibration set. Nevertheless, once this learning stage is concluded, the final result is perhaps close to the result of an ideal analytical method (Pieris *et al.*, 1999).

Basically, spectroscopic methods provide detailed fingerprints of the biological sample to be analysed using physical characteristics of the interaction between electromagnetic radiation and the sample material, such as reflectance, transmittance, absorbance, phosphorescence, fluorescence, and radioactive decay. *Spectroscopic analysis* exploits the interaction of electromagnetic radiation with atoms and molecules to provide qualitative and quantitative chemical and physical information contained within the wavelength spectrum that is either absorbed or emitted. Among these spectroscopic techniques, NIR spectroscopy is one of the most successful within the food industry. The absorption bands seen in this spectral range arise from overtones and combination bands of O–H, N–H, C–H, and S–H stretching and bending vibrations that enable qualitative and quantitative assessment of chemical and physical features. Therefore, NIR could be applied to all organic compounds rich in O–H bonds (such as moisture, carbohydrate and fat), C–H bonds (such as organic compounds and petroleum derivatives), and N–H bonds (such as proteins and amino acids). In a given wavelength range, some frequencies will be absorbed, others (that do not match any of the energy differences between vibration response energy levels for that molecule) will not be absorbed, while some will be partially absorbed. This complex relation between the intensity of absorption and wavelength constitutes the absorption spectra of a substance or sample (Pasquini, 2003). Since all biological substances contain thousands of C–H, O–H, and N–H molecular bonds, the exposure of a sample to NIR radiation results in a complex *spectrum* that contains qualitative and quantitative information about the physical and chemical computational changes of that sample.

Indeed, modern NIR spectroscopy technique requires a low-noise spectrometer, computerized control of the spectrometer and data acquisition, and the use of multivariate mathematical and statistical computer algorithms to analyse the data. The bonds of organic molecules change their vibration response energy when irradiated by NIR frequencies and exhibit absorption peaks through the spectrum. Thus, qualitative and quantitative chemical and physical information is contained within the wavelength spectrum of absorbed energy (Carlomagno *et al.*, 2004). However, NIR spectroscopic techniques rely on measuring only the aggregate amount of light reflected or transmitted from a specific area of a sample (point measurement where the sensor is located), and do not give information on the spatial distribution of light in the sample. Besides, when the samples are presented to the spectrometers, their homogeneity is an important issue, since a traditional spectrometer integrates the spatial information present, e.g. in a cuvette. This fact does not influence the measurements when the sample is in the liquid or gaseous phase, but in the case of a solid sample (like all agro-food products), this means losing a great deal of information since there are many cases in which the mapping of some characteristic property spectrally identifiable is of the utmost importance. This greatly limits the ability of NIR spectroscopy to quantify structurally related properties and spatial-related distribution. The logical solution would be the use of hyperspectral imaging, but such a technique imposes major technological challenges both from the hardware and software point of view that should be carefully evaluated before starting any research project.

1.3.2. Importance of Light in Hyperspectral Imaging

In modern physics, the discipline of studying light and interaction of light with matter is called optics. Yet while light enables us to see, we cannot see light itself. In fact, what we see depends fundamentally on the properties of light as well as the physical and physiological processes of our interpretation of the scenes. By the end of the nineteenth century, it seemed that the question of the nature of light had been conclusively settled. *Light* is a nonmaterial wave composed of oscillating electric and magnetic fields and, being nonmaterial, the wave can travel through a vacuum without the aid of a material substance (medium). Through the development of quantum theory during the twentieth century, it has been proved by several investigations that under certain circumstances light behaves as a wave, while under different circumstances it behaves as a stream of massless particles. Thus, light has a dual nature. It displays a wave nature in some experiments and particle-like behavior in others. Therefore, it was also neatly and precisely assumed that light consists of a stream of particles, called *photons*, that travel at the speed of light and carry an amount of energy proportional to the light frequency. Depending on the circumstances, when light behaves as a wave it is characterized by a speed, wavelength, and frequency; when considered as particles, each particle has an energy related to the frequency of the wave, given by the following *Planck's relation*:

$$E = hf \tag{1.1}$$

where E is the energy of the photon, h is Planck's constant (6.626×10^{-34} J.s), and f is the frequency. When light interacts with a single atom and molecule, its behavior depends on the amount of energy per quantum it carries.

During the nineteenth century there was an explosive increase in our understanding of the properties of light and its behaviors. Wave interference and polarization were discovered, and the speed of light was measured in different media. Instruments using prism and diffraction gratings gave rise to analysis of light spectra from various sources and the field of spectroscopy was born. These spectra became the key to understanding the structure of the atom and discovering numerous characteristics of molecules. In hyperspectral imaging, *light* plays a crucial role in the system in order to see clearer, farther, and deeper and to gain detailed information about different objects under investigation. A hyperspectral imaging system can capture light from frequencies beyond the visible light range. This can allow extraction of additional information that the human eye fails to capture.

1.3.3. Electromagnetic Spectrum

Electromagnetic radiation is a unique phenomenon that takes the form of self-propagating waves in a vacuum or in matter. It consists of electric and magnetic field components that oscillate in phase perpendicular to each other and perpendicular to the direction of energy propagation. The *electromagnetic spectrum*, as shown in Figure 1.1, consists of several categories (or regions), including gamma rays, X-rays, ultraviolet radiation (UV), visible light (VIS), infrared radiation (IR)—divided into near-infrared (NIR), mid-infrared (MIR), and far-infrared (FIR) regions—microwaves and radio waves (FM and AM). Each region corresponds to a specific kind of atomic or molecular transition corresponding to different energies. It is important to indicate that wavelength increases to the right and the frequency increases to the left. These categories are classified in the order of increasing wavelength and decreasing frequency. It has been convenient to divide the spectrum into these categories, even though the division is arbitrary and the categories sometimes overlap. The small region of frequencies with an extremely small range of wavelengths between 400 and 700 nm is sensed by the eyes of

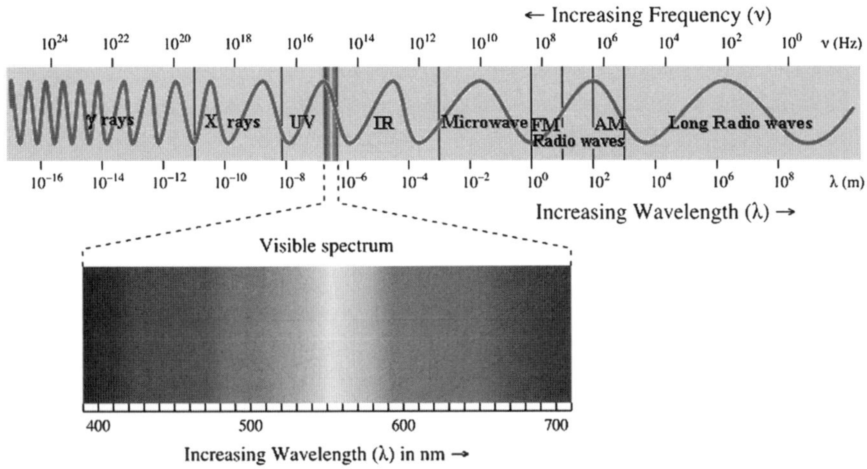

FIGURE 1.1 *Electromagnetic spectrum with visible spectrum (light) magnified. (Full color version available on http://www.elsevierdirect.com/companions/9780123747532/)*

humans and various organisms and is what we call the *visible spectrum*, or *light*.

Light waves are electromagnetic and thus consist of an oscillating electric field perpendicular to and in phase with an oscillating magnetic field. As with all types of waves, the frequency f of an electromagnetic wave is determined by the frequency of the source. The speed of light in a vacuum is defined to be exactly $c = 299{,}792{,}458\ \mathrm{m\,s^{-1}}$ (about 186,282.397 miles per second) which is usually rounded to $3.0 \times 10^8\ \mathrm{m\,s^{-1}}$. In general, an electromagnetic wave consists of successive troughs and crests, and the distance between two adjacent crests or troughs is called the *wavelength*. Waves of the electromagnetic spectrum vary in size, from very long radio waves like the size of a building to very short gamma rays smaller than atom nuclei. *Frequency* (f) is inversely proportional to wavelength (λ), according to the equation of the speed of the wave (υ) which is equal to c in a vacuum:

$$\upsilon = f\lambda \tag{1.2}$$

As waves cross boundaries between different media, their speeds change but their frequencies remain constant. All forms of waves, such as sound waves, water waves, and waves on a string, involve vibrations that need some material to support the wave or media to be conveyed. In the case of electromagnetic waves travelling through empty space, however, no material is needed to support the wave.

1.3.4. Interaction of Light with the Sample

The rationale for the development of a hyperspectral imaging system as a tool for nondestructive food analysis is based on the physical understanding of the interaction of light photons with the molecular structure of food samples. Indeed, studying the subject of interaction of light with biological materials and food samples is of paramount importance in identifying molecules based on their intrinsic properties in order to find their functions, to monitor interactions between different molecules, to detect morphological changes within biological materials, and to correlate changes that occur in the samples with the relevant physiological disorders or disease. In fact, all materials, including food samples, continuously emit and absorb energy by lowering or raising their molecular energy levels. The strength and wavelengths of emission and absorption depend on the nature of the material. Basically, when an electromagnetic wave (from an illumination unit) strikes the surface of a sample, the wave may be partly or totally reflected, and any nonreflected part will penetrate into the material. If a wave passes through a material without any attenuation, the material is called transparent. A material with partial attenuation is known as semitransparent, and a material through which none of the incoming radiation penetrates is called opaque. Most gases are rather transparent to radiation, while most solids (like raw food samples) tend to be strong absorbers for most wavelengths, making them opaque over a distance of a few nanometres to a few micrometres.

Visible light reflected, emitted or transmitted from a product carries information used by inspectors and consumers to judge several aspects of its quality. However, human vision is limited to a small region of the spectrum (as shown in Figure 1.1), and some quality features respond to wavelengths in regions outside the visible spectrum. The characteristics of the radiation that leaves the surface of the product depend on the properties of the product and the incident radiation. When radiation from the lighting system illuminates an object, it is transmitted through, reflected or absorbed. These phenomena are referred to as optical properties. Thus, determining such *optical* characteristics of an agricultural product can provide information related to quality factors of the product. When a sample is exposed to light, some of the incident light is reflected at the outer surface, causing specular reflectance (mirror-like reflectance), and the remaining incident energy is transmitted through the surface into the cellular structure of the sample where it is scattered by the small interfaces within the tissue or absorbed by cellular constituents (Birth, 1976). This is called diffuse reflection, where incoming light is reflected in a broad range of directions. Even when a surface exhibits only specular reflection with no diffuse reflection, not all of the light is necessarily reflected. Some of the light may be absorbed by the materials. Additionally, depending on the type of material behind the surface, some of the light may be transmitted through the surface. For opaque objects such as most food products, there is no transmission. The detected energy is converted by the spectrometers into *spectra*. These spectra are sensitive to the physical and chemical states of individual constituents. The high spectral signal-to-noise ratio obtained from modern instruments means that even constituents present in quite low concentrations can be detected (Gao *et al.*, 2003).

Most light energy penetrates only a very short distance and exits near the point of entry; this is the basis for color. However some penetrates deeper into the tissues and is altered by differential absorbance of various wavelengths before exiting and therefore contains useful chemometric information. Such light may be called diffuse reflectance, body reflectance, diffuse transmittance, body transmittance or interactance (Abbott, 1999). Meanwhile, the interactions of constituents within product cells alter the characteristic absorbance wavelength and cause many overlapping absorbances (Park *et al.*, 2002). In an attempt to determine the light penetration depth in fruit tissue for each wavelength in the range from 500 to 1900 nm, Lammertyn *et al.* (2000) found that the penetration depth in apple fruit is wavelength-dependent: up to 4 mm in the 700–900 nm range and between 2 and 3 mm in the 900–1900 nm range. In addition, the absorbed light can also be re-emitted (fluorescence), usually at longer wavelengths. A number of compounds emit fluorescence in the VIS region of the spectrum when excited with UV radiation; these compounds are called *fluorophores*. Fluorophores are a functional group in a molecule that will absorb energy of a specific wavelength and re-emit energy at a different, specific wavelength. The amount of the emitted energy and the wavelength at which the energy emits depend on both the fluorophore and the chemical environment of the fluorophore. The optical properties and fluorescence emission from the object are integrated functions of the angle and wavelength of the incident light and chemical and physical composition of the object (Chen *et al.*, 2002). Fluorescence refers to the phenomenon that light of short wavelengths is being absorbed by molecules in the sample tissue with subsequent emission of longer wavelength light. The fluorescence technique has been used for investigating biological materials, detecting environmental, chemical, and biological stresses in plants, and monitoring food quality and safety (Noh & Lu, 2005).

On the other hand, absorption and scattering are two basic phenomena as light interacts with biological materials. Light absorption is related to certain chemical constituents in agro-food samples, such as sugar, acid, water, etc. Modern reflectance NIR spectrometers measure an aggregate amount of light reflected from a sample, from which light absorption may be estimated and then related to certain chemical constituents. However, scattering is a physical phenomenon that is dependent on the density, cell structures, and cellular matrices of fruit tissue. NIR does not provide quantitative information on light scattering in the sample (Lu, 2004; Peng & Lu, 2005). If both absorption and scattering are to be measured, more significant information about the chemical and physical/mechanical properties of food products could be gained (Lu, 2003a).

1.3.5. Terminology

In dealing with a hyperspectral imaging system, some familiarity with technical information, essential expressions, and definitions will be useful. In this section, basic terminologies normally used in hyperspectral imaging will be highlighted and differentiation among them will be discussed.

1.3.5.1. Spectral range

The *spectral range* describes the wavelength regions covered by the hyperspectral imaging system. Spectral imaging instruments could cover either the ultraviolet, visible, near-infrared or infrared wavelengths based on the required application. Hyperspectral imaging system in the visible and very near-infrared range 380–800 nm or 400–1000 nm is the most widely used in food analysis applications. Nowadays, hyperspectral imaging systems in the range 900–1700 nm that provide the accuracy required in today's most challenging applications in food analysis are available. Moreover, some hyperspectral imaging systems that cover the shortwave-infrared (SWIR) region (900–2500 nm) are currently produced by many manufacturers to serve as significant tools in numerous applications in food and agricultural analyses, chemical imaging, and process analytical technologies.

1.3.5.2. Spectral resolution

The *spectral resolution* of the hyperspectral imaging system is related to its spectrograph as a measure of its power to resolve features in the electromagnetic spectrum. Spectral resolution is defined as the absolute limit of the ability of a hyperspectral imaging system to separate two adjacent monochromatic spectral features emitted by a point in the image. Spectral resolution is a measure of the narrowest spectral feature that can be resolved by a hyperspectral imaging system. The magnitude of spectral resolution is determined by the wavelength dispersion of the spectrograph and the sizes of the entrance and exit apertures. The goal of any spectral imaging system should be to accurately reconstruct the true spectral profile of an emitting light from all points in the tested sample.

1.3.5.3. Spatial resolution

The *spatial resolution* of the hyperspectral imaging system determines the size of the smallest object that can be seen on the surface of the specimen by the sensor as a distinct object separate from its surroundings. *Spatial resolution* also determines the ability of a system to record details of the objects under study. Higher spatial resolution means more image detail explained. In other words, spatial resolution is defined as the area in the scene that is represented by one image pixel. For practical purposes the clarity of the image is decided by its spatial resolution, not the number of pixels in an image. The parameter most commonly used to describe spatial resolution is the *field of view* (FOV). In effect, spatial resolution refers to the number of pixels per unit length. The spatial resolution is determined by the pixel size of the two-dimensional camera and the objective lens as the spectrograph is designed with a unity magnification.

1.3.5.4. Band numbers

The number of bands is one of the main parameters that characterize hyperspectral imaging systems. Based on the type of spectral imaging system, i.e. multispectral or hyperspectral, the number of spectral bands could vary from a few (usually fewer than 10) in multispectral imaging to about 100–250 spectral bands in the electromagnetic spectrum in the case of hyperspectral imaging. However, the band number is not the only and decisive criterion for choosing a hyperspectral system for certain applications; the second important criterion is the bandwidth.

1.3.5.5. Bandwidth

The *bandwidth* is a parameter that is defined as the full width at half maximum (FWHM) response to a spectral line, describing the narrowest spectral feature that can be resolved by spectrography. Bandwidth should not be interchanged with the spectral sampling intervals, indicating that the spectral distance between two contiguous bands is the same without referring to their bandwidth.

1.3.5.6. Signal-to-noise ratio (SNR or S/N)

The *signal-to-noise ratio* (SNR) is the ratio of the radiance measured to the noise created by the detector and instrument electronics. In other words, signal-to-noise ratio compares the level of a desired signal to the level of background noise. In hyperspectral imaging systems, the SNR is always wavelength-dependent because of overall decreasing radiance towards longer wavelengths. The higher the ratio, the less obtrusive the background noise is.

1.3.5.7. Spectral signature

Hyperspectral imaging exploits the fact that all materials, due to the difference of their chemical composition and inherent physical structure, reflect, scatter, absorb, and/or emit electromagnetic energy in distinctive patterns at specific wavelengths. This characteristic is called *spectral signature* or *spectral fingerprint*, or simply *spectrum*. Every image element (*pixel*) in the hyperspectral image contains its own spectral signature. Briefly, spectral signature is defined as the pattern of reflection, absorbance, transmittance, and/or emitting of electromagnetic energy at specific wavelengths. In principle, the spectral signature can be used to uniquely characterize, identify, and discriminate by class/type any given object(s) in an image over a sufficiently broad wavelength band (Shaw & Manolakis, 2002).

1.3.6. Hyperspectral Image and Hyperspectral Data

Hyperspectral image data consist of several congruent images representing intensities at different wavelength bands composed of vector pixels (*voxels*) containing two-dimensional spatial information (of m rows and n columns) as well as spectral information (of K wavelengths). These data are known as a three-dimensional hyperspectral cube, or *hypercube, datacube, data volume, spectral cube* or *spectral volume*, which can provide physical and/or chemical information of a material under test (Cogdill *et al.*, 2004). This information can include physical and geometric observations of size, orientation, shape, color, and texture, as well as chemical/molecular information such as water, fat, proteins, and other hydrogen-bonded constituents (Lawrence *et al.*, 2003). However, the combination of these two features (spectral and spatial) is not trivial, mainly because it requires creating a three-dimensional (3D) data set that contains many images of the same object, where each one of them is measured at a different wavelength. Because pixels are digitalized gray values or intensities at a certain wavelength, they may be expressed as integers. Intensity values of a spatial image in the *hypercube* at one wavelength may have 8-bit gray values meaning that 0 is the black and 255 is the white. In more precise systems, the intensity values of each pixel having 12-bit (2^{12} gradations, i.e., 0–4095), 14-bit (2^{14} gradations, i.e., 0–16383) or 16-bit (2^{16} gradations, i.e., 0–65535) gray levels are used. For many applications, 12-bit dynamic range is adequate and can provide high frame rates. For more demanding scientific applications such as cell, fluorescence or Raman imaging, a higher performance 16-bit cooled camera may be advantageous.

Figure 1.2 illustrates one example of the *hypercube* extracted from a hyperspectral image acquired for a piece of meat. The raw hyperspectral image consists of a series of contiguous sub-images; each one represents the intensity and spatial distribution of the tested object at a certain waveband. All individual spatial images could be picked up from the *hypercube* at any wavelength(s) covering the spectral sensitivity of the system. Therefore, a *hyperspectral image* described as I(x, y, λ) can be viewed either as a separate spatial image I(x, y) at each wavelength (λ), or as a spectrum I(λ) at every pixel (x, y). Each pixel in a hyperspectral image contains the spectrum of that specific position. The resulting spectrum acts like a fingerprint which can be used to characterize the composition of that particular pixel. Since hyperspectral imaging acquires spatially distributed spectral responses at pixel levels, this allows flexible selection of any regions of interest on a target object,

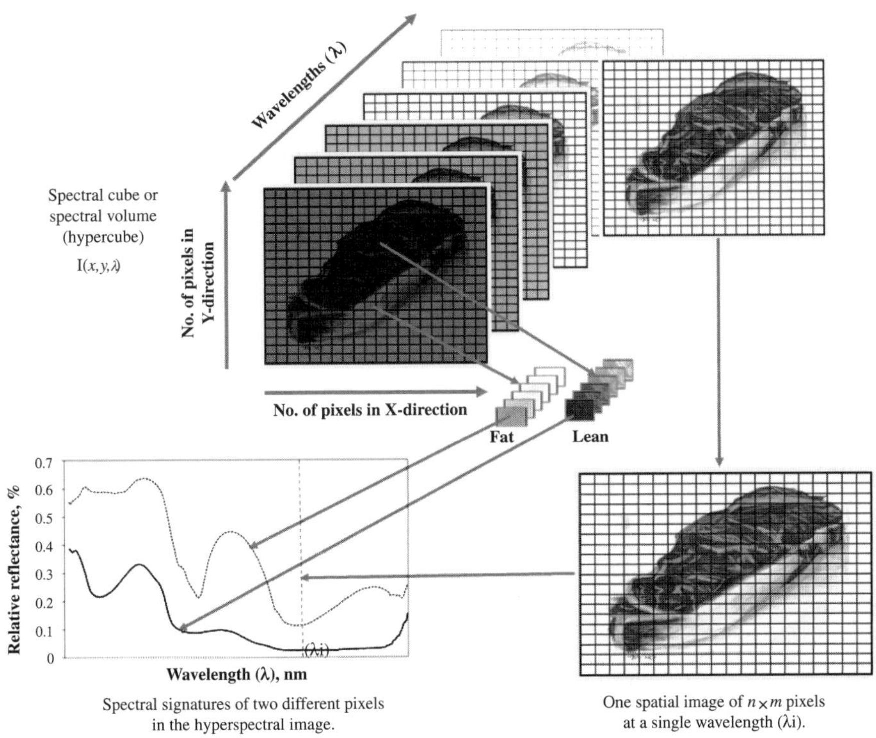

FIGURE 1.2
*Schematic diagram of hyperspectral image (*hypercube*) for a piece of meat showing the relationship between spectral and spatial dimensions. Every pixel in the hyperspectral image is represented by an individual spectrum containing information about chemical composition at this pixel. (Full color version available on http://www.elsevierdirect.com/companions/9780123747532/)*

e.g. variable sizes and locations. For instance, if two different pixels from two different compositional locations in the *hypercube* are extracted, they will show different fingerprints or different *spectral signatures*. Therefore, without any further manipulation or preprocessing treatments of these spectral data, the difference in spectral signatures between lean meat pixel and fat pixel of the tested piece of meat shown in Figure 1.2 are noticeably distinguished.

Technically speaking, the hyperspectral data are characterized by the following features:

Hyperspectral data volumes are very large and suffer from colinearity problems. This has implications for storage, management, and further image processing and analyses. The amount of data is the greatest problem that has to be coped with. Assuming collection of an image of 160 wavebands between 900 and 1700 nm (with 5 nm bandwidth) with spatial dimensions of 512 × 512 pixels and 8 bits precision (1 byte), the size of the image would be 512 × 512 × 160 bytes = 41.94 Mega bytes. The primary goal of data analysis is therefore a reduction step to decrease the data size.

Hyperspectral data are inherently high dimensional since they are, by definition, composed of large numbers of spectral bands. For example, the hyperspectral imaging system that ElMasry *et al.* (2009) used in their experiment for chilling injury detection in apples and for predicting quality attributes in strawberries (ElMasry *et al.*, 2007) recorded 826 spectral bands in the VIS and NIR region between 400 and 1000 nm with about 0.73 nm between contiguous bands. Even though these high *dimensionality* data offer access to rich information content they also represent a dilemma in themselves for data processing especially when the major purpose is to use the system in a real-time application.

The *hypercube* can be viewed in the spatial domain as images ($m \times n$) at different wavelengths or in the spectral domain as spectral vectors at all wavelengths, as shown in Figure 1.3. Both representations are essential for analyzing the hyperspectral data with the suitable *chemometric* tools using one or more of the *multivariate analysis* techniques. For instance, if one hyperspectral image has dimensions of 256 × 320 × 128, this *image cube* can be interpreted as 128 single channel images each with 256 × 320 pixels. Alternatively, the same hypercube can be viewed as 81,920 spectra, each with 128 wavelength points. This huge amount of data poses data mining challenges, but also creates new opportunities for discovering detailed hidden information.

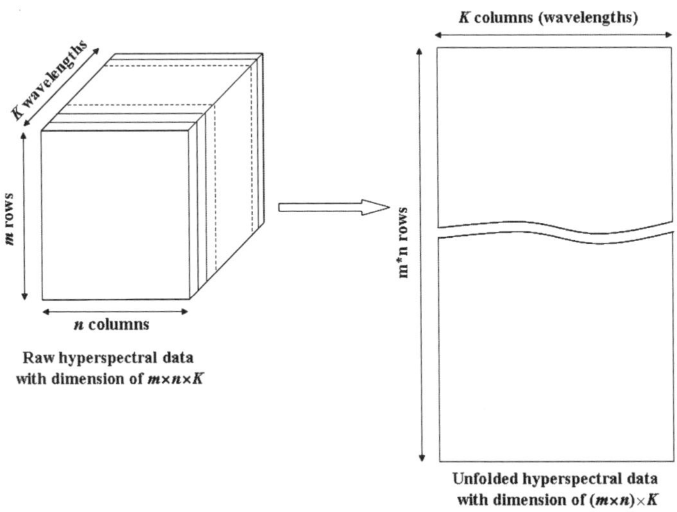

FIGURE 1.3 *Unfolding the hyperspectral data "hypercube" to facilitate multivariate analysis*

As explained in the previous sections, the product of a spectral imaging system is a stack of images of the same object, each at a different spectral narrow band. However, the field of spectral imaging is divided into three techniques called *multispectral*, *hyperspectral*, and *ultraspectral*. The concept of multispectral, hyperspectral, and ultraspectral imaging is similar. It is believed by many researchers that the only difference between them is the number of wavebands used during image acquisition. If an image is acquired with very few separated wavelengths, the system is called *multi-spectral imaging*. If the spectral image is acquired with an abundance of contiguous wavelengths, the system is then called *hyperspectral imaging*. While no formal definition exists, the difference is not based on the number of bands, contrary to various popular notations by many scientists working in this field. Multispectral deals with several images at discrete and somewhat narrow bands. The simplest method to obtain images at a discrete wavelength region is by using band-pass filters (or interference filter) in front of a monochrome camera lens. Multispectral images can be obtained by capturing a series of spectral images by using either a liquid crystal tunable filter (LCTF) or an acousto–optic tunable filter (AOTF), or by sequentially changing filters in front of the camera (Chen *et al.*, 2002). Regrettably, multispectral images do not produce the "spectrum" of an object. On the other hand, hyperspectral deals with imaging at narrow bands over a contiguous wavelength range, and produces the "*spectra*" of all pixels in the scene. Therefore a system with only 20 wavebands can also be a hyperspectral system if it covers a certain spectral range (VIS, NIR, SWIR, IR, etc.) to produce spectra of all pixels within this range. Given that the visible range spectrum spans a wavelength range of approximately 300 nanometres (400–700 nm), the system of only 20 wavebands of 15 nm bandwidth can be named as hyperspectral. *Ultraspectral imaging* is typically used for spectral imaging systems with a very fine spectral resolution. These systems often have a low *spatial resolution* of several pixels only.

1.4. CONFIGURATION OF HYPERSPECTRAL IMAGING SYSTEM

The optical and spectral characteristics of a hyperspectral imaging system are determined largely by the application requirements. However, all systems have the same basic components in common: a means to image the object, a means to provide both spectral and spatial resolution, and a means to detect. The complete optical system for a hyperspectral imaging system consists of a suitable objective lens matched to the spatial and spectral requirements of the application, a wavelength dispersion device such as an imaging spectrograph and a two-dimensional detector such as a CCD or CMOS camera to simultaneously collect the spectral and spatial information. The main part of this system is the spectrograph. A spectrograph is a system for delivering multiple images of an illuminated entrance slit onto a photosensitive surface (detector). The location of the images is a function of wavelength. It is normally characterized by an absence of moving parts.

1.4.1. Acquisition Modes of Hyperspectral Images

There are three conventional ways to build one spectral image: area scanning, point scanning, and line scanning. These instruments capture a one- or two-dimensional subset of the datacube, and thus require the temporal scanning of the remaining dimension(s) to obtain the complete datacube. The *area-scanning* design, also known as *staring imaging* or *focal plane scanning imaging* or the *tunable filter*, involves keeping the image field of view fixed, and obtaining images one wavelength after another, therefore it is conceptually called the *wavelength-scanning* method or band sequential method. Acquiring an image at different wavelengths using this configuration requires a tunable filter, and the resulting hypercube data is stored in Band Sequential (BSQ) format. The *point-scanning* method, also known as *whiskbroom*, produces hyperspectral images by measuring the spectrum of a single point and then the sample is moved and another spectrum is taken. Hypercube data obtained using this configuration are stored as Band Interleaved by Pixel (BIP) format. The third method is *line scanning*, also called *pushbroom*, involving acquisition of spectral measurements from a line of sample which are simultaneously recorded by an array detector; and the resultant hypercube is stored in the Band Interleaved by Line (BIL) format. This method is particularly well suited to conveyor belt systems, and may therefore be more practicable than the former ones for food industry applications.

In point scanning the sample is moved in the *x* and *y* directions point-by-point using a computer-controlled stage; meanwhile it is moved line-by-line in the case of line scanning. In imaging by area scanning, data are collected with a two-dimensional detector, hence capturing the full desired field-of-view at one time for each individual wavelength, without having to move the sample. The point-scanning and line-scanning methods are conceptually called *spatial-scanning* methods since they depend on scanning the specimen in the spatial domain by moving the specimen either point-by-point or line-by-line respectively, while area scanning is a spectral-scanning method. These three configurations of acquisition modes—based on the spectral imaging sensors—are explained in more detail below.

1.4.1.1. Staring imaging (area-scanning imaging, focal plane-scanning imaging or tunable filter or wavelength scanning)

The detector in an *area-scanning imaging* configuration is located in a plane parallel to the surface of the sample and the sample is imaged on the focal plane detector. The camera, lens, spectrograph, and the sample itself (field of view) remain fixed in position relative to the detector. The spectral domain is electronically scanned and the image is collected one spectral plane (wavelength) after another. One of the simplest methods for gathering the images at one wavelength at a time can be performed by collecting images using interchangeable narrow bandpass interference filters at distinct wavelengths. The bandpass size of the filters determines the number of wavelengths in the spectral range. The filters are positioned in front of the camera and a filter wheel rotates a bandpass filter into the optical path to acquire wavelength bands of equal bandwidth. This technique is usually preferred only where a limited number of wavebands are required because this process is inherently slow, which is considered one of its disadvantages. The disadvantage of using this configuration is the requirement for repetitive scanning of the same specimen at several wavelengths. Such repetition in scanning is necessary so that successive images at each wavelength increment can be gathered. An alternative mechanism for obtaining wavelength scanning is to use *tunable filters*. Typically, this is achieved by using electronically tunable filters or imaging interferometers. In this configuration, the most predominantly employed filters are Liquid Crystal Tunable Filters (LCTFs), Acousto–Optic Tunable Filters (AOTFs), and interferometers either between the illumination source and specimen or between the specimen and the detector. The staring image acquisition is suitable for many applications where a moving tested sample is not required, such as florescence imaging using an excitation–emission matrix in which the wavelengths of both excitation and emission are controlled by the tunable filters where the filter change is done electronically. Lengthy image acquisition times can also be an issue for biological samples, which may be sensitive to heating caused by the continuous illumination from source lamps. Furthermore, staring imaging is not effective for either a moving target or for real-time delivery of information concerning a particular specimen.

1.4.1.2. Whiskbroom (point-scan imaging or Raster-scanning imaging)

It is obvious that the easiest way to acquire a particular spectral image of an object is to use a filter-based imaging system (i.e., area-scanning imaging). This is mostly due to the poor optical quality and transmission efficiency of wavelength dispersive systems such as those based on a diffraction *grating*. The use of newer, highly specialized *prism* spectrometers has enabled the design of spectral imaging systems with high efficiency. The *whiskbroom* is an example of this technology which operates as an electromechanical scanner with a single detector. Whiskbroom scans a single pixel at a time, with the scanning element moving continuously. Light coming from the specimen is dispersed using an optical grating, prism or a similar *dispersing element* and is detected wavelength by wavelength by a line detector array. Thus whiskbroom scanners have one detector element for each wavelength (spectral band) recorded. A single, small sensor can be moved in a zigzag or raster fashion to sense the light intensity on a grid of points covering the whole image. The image is recorded with a double scanning step: one in the wavelength domain and the other in the spatial domain. This design is commonly used for microscopic imaging where the acquisition time is usually not a problem since a double scan (i.e., spatial and spectral) is required. By moving the sample systematically in two spatial dimensions, a complete hyperspectral image can be obtained. This system provides very stable high resolution spectra; however, positioning the sample is very time-consuming and has high demands on repositioning hardware to ensure repeatability. The spatial size dimensions of the hyperspectral image are limited only by the sample positioning hardware.

1.4.1.3. Pushbroom (line-scan imaging)

Line-scanning devices record a whole line of an image rather than a single pixel at a time using a two-dimensional *dispersing element* (*grating*) and a two-dimensional *detector* array. A narrow line of the specimen is imaged onto a row of pixels on the sensor chip and the spectrograph generates a spectrum for each point on the line, spread across the second dimension of the chip. Therefore, hyperspectral images are acquired by a wavelength dispersive system that incorporates a diffraction *grating* or *prism*. These instruments typically require an entrance *aperture*, usually a *slit*, which is imaged onto the focal plane of a spectrograph at each wavelength simultaneously. Therefore, an object imaged on the slit will be recorded as a function of its entire spectrum and its location in the sample. In this design an array of detectors is used to scan over a two dimensional scene using a two dimensional detector perpendicular to the surface of the specimen. This configuration is normally used when either the specimen or the imaging unit is moving one in respect to the other, such as those used in industrial applications. The sensor detectors in a pushbroom scanner are lined up in a row called a linear array. Instead of sweeping from side to side as the sensor system moves forward, the one-dimensional sensor array captures the entire scan line at once. Since no filter change is required, the speed of image acquisition is limited only by camera read-out speeds.

The difference between *wavelength scanning* (implemented in tunable filter systems) and *spatial scanning* (implemented in pushbroom systems) approaches to acquire a cube of spatial and spectral data is shown in Figure 1.4. One approach is used to acquire a sequence of two-dimensional images at different wavelengths (from λ_1 to λ_n) and the other approach is used to acquire a sequence of line images in which a complete spectrum is captured for each pixel on the line. In the first approach (*wavelength scanning*), illustrated in Figure 1.4a, the detector sequentially captures a full spatial scene at each spectral band (wavelength) to form a three-dimensional image cube. This approach is preferable if the number of bands needed is limited and the object can be held fixed in front of the camera during capturing. In the second approach (*spatial scanning*), shown in Figure 1.4b, a line of spatial information with a full spectral range per spatial pixel is captured sequentially to complete a volume of spatial–spectral data (Kim *et al.*, 2001). Since the *spatial-scanning* mode requires moving the specimen line by line, this method is particularly well suited to conveyor belt systems

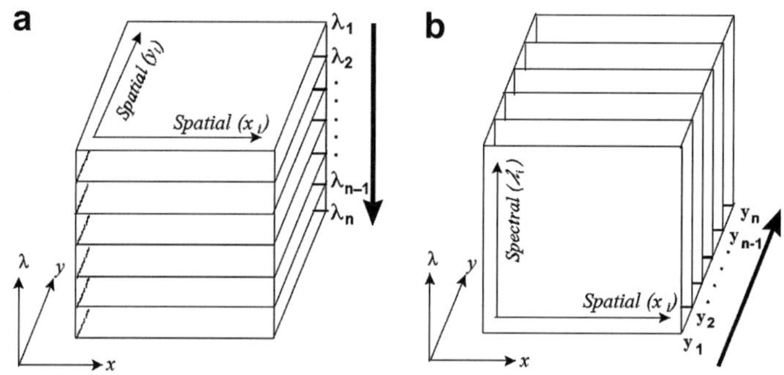

FIGURE 1.4 *Conceptual representations of image acquisition modes. Data arrows indicate directions for sequential acquisition to complete the volume of spatial and spectral data "hypercube". (a) Wavelength-scanning mode; (b) spatial-scanning mode*

and is more practicable than the *wavelength scanning* for real-time applications (Chen *et al.*, 2002; Mehl *et al.*, 2004; Polder *et al.*, 2002).

1.4.2. Detectors in Hyperspectral Imaging Systems

The two-dimensional *detector* (i.e., the area detector) for the spectrograph of the hyperspectral imaging system plays an important role in recording the spatial and spectral signals. The detectors used in hyperspectral imaging systems are generally photovoltaic semiconductor detectors, so-called *charge-coupled devices* (CCDs). Semiconductor devices are electronic components that exploit the electronic properties of semiconductor materials, principally silicon (Si), germanium (Ge), and gallium arsenide (GaAr). Silicon (Si) is the most widely used material in semiconductor devices. The many advantages such as low raw material cost, relatively simple processing, and a useful temperature range makes it currently the best compromise among the various competing materials. Semiconductor line or area arrays typically used in most spectral imaging systems include silicon (Si) arrays, indium antimonide (InSb) arrays, mercury cadmium telluride (HgCdTe) arrays, and indium gallium arsenide (InGaAs) arrays. Silicon arrays are sensitive to radiation in the 400–1000 nm wavelength range, InSb, HgCdTe, and InGaAs arrays at longer wavelengths between 1000 and 5000 nm. In some instruments, several different and overlapping detector elements are used for optimized sensitivity in different wavelength regions (Goetz, 2000). To increase detection efficiency especially in the infrared regions, the detector should be cooled. Cooling reduces the array's dark current, thus improving the sensitivity of the detector to low light intensities, even for ultraviolet and visible wavelengths, and hence reducing the thermal noise to a negligible level.

1.4.3. Main Components of Hyperspectral Imaging System

In food analysis applications it is desirable to know what is the main components of the most acceptable hyperspectral system in this field. Therefore, in this section the main components of a hyperspectral imaging system employing the pushbroom design will be explained due to the fact that it uses the line-scan method and therefore is more consistent for on-line application. An image of a specimen located in the field of view (FOV) is collected by translating the specimen across the slit aperture of the spectrograph in a pushbroom acquisition method. Thus the spectral data are measured simultaneously and the image or FOV is generated sequentially. The prime advantage of this method is that all the wavelength data needed to identify an object or objects, even if the spectra are highly convoluted, are acquired simultaneously and are immediately available for processing. Consequently, this technique is ideal for kinetic studies on samples that exhibit movement, for studies of time-based changes in molecular characteristics, and for any condition that benefits from real-time spectral analysis. As stated by many researchers (e.g. Kim *et al.*, 2002; Polder *et al.*, 2002), the pushbroom hyperspectral imaging system consists of five main components: camera containing a cooled two-dimensional (2D) light detector, spectrograph, translation stage, illumination units, and a computer. Each of these components has its own characteristics that influence the total accuracy of the system. To characterize the performance of the whole system, it is important to measure and optimize all parameters that influence the quality of the obtained spectral image. For instance, the ideal illumination should be homogeneous illumination over a large area without radiation damage to the samples. By scanning the object by moving the linear translation stage, the second spatial dimension is incorporated, resulting in a three-dimensional (3D) datacube of (x, y, K) dimensions. The main components of a pushbroom hyperspectral imaging system used for nondestructive meat quality assessment in University College Dublin (UCD), Ireland, are depicted in Figure 1.5.

The wavelength dispersing unit in the hyperspectral imaging system is essentially a grating spectrograph with a 2D detector array. It utilizes a field-limiting entrance slit and an imaging spectrometer with a dispersive element to allow the 2D detector to sample the spectral dimension and one spatial dimension simultaneously. The imaging lens focuses the light onto an entrance slit, the light is then collimated, dispersed by a grating and focused on the detector. The second spatial dimension, y, is typically generated by moving or scanning the camera's field of view relative to the scene. The spectral resolution of the system depends on both the slit width and the

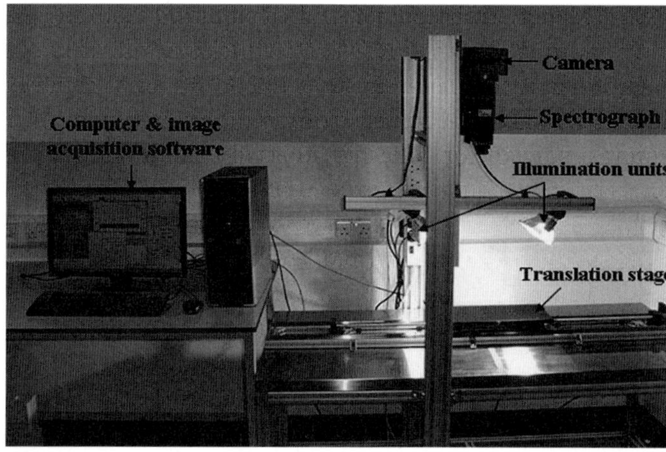

FIGURE 1.5 *Main components of a pushbroom hyperspectral imaging system. (Full color version available on http://www.elsevierdirect.com/companions/9780123747532/)*

optical aberration. As the light beam enters the spectrograph, it is dispersed into different directions according to wavelength while preserving its spatial information. The dispersed light is then mapped onto the detector array, resulting in a 2D image, one dimension representing the spectral axis and the other containing the spatial information for the scanning line. By scanning the entire surface of the specimen, a complete 3D hyperspectral image cube is created, where two dimensions represent the spatial information and the third represents the spectral information (Lu, 2003b). Figure 1.6 shows an implementation of this principle from Specim Ltd (Finland).

Technically speaking in the context of system integration, the basic elements of a hyperspectral imaging spectrograph are shown in Figure 1.7. The light source, such as a halogen lamp, illuminates the object to be measured, and the entrance optics, e.g. a camera lens, collects the radiation from the object and forms an image on the image plane (image plane 1 in Figure 1.7), where the entrance slit of the imaging spectrograph is located. The slit acts as a field-stop to determine the instantaneous FOV in spatial directions to a length of Δx and a width of Δy, marked as the measured area in Figure 1.7. Each point A in the spatial **x**-direction of the measured area has its image A′ on the entrance slit. The radiation from the slit is collimated by either a lens or a mirror and then dispersed by a dispersing element, which is typically a prism or grating, so that the direction of propagation of the radiation depends on its wavelength. It is then focused on image plane 2 by the focusing optics, i.e. a lens or mirror. Every point A is represented on image plane 2 by a series of monochromatic images forming a continuous spectrum

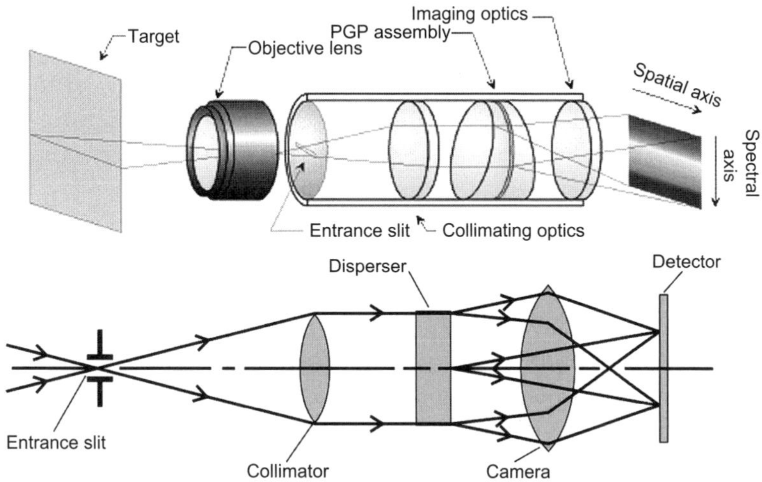

FIGURE 1.6 *Working principle of prism-grating-prism (PGP) spectrograph (courtesy of Specim Ltd). (Full color version available on http://www.elsevierdirect.com/companions/9780123747532/)*

in the direction of the spectral axis, marked with different sizes of A″. The focused radiation is detected by a 2D detector array such as charge-coupled device (CCD) or a complementary metal-oxide-semiconductor (CMOS) detector. The imaging spectrograph allows a 2D detector array to sample one spatial dimension of length Δx and infinite width Δy and the spectral dimension of the 3D cube simultaneously. The width Δy also defines the spectral resolution, which can be seen as $\Delta y''$ in the direction of the spectral

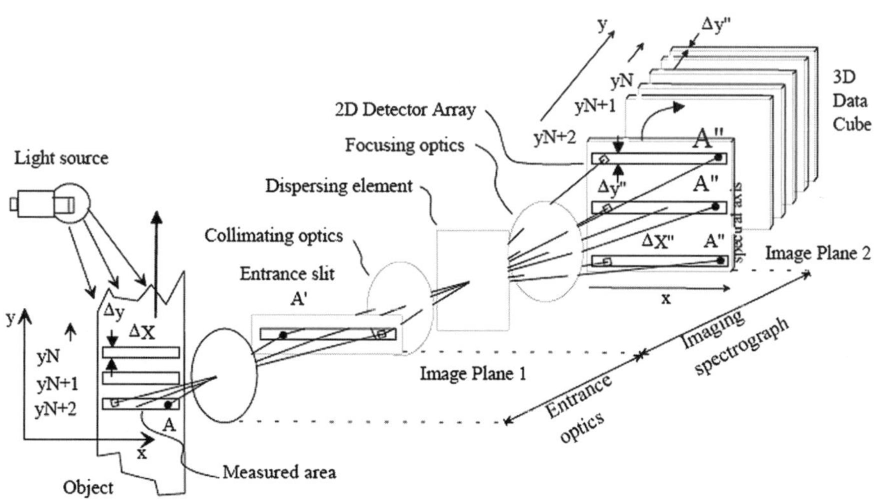

FIGURE 1.7 *The basic elements of a hyperspectral imaging spectrograph, with the entrance optics and generation of the 3D datacube: spatial (x and y) and spectral (K) dimensions (reproduced from Aikio, 2001 by permission of the author)*

axis in Figure 1.7. In addition to defining spectral resolution, slit width controls the amount of light entering the spectrograph. Also, the collimator makes this light parallel so that the disperser (a grating or prism) disperses it. The second spatial dimension of the object, *y*, is generated by scanning or moving the FOV of the instrument relative to the scene, corresponding to the positions yN, yN + 1, yN + 2 in Figure 1.7.

1.5. CALIBRATION OF HYPERSPECTRAL IMAGING SYSTEM

Hyperspectral imaging systems are basically able to delineate multiple mapping of essential chemical constituents such as moisture, fat, and protein on most biological specimens by performing spectral characterizations of these constituents. However, some systems may present inconsistent spectral profiles of reference spectra even under controlled conditions. This variability confirms that there is a need for a standardized, objective *calibration* and a validation protocol. In all hyperspectral imaging systems the *spectrograph* and its dispersive element is the most important component for the determination of its optical properties because it determines the spectral range and the spectral resolution. The dispersive element separates the light depending on its wavelengths and projects these fractions on different spatial positions. Therefore, the goals of the calibration process are to (a) standardize the spectral axis of the hyperspectral image, (b) determine whether a hyperspectral imaging system is operating properly, (c) provide information about the accuracy of the extracted spectral data and thus validate their acceptability and credibility, and (d) diagnose instrumental errors, measurement accuracy, and reproducibility under different operating conditions.

In essence, calibrating a spectral imaging system is vital before acquiring the images. A system calibration test is always a prudent step when doing qualitative and quantitative analyses. This procedure is performed after assembling all the components of the hyperspectral imaging system to ensure both spectral and spatial dimensions are projected in their right directions. The manufacturers are obliged to produce calibrated systems to guarantee trustworthy results. Recalibration is generally not required unless the physical arrangement of the components of the imaging system is disturbed. The first precaution in the calibration process is to cool the imaging system to its initial operating temperature, which is usually between −80 and −120 °C in most modern systems. Also, the combination of lamp intensity and detector integration time has to be adjusted to avoid saturation of the analog to digital (A/D) converter.

Another precaution that requires consideration is to set image binning, which is determined by the spectral distribution of useful wavelengths and the size of spatial image features to be processed for the application. In the case of line-scanning mode (pushbroom), one of the dimensions is assigned to one spatial axis and the other is used for projecting the spectral axis as a spectral dispersion plane. For instance, if the image resolution is of $x \times y$ pixels, x pixels will be used for projecting the spatial resolution of the scanned line and y pixels will be used for projecting the spectral resolution of K wavelengths. Moreover, wavelength dispersion controls the physical distance that separates one wavelength from another on the spectral axis and is a key parameter in determining the limits of spectral resolution. The binning in both spatial and spectral directions will lead to a reduction in the resolution of both axes. The new resolution will be the initial number of pixels of this axis over the binning factor. Therefore, the new resolution would be x/b_1 pixels for spatial resolution and y/b_2 pixels for the spectral resolution, where b_1 and b_2 are the binning factors in the spatial and spectral axes respectively. To make this sophistication much clearer, it can be considered that the spatial and spectral resolution in most widely used hyperspectral imaging systems implemented in food quality assessment are of 512×320 pixels. If under certain applications a unity binning factor ($b_1 = 1$) is required in the spatial direction, this will result in line-scan images with a spatial resolution of 512 pixels (512 divided by 1). On the other hand, if a binning factor of value $b_2 = 2$ is used the resulting spectral resolution would be 160 pixels (320 divided by 2) in the spectral axis. This will lead to a total number of 160 contiguous wavebands (channels) in the spectral axis. Strictly speaking, the binning process in the spectral direction adds together photons from adjacent pixels in the detector array which will produce a reduced number of pixels to be digitized by the A/D converter for the computer to process. Reducing total pixel readout time decreases the acquisition time of each line-scan image, which allows a higher image acquisition speed for the imaging device.

The most significant step in the calibration process is the spectral waveband calibration (wavelength calibration) that identifies each spectral channel with a specific wavelength. Each wavelength on the spectral axis is identified as a function of its physical location on this axis. To determine the relation between distance (in pixels) on the spectral axis and wavelength, the spectral axis must be calibrated by using a standard emission lamp as a light source. A specific wavelength will then be assigned to a specific column of CCD pixels. The most acceptable calibration protocol involves the use of a single or multi-ion discharge lamp of mercury (Hg), helium (He), argon (Ar), neon (Ne), and/or cadmium (Cd) that emits distinct, stable, spectral features in place of a sample. These reference spectra from this lamp will be used to accurately predict the spectral resolution of the system and adjust the spectral axis. Therefore, using these reference light sources that emit absolute standard "reference spectra" is a sensible tool for diagnosing instrumental errors and measurement accuracy and reproducibility under different operating conditions. With this information on one hand, the researcher can determine whether the spectral imaging system is working optimally and make objective comparisons with the performance of other spectral imaging systems. On the other hand, if spectral imaging systems are standardized to produce the same spectral profile of a reference lamp, the researcher can be confident that the experimental findings are comparable with those obtained from other spectral imaging systems. Different light sources of known spectrum should be used for this task, such as mercury, helium, and/or cadmium calibration lamps, as shown in Figure 1.8. One example of a single ion discharge calibration lamp is the cadmium lamp that has five distinct peaks in the visible range of the electromagnetic spectrum at 467.8, 479.9, 508.58, 607.2, and 643.8 nm. as depicted in Figure 1.8.

In addition, there are several readily available calibration sources of a multi-ion discharge type, the most common of which is a low-pressure Hg^{+}/Ar^{+} discharge lamp that covers the wavelength range of 400 to 840 nm. The emission spectrum of this lamp is shown in Figure 1.9 (Oriel Instruments, Stratford, CT, USA). The benefit of this spectrum is that the spectrum acts as a spectral fingerprint that can be used to calibrate the performance of any spectroscopic system.

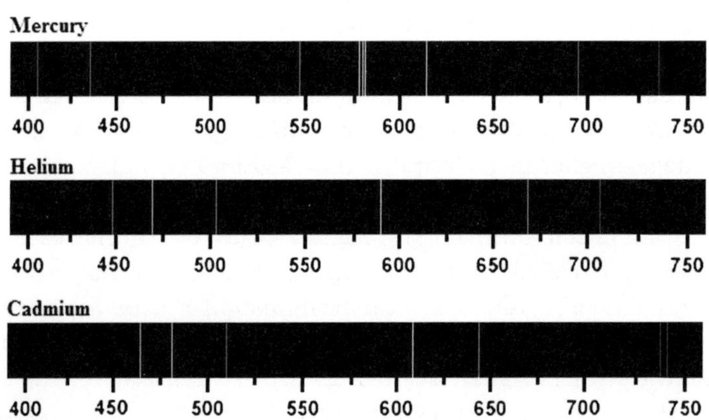

FIGURE 1.8 *Emission (bright lines) spectra of different calibration lamps. (Full color version available on http://www.elsevierdirect.com/companions/9780123747532/)*

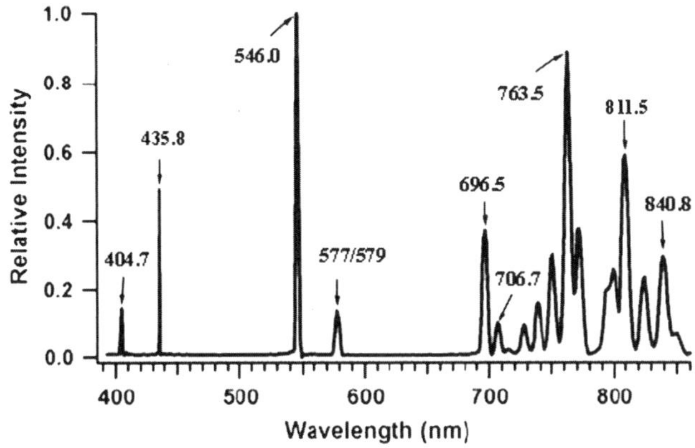

FIGURE 1.9 *Spectrum of calibration light source of pure Hg^{+}/Ar^{+} low-pressure discharge lamp*

In practice, the calibration lamp is first scanned by the hyperspectral imaging system under controlled operating conditions. Once the calibration lamp is scanned, its peaks are then assigned to standardize the spectral axis. Then, a polynomial regression of first or second order can be established to convert spectral axis (in pixels) to its corresponding wavelength using the reference wavelength peaks of the calibration lamp. Following system calibration, the spectral imaging system will be ready to use for the acquisition of real line-scan images. The extracted data from such images before calibration will be in the pure form (nonlinearized pixel versus intensity), and after calibration will be wavelength versus intensity. However, if some error occurs in the physical arrangement of the hyperspectral imaging system or if some of its components have to be reassembled, the system should be recalibrated with the calibration lamp. The system can be used safely provided that it gives the same peaks of the calibration lamp with an acceptable error. This step must be repeated several times to diagnose the level of this error and to judge the reproducibility of the system under different operating conditions.

Finally, after acquiring hyperspectral images of real samples, another calibration step, called *reflectance calibration*, should be performed to account for the background spectral response of the instrument and the 'dark' current of the camera. The background is obtained by acquiring a spectral image from a uniform, high reflectance standard or white ceramic (~100% reflectance), and the dark response (~0 % reflectance) is acquired by recording an image when the light source is turned off and the camera lens is completely covered with its nonreflective opaque black cap. These two reference images are then used to calculate the pixel-based relative reflectance for the raw line-scan images using the following formula:

$$I = \frac{I_0 - D}{W - D} \tag{1.3}$$

where I is the relative reflectance image, I_0 is the raw reflectance image, D is the dark reference image, and W is the white reference image.

The corrected hyperspectral image can also be expressed in absorbance (A) by taking logarithms of the above equation as:

$$A = -\log_{10}\left(\frac{I_0 - D}{W - D}\right) \tag{1.4}$$

1.6. SPECTRAL DATA ANALYSIS AND CHEMOMETRICS

Hyperspectral imaging systems cannot stand alone without the help of some software for gaining high performance in acquisition, controlling, and analyses. It is essential to support the system with software for image acquisition, software for controlling the motor to move the sample line by line, software for extracting spectral data and preprocessing steps, software for multivariate analysis, and software for final image processing. Integration of image acquisition, spectral analysis, chemometric analysis, and digital image analysis in single software has not been explored yet. In fact, some of these processes are integrated in one software package to perform some of these operations. Alternatively, professional researchers can develop their own software routines or build a comprehensive graphical user interface (GUI) to perform each of the key steps of these processes. Typically, routines can be developed by using packages that support scripting capability, such as C^{++}, Matlab, IDL or LabView. However, researchers should be familiar with the main fundamentals of the necessary steps required to obtain the key information about the process or about the samples being monitored for achieving the final goals of the tests. Typical steps usually undertaken in hyperspectral imaging experiments are outlined in the flowchart described in Figure 1.10.

The first step is the collection of a hyperspectral image by utilizing ideal acquisition conditions in terms of illumination, spatial and spectral resolution, motor speed, frame rate, and exposure/integration time. After acquiring a hyperspectral image for the tested sample, this image should be calibrated with the help of white and dark hyperspectral images as mentioned earlier in

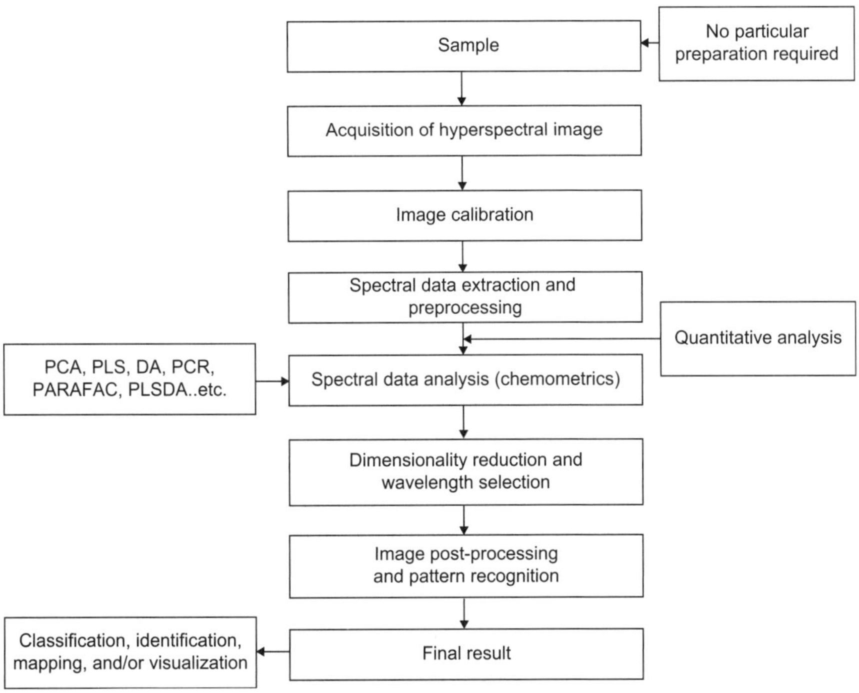

FIGURE 1.10 *Flowchart of the key steps involved in hyperspectral imaging analyses*

this chapter. The spectral data are then extracted from different regions of interest (ROIs) that present different quality features in the calibrated image. Extracted spectral data should be preprocessed to reduce noise, improve the resolution of overlapping data, and to minimize contributions from imaging instrument responses that are not related to variations in the composition of the imaged sample itself. *Preprocessing* of spectral data is often of vital importance if reasonable results are to be obtained from the spectral analysis step. Preprocessing includes spectral and spatial operations. Spectral preprocessing includes some operations such as spectral filters, normalization, mean centering, auto scaling, baseline correction, differentiation (Savitsky-Golay), standard normal variate (SNV), multiplicative scatter correction (MSC), and smoothing. On the other hand, spatial operations include low-pass filters, high-pass filters, and a number of other spatial filters. Detailed overviews of the most admired preprocessing operations are further explained in subsequent relevant chapters in the book.

Once instrument response has been suppressed by means of preprocessing, qualitative analysis can be employed. Qualitative analysis attempts to address what different components are present in the sample and how these components are distributed. Many *chemometric* tools fall under this category. Strictly speaking, the cornerstone of this process is the data analysis using multivariate analysis by one or more chemometrics tools, including correlation techniques such as cosine correlation and Euclidean distance correlation; classification techniques such as principal components analysis (PCA), cluster analysis, discriminant analysis (DA), and multi-way analysis; and spectral deconvolution techniques. To build concentration maps for determining the estimated concentrations of different components present in the tested sample and their spatial distribution, a quantitative assessment should be performed using a standard analytical means. In quantitative spectral analysis, a number of multivariate chemometric techniques can be used to build the calibration models to relate spectral data to the actual quantitative data. Depending on the quality of the models developed, the results can range from semi-quantitative concentration maps to rigorous quantitative measurements.

Moreover, with the aid of multivariate analysis, the huge *dimensionality* and *colinearity* problems of hyperspectral data can be reduced or eliminated by selecting the spectral data at some important wavelengths. In most cases, not all the spectral bands are required to address a particular attribute. Selection of *important wavelength* is an optional step based on the speed requirements of the whole process. Generally, the selection of these optimal wavelengths reduces the size of the required measurement data while preserving the most important information contained in the data space. The wavelength preserving the largest amount of energy among the hyperspectral data carries the most important spectral information and maintains any valuable details about the tested samples. The selected *essential wavelengths* should not only maintain any valuable required details, but also simplify the successive discrimination and classification procedures (Cheng *et al.*, 2004). Indeed, the selection of the most *efficient wavelength* can be done off-line and then the on-line process consisting of image acquisition and analyses may be executed at acceptable speeds (Kleynen *et al.*, 2005). Several essential wavelengths could be sorted from the whole spectral cube through a variety of strategies, such as general visual inspection of the spectral curves and correlation coefficients (Keskin *et al.*, 2004; Lee *et al.*, 2005), analysis of spectral differences from the average spectrum (Liu *et al.*, 2003), stepwise regression (Chong & Jun, 2005), discriminant analysis (Chao *et al.*, 2001), principal component analysis (PCA) (Mehl *et al.*, 2004; Xing & De Baerdemaeker, 2005), partial least square (PLS), and others (ElMasry *et al.*, 2009; Hruschka, 2001). The mathematical principles of these approaches are given in subsequent relevant chapters in the book.

Results obtained from preprocessing, qualitative analysis, and quantitative analysis must be visualized either by scaling, surface mapping or pseudocolor representation. Once the final digital concentration images have been generated, traditional postprocessing of these images, such as segmentation, enhancement, and morphological feature extraction can be applied as a final step of the work flow. The final image processing step is carried out to convert the contrast developed by the classification step into a picture depicting component distribution. Grayscale or color mapping with intensity scaling is commonly used to display compositional contrast between pixels in an image. Final results of these calculations are used to develop key quantitative image parameters to characterize various traits in the tested samples in different categories by performing classification, identification, mapping and/or visualization.

1.7. CONCLUSIONS

Hyperspectral imaging is a complex, highly multidisciplinary field that can be defined as the simultaneous acquisition of spatial images in many spectrally contiguous bands. It is quite clear that measurement in contiguous spectral bands throughout the visible, near-infrared and/or shortwave regions of the electromagnetic spectrum makes it possible to collect all the necessary information about the tested objects. Each pixel in the hyperspectral image contains a complete spectrum. Therefore hyperspectral imaging is a very powerful technique for characterizing and analyzing biological and food samples. The strong driving force behind the development of hyperspectral imaging systems in food quality evaluation is the integration of spectroscopic and imaging techniques for discovering hidden information nondestructively for direct identification of different components and their spatial distribution in food samples. As a result, hyperspectral imaging represents a major technological advance in the capturing of morphological and chemical information from food and food products. Although effective use of hyperspectral imaging systems requires an understanding of the nature and limitations of the data and of various strategies for processing and interpretation, the wealth of additional information available and the application benefits that hyperspectral imaging produce are almost without limit in monitoring, control, inspection, quantification, classification, and identification purposes. It is therefore anticipated that work in this area will gain prominence over the coming years and its potentialities present significant challenges to food technologists and food engineers.

NOMENCLATURE

Symbols

E	energy of the photon (J)
h	Planck's constant (6.626×10^{-34} J.s)
f	frequency (Hz)
c	speed of light in vacuum (299 792 458 ms^{-1})
υ	speed of the wave, ms^{-1} (equals c in a vacuum)
I	relative reflectance image (calibrated image)
I_0	raw reflectance image
D	dark reference image
W	white reference image
A	absorbance calibrated spectral image

Abbreviations

AM	amplitude modulation of radio waves
AOTF	acousto–optic tunable filter
BIL	band interleaved by line
BIP	band interleaved by pixel
BSQ	band sequential
CCD	charge-coupled device
CMOS	complementary metal-oxide-semiconductor
DA	discriminant analysis
FIR	far-infrared
FLIM	fluorescence lifetime imaging microscopy
FM	frequency modulation of radio waves
FOV	field of view
FWHM	full width at half maximum
HACCP	hazard analysis critical control point
IR	infrared
LCTF	liquid crystal tunable filter
MSC	multiplicative scatter correction
NIR	near-infrared
NIRS	near-infrared spectroscopy
PCA	principal component analysis
PCR	principal component regression
PLS	partial least square
RGB	red, green, blue (components of a color image)
ROI	region of interest
SNR	signal-to-noise ratio

SNV	standard normal variate
SWIR	shortwave-infrared
UV	ultraviolet
VIS	visible light

REFERENCES

Abbott, J. A. (1999). Quality measurement of fruits and vegetables. *Postharvest Biology and Technology, 15*, 207–225.

Abdullah, M. Z., Guan, L. C., Lim, K. C., & Karim, A. A. (2004). The applications of computer vision system and tomographic radar imaging for assessing physical properties of food. *Journal of Food Engineering, 61*, 125–135.

Aikio, M. (2001). Hyperspectral prism–grating–prism imaging spectrograph. In *Technical Research Centre of Finland (VTT) Publications, Vol. 435*. Espoo, Finland: Department of Electrical Engineering of the University of Oulu.

Birth, G. S. (1976). How light interacts with foods. In J. J. Gafney (Ed.), *Quality detection in foods* (pp. 6–11). St Joseph, MI: ASAE.

Carlomagno, G., Capozzo, L., Attolico, G., & Distante, A. (2004). Non-destructive grading of peaches by near-infrared spectrometry. *Infrared Physics & Technology, 46*, 23–29.

Chao, K., Chen, Y. R., Hruschka, W. R., & Park, B. (2001). Chicken heart disease characterization by multi-spectral imaging. *Applied Engineering in Agriculture, 17*, 99–106.

Chao, K., Mehl, P. M., & Chen, Y. R. (2002). Use of hyper- and multi-spectral imaging for detection of chicken skin tumors. *Applied Engineering in Agriculture, 18*(1), 113–119.

Chen, Y.-R., Chao, K., & Kim, M. S. (2002). Machine vision technology for agricultural applications. *Computers and Electronics in Agriculture, 36*(2), 173–191.

Cheng, X., Chen, Y. R., Tao, Y., Wang, C. Y., Kim, M. S., & Lefcourt, A. M. (2004). A novel integrated PCA and FLD method on hyperspectral image feature extraction for cucumber chilling damage inspection. *Transactions of the ASAE, 47*(4), 1313–1320.

Chong, I.-G., & Jun, C. H. (2005). Performance of some variable selection methods when multicollinearity is present. *Chemometrics and Intelligent Laboratory Systems, 78*(1), 103–112.

Cogdill, R. P., Hurburgh, C. R., Jr., & Rippke, G. R. (2004). Single-kernel maize analysis by near-infrared hyperspectral imaging. *Transactions of the ASAE, 47*(1), 311–320.

Du, C. J., & Sun, D.-W. (2004). Recent developments in the applications of image processing techniques for food quality evaluation. *Trends in Food Science & Technology, 15*, 230–249.

Du, C.-J., & Sun, D.-W. (2006). Learning techniques used in computer vision for food quality evaluation: A review. *Journal of Food Engineering, 72*(1), 39–55.

ElMasry, G., Wang, N., ElSayed, A., & Ngadi, M. (2007). Hyperspectral imaging for nondestructive determination of some quality attributes for strawberry. *Journal of Food Engineering, 81*(1), 98–107.

ElMasry, G., Wang, N., & Vigneault, C. (2009). Detecting chilling injury in Red Delicious apple using hyperspectral imaging and neural networks. *Postharvest Biology and Technology, 52*(1), 1–8.

Gao, X., Heinemann, P. H., & Irudayaraj, J. (2003). Non-destructive apple bruise on-line test and classification with Raman spectroscopy. Paper No. 033025, ASAE Annual International Meeting, Las Vegas, Nevada, USA.

Goetz, A. F. H. (2000). Short course in hyperspectral imaging and data analysis. In J. W. Boardman (Ed.), *Center for the Study of Earth from Space*. Boulder, CO: University of Colorado.

Gunasekaran, S. (1996). Computer vision technology for food quality assurance. *Trends in Food Science & Technology, 7*(8), 245–256.

Hecht, E. (2002). *Optics* (4th ed.). San Francisco, CA and London, UK: Addison–Wesley.

Hruschka, W. R. (2001). Data analysis: wavelength selection methods. In: P. Williams and K. Norris (Eds.), *Near infrared technology in the agricultural and food industries* (2nd ed., pp. 39–58). St Paul, MN: American Association of Cereal Chemists.

Keskin, M., Dodd, R. B., Han, Y. J., & Khalilian, A. (2004). Assessing nitrogen content of golf course turfgrass clippings using spectral reflectance. *Applied Engineering in Agriculture, 20*(6), 851–860.

Kim, M. S., Chen, Y. R., & Mehl, P. M. (2001). Hyperspectral reflectance and fluorescence imaging system for food quality and safety. *Transactions of the ASAE, 44*(3), 721–729.

Kim, M. S., Lefcourt, A. M., Chao, K., Chen, Y. R., Kim, I., & Chan, D. E. (2002). Multispectral detection of fecal contamination on apples based on hyperspectral imagery: Part I. Application of visible and near-infrared reflectance imaging. *Transactions of the ASAE, 45*, 2027–2037.

Kleynen, O., Leemans, V., & Destain, M.-F. (2005). Development of a multi-spectral vision system for the detection of defects on apples. *Journal of Food Engineering, 69*(1), 41–49.

Lammertyn, J., Peirs, A., De Baerdemaeker, J., & Nicolaï, B. (2000). Light penetration properties of NIR radiation in fruit with respect to non-destructive quality assessment. *Postharvest Biology and Technology, 18*(1), 121–132.

Lawrence, K. C., Park, B., Windham, W. R., & Mao, C. (2003). Calibration of a pushbroom hyperspectral imaging system for agricultural inspection. *Transactions of the ASAE, 46*(2), 513–521.

Lee, K.-J., Kang, S., Kim, M., & Noh, S.-H. (2005). Hyperspectral imaging for detecting defect on apples. Paper No. 053075, ASAE Annual International Meeting, Tampa Convention Center, Tampa, Florida, USA.

Liu, Y., Windham, W. R., Lawrence, K. C., & Park, B. (2003). Simple algorithms for the classification of visible/near-infrared and hyperspectral imaging spectra of chicken skins, feces, and fecal contaminated skins. *Applied Spectroscopy, 57*(12), 1609–1612.

Lu, R. (2003a). Imaging spectroscopy for assessing internal quality of apple fruit. Paper No. 036012, ASAE Annual International Meeting, Las Vegas, Nevada, USA.

Lu, R. (2003b). Detection of bruises on apples using near-infrared hyperspectral imaging. *Transactions of the ASAE, 46*(2), 523–530.

Lu, R. (2004). Multispectral imaging for predicting firmness and soluble solids content of apple fruit. *Postharvest Biology and Technology, 31*(2), 147–157.

Mehl, P. M., Chen, Y. R., Kim, M. S., & Chan, D. E. (2004). Development of hyperspectral imaging technique for the detection of apple surface defects and contaminations. *Journal of Food Engineering, 61*, 67–81.

Nagata, M., Tallada, J. G., Kobayashi, T., & Toyoda, H. (2005). NIR hyperspectral imaging for measurement of internal quality in strawberries. Paper No. 053131, ASAE Annual International Meeting, Tampa, Florida, USA.

Noh, H. K., & Lu, R. (2005). Hyperspectral reflectance and fluorescence for assessing apple quality. Paper No. 053069, ASAE Annual International Meeting, Tampa, Florida, USA.

Park, B., Abbott, A. J., Lee, K. J., Choi, C. H., & Choi, K. H. (2002). Near-infrared spectroscopy to predict soluble solids and firmness in apples. Paper No. 023066, ASAE/CIGR Annual International Meeting, Chicago, Illinois, USA.

Pasquini, C. (2003). Near infrared spectroscopy: fundamentals, practical aspects and analytical applications. *Journal of the Brazilian Chemical Society, 14*, 198–219.

Peng, Y., & Lu, R. (2005). Modeling multispectral scattering profiles for prediction of apple fruit firmness. *Transactions of the ASAE, 48*(1), 235–242.

Pieris, K. H. S., Dull, G. G., Leffler, R. G., & Kays, S. J. (1999). Spatial variability of soluble solids or dry matter in fruits, bulbs and tubers using NIRS. *Postharvest Biology and Technology, 34*, 114–118.

Polder, G., van der Heijden, G. W. A. M., & Young, I. T. (2002). Spectral image analysis for measuring ripeness of tomatoes. *Transactions of the ASAE, 45*(4), 1155–1161.

Schweizer, S. M., & Moura, J. M. F. (2001). Efficient detection in hyperspectral imagery. *IEEE Transactions on Image Processing, 10*, 584–597.

Shaw, G., & Manolakis, D. (2002). Signal processing for hyperspectral image exploitation. *IEEE Signal Processing Magazine, 19*(1), 12–16.

Xing, J., & De Baerdemaeker, J. (2005). Bruise detection on 'Jonagold' apples using hyperspectral imaging. *Postharvest Biology and Technology, 37*(1), 152–162.

Xing, J., Ngadi, M., Wang, N., & De Baerdemaeker, J. (2006). Wavelength selection for surface defects detection on tomatoes by means of a hyperspectral imaging system. Paper No. 063018, *ASABE Annual International Meeting*, Portland, Oregon, USA.

CHAPTER 2

Spectral Preprocessing and Calibration Techniques

Haibo Yao [1], David Lewis [2]

[1] *Mississippi State University, Stennis Space Center, Mississippi, USA*
[2] *Radiance Technologies, Inc., Stennis Space Center, Mississippi, USA*

CONTENTS

2.1. INTRODUCTION

The food industry and its associated research communities continually seek sensing technologies for rapid and nondestructive inspection of food products and for process control. In the past decade, significant progress has been made in applying hyperspectral imaging technology in such applications. Hyperspectral imaging technology integrates both imaging and spectroscopy into unique imaging sensors. Thus, imaging spectrometers or hyperspectral imagers can produce hyperspectral images with exceptional spectral and spatial resolution. A single hyperspectral image has a contiguous spectral resolution between one and several nanometers, with the number of bands ranging from tens to hundreds. Generally, high spectral resolution images can be used to study either the physical characteristics of an object at each pixel by looking at the shape of the spectral reflectance curves or the spectral/spatial relationships of different classes using pattern recognition and image processing methods.

Traditionally, hyperspectral imagery was employed in earth remote sensing applications using aerial or satellite image data. More recently, low cost portable hyperspectral sensing systems became available for laboratory-based research. The literature reports food-related studies where hyperspectral technology was applied for detection of fungal contamination, bruising in apples, fecal contamination, skin tumors on chicken carcasses, grain inspections, and so on. The generic approach for applying hyperspectral technology in food-related research includes experiment design, sampling

Hyperspectral Imaging for Food Quality Analysis and Control

preparation, image acquisition, spectral preprocessing/calibration, sample ground truth characterization, data analysis, and information extraction.

The need for spectral preprocessing and calibration of image data is due to the fact that hyperspectral imaging systems are an integration of many different optical and electronic components. Such systems generally require correction of systematic defects or undesirable sensor characteristics before performing reliable data analysis. In addition, random errors and noise can be introduced in the experimenting and image acquisition process. Consequently, spectral preprocessing and calibration is always needed before data analysis. Specifically, the main goals for calibration include (1) wavelength alignment and assignment, (2) converting from radiance values received at the sensor to reflectance values of the target surface, and (3) removing and reduction of random sensor noise.

The objective of this chapter is to discuss image preprocessing techniques to fulfill these stated calibration goals. First, methods and materials are presented which can be used for hyperspectral image wavelength calibration. This includes the introduction of an example hyperspectral imaging system for a case study. Secondly, radiometric reflectance/transmittance calibration will be discussed including calibration to percentage reflectance, relative reflectance calibration, calibration of hyperspectral transmittance data, and spectral normalization. The last part of the chapter is on noise reduction and removal. Techniques such as dark current removal, spectral low pass filter, Savitzky–Golay filtering, noisy band removal, and minimum noise fraction transformation will also be discussed.

2.2. HYPERSPECTRAL IMAGE SPECTRAL PREPROCESSING

2.2.1. Wavelength Calibration

2.2.1.1. Purpose of wavelength calibration

The purpose of wavelength calibration is to assign a discrete wavelength to the hyperspectral image band. This will enable data analysis and information extraction from the hyperspectral images to associate the correct wavelengths to the observed target. As mentioned previously, an imaging spectrometer or hyperspectral imager can produce hyperspectral images with exceptional spectral and spatial resolution. For example, when a hyperspectral image is acquired with a line-scan mechanism using a pushbroom scanner as shown in Figure 2.1 (Schowengerdt, 1997), one line of target reflectance is dispersed by a prism to generate full spectral information on the

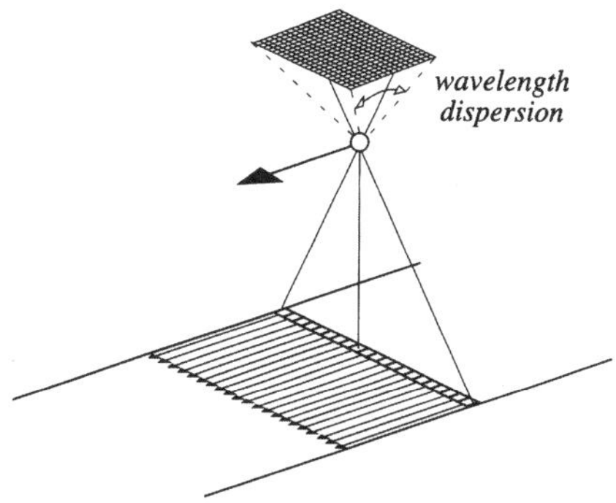

FIGURE 2.1 *Pushbroom scanning and data acquisition on a camera's detector array (reproduced from Schowengerdt (1997), figure 1.11, p. 23. © Elsevier 1997)*

camera's detector array such as a charge-coupled device (CCD). Successive line scans eventually create the three-dimensional hyperspectral cube. Thus, for each line of target reflectance, the prism disperses the target spectral information along the vertical dimension of the detector array. The horizontal dimension of the detector array represents the spatial information of each line of the target. Every column of the detector array's pixels represents the full spectral information of one target pixel. Therefore each row or line of the detector array records the target's spectral information at one discrete wavelength. This one row of the detector array's information is stored as one band of the hyperspectral image. Since each row of the detector array's pixels represents a different wavelength, wavelength calibration is needed to assign each row to its corresponding wavelength. This wavelength calibration basically establishes the wavelength to detector array row assignment for the sensor.

Wavelength calibration is needed in the initial instrumentation stage when a hyperspectral imager is manufactured and tested. Re-calibration of the instrument is also necessary after some physical changes in the instrument, such as when sensor maintenance, upgrading or repairing has been performed. The upgrade may cause misalignment between components of the sensor. Furthermore, for a hyperspectral camera, the wavelengths will drift slightly due to time and environmental conditions. Wavelength calibration is thus needed at certain time intervals, e.g., after several months or a year of significant operation of the sensor. There could be a significant difference between these two types of misalignments. Sensor misalignment due to maintenance, upgrading or repairing may cause the alignment between the camera's detector array and the spectrograph (where the prism locates) to change significantly. This could shift the response of the wavelength currently assigned to a specific detector row. This, in turn, could result in the wavelength to detector array line assignment to be offset by possibly tens of lines. For the latter case, sensor drift might only change the wavelength to detector array assignment a few lines or less. In either case, wavelength calibration is required to keep the sensor in proper working condition.

Generally, wavelength calibration can be accomplished by using calibration light sources with known accurate, narrow emission peaks covering the usable wavelength range of a hyperspectral imaging system and following a predefined calibration procedure (Lawrence, Park *et al.*, 2003; Lawrence, Windham *et al.*, 2003). The procedure basically collects image data of the calibration lights and then associates the lines in the detector array with peak signals to the wavelength known to be associated with the light source. Then a simple linear (Kim *et al.*, 2008; Mehl *et al.*, 2002), a quadratic (Chao *et al.*, 2007a; Yang *et al.*, 2006), or a cubic (Park *et al.*, 2006) regression is performed to fill in the wavelength assignment for the detector lines between those which are associated with the emission peaks of the light sources. The wavelength calibration can use data collected from:

1. a center column of the detector if only one line (one frame) of image is taken, or

2. an average of a region of interest (ROI) if a datacube is acquired.

2.2.1.2. A typical hyperspectral image system for wavelength calibration

Hyperspectral image data can be conceptualized as a three-dimensional datacube. In practice, this three-dimensional datacube is acquired through using a two-dimension focal plane array. There are two main hyperspectral imaging techniques used for three-dimensional datacube acquisition. One approach involves the use of tunable wavelength devices such as a acousto–optic tunable filter (AOTF) (Suhre *et al.*, 1999) or a liquid crystal tunable filter (LCTF) (Evans *et al.*, 1998; Zhang *et al.*, 2007). In this approach, each image frame represents a two-dimensional spatial image of a target for a given wavelength, or image band. The three-dimensional datacube is thus acquired through sequentially varying wavelength via the wavelength tuning device. The other approach involves a line-scanning mechanism such as the one mentioned in the previous section. An actual system of the latter approach is described in the following paragraphs to show how a typical hyperspectral imaging system is used for wavelength calibration.

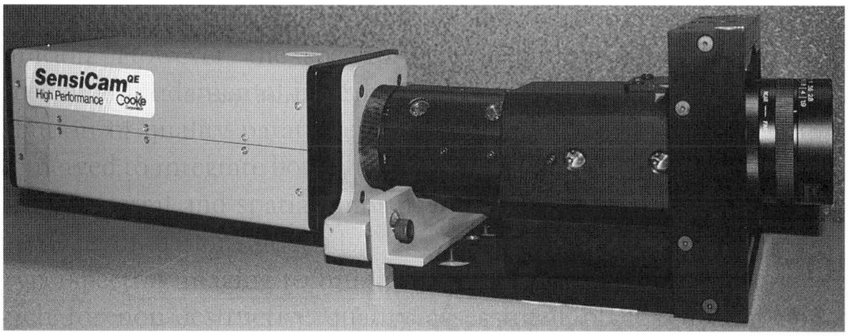

FIGURE 2.2 *ITD's VNIR 100E hyperspectral imaging system. (Full color version available on http://www.elsevierdirect.com/companions/9780123747532)*

The VNIR 100E hyperspectral imaging system (Figure 2.2) developed by the Institute for Technology Development (ITD, Stennis Space Center, MS 39529, USA) is a pushbroom line-scanning hyperspectral imaging system. The VNIR 100E incorporates a patented line-scanning technique (Mao, 2000) that requires no relative movement between the target and the sensor. The scanning motion for the data collection is performed by moving the lens across the focal plane of the camera on a motorized stage. The hyperspectral focal plane scanner eliminates the requirement of a mobile platform in a pushbroom scanning system. For this system, the front lens is driven by a Model Stage A-10 motor with a NCS-1S Motor controller (Newmark Systems Inc., Mission Viejo, CA, USA).

The hyperspectral imaging system uses a prism–grating–prism to separate incoming light into its component wavelengths with a high signal-to-noise ratio. The prism is located in an ImSpector V10E spectrograph from Specim (Spectral Imaging Ltd, Oulu, Finland) with a 30 μm entrance slit. The spectral range of the spectrograph is from 400 to 1000 nm. In this system, image data are recorded by a 12-bit CCD SensiCam QE (The Cooke Corporation, Romulus, MI, USA) digital camera with a 1376×1040 pixel array (Yao *et al.*, 2008). The system uses thermo–electrical cooling to cool the image sensor down to $-12\,°C$. The variable binning capability of the camera allows image acquisition at user-specified spatial and spectral resolutions. Each output image contains a complete reflectance spectrum from 400 to 1000 nm. Even though several lines of data from the detector can be binned together, wavelength calibration is always implemented at the maximum detector resolution (1×1 binning) along the vertical dimension on the CCD array. This provides wavelength to detector array line assignments no matter what type of binning is used.

To calibrate the system, the following items are needed:

1. a light source that produces spectral lines at fixed wavelengths,
2. regression programs, and
3. (optional) integrating sphere, or standard white reflectance surface such as Spectralon® surface.

2.2.1.3. Wavelength calibration procedure

The light source used to produce spectral lines at fixed wavelengths can be a spectral calibration lamp such as a mercury–argon lamp or a laser. This is because the calibration lamps and lasers can provide emission peaks at known wavelengths. For example, Park *et al.* (2002) and Lawrence *et al.* (Lawrence, Park *et al.*, 2003; Lawrence, Windham *et al.*, 2003) used mercury–argon (Hg–Ar) and krypton (Kr) calibration lamps (Oriel Model 6035 and 6031, Oriel Instruments, Stratford, CT, USA) together with an Oriel 6060 DC power supply to provide calibration wavelengths from about 400 to 900 nm. In addition, a Uniphase Model 1653 helium–neon laser and a Melles Griot Model 05-LHR-151 helium–neon laser were also used as spectral standards at 543.5 and 632.8 nm. Other studies mentioned slightly different types of wavelength calibration lamps such as a custom-made Ne lamp (Tseng *et al.*, 1993), an Oriel lamp set including mercury–neon (Hg–Ne), krypton, helium (He), and neon (Ne) lamps (Mehl *et al.*, 2002), a mercury vapor lamp from Pacific Precision Instruments (Concord, CA, USA) (Cho *et al.*, 1995), and a mercury–neon lamp from Oriel Instrument (Chao *et al.*, 2007a; Kim *et al.*, 2008). In general, these calibration lamps produce narrow, intense lines from the excitation of various rare gases and metal vapors at different fixed known wavelengths. They are widely used for wavelength calibration of spectroscopic instruments such as monochromators, spectrographs, spectral radiometers, and imaging spectrometers. Figure 2.3 shows a calibration pencil lamp from Oriel and the emission peaks for a mercury–argon (Hg–Ar) lamp.

There are three instrument setups that can be used to perform wavelength calibration data with the calibration lamps. The goal is to obtain uniformly distributed spectral data for wavelength calibration. The first setup requires the use of an integrating sphere. An integrating sphere is an optical device with a hollow cavity. Its interior is coated white to create highly diffuse reflectivity. An integrating sphere can provide spatially-uniform diffuse light. Consequently, when acquiring calibration data with the hyperspectral camera, the integrating sphere can disperse the spectral peaks uniformly across the length of the spectrograph slit. Lawrence *et al.* (Lawrence, Park *et al.*, 2003; Lawrence, Windham *et al.*, 2003) used a 30.5 cm (12 inch)

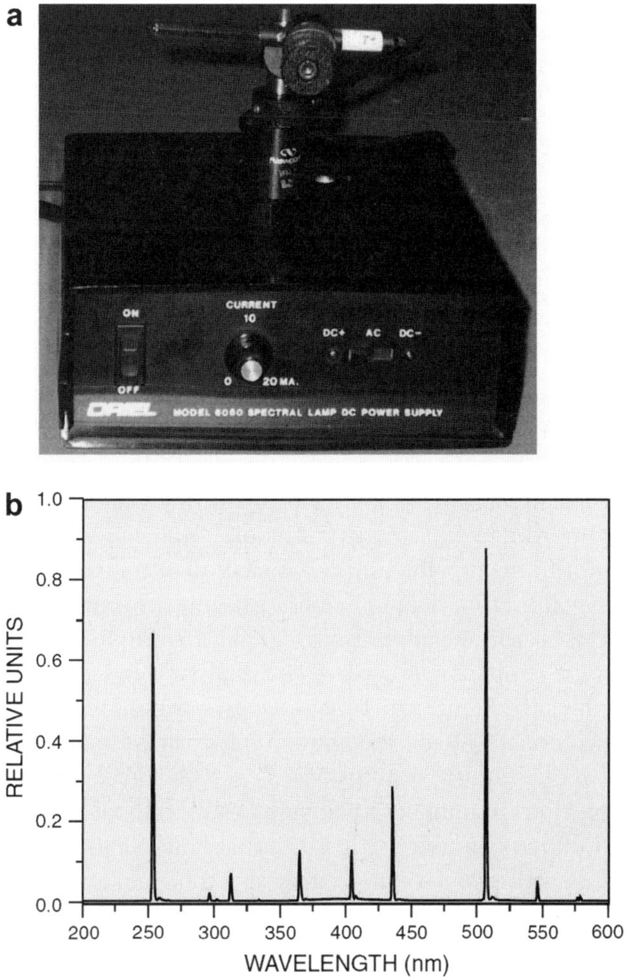

FIGURE 2.3 *Wavelength calibration: (a) calibration pencil light (Hg–Ar, Oriel Model 6035) with power supply; (b) output spectrum of 6035 Hg-Ar Lamp, run at 18 mA, measured with MS257™ 1/4 m Monochromator with 50 μm slits (Oriel Instruments, Stratford, CT) (Full color version available on http://www.elsevierdirect.com/companions/ 978012374753)*

integrating sphere (Model OL-455-12-1, Optronic Laboratories, Inc., USA). The sphere had a 1.27 cm (0.5 inch) input port behind the integrating sphere baffle for the insertion of additional calibration sources such as the calibration lamps. The second setup is to place the calibration lamp above a standard reference surface (Kim *et al.*, 2008). The standard reference surface used by Kim *et al.* (2008) was a $30 \times 30\ cm^2$, 99% diffuse reflectance polytetrafluoroethylene (Spectralon™) reference panel (SRT-99-120) from Labsphere

(North Sutton, NH, USA). In this study, an Hg–Ne pencil light was placed 25 cm above and at 5° forward angle over the reference surface. The pencil light was positioned horizontally. The third setup is to place the calibration pencil light directly underneath the entrance slit of the spectrograph with a distance of approximately 5 cm. Calibration data are then acquired with all ambient light off. In a similar setup to calibrate wavelength of a spectrometer, Chen *et al.* (1996) used a high intensity short wave ultraviolet light source (Hg (Ar) Penray™, UVP Inc., San Gabriel, CA, USA). It was placed near the probe receptor to ensure the accuracy of the spectral calibration.

Actual data acquisition can be started after the calibration lamp is turned on for several minutes to allow time for the lamp to reach a stable condition. For example, when using a mercury–neon (Hg–Ne) pencil light, neon is a starter gas. Light output from the pencil light in the first minute is influenced by the neon. The pencil light then automatically switches to mercury after the first minute and then the influence of mercury will dominate the output spectrum (Kim *et al.*, 2008; Yang *et al.*, 2009). Thus, data acquisition should begin at this stage if the purpose is to acquire mercury lines. Another issue in taking calibration data is camera integration time. The integration time for the hyperspectral camera is adjusted to ensure that the highest peak of the calibration lamps is not saturated. Finally, a 1 × 1 binning is used in the wavelength calibration process in order to assign a wavelength to each line of the detector array. Band wavelength information can be subsequently calculated for other binning settings based on these discrete values.

Once calibration data are obtained, a program such as ENVI (ITT Visual Information Solutions, Boulder, CO, USA) that has been designed to process hyperspectral data can be used to extract spectral information. A region of interest (ROI), preferably from the center of the image, is normally generated to obtain mean spectral information. A spectral profile of different pixels in the image can then be produced. This profile should appear similar to the spectral profile in Figure 2.3b. Peak values in the spectral profile can be assigned to the known peaks of the target light sources. These assignments are then used in the subsequent regression process to calculate a wavelength for each line of the detector array. When selecting peak features, Bristow & Kerber (2008) have set up several guidelines:

- *They will not be blended at the resolution of the instrument in question.*
- *They are bright enough to be seen in realistic calibration exposures.*
- *They provide adequate coverage (baseline and density) across the wavelength range, detector co-ordinates and spectral orders.*

The last step in the calibration process is to run a regression using the selected peak features. The regression can be based on linear, quadratic, cubic, and trigonometric equations. The key point at this step is not to overfit the regression model. Past studies have used a broad distribution in applying these equations. Below, each equation will be presented with a list of related works:

Linear (Kim *et al.*, 2008; Mehl *et al.*, 2002; Naganathan *et al.*, 2008; Xing *et al.*, 2008):

$$\lambda_i = \lambda_0 + C_1 X_i \tag{2.1}$$

Quadratic (Chao *et al.*, 2007a, 2007b, 2008; Yang *et al.*, 2006, 2009)

$$\lambda_i = \lambda_0 + C_1 X_i + C_2 X_i^2 \tag{2.2}$$

Cubic (Lawrence, Park *et al.*, 2003; Lawrence, Windham *et al.*, 2003; Park *et al.*, 2006):

$$\lambda_i = \lambda_0 + C_1 X_i + C_2 X_i^2 + C_3 X_i^3 \tag{2.3}$$

Trigonometric 1 (Cho *et al.*, 1995):

$$\lambda_i = \lambda_0 + C_1 X_i + C_2 \sin\left(X_i \frac{\pi}{n_p}\right) \tag{2.4}$$

Trigonometric 2 (Cho *et al.*, 1995):

$$\lambda_i = \lambda_0 + C_1 X_i + C_2 \sin\left(X_i \cdot \frac{\pi}{n_p}\right) + C_3 \cos\left(X_i \cdot \frac{\pi}{n_p}\right) \tag{2.5}$$

where λ_i is the wavelength in nm of band *i*, λ_0 is the wavelength of band *0*. The coefficient C_1 is the first coefficient (nm/band), C_2 is the second coefficient (nm/band2), and C_3 is the third coefficient (nm/band3) (if any) for the first three models. The coefficients C_1, C_2, and C_3 in trigonometric models (1) and (2) are the first, second, and third coefficients of a Fourier series expansion. X_i is peak position in band number (or pixel number). n_p is the number of bands within a given spectral range.

As an example, Table 2.1 presents some selected peak wavelengths along with their corresponding band numbers. Data were acquired using an Hg–Ar lamp with the hyperspectral imaging system described in section 2.2.1.2. Both mercury and argon lines were used in the calibration. The first two columns are the selected peak wavelength and the corresponding band numbers. The selected wavelength for band 36, 87, 264, 316, 502, and

Table 2.1 Example data for wavelength calibration using Hg–Ar lamp

Peak wavelength (nm)	Band number	Calibrated wavelength (nm)
404.66	36	404.61
435.84	87	435.99
546.08	264	546.08
579.07	316	578.77
696.54	502	696.98
763.51	605	763.30

605 is 404.66 nm, 435.84 nm, 546.08 nm, 578.07 nm, 696.54 nm, and 763.51 nm, respectively. To run a regression analysis, the peak wavelength is used as the dependent variable and the band number is used as the independent variable. In this case, a quadratic regression function is generated as:

$$\lambda_i = 382.54 + 0.61X_i + 2.90\text{E} - 05X_i^2 \quad (2.6)$$

The resulted wavelength for each selected band after calibration is listed in column three in Table 2.1. The calibrated wavelength for band 36, 87, 264, 316, 502, and 605 is 404.61 nm, 435.99 nm, 546.08 nm, 578.77 nm, 696.98 nm, and 763.30 nm, respectively. Once the regression equation is established, wavelength information for every band can be subsequently calculated. The resulting average bandwidth is 0.63 nm. The regression results are also plotted in Figure 2.4 with regression coefficient of determination R^2 being equal to 0.999996. The rule of thumb is that this number should be very close to 1. If it is not the case, the assignment of wavelength

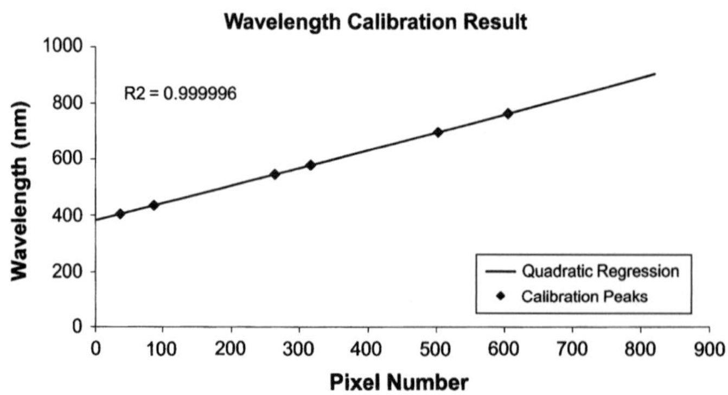

FIGURE 2.4 *Quadratic regression curve for wavelength calibration. The pixel number is also known as band number*

might be incorrect. In this case it is possible that another regression equation that fits the data better should be used. Cho *et al.* (1995) also used standard error of estimate (*SEE*) as a criterion for the goodness of fit when comparing regression Equations (2.1) through (2.5). *SEE* is described as:

$$SEE = \sqrt{\frac{\sum_1^n (\widehat{\lambda}_i - \lambda_i)^2}{n - p}} \quad (2.7)$$

where n is the number of calibration wavelengths, p is the number of coefficients in the regression models, and $\widehat{\lambda}_i$ and λ_i are the regression estimated and actual wavelengths of known mercury lines, respectively.

Instead of using all available peaks to run a regression across the wavelength range, an alternative approach is to perform a segmented linear regression. In the segmented linear regression, a linear regression is implemented only between two adjacent wavelength peaks. Compared with the previous approach, the segmented linear regression guarantees wavelengths for the selected band numbers with emission peaks staying the same after the regression is completed. The latter approach also results in variable bandwidths for different regression segment regions. Difference between the two regression approaches within the regression wavelength range is plotted in Figure 2.5. It can be seen that the difference is generally within 0.3 nm. The largest difference within the regression peak wavelength range is about 0.4 nm at 696.54 nm. Another observation is that outside the regression peak wavelength range the difference gradually increases.

2.2.2. Radiometric Calibration

The detector array of a hyperspectral imaging system's camera, such as the one mentioned previously, records digital counts (DN) of at-sensor radiance from the target. This radiance is called uncorrected radiance for the

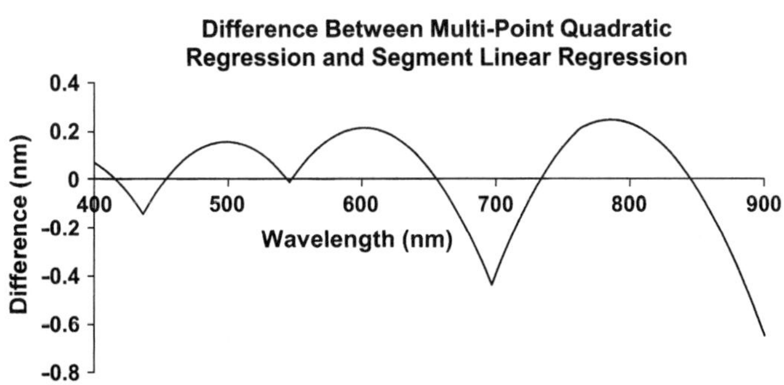

FIGURE 2.5 *Difference between two regression approaches*

hyperspectral imaging system. Because of the differences in camera quantum efficiency and physical configuration of hyperspectral imaging systems, the uncorrected radiance for different hyperspectral imaging system may not be the same even when imaging the same target under the same imaging conditions. In order to perform cross sensor comparison, radiometric calibration of hyperspectral image data is required. Radiometric calibration also makes it easier to adopt results and knowledge learned from one study to other similar investigations. In addition, the radiometric calibration process reduces errors from uncorrected data. Furthermore, there are other advantages (Clark *et al.*, 2002) from calibrated surface reflectance spectra over uncorrected radiance data based on the United State Geological Survey (USGS). First, the shapes of the calibrated spectra are mainly affected by the chemical and physical properties of surface materials. Secondly, the calibrated spectra can be compared with other spectra measurements of known materials. Lastly, spectroscopic methods may be used to analyze the calibrated spectra to isolate absorption features and relate them to chemical bonds and physical properties of materials.

Several radiometric calibration techniques are discussed here including: radiometric calibration to percentage reflectance; radiometric calibration to relative reflectance; radiometric calibration of transmittance; and radiometric normalization.

2.2.2.1. Radiometric calibration to percentage reflectance

The radiometric reflectance calibration process involves a pixel-by-pixel calibration of the hyperspectral image data to percentage reflectance. This is the most common approach for radiometric calibration and is widely used in spectral-based food safety and quality assessment research. Some of these research activities include apple bruise and stem-end/calyx regions detection (Xing *et al.*, 2007), citrus canker detection (Qin *et al.*, 2008), defect detection on apples (Mehl *et al.*, 2002), apple bruise detection (Lu, 2003), fecal contamination on apples (Kim *et al.*, 2002), assessment of chilling injury in cucumbers (Liu *et al.*, 2006), grain attribute measurements (Armstrong, 2006), corn genotype differentiation (Yao *et al.*, 2004), *Fusarium* head blight (SCAB) detection in wheat (Delwiche & Kim, 2000), optical sorting of pistachio nut with defects (Haff & Pearson, 2006), differentiation of wholesome and systemically diseased chicken carcasses (Chao *et al.*, 2007a, 2007b, 2008), fecal contamination detection on poultry carcasses (Heitschmidt *et al.*, 2007), identification of fecal and ingesta contamination on poultry carcasses (Lawrence, Windham *et al.*, 2003b), chicken inspection (Yang *et al.*, 2006), beef tenderness prediction (Naganathan *et al.*, 2008), differentiation of toxigenic fungi (Yao *et al.*, 2008), and contamination detection on the surface of processing equipment (Cho *et al.*, 2007), etc.

Using hyperspectral imagery for food quality and safety inspections is a natural extension from using such data in space or terrestrial remote sensing. Different from traditional earth-based hyperspectral remote sensing applications where solar radiation is the sole source for target illumination, the aforementioned research activities all utilized artificial light. The artificial light can be fiber light (Armstrong, 2006; Cho *et al.*, 2007; Kim *et al.*, 2001; Lawrence, Windham *et al.*, 2003; Lu, 2003; Pearson & Wicklow, 2006), tungsten halogen light (Haff & Pearson, 2006; Yao *et al.*, 2008), tungsten halogen light in a diffuse lighting chamber (Naganathan *et al.*, 2008), light emitting diode (LED) (Chao *et al.*, 2007a; Lawrence *et al.*, 2007). These lab-based research experiments are normally implemented in an indoor environment in close distance. Thus, atmospheric effect correction, which is a major part in calibrating space or airborne-based hyperspectral imagery, is not necessary for lab-based hyperspectral applications. Still, a pixel-by-pixel radiometric calibration to convert at-sensor radiance to percent reflectance is necessary. The calibration can minimize or eliminate the inherent spatial nonuniformity in the artificial light intensity on the target area. In addition, the intensity of the artificial light source also varies over time and the radiometric calibration process can compensate for such variations.

For radiometric reflectance calibration, the general approach includes collecting reference image, dark current image, and sample images. Then percent reflectance can be computed on a pixel-by-pixel basis using a transformation equation, usually through a computer program that runs in batch mode.

Reference Image and White Diffuse Reflectance Standard

Reference image is taken normally when the imaging system can collect data from a standard reflectance surface in the same image with the target phenomenon. Ideally, a standard reflectance surface should represent 100% uniform reflectance to enable proper conversion of sample images from at-sensor radiance to percent reflectance. Currently, the widely used standard reflectance surface is the NIST (National Institute of Standards and Technology) certified 99% Spectralon® White Diffuse Reflectance (SRT-99) target from Labsphere, Inc. (North Sutton, NH, USA).

To make the 99% Spectralon® White Diffuse Reflectance target, Labsphere uses their patented diffuse reflectance material, Spectralon. Spectralon is claimed to have the highest diffuse reflectance of any known material or coating over the ultraviolet (UV)–visible (VIS)–near-infrared (NIR) region of the spectrum. It is hydrophobic and is thermally stable to

350 °C. The material exhibits nearly Lambertian (perfectly diffuse) properties and provides consistent uniform reflectance. For its performance, the reflectance is generally >99% reflective over a range from 400 nm to 1500 nm and >95% reflective from 250 nm to 2500 nm. Its calibration is traceable with NIST. Because of the diffuse reflectance properties of Spectralon, the Spectralon® White Diffuse Reflectance target can maintain a constant contrast over a wide range of lighting conditions. Thus it is ideal for field spectral calibration as well as for lab spectral calibration. Spectralon is also a durable material that provides highly accurate, reproducible data. It is durable and optically stable over time, and is resistant to UV degradation. Because Spectralon is a thermoplastic resin, it can be made into different shapes for different application purposes. The Spectralon material is normally pressed into a rugged anodized aluminum frame. Spectralon® White Diffuse Reflectance target is available from Labsphere at different sizes (from SRT-99-020, 2 × 2 inch to SRT-99-240, 24 × 24 inch). The more practical sizes used for food quality and safety research are 10 × 10 inch and 12 × 12 inch to cover the target viewing area of hyperspectral imaging systems. Figure 2.6 shows typical Spectralon® White Diffuse Reflectance target panels with its reflectance measurement. Further details on reflectance standards can also be found from Springsteen (1999).

In addition to Spectralon® White Diffuse Reflectance target, other targets such as the WS-1 Diffuse Reflectance Standard from Ocean Optics (Dunedin, FL, USA) is also available for food quality research using hyperspectral imagery (Lin *et al.*, 2006). The WS-1 Diffuse Reflectance Standard is made of PTFE, a diffuse white plastic that provides a Lambertian reference surface. The material is hydrophobic, chemically inert, and stable. For its performance, the reflectance is generally > 98% reflective from 250 to 1500 nm and > 95% reflective from 250 to 2200 nm.

The integration time is normally adjusted when taking the 99% reference image. The goal is to keep the magnitude of the spectral response of a camera within the maximum range of a camera's detector array. Different intensity levels such as 30% (Cho *et al.*, 2007) or 90% (Delwiche & Kim, 2000; Kim *et al.*, 2001) of the full dynamic range of the detector array were reported to be used in different applications. A sample reference mean spectral curve is presented in Figure 2.6(b) for the camera system presented in Section 2.2.1.2.

Dark Current Image

Modern hyperspectral imaging systems typically use InGaAs (indium gallium arsenide) or CCD arrays for image acquisition. For such image sensors, there is an electronic current flowing in the detector arrays even without light shining on it. This current is called the electronic dark current

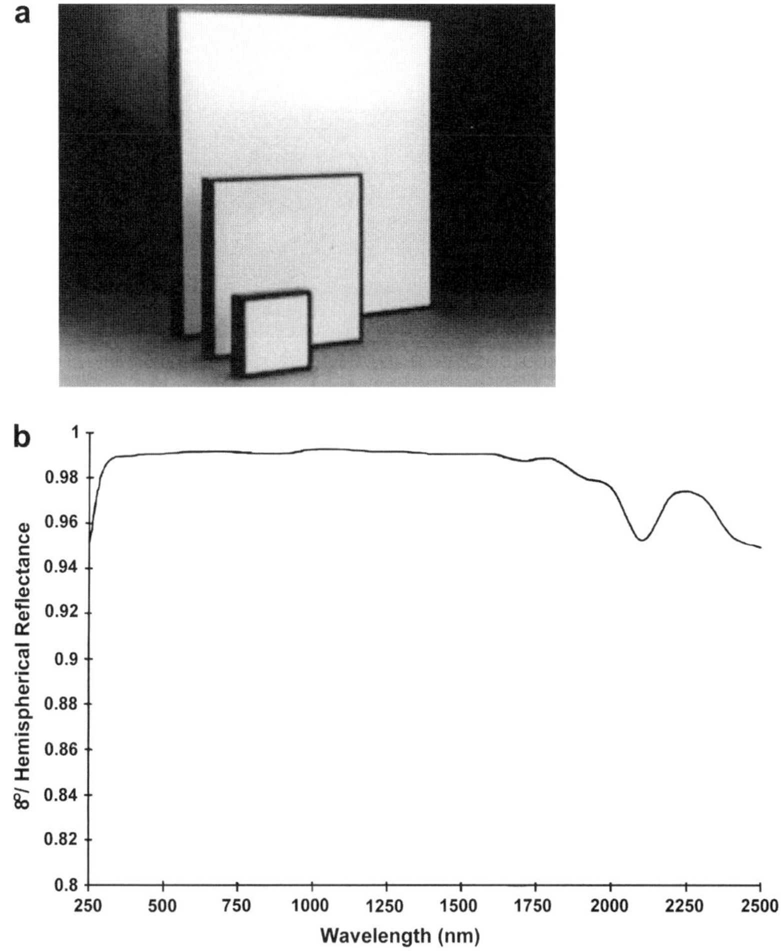

FIGURE 2.6 *White diffuse reflectance standard: (a) typical 99% Spectralon® White Diffuse Reflectance targets; (b) reflectance curve (courtesy of Labsphere, Inc.)*

or simply dark current. Dark current is generated from thermally induced electron hole pairs. Thus, dark current is dependent on temperature. Dark current is also proportional to integration time. For these reasons, imaging devices for scientific applications are normally cooled to minimize dark current level. For example, a SensiCam QE (The Cooke Corporation, Romulus, MI, USA) is cooled to $-12\,^{\circ}\mathrm{C}$. The cooling mechanism is thermoelectrical and it uses a two-stage Peltier cooler with forced air cooling. This type of camera is used by Delwiche & Kim (2000), Kim *et al.* (2001), Lawrence *et al.* (Lawrence, Park *et al.*, 2003; Lawrence, Windham *et al.*, 2003; Lawrence *et al.*, 2007), and Yao *et al.* (2008) for their research. A sample dark

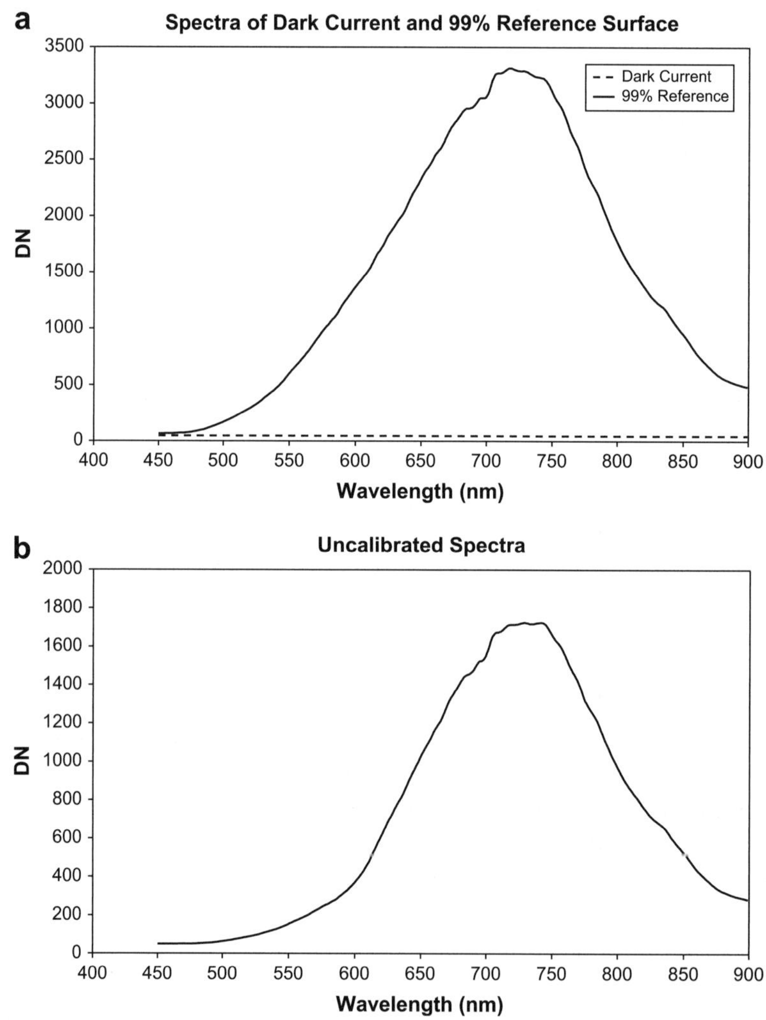

FIGURE 2.7 *Dark current image: (a) typical mean reference spectra (99%) and mean dark current curve for a SensiCam QE camera (taken by ITD VNIR-100E hyperspectral imaging system); (b) uncalibrated mean spectra of corn kernel samples*

current spectral curve is presented in Figure 2.7(a). Uncalibrated mean spectra collected from corn kernels are presented in Figure 2.7(b).

A relatively new type of CCD camera called electron-multiplying CCD (EMCCD) (Chao *et al.*, 2007a; Cho *et al.*, 2007; Qin *et al.*, 2008) uses a three-stage Peltier cooler with adjustable cooling temperature to further reduce sensor dark current. For an EMCCD camera the lowest temperature can go as low as −60 °C depending on application (Photometrics, Tucson, AZ, USA).

To take a dark current image, the same integration time is used as for acquiring the target image. Many practices have been employed to reduce the ambient light, such as blocking the light entrance of fiber-optic cables (Armstrong, 2006), covering the lens with a lens cap and turning off all other light sources (Delwiche & Kim, 2000; Mehl *et al.*, 2002; Naganathan *et al.*, 2008; Qin *et al.*, 2008), or covering the lens with a non-reflective opaque black fabric (Chao *et al.*, 2007a, 2007b, 2008).

Normally, reference and dark current images are taken before acquiring sample images. Some researchers (Delwiche & Kim, 2000; Kim *et al.*, 2001) used an average of 20 reference and 20 dark current images for calibration purposes. Because imaging system and lighting conditions are relatively stable within a short period of time in lab conditions, it is not required to take calibration data for each sample image and the calibration data could be used for the same imaging day (Chao *et al.*, 2007b). Repetitive acquisition of calibration images can also be made after a fixed number of samples (Haff & Pearson, 2006; Peng & Lu, 2006) or at certain time intervals (Naganathan *et al.*, 2008).

Sample Image and Calibration

When taking sample images, the same integration time and imaging settings as used for acquiring the reference and dark images should be used. An uncalibrated sample mean spectral curve for corn kernel is presented in Figure 2.7(b). The following equation can be used to convert raw digital counts of reflectance into percent reflectance:

$$Reflectance_\lambda = \frac{S_\lambda - D_\lambda}{R_\lambda - D_\lambda} \times 100\% \tag{2.8}$$

where $Reflectance_\lambda$ is the reflectance at wavelength λ, S_λ is the sample intensity at wavelength λ, D_λ is the dark intensity at wavelength λ, and R_λ is the reference intensity at wavelength λ. Eventually, the calibrated reflectance value lies in the range from 0% to 100%. The image in Figure 2.8a is a true color representation of the calibrated corn sample, while Figure 2.8b shows the mean calibrated spectral reflectance curve from the corn kernels.

There also exists a variation for Equation (2.8), when the reflectivity of the reference surface is considered. The variation is as follows:

$$Reflectance_\lambda = \frac{S_\lambda - D_\lambda}{R_\lambda - D_\lambda} RC_\lambda \times 100\% \tag{2.9}$$

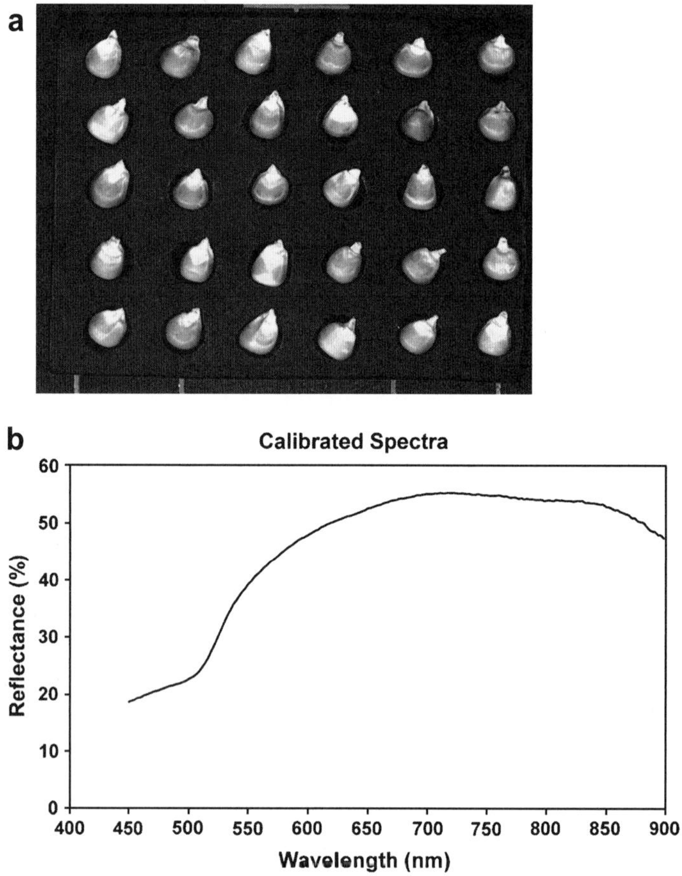

FIGURE 2.8 *Corn sample and its calibrated spectra: (a) corn sample images; (b) mean calibrated spectra of corn samples. (Full color version available on http://www.elsevierdirect.com/companions/9780123747532)*

Here RC_λ is the correction factor for the reference panel. For the white Spectralon panel mentioned previously, it can be assumed that the white Spectralon panel has a correction factor of 0.99 in the spectral range covered by some hyperspectral imaging systems. Thus, $RC_\lambda = 1.0$ was used in these studies (Delwiche & Kim, 2000; Kim *et al.*, 2001). It can be seen that Equations (2.8) and (2.9) have the same representation if the reference surface has a correction factor close to 1.

Calibration Verification

In order to validate the reflectance calibration results, a NIST certified gradient reference panel with known reflectance values can be used.

Lawrence *et al.* (Lawrence, Park *et al.*, 2003; Lawrence, Windham *et al.*, 2003) used a gradient Spectralon panel consisting of four vertical sections with nominal reflectance values of 99%, 50%, 25%, and 12% from Labsphere (Model SRT-MS-100). The studies pointed out that the calibration can reduce errors across the panel, especially along the edge and at high reflectance values. For example, the raw data values for the 99% reflectance portion of the gradient panel displayed drops near the detector edge. The calibration can correct the drop and the effect of calibration is quite evident (Lawrence, Park *et al.*, 2003). Mean and standard deviation of percentage reflectance values are constant within the middle wavelength region and vary significantly at the extremes. The studies further reported that the observed trend follows the errors reported by the spectrograph manufacturer.

2.2.2.2. Relative reflectance calibration

A sensor's raw digital count can also be calibrated in a relative way. Similar to the previous percentage reflectance approach, the relative reflectance calibration method requires image acquisition of reference, dark current, and sample images. The same equation (Eq. 2.8) presented in the previous section is also used for relative reflectance calculation. However, because this approach only calibrates the sample image to a relative reference standard, it is not necessary to use a 99% or 100% white diffuse reflectance standard. Some researchers (Ariana *et al.*, 2006; Ariana & Lu, 2008; Lu, 2007; Peng & Lu, 2006) used a Teflon surface as reference standard. On the other hand, Gowen *et al.* (2008) used a uniform white ceramic surface which was calibrated against a tile with known reflectance. Meanwhile, Ariana & Lu (2008) found that other materials such as PVC (polyvinyl chloride) could also be used for relative reflectance calibration in quality evaluation of pickling cucumbers. One consideration for choosing PVC as the reference surface is because of its low reflecting property. This property matched the low reflectance of cucumbers in the visible region in its specific application.

The relative reflectance calibration method has been used in several applications such as bruise detection on pickling cucumbers (Ariana *et al.*, 2006), apple firmness estimation (Peng & Lu, 2006), nondestructive measurement of firmness and soluble solids content for apple (Lu, 2007), pickling cucumber quality evaluation (Ariana & Lu, 2008), and definition of quality deterioration in sliced mushrooms (Gowen *et al.*, 2008). One advantage of the method is it avoids the use of expensive 99% or 100% white diffuse reflectance standards and still achieves the research goals. The calibration process can still compensate for the spatial nonuniformity from light, aging of light, and other factors such as power supply fluctuation, etc. The drawback is that it is difficult to compare results generated from this calibration with other approaches, especially when a direct spectral comparison is needed.

2.2.2.3. Calibration of hyperspectral transmittance image

Hyperspectral reflectance imagery has proven to be a good tool for external inspection and evaluation for food quality and safety applications. For studying internal properties of food, hyperspectral images of transmittance can be useful. It was reported that NIR spectroscopy in transmittance mode can penetrate the deeper region of fruit (>2 mm) compared with that in reflectance mode (McGlone & Martinsen, 2004). The internal property of targets can be analyzed using light absorption within the detector's spectral range. One drawback of transmittance imaging is the low signal level from light attenuation due to light scattering and absorption.

Hyperspectral transmission measurement involves projecting light at one side of the target and recording light transmitted through the target at the opposite side with a hyperspectral imager. Recently research activity using hyperspectral transmittance image for food quality and safety have been reported in corn kernel analysis (Cogdill *et al.*, 2004), detection of pits in cherries (Qin & Lu, 2005), egg embryo development detection (Lawrence *et al.*, 2006), quality assessment of pickling cucumbers (Kavdir *et al.*, 2007), bone fragment detection in chicken breast fillets (Yoon *et al.*, 2008), detection of insects in cherries (Xing *et al.*, 2008), and defect detection in cucumbers (Ariana & Lu, 2008). These studies demonstrated that hyperspectral transmittance imagery has the potential for food quality evaluation and detection of defects in food.

To calibrate hyperspectral transmittance images, Equation (2.8) used in reflectance calibration is also applicable to calculate the calibrated relative transmittance. Similarly, a dark current image and a reference transmittance image are needed in the calibration equation. It was reported (Ariana & Lu, 2008; Qin & Lu, 2005) that the reference transmittance image could be collected using a white Teflon disk due to its relatively flat transmittance responses over the spectral range of 450–1000 nm. In addition, an absorption transformation (Clark *et al.*, 2003) is sometimes used to convert the relative transmittance into absorbance unit based on the equation below (Cogdill *et al.*, 2004):

$$A = \log\left(\frac{1}{I}\right) \tag{2.10}$$

where I is the transmittance intensity, and A is the calculated absorbance spectrum.

2.2.2.4. Radiometric normalization

One spectral preprocessing technique known as image normalization can be used to standardize input data and reduce light variations in the reflectance data (Kavdir & Guyer, 2002). For example, one study (Cheng *et al.*, 2003) on apples found that a dark-colored apple has a lower light reflectance than a bright-colored apple in the near-infrared spectrum from 700 to 1000 nm. This difference in brightness levels could cause detection errors, especially for bright-colored defective apples and dark-colored good apples. Thus, data normalization was applied to the original NIR image to avoid these kinds of errors by eliminating the effect of the brightness variations in the original data. Generally, normalized data can be insensitive to surface orientation, illumination direction, and intensity. Consequently, normalized data could be regarded as independent of the illumination spectral power distribution, illumination direction (Polder *et al.*, 2002), and object geometry (Lu, 2003; Polder *et al.*, 2002). Normalization has been found in applications such as measurement of tomato ripeness (Polder *et al.*, 2002), detection of apple bruise (Lu, 2003), recognition of apple stem-end/calyx, prediction of firmness and sugar content of sweet cherries (Lu, 2001), apple sorting (Kavdir & Guyer, 2002), and prediction of beef tenderness (Cluff *et al.*, 2008).

For normalization implementation, many approaches may be used. Some equations appearing in literatures are shown below:

Normalizing reflectance data for each band to the average of each scanning line of the same image band (Lu, 2003):

$$\boldsymbol{R}_\lambda = \frac{R_\lambda}{\sum R_\lambda / N} \tag{2.11}$$

where $\boldsymbol{R}_\lambda$ is the resulted relative reflectance, R_λ is the reflectance measurement, and N is the number of pixels for the scanning.

Normalizing reflectance data for each band of each pixel to the sum of all bands of the same pixel (Polder *et al.*, 2002):

$$\boldsymbol{R}_\lambda = \frac{R_\lambda}{\sum_\lambda R_\lambda} \tag{2.12}$$

Normalizing reflectance data to the largest intensity within the image (Cheng *et al.*, 2003):

$$NNI(x,y) = c_0 \frac{ONI(x,y)}{I_{\max}(x,y)} \tag{2.13}$$

where $ONI(x, y)$ is original NIR image, $NNI(x, y)$ is normalized NIR image, $I_{max}f(x, y) = max[ONI(x, y)]$ for all (x, y), and $C_0 = constant$ equals to 255 in the paper (Cheng *et al.*, 2003).

The internal average relative reflectance (IARR) normalization procedure described by Schowengerdt (1997) is another approach for normalization. It attempts to normalize each pixel's spectrum by the average spectrum of the entire scene. The procedure was used by Yao *et al.* (2006) to study aflatoxin-contaminated corn kernels.

2.2.3. Noise Reduction and Removal

For a hyperspectral imaging system, there exist many different types of random noise including camera read-out noise, wire connection and data transfer noise between camera and computers, electronic noise inherent to the camera such as dark current, and noise from digitizing while doing analog to digital (A/D) conversion. These noise values will obviously impact results produced from subsequent image analysis. In the spectral preprocessing stage, the random noise needs to be dealt with through specific processing steps. Five techniques for noise reduction and removal will be introduced here: 1. dark current subtraction; 2. spectral low pass filtering; 3. Savitzky–Golay filtering; 4. noisy band removal; and 5. minimum noise fraction transformation.

2.2.3.1. Dark current subtraction

In the previous section the temperature-dependent dark current was introduced as an inherent property of a hyperspectral imaging system. Dark current data are normally collected together with a reference data set and then later used in a reflectance/transmittance calibration process. In some cases where reference data are not available, a reflectance calibration cannot be implemented. Instead of just using the raw sample data for data analysis, dark current can be subtracted from the sample data prior to further data analysis (Cluff *et al.*, 2008; Singh *et al.*, 2007; Wang and Paliwal., 2006). Although this simplified approach cannot achieve results obtained from a more stringent reflectance calibration by transforming the data with Equation (2.8), it will still be able to remove some inherent noise generated from a hyperspectral imaging system and is better than doing nothing for calibration. The equation for dark current subtraction is straightforward:

$$DN_\lambda = S_\lambda - D_\lambda \tag{2.14}$$

where DN_λ is the dark current removed sample data digital number at wavelength λ, S_λ is the raw sample intensity at wavelength λ, and D_λ is the dark intensity at wavelength λ.

2.2.3.2. Spectral low pass filtering

The most common and simplest way to smooth random noise from raw data is through a moving average process or spectral low pass filtering. Theoretically, a low pass filter preserves the local means and smoothes the input data signal. Generally, a low pass filter has a window size of an odd number and is running a moving average along the wavelength for each pixel based on:

$$Y_j^* = \frac{\sum_{i=-m}^{m} Y_{j+i}}{N} \tag{2.15}$$

where Y_j^* is the smoothed data at wavelength j, j is also the center location of the smoothing operation, $N = 2m + 1$ is the window size, m is half of the window size minus 1, and $Yj + i$ is the data point at band $j + i$ within the window. In Equation (2.15), it can be seen that the larger the window, the more smoothing the data experience. Various smoothing window sizes have been reported in past researches, such as five (Yao *et al.*, 2008) and nine (Heitschmidt *et al.*, 2007).

Alternatively, a spectral Gaussian filter can be used to reduce random noise and smooth data. Theoretically, a Gaussian filter smoothes the input signal by convolution with a Gaussian function. In studies of using hyperspectral data for fecal contamination detection (Park, *et al.*, 2007; Yoon *et al.*, 2007a, 2007b), a Gaussian filter with a 10 nm bandwidth as the full width at half maximum (FWHM) was applied as an optimal trim filter.

2.2.3.3. Savitzky–Golay filtering

Similar to the spectral low pass filtering method, the Savitzky–Golay filtering technique (Savitzky & Golay, 1964) also used a moving window of different odd-numbered window sizes in the process. However, unlike spectral low pass filtering, which uses an averaging approach, the Savitzky–Golay filtering technique uses a convolution approach to do the filtering calculation. It is stated mathematically as:

$$Y_j^* = \frac{\sum_{i=-m}^{m} C_i Y_{j+i}}{N} \tag{2.16}$$

where Y is the original spectral data, Y^* is the filtered spectral data, C_i is the convolution coefficient for the ith spectral value of the filter within the filter window, and N is the number of convolution integers. The filter consists of $2m + 1$ points, which is called filter size. Thus, m is half-width of the filter window. The index j is the running index of the original ordinate data table.

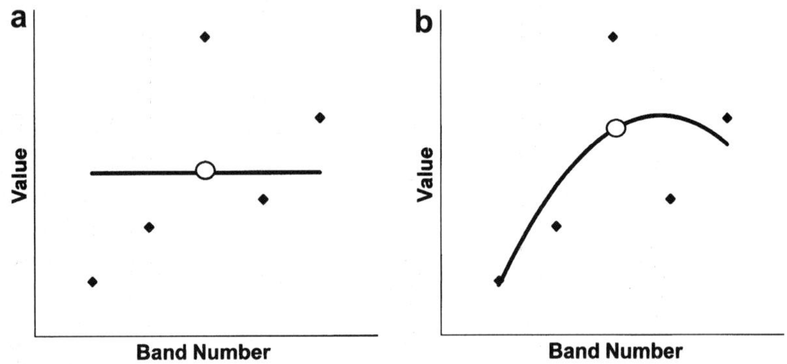

FIGURE 2.9 *Example of zeroth-order linear least-square smoothing, the resulted convolution point is marked as a circle: (a) simple moving average; (b) polynomial least-square smoothing*

The convolution is solved through fitting a polynomial equation based on the least-square concept. This polynomial least-square fitting is different from the linear least-square principle. The coefficients in the zeroth-order linear least-square fitting are all the same and the application of such fitting is essentially the same as the application of a simple moving window average. The coefficients in polynomial least-square fitting are different, thus they provide shaped filter windows for data smoothing. For example, Figure 2.9 provides smoothing results of the two approaches using a five-point filter window for comparison.

In the above five-point filter window, a quadratic polynomial can be approximated to describe the data curve through:

$$Y(x) = a_0 + a_1 x + a_2 x^2 \tag{2.17}$$

where a_0, a_1, and a_2 are coefficients for the polynomial fitting and x, y are spectral data points. Because this polynomial has three unknowns and five equations, it can be solved in a least-square way. Upon substituting results back to the center point of the convolution window, the spectral smoothing process is complete. Furthermore, instead of solving the least-square equation at every filter window, Savitzky & Golay (1964) provided several tables of coefficients for convolution calculation for various sizes of filter windows. The lookup tables were later corrected (Steinier *et al.*, 1972) for some errors presented in the original tables. These tables provide window size to as much as 25 points.

The advantage of the Savitzky–Golay filtering approach is that it greatly improves speed through the use of convolution instead of the more computationally demanding least-square calculation. One of the major drawbacks of the Savitzky–Golay filtering approach is that it truncates the data by m points at both ends. The reason is because the convolution process needs m points at both ends to calculate the required least-square values. So this method is not applicable to data with limited spectral sampling points but should not be a problem for large data sets. Savitzky & Golay (1964) also listed some requirements for using this method: (1) the points must be arranged in a way to have fixed, uniform intervals along the abscissa (spectral dimension); in the spectral image data, the intervals should represent image bandwidth for each adjacent band and in most cases is stated in "nanometer (nm)"; and (2) the sampling points under processing along the spectral dimension should form curves that must be continuous and smooth.

In recent years, the Savitzky–Golay filtering technique has been applied in food quality and safety related research using hyperspectral imaging technology. An incomplete list of these applications is: prediction of cherry firmness and sugar content (Lu, 2001), aflatoxin detection in single corn kernel (Pearson *et al.*, 2001), on-line measurement of grain quality (Maertens *et al.*, 2004), apple firmness estimation (Peng & Lu, 2006), quality assessment of pickling cucumbers (Kavdir *et al.*, 2007), detection of fecal/ingesta on poultry processing equipment (Chao *et al.*, 2008), paddy seeds inspection (Li *et al.*, 2008), quality evaluation of fresh pork (Hu *et al.*, 2008), and foodborne pathogen detection (Yoon *et al.*, 2009). When applying this method, special attention should be given to the filter size. Tsai and Philpot (1998) showed that the size of the convolved filter had the greatest effect on the degree of spectral smoothing. Different filter sizes should be tested to determine the size that provides the optimum noise removal without significant elimination of useful signal.

2.2.3.4. Noisy band removal

One feature of a hyperspectral camera such as the SensiCam QE camera mentioned previously is that the quantum efficiency of the camera drops significantly around the detector edges. This introduces high noisy bands at both ends of the camera's wavelength range. In addition, the effective spectral range of the spectrograph is also limited (Lawrence, Park *et al.*, 2003). The effective spectral range is also affected by the wavelength calibration process when known wavelength peaks from calibration lamps are selected. Thus, some image bands at both ends of the spectral range should be removed in the spectral preprocessing step. For example, it was reported that because image data from 400 nm to 450 nm and from 900 nm to 1000 nm contain relatively high levels of background noise (Yao *et al.*, 2008), image bands within the above spectral regions were discarded during the noisy band removal step.

2.2.3.5. Minimum noise fraction transformation

Minimum noise fraction (MNF) transformation is a procedure to remove noise in the image caused by the image sensor (ENVI, 2000; Green *et al.*, 1988). This procedure was used to enhance bruise feature and reduce data dimensionality (Lu, 2003). Certain features such as bruises on apples also show up in one MNF image band. It normally includes a forward minimum noise fraction and an inverse MNF transformation. The forward MNF transformation, which uses the original image and the dark current image, transforms the original image into data space with one part holding the large eigenvalues and coherent eigenimages, and a complementary part holding the near-unity eigenvalues and noise-dominated images. The transformation uses a noise covariance matrix which is computed with the dark current image. The inverse MNF transformation normally selects a group of the high ranking bands from the forward MNF transformed image (Yao & Tian, 2003). In order to avoid the potential to remove a signal when too few bands are used in the inverse MNF transformation, the eigenimages and eigenvalues should be examined to determine the best spectral subset for removing noise and minimizing signal loss.

2.3. CONCLUSIONS

As discussed throughout the chapter, hyperspectral imagery has been increasingly used in food quality and safety-related research and applications in recent years. In order to correctly understand the image data, it is important to properly preprocess the hyperspectral image prior to enhancing the quality of the data analysis. There are many different methods available for image spectral preprocessing. In summary, a systematic approach includes spectral wavelength calibration, radiometric calibration, and noise reduction and removal. Different techniques for implementing each calibration approach were discussed. Because the cost, time, and complexity associated with each preprocessing technique and calibration method varies significantly, it is the user's decision to choose the right spectral preprocessing method or combination of methods to respond to the needs of each food safety and food security application.

NOMENCLATURE

Symbols

a_0, a_1, a_2	coefficients for the polynomial fitting in Savitzky–Golay filtering equation
A	calculated absorbance spectrum
C_0	constant
C_1	first coefficient of wavelength regression, nm band^{-1}
C_2	second coefficient of wavelength regression, nm band^{-2}
C_3	third coefficient of wavelength regression, nm band^{-3}
C_i	convolution coefficient for the ith spectral value in Savitzky–Golay filtering equation
D_λ	dark intensity at wavelength λ
DN_λ	dark current removed sample data digital number at wavelength λ
I	transmittance intensity
$I_{max} f(x, y)$	equal to $max[ONI(x, y)]$ for all (x, y)
m	half of the window size minus 1 in Savitzky–Golay filtering equation
N	equal to $2m + 1$, window size in Savitzky–Golay filtering equation
N	number of pixels
$NNI(x, y)$	normalized NIR image
n_p	number of bands within a given spectral range
$ONI(x, y)$	original NIR image
$\boldsymbol{R}_\lambda$	resulted relative reflectance
R_λ	reference intensity at wavelength λ
RC_λ	correction factor for the reference panel
$Reflectance_\lambda$	reflectance at wavelength λ
S_λ	sample intensity at wavelength λ
x, y	spectral data for the polynomial fitting in Savitzky–Golay filtering equation
X_i	peak position
Y^*	smoothed data
Y	data point within the filter window
λ_i	wavelength of band i, nm
λ_0	wavelength of band 0, nm
$\widehat{\lambda}_i$	regression estimated wavelength, nm

Abbreviations

AOTF	acousto–optic tunable filter
A/D	analog to digital
CCD	charge-coupled device
DN	digital counts
EMCCD	electron-multiplying CCD
FWHM	full width at half maximum
He	helium
Hg–Ar	mercury–argon
Hg–Ne	mercury–neon
IARR	internal average relative reflectance
InGaAs	indium gallium arsenide
ITD	Institute for Technology Development
Kr	krypton
LCTF	liquid crystal tunable filter
LED	light emitting diode
MNF	minimum noise fraction
Ne	neon
NIR	near-infrared
NIST	National Institute of Standards and Technology
nm	nanometer
PVC	polyvinyl chloride
ROI	region of interest
SEE	standard error of estimate
USGS	United State Geological Survey
VNIR	visible near-infrared
VIS	visible
UV	ultraviolet

REFERENCES

Ariana, D. P., Lu, R., & Guyer, D. E. (2006). Near-infrared hyperspectral reflectance imaging for detection of bruises on pickling cucumbers. *Computers and Electronics in Agriculture, 53*, 60–70.

Ariana, D. P., & Lu, R. (2008). Quality evaluation of pickling cucumbers using hyperspectral reflectance and transmittance imaging: Part I. Development of a prototype. *Sensing and Instrumentation for Food Quality and Safety, 2*, 144–151.

Armstrong, P. R. (2006). Rapid single-kernel NIR measurement of grain and oil-seed attributes. *Applied Engineering in Agriculture, 22*(5), 767–772.

Bristow, P., & Kerber, F. (2008). Selection of wavelength calibration features for automatic format recovery in astronomical spectrographs. In Ground-based and Airborne Instrumentation for Astronomy II. *Proceedings of SPIE*, 70145V.

Chao, K., Yang, C. C., Chen, Y. R., Kim, M. S., & Chan, D. E. (2007a). Fast line-scan imaging system for broiler carcass inspection. *Sensing and Instrumentation for Food Quality and Safety, 1*, 62–71.

Chao, K., Yang, C. C., Chen, Y. R., Kim, M. S., & Chan, D. E. (2007b). Hyperspectral–multispectral line-scan imaging system for automated poultry carcass inspection applications for food safety. *Poultry Science, 86*, 2450–2460.

Chao, K., Yang, C. C., Kim, M. S., & Chan, D. E. (2008). High throughput spectral imaging system for wholesomeness inspection of chicken. *Applied Engineering in Agriculture, 24*(45), 475–485.

Chen, Y. R., Huffman, R. W., & Park, B. (1996). Changes in the visible/near-infrared spectra of chicken carcasses in storage. *Journal of Food Processing Engineering, 19*, 121–134.

Cheng, X., Tao, Y., Chen, Y. R., & Luo, Y. (2003). NIR/MIR dual-sensor machine vision system for online apple stem-end/calyx recognition. *Transactions of the ASAE, 46*(2), 551–558.

Cho, B. K., Chen, Y. R., & Kim, M. S. (2007). Multispectral detection of organic residues on poultry processing plant equipment based on hyperspectral reflectance imaging technique. *Computers and Electronics in Agriculture, 57*, 177–180.

Cho, J., Gemperline, P. J., & Walker, D. (1995). Wavelength calibration method for a CCD detector and multichannel fiber-optic probes. *Applied Spectroscopy, 49*(12), 1841–1845.

Clark, C. J., McGlone, V. A., & Jordan, R. B. (2003). Detection of brownheart in "braeburn" apple by transmission NIR spectroscopy. *Postharvest Biology and Technology, 28*, 87–96.

Clark, R. N., Swayze, G. A., Livo, K. E., Kokaly, R. F., King, T. V. V., Dalton, J. B., *et al.* (2002). Surface reflectance calibration of terrestrial imaging spectroscopy data: a tutorial using AVIRIS. US Geological Survey, Mail Stop 964, Box 25046 Federal Center, Lakewood, Colorado 80225, USA.

Cluff, K., Naganathan, G. K., Subbiah, J., Lu, R., Calkins, C. R., & Samal, A. (2008). Optical scattering in beef steak to predict tenderness using hyperspectral imaging in the VIS-NIR region. *Sensing and Instrumentation for Food Quality and Safety, 2*, 189–196.

Cogdill, R. P., Hurburgh, C. R., Jr., Rippke, G. R., Bajic, S. J., Jones, R. W., McClelland, J. F., *et al.* (2004). Single-kernel maize analysis by near-infrared hyperspectral imaging. *Transactions of the ASAE, 47*(1), 311–320.

Delwiche, S., Kim, M. (2000). Hyperspectral imaging for detection of scab in wheat. In Biological Quality and Precision Agriculture II. *Proceedings of SPIE*, Vol. 4203, 13–20.

ENVI. (2000). *The environment for visualizing images, user's guide*. Boulder, CO: Research Systems.

Evans, M. D., Thai, C. N., & Grant, J. C. (1998). Development of a spectral imaging system based on a liquid crystal tunable filter. *Transactions of the ASAE, 41*(6), 1845–1852.

Gowen, A. A., O'Donnell, C. P., Taghizadeh, M., Gaston, E., O'Gorman, A., Cullen, P. J., Frias, J. M., Esquerre, C., & Downey, G. (2008). Hyperspectral imaging for the investigation of quality deterioration in sliced mushrooms (*Agaricus bisporus*) during storage. *Sensing and Instrumentation for Food Quality and Safety, 2*, 133–143.

Green, A. A., Berman, M., Swetzer, P., & Graig, M. D. (1988). A transformation of ordering multispectral data in terms of image quality with implications for noise removal. *IEEE Transactions on Geoscience and Remote Sensing, 26*, 65–74.

Haff, R. P., & Pearson, T. (2006). Spectral band selection for optical sorting of pistachio nut defects. *Transactions of the ASABE, 49*(4), 1105–1113.

Heitschmidt, G. W., Park, B., Lawrence, K. C., Windham, W. R., & Smith, D. P. (2007). Improved hyperspectral imaging system for fecal detection on poultry carcasses. *Transactions of the ASABE, 50*(4), 1427–1432.

Hu, Y., Guo, K., Suzuki, T., Noguchi, G., & Satake, T. (2008). Quality evaluation of fresh pork using visible and near-infrared spectroscopy with fiber optics in interactance mode. *Transactions of the ASABE, 51*(3), 1029–1033.

Kavdir, I., & Guyer, D. E. (2002). Apple sorting using artificial neural networks and spectral imaging. *Transactions of the ASAE, 45*(6), 1995–2005.

Kavdir, I., Lu, R., Ariana, D., & Ngouajio, M. (2007). Visible and near-infrared spectroscopy for nondestructive quality assessment of pickling cucumbers. *Postharvest Biology and Technology, 44*, 165–174.

Kim, M. S., Chen, Y. R., & Mehl, P. M. (2001). Hyperspectral reflectance and fluorescence imaging system for food quality and safety. *Transactions of the ASAE, 44*(3), 721–729.

Kim, M. S., Lee, K., Chao, K., Lefcourt, A. M., Jun, W., & Chan, D. E. (2008). Multispectral line-scan imaging system for simultaneous fluorescence and reflectance measurements of apples: multitask apple inspection system. *Sensing and Instrumentation for Food Quality and Safety, 2*, 123–129.

Kim, M. S., Lefcourt, A. M., Chen, Y. R., Kim, I., Chan, D. E., & Chao, K. (2002). Multispectral detection of fecal contamination on apples based on hyperspectral imagery. Part II: Application of hyperspectral fluorescence imaging. *Transactions of the ASAE, 45*(6), 2039–2047.

Lawrence, K. C., Park, B., Heitschmidt, G. W., Windham, W. R., & Mao, C. (2003). Calibration of a pushbroom hyperspectral imaging system for agricultural inspection. *Transactions of the ASAE, 46*(2), 513–521.

Lawrence, K. C., Park, B., Heitschmidt, G. W., Windham, W. R., & Thai, C. N. (2007). Evaluation of LED and tungsten-halogen lighting for fecal contaminant detection. *Applied Engineering in Agriculture, 23*(6), 811–818.

Lawrence, K. C., Smith, D. P., Windham, W. R., Heitschmidt, G. W., & Park, B. (2006). Egg embryo development detection with hyperspectral imaging. *International Journal of Poultry Science, 5*(10), 964–969.

Lawrence, K. C., Windham, W. R., Park, B., & Jeff Buhr, R. (2003). A hyperspectral imaging system for identification of faecal and ingesta contamination on poultry carcasses. *Journal of Near Infrared Spectroscopy, 11*, 269–281.

Li, X. L., He, Y., & Wu, C. Q. (2008). Least square support vector machine analysis for the classification of paddy seeds by harvest year. *Transaction of the ASABE, 51*(5), 1793–1799.

Lin, L. L., Lu, F. M., & Chang, Y. C. (2006). Development of a near-infrared imaging system for determination of rice moisture. *Cereal Chemistry, 83*(5), 498–504.

Liu, Y., Chen, Y. R., Wang, C. Y., Chan, D. E., & Kim, M. S. (2006). Development of hyperspectral imaging technique for the detection of chilling injury in cucumbers spectral and image analysis. *Applied Engineering in Agriculture, 22*(1), 101–111.

Lu, R. (2001). Predicting firmness and sugar content of sweet cherries using near-infrared diffuse reflectance spectroscopy. *Transaction of the ASAE, 44*(5), 1265–1271.

Lu, R. (2003). Detection of bruises on apples using near-infrared hyperspectral imaging. *Transactions of the ASAE, 46*(2), 523–530.

Lu, R. (2007). Nondestructive measurement of firmness and soluble solids content for apple fruit using hyperspectral scattering images. *Sensing and Instrumentation for Food Quality and Safety, 1*, 19–27.

Maertens, K., Reyns, P., & De Baerdemaeker, J. (2004). On-line measurement of grain quality with NIR technology. *Transactions of the ASAE, 47*(4), 1135–1140.

Mao, C. (2000). Focal plane scanner with reciprocating spatial window. US Patent No. 6,166,373.

McGlone, V. A., & Martinsen, P. J. (2004). Transmission measurements on intact apples moving at high speed. *Journal of Near Infrared Spectroscopy, 12*, 37–42.

Mehl, P. M., Chao, K., Kim, M., & Chen, Y. R. (2002). Detection of defects on selected apple cultivars using hyperspectral and multispectral image analysis. *Applied Engineering in Agriculture, 18*(2), 219–226.

Naganathan, G. K., Grimes, L. M., Subbiah, J., Calkins, C. R., Samal, A., & Meyer, G. E. (2008). Partial least squares analysis of near-infrared hyperspectral images for beef tenderness prediction. *Sensing and Instrumentation for Food Quality and Safety, 2*, 178–188.

Park, B., Lawrence, K. C., Windham, W. R., & Buhr, R. J. (2002). Hyperspectral imaging for detecting fecal and ingesta contaminants on poultry carcasses. *Transactions of the ASAE, 45*(6), 2017–2026.

Park, B., Lawrence, K. C., Windham, W. R., & Smith, D. P. (2006). Performance of hyperspectral imaging system for poultry surface fecal contamination detection. *Journal of Food Engineering, 75*, 340–348.

Park, B., Yoon, S. C., Lawrence, K. C., & Windham, W. R. (2007). Fisher linear discriminant analysis for improving fecal detection accuracy with hyperspectral images. *Transactions of the ASABE, 50*(6), 2275–2283.

Pearson, T. C., & Wicklow, D. T. (2006). Detection of corn kernels infected by fungi. *Transactions of the ASABE, 49*(4), 1235–1245.

Pearson, T. C., Wicklow, D. T., Maghirang, E. B., Xie, F., & Dowell, F. E. (2001). Detecting aflatoxin in single corn kernels by transmittance and reflectance spectroscopy. *Transaction of the ASAE, 44*(5), 1247–1254.

Peng, Y., & Lu, R. (2006). An LCTF-based multispectral imaging system for estimation of apple fruit firmness. Part 1: Acquisition and characterization of scattering images. *Transactions of the ASABE, 49*(1), 259–267.

Polder, G., van der Heijden, G. W. A. M., & Young, I. T. (2002). Spectral image analysis for measuring ripeness of tomatoes. *Transactions of the ASAE, 45*(4), 1155–1161.

Qin, J., & Lu, R. (2005). Detection of pits in tart cherries by hyperspectral transmission imaging. *Transactions of the ASAE, 48*(5), 1963–1970.

Qin, J., Burks, T. F., Kim, M. S., Chao, K., & Ritenour, M. A. (2008). Citrus canker detection using hyperspectral reflectance imaging and PCA-based image classification method. *Sensing and Instrumentation for Food Quality and Safety, 2*, 168–177.

Savitzky, A., & Golay, M. J. E. (1964). Smoothing and differentiation of data by simplified least squares procedures. *Analytical Chemistry, 36*, 1627–1639.

Schowengerdt, R. A. (1997). *Remote sensing: models and methods for image processing* (2nd ed.). San Diego, CA: Academic Press.

Singh, C. B., Jayas, D. S., Paliwal, J., & White, N. D. G. (2007). Fungal detection in wheat using near-infrared hyperspectral imaging. *Transactions of the ASABE, 50*(6), 2171–2176.

Springsteen, A. (1999). Standards for the measurement of diffuse reflectance—an overview of available materials and measurement laboratories. *Analytica Chimica Acta, 380*(2–3), 379–390.

Steinier, J., Termonia, Y., & Deltour, J. (1972). Comments on smoothing and differentiation of data by simplified least squares procedure. *Analytical Chemistry, 44*(11), 1906–1909.

Suhre, D. R., Taylor, L. H., Singh, N. B., Rosch, W. R. (1999). Comparison of acousto-optic tunable filters and acousto-optic dispersive filters for hyperspectral imaging. In R.J. Mericsko (Ed.), 27th AIPR Workshop: Advances in Computer-Assisted Recognition. *Proceedings of SPIE*, Vol. 3584, 142–147.

Tsai, F., & Philpot, W. (1998). Derivative analysis of hyperspectral data. *Remote Sensing of Environment, 66*, 41–51.

Tseng, C. H., Ford, J. F., Mann, C. K., & Vickers, T. J. (1993). Wavelength calibration of a multichannel spectrometer. *Applied Spectroscopy, 47*(11), 1808–1813.

Wang, W., & Paliwal, J. (2006). Spectral data compression and analyses techniques to discriminate wheat classes. *Transactions of the ASABE, 49*(5), 1607–1612.

Xing, J., Karoui, R., & Baerdemaeker, J. D. (2007). Combining multispectral reflectance and fluorescence imaging for identifying bruises and stem-end/calyx regions on Golden Delicious apples. *Sensing and Instrumentation for Food Quality and Safety, 1*, 105–112.

Xing, J., Guyer, D., Ariana, D., & Lu, R. (2008). Determining optimal wavebands using genetic algorithm for detection of internal insect infestation in tart cherry. *Sensing and Instrumentation for Food Quality and Safety, 2*, 161–167.

Yang, C. C., Chao, K., Chen, Y. R., Kim, M. S., & Chang, D. E. (2006). Development of fuzzy logic based differentiation algorithm and fast line-scan imaging system for chicken inspection. *Biosystems Engineering, 95*(4), 483–496.

Yang, C. C., Chao, K., & Kim, M. S. (2009). Machine vision system for online inspection of freshly slaughtered chickens. *Sensing and Instrumentation for Food Quality and Safety, 3*, 70–80.

Yao, H., & Tian, L. (2003). A genetic-algorithm-based selective principal component analysis (GA-SPCA) method for high-dimensional data feature extraction. *IEEE Transactions on Geoscience and Remote Sensing, 41*(6), 1469–1478.

Yao, H., Hruska, Z., Brown, R. L., Cleveland, T. E. (2006). Hyperspectral bright greenish-yellow fluorescence (BGYF) imaging of aflatoxin contaminated corn kernels. In Optic East, a SPIE Conference on Nondestructive Sensing for Food Safety, Quality, and Natural Resource. *Proceedings of SPIE*, 63810B.

Yao, H., Hruska, Z., DiCrispino, K., Lewis, D., Beach, J., Brown, R. L., Cleveland T. E. (2004). Hyperspectral imagery for characterization of different corn genotypes. In Optic East, a SPIE Conference on Nondestructive Sensing for Food Safety, Quality, and Natural Resources. *Proceedings of SPIE*, Vol. 5587, 144–152.

Yao, H., Hruska, Z., Kincaid, R., Brown, R. L., & Cleveland, T. E. (2008). Differentiation of toxigenic fungi using hyperspectral imagery. *Sensing and Instrumentation for Food Quality and Safety, 2*, 215–224.

Yoon, S. C., Lawrence, K. C., Park, B., & Windham, W. R. (2007a). Optimization of fecal detection using hyperspectral imaging and kernel density estimation. *Transactions of the ASABE, 50*(3), 1063–1071.

Yoon, S. C., Lawrence, K. C., Park, B., & Windham, W. R. (2007b). Statistical model-based thresholding of multispectral images for contaminant detection on poultry carcasses. *Transactions of the ASABE, 50*(4), 1433–1442.

Yoon, S. C., Lawrence, K. C., Siragusa, G. R., Line, J. E., Park, B., & Feldner, P. W. (2009). Hyperspectral reflectance imaging for detecting a foodborne pathogen: Campylobacter. *Transaction of the ASABE, 52*(2), 651–662.

Yoon, S. C., Lawrence, K. C., Smith, D. P., Park, B., & Windham, W. R. (2008). Bone fragment detection in chicken breast fillets using transmittance image enhancement. *Transactions of the ASABE, 50*(4), 1433–1442.

Zhang, H., Paliwal, J., Jayas, D. S., & White, N. D. G. (2007). Classification of fungal infected wheat kernels using near-infrared reflectance hyperspectral imaging and support vector machine. *Transactions of the ASABE, 50*(5), 1779–1785.

CHAPTER 3

Hyperspectral Image Classification Methods

Lu Jiang, Bin Zhu, Yang Tao

Bio-imaging and Machine Vision Lab, The Fischell Department of Bioengineering, University of Maryland, USA

CONTENTS

3.1. HYPERSPECTRAL IMAGE CLASSIFICATION IN FOOD: AN OVERVIEW

Hyperspectral imaging techniques have received much attention in the fields of food processing and inspection. Many approaches and applications have shown the usefulness of hyperspectral imaging in food safety areas such as fecal and ingesta contamination detection on poultry carcasses, identification of fruit defects, and detection of walnut shell fragments, and so on (Casasent & Chen, 2003, 2004; Cheng *et al.*, 2004; Jiang *et al.*, 2007a, 2007b; Kim *et al.*, 2001; Lu, 2003; Park *et al.*, 2001; Pearson *et al.*, 2001; Pearson & Young, 2002).

Because hyperspectral imaging technology provides a large amount of spectral information, an effective approach for data analysis, data mining, and pattern classification is necessary to extract the desired information, such as defects, from images. Much work has been carried out in the literature to present the feature extraction and pattern recognition methods in hyperspectral image classification. Several main approaches can be identified:

1. A general two-step strategy, which is feature extraction followed by pattern classification. The feature extraction step is also called optimal band selection or extraction, whose aim is to reduce or transform the original feature space into another space of a lower dimensionality. Principal component analysis (PCA) followed by K-means clustering is the most popular technique in this method.

2. Sample regularization of the second-order statistics, such as the covariance matrix. This approach uses the multivariate normal (Gaussian) probability density model, which is widely accepted for hyperspectral image data. The Gaussian Mixture Model (GMM) is a classic method in this category.

3. The artificial neural network, which is a pattern classification method used in hyperspectral image processing. The neural network is considered to be a commonly used pattern recognition tool because of its nonlinear property and the fact that it does not need to make assumptions about the distribution of the data.

4. Kernel-based methods for hyperspectral image classification. This approach is designed to tackle the specific characteristics of hyperspectral images, which are the high number of spectral channels and relatively few labeled training samples. One popular kernel-based method is the support vector machine (SVM). In this chapter, several main approaches to feature extraction and pattern classification in hyperspectral image classification are illustrated.

The image data acquired by the hyperspectral system are often arranged as a three-dimensional image cube $f(x, y, \lambda)$, with two spatial dimensions x and y, and one spectral dimension λ, as shown in Figure 3.1.

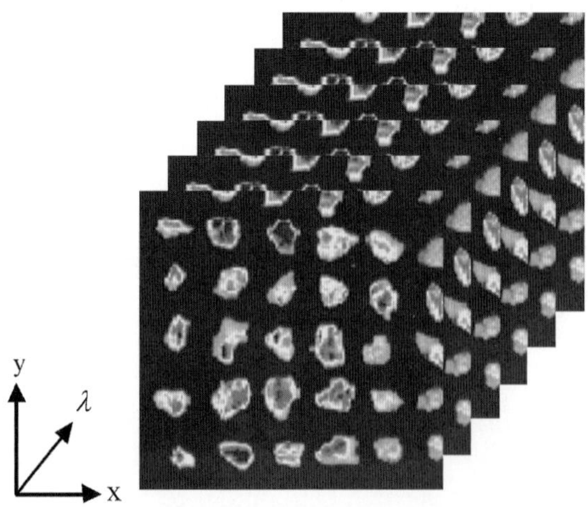

FIGURE 3.1 *A typical image cube acquired by hyperspectral imager, with two spatial dimensions and one spectral dimension (x, y, λ). (Full color version available on http://www.elsevierdirect.com/companions/9780123747532/)*

3.2. OPTIMAL FEATURE AND BAND EXTRACTION

In hyperspectral image analysis the data dimension is high. It is necessary to reduce the data redundancy and efficiently represent the distribution of the data. Feature selection techniques perform a reduction of spectral channels by selecting a representative subset of original features.

3.2.1. Feature Selection Metric

The feature selection problem in pattern recognition may be stated as follows: Given a set of n features (e.g. hyperspectral bands or channels information measured on an object to be classified), find the best subset consisting of k features to be used for classification. Usually the objective is to optimize a trade-off between the classification accuracy (which is generally reduced when fewer than the n available features are used) and computational speed. The feature selection criterion aims at assessing the discrimination capabilities of a given subset of features according to a statistical distance metric among classes.

As a start, the simplest and most frequently used distance metric in feature extraction is Euclidean distance (Bryant, 1985; Searcóid, 2006). The definition of Euclidean distance between feature points $P = (p_1, p_2, \ldots p_n)$ and $Q = (q_1, q_2, \ldots q_n)$ in Euclidean n-space is $\sqrt{\sum_{i=1}^{n}(p_i - q_i)^2}$, which is based on L_2 norm. Another distance metric that has been used in feature selection is the L_1 norm-based metric. It is also called Manhattan distance (Krause, 1987), and is defined as $\sum_{i=1}^{n}|p_i - q_i|$. More generally, an L_p norm-based distance metric can be used in feature selection, and is defined as $\left(\sum_{i=1}^{n}(p_i - q_i)^p\right)^{\frac{1}{p}}$, which can be found in many classical literatures (Bryant, 1985; Searcóid, 2006).

Some other more complicated statistical distance measures among classes have been reported, such as Bhattacharyya distance (Bhattacharyya, 1943), Jefferies–Matusita (JM) distance (Richards, 1986), and the divergence measure (Jeffreys, 1946) in hyperspectral data analysis. The JM distance between a pair of probability distributions (spectral classes) is defined as:

$$J_{ij} = \int_X \left[\sqrt{p_i(x)} - \sqrt{p_j(x)}\right]^2 dx \qquad (3.1)$$

where $p_i(x)$ and $p_j(x)$ are two class probability density functions. For normally distributed classes, the JM distance becomes:

$$J_{ij} = 2(1 - e^{-B}) \qquad (3.2)$$

where

$$B = \frac{1}{8}(m_i - m_j)^T \left\{ \frac{\Sigma_i + \Sigma_j}{2} \right\}^{-1} (m_i - m_j) + \frac{1}{2} \ln \left\{ \frac{\frac{1}{2}|\Sigma_i + \Sigma_j|}{|\Sigma_i|^{1/2} |\Sigma_j|^{1/2}} \right\} \tag{3.3}$$

in which m_i is the mean of ith class, Σ_i is the covariance of the ith class, and B is referred to as the Bhattacharyya distance. For multiclass problems, an average J among multiclasses can be achieved.

Divergence is another measure of the separability of a pair of probability distributions that has its basis in their degree of overlap. The divergence D for two densities $p_i(x)$ and $p_j(x)$ can be defined as:

$$D_{ij} = \int_x [p_i(x) - p_j(x)] \ln \frac{p_i(x)}{p_j(x)} dx \tag{3.4}$$

If the $p_i(x)$ and $p_j(x)$ are multivariate Gaussian densities with mean m_i and m_j, covariance matrices Σ_i and Σ_j, respectively, then:

$$D_{ij} = \frac{1}{2} tr[\Sigma_i - \Sigma_j][\Sigma_j^{-1} - \Sigma_i^{-1}] + \frac{1}{2} tr[\Sigma_i^{-1} + \Sigma_j^{-1}][m_i - m_j][m_i - m_j]^T \tag{3.5}$$

where trA denotes trace of matrix A, A^{-1} is the inverse of A, and A^T is the transpose of A. Similarly with JF distance, an average D among multiclasses can be obtained in more than two classes case.

3.2.2. Feature Search Strategy

Optimal feature search algorithms identify a subset that contains a predetermined number of features and is the best in terms of the adopted criterion function. The most straightforward ways to realize feature search are sequential forward/backward selections. The sequential forward selection method (SFS) (Marill & Green, 1963) starts with no features and adds them one by one, at each step adding the one that decreases the error the most, until any further addition does not significantly decrease the error. The sequential backward selection method (SBS) (Whitney, 1971) starts with all the features and removes them one by one, at each step removing the one that decreases the classification error most (or increases it only slightly), until any further removal increases the error significantly. A problem with this hill-climbing search technique is that when a feature is deleted in SBS, it cannot be picked up again in the following selection and when a feature is added in SFS, it cannot be deleted.

More generalized than SFS/SBS, the plus-Z-minus-R method (Stearns, 1976) utilizes a more complex sequential search approach to select optimal features. The settings of parameters Z in forward selection and R in backward selection are fixed and cannot be changed during the selection process. Pudil *et al.* (1994) introduced the sequential forward floating selection (SFFS) method and the sequential backward floating selection (SBFS) method as feature selection strategies. They improve the standard SFS and SBS techniques by dynamically changing the number of features included (SFFS) or removed (SBFS) at each step and by allowing the reconsideration of the features included or removed at the previous steps. According to the comparisons made in the literature (Jain, 2000; Kudo & Sklansky, 2000), the sequential floating search methods (SFFS and SBFS) can be regarded as being the most effective ones, when one deals with very high-dimensional feature spaces.

A random search method such as genetic algorithm can also be used in the hyperspectral feature selection strategy. Yao & Tian (2003) proposed a genetic-algorithm-based selective principal component analysis (GA-SPCA) method to select features using hyperspectral remote sensing data and ground reference data collected within an agricultural field. Compared with a sequential feature selection method, a genetic algorithm helps to escape from a local optimum in the search procedure.

3.2.3. Principal Component Analysis (PCA)

The focus of the preceding sections has been on the evaluation of existing sets of features of the hyperspectral data with regard to selecting the most differentiable and discarding the rest. Feature reduction can also be achieved by transforming the data to a new set of axes in which differentiability is higher in a subset of the transformed features than in any subset of the original data. The most commonly used image transformations are principal component and Fisher's discriminant analyses.

As a classical projection-based method, PCA is often used for feature selection and data dimension reduction problems (Campbell, 2002; Fukunaga, 1990). The advantage of PCA compared with other methods is that PCA is an unsupervised learning method. The PCA approach can be formulated as the following. The scatter matrix of the hyperspectral samples, S_T is given by:

$$S_T = \sum_{k=1}^{n} (\mathrm{x}_k - \mu)(\mathrm{x}_k - \mu)^T \tag{3.6}$$

where S_T is an N by N covariance matrix, x_k is an N-dimensional hyperspectral grayscale vector, μ is the sample's mean vector, and n is the total number of training samples. In PCA the projection W_{opt} is chosen to maximize the determinant of the total scatter matrix of the projected samples. That is:

$$W_{opt} = \arg\max_{\eta} |W^T S_T W| = [w_1\ w_2\ \ldots\ w_m] \tag{3.7}$$

where $\{w_i | i = 1, 2, \ldots, m\}$ is the set of N-dimensional eigenvector of S_T corresponding to the m largest eigenvalues (Fukunaga, 1990). In general the eigenvectors of S_T corresponding to the first three largest eigenvalues have preserved more than 90% energy of the whole dataset. However, the selection of the parameter m is still an important problem because the performance of the classifier becomes better as the principal components increase to some extent; on the other hand, the computation time also increases as the principal components increase. As a result, there is a balance among the number of selected principal components, the performance of the classifier and the computation time. A cross-validation method could be used to select optimal m in PCA analysis (Goutte, 1997).

3.2.4. Fisher's Discriminant Analysis (FDA)

Fisher's discriminant analysis (FDA) is another method of feature extraction in hyperspectral image classification (Fukunaga, 1990). It is a supervised learning method. This method selects projection W in such a way that the ratio of the between-class scatter S_B and the within-class scatter S_W is maximized. Let the between-class scatter matrix be defined as:

$$S_B = \sum_{i=1}^{c} (u_i - u)(u_i - u)^T \tag{3.8}$$

and the within-class scatter matrix S_W be defined as:

$$S_W = \sum_{i=1}^{c} \sum_{x_k \in X_i} (x_k - u_i)(x_k - u_i)^T \tag{3.9}$$

where x_k is an N-dimensional hyperspectral grayscale vector, u_i is the mean vector of class X_i, μ is the sample's mean vector, c is the number of classes. If S_W is nonsingular, the optimal projection W_{opt} is chosen as the matrix with orthonormal columns that maximize the ratio of the determinant of the between-class scatter matrix of the projected samples over the determinant of the within-class scatter matrix of the projected samples, i.e.

$$W_{opt} = \arg\max_{W} \frac{|W^T S_B W|}{|W^T S_W W|} \tag{3.10}$$

where $\{w_i | i = 1, 2, \ldots, m\}$ is the set of generalized eigenvector of S_B and S_W corresponding to the m largest generalized eigenvalues $\{\lambda_i | i = 1, 2, \ldots, m\}$ .i.e.,

$$S_B w_i = \lambda_i S_W w_i \qquad i = 1, 2, \ldots, m \tag{3.11}$$

In hyperspectral image classification, sometimes S_W is singular when there are a small number of training data. This will lead to the rank of S_W being at most *N-c*. In order to overcome the complication of a singular S_W, one method (Turk & Pentland, 1991) is to project the image set to a lower dimensional space so that the result in S_W is nonsingular, i.e. W_{opt} is given by

$$W_{opt} = W_{fld}^T W_{pca}^T \tag{3.12}$$

where

$$W_{pca} = \arg\max_{W} |W^T S_T W| \tag{3.13}$$

$$W_{fld} = \arg\max_{W} \frac{|W^T W_{pca}^T S_B W_{pca} W|}{|W^T W_{pca}^T S_W W_{pca} W|} \tag{3.14}$$

3.2.5. Integrated PCA and FDA

The PCA method is believed to be one of the best methods to represent band information in hyperspectral images, but does not guarantee the feature class separability of the selected band. On the other hand, the FDA method, though effective in class segmentation, is sensitive to noise and may not convey enough energy from the original data. In order to design a set of projection vector-bases that can provide supervised classification information well, and at the same time preserve enough information from the original hyperspectral data cube, a novel method is presented in Cheng *et al.* (2004) to combine Equations (3.7) and (3.10) to construct an evaluation equation called Integrated PCA–FDA method. A weight factor k is introduced to adjust the degree of classification and energy preservation as desired. The constructed evaluation equation is given as:

$$W_{evl} = \arg\max_{W} \frac{\left|W^T[kS_T + (1-k)S_B]W\right|}{\left|W^T[kI + (1-k)S_W]W\right|} \tag{3.15}$$

where $0 \leq k \leq 1$, and I is the identity matrix. In Equation (3.11), if the within-scatter matrix S_W becomes very small, the eigen-decomposition becomes inaccurate. Equation (3.15) overcomes this problem by adjusting the weight factor k toward 1, the effects of S_W can then be ignored, which means that the principal components are more heavily weighted. On the other hand, if the k value is chosen small, which means more differential information between classes is taken into account, the ratio between S_B and S_W dominates.

The integrated method magnifies the advantages of PCA and FDA and compensates the disadvantages of the two at the same time. In fact, the FDA and PCA methods represent the extreme situation of Equation (3.15). When $k = 0$, only the discrimination measure is considered, and the equation is in fact equal to FDA (Equation 3.10). Meanwhile, when $k = 1$, only the representation measure is presented, and the evaluation equation is equivalent to PCA method (Equation 3.7). An optimal projection W_{opt} is chosen as the matrix with orthonormal columns that maximizes Equation (3.15) when $k = 0.5$ in order to find a projection transform that provides equally well both representation and discrimination. The solution of Equation (3.15) is the set of generalized eigenvector that can be obtained by:

$$[kS_T + (1-k)S_B]w_i = \lambda_i[kI + (1-k)S_W]w_i \quad i = 1, 2, \ldots, m \tag{3.16}$$

where, λ_i represents m largest eigenvalues, and w_i is the generalized eigenvector corresponding to m largest eigenvalues.

3.2.6. Independent Component Analysis (ICA)

Another method used often in hyperspectral image feature selection is the independent component analysis (ICA). It is well known that the ICA has become a useful method in blind source separation (BSS), features extraction, and other pattern recognition related areas. The ICA method was first introduced by Herault & Jutten (1986) and was fully fledged by Comon (1994). It extracts independent source signals by looking for a linear or nonlinear transformation that minimizes the statistical dependence between components.

Given the observed signal $X = (X_1, X_2, \ldots, X_n)^T$, which is the spectral profile of the hyperspectral image pixel vector, and the source signal $S = (S_1, S_2, \ldots, S_m)^T$ with each component corresponding to the existing classes in the hyperspectral image, a linear ICA unmixing model can be shown as:

$$S_{m\times p} = W_{m\times n} X_{n\times p} \tag{3.17}$$

where W is the weight matrix in the unmixing model, and p is the number of pixels in the hyperspectral images.

From Equation (3.17), the system mixing model with additive noise may be written as:

$$X_{n\times p} \equiv Y_{n\times p} + N_{n\times p} = A_{n\times m} S_{m\times p} + N_{n\times p} \tag{3.18}$$

Assume the additive noise $N_{n\times p}$ is a stationary, spatially white, zero-mean complex random process independent of source signal. Also assume the matrix A has full column rank and the component of source S is statistically independent, and no more than one component is Gaussian distributed. The weight matrix A can be estimated by the second-order blind identification ICA (SOBIICA) algorithm which was introduced by Belouchrani *et al.* (1997) and Ziehe & Miller (1998).

SOBI is defined as the following procedure:

(1) Estimate the covariance matrix R_0 from p data samples. R_0 is defined as

$$R_0 = E(XX^*) = AR_{s0}A^H + \sigma^2 I \tag{3.19}$$

where R_{s0} is the covariance matrix of source S at initial time, and H denotes the complex conjugate transpose of the matrix. Denote by $\lambda_1, \lambda_2 \ldots\ldots \lambda_l$ being the l largest eigenvalues and being $u_1, u_2 \ldots\ldots u_l$ the corresponding eigenvectors of R_0.

(2) Calculate the whitened signal $Z = [z_1, z_2, \ldots\ldots z_l] = BX$, where $z_i = (\lambda_i - \sigma^2)^{-\frac{1}{2}} u_i^* x_i$ for $1 \le i \le l$. This equals to forming a whitening matrix B by:

$$B = \left[(\lambda_1 - \sigma^2)^{-\frac{1}{2}} u_1, (\lambda_2 - \sigma^2)^{-\frac{1}{2}} u_2, \ldots\ldots, (\lambda_l - \sigma^2)^{-\frac{1}{2}} u_l \right] \tag{3.20}$$

(3) Estimate the covariance matrix R_τ from p data samples by calculating the covariance matrix of Z for a fixed set of time lag, such as $\tau = [1, 2, \ldots\ldots, K]$.

(4) A unitary matrix U is then obtained as joint diagonalizer of the set $\{R_\tau | \tau = 1, 2, \ldots\ldots, K\}$.

(5) The source signals are estimated as $S = U^H W X$ and the mixing matrix A is estimated by $A = W^{\#} U$, where # denotes the Moore–Penrose pseudoinverse.

If the number of categories in the n-band hyperspectral images is m, the related weight matrix W is approximated by the SOBIICA algorithm. The source component S_{ij} with $i = 1, \ldots, m$ can be expressed as the following equation according to the ICA unmixing model.

$$\begin{bmatrix} s_{11} & \cdot & \cdot & \cdot & s_{1p} \\ \cdot & \cdot & \cdot & \cdot & \cdot \\ \cdot & \cdot & s_{ij} & \cdot & \cdot \\ \cdot & \cdot & \cdot & \cdot & \cdot \\ s_{m1} & \cdot & \cdot & \cdot & s_{mp} \end{bmatrix} = \begin{bmatrix} w_{11} & \cdot & \cdot & \cdot & w_{1n} \\ \cdot & \cdot & \cdot & \cdot & \cdot \\ \cdot & \cdot & w_{ik} & \cdot & \cdot \\ \cdot & \cdot & \cdot & \cdot & \cdot \\ w_{m1} & \cdot & \cdot & \cdot & w_{mn} \end{bmatrix} \times \begin{bmatrix} x_{11} & \cdot & \cdot & \cdot & x_{1p} \\ \cdot & \cdot & \cdot & \cdot & \cdot \\ \cdot & \cdot & x_{kj} & \cdot & \cdot \\ \cdot & \cdot & \cdot & \cdot & \cdot \\ x_{n1} & \cdot & \cdot & \cdot & x_{np} \end{bmatrix} \tag{3.21}$$

That is,

$$s_{ij} = \sum_{k=1}^{n} w_{ik} x_{kj} \tag{3.22}$$

From Equation (3.22), the ith class material in the source is the weighted sum of the kth band in the observed hyperspectral image pixel X with corresponding weight w_{ik}, which means the weight w_{ik} shows how much information the kth band contributes to the ith class material. Therefore, the significance of each spectral band for all the classes can be calculated as the average absolute weight coefficient $\overline{w}_k$, which is written as (Du *et al.*, 2003):

$$\overline{w}_k = \frac{1}{m} \sum_{i=1}^{m} |w_{ik}| \quad k = 1, 2, \ldots, n \tag{3.23}$$

As a result, an ordered band weight series as

$$[\overline{w}_1, \overline{w}_2, \overline{w}_3, \ldots.., \overline{w}_n] \text{ with } \overline{w}_1 > \overline{w}_2 > \overline{w}_3, \ldots.. > \overline{w}_n \tag{3.24}$$

can be obtained by sorting the average absolute coefficients for all the spectral bands. In this sequence the band with the higher averaged absolute weights contributes more to ICA transformation. In other words, the band with the higher averaged absolute weights contains more spectral information than the other band. Therefore, the bands with the top highest averaged absolute weights will be selected as the optimal bands for hyperspectral feature extraction.

3.3. CLASSIFICATIONS BASED ON FIRST- AND SECOND-ORDER STATISTICS

This approach applies the multivariate Gaussian probability density model, which has been widely accepted for hyperspectral sensing data. The model requires the correct estimation of first- and second-order statistics for each category.

The Gaussian Mixture Model (GMM) is a classical first- and second-order-based classification method. GMM (Duda *et al.*, 2001) has been widely used in many data modeling applications, such as time series classification (Povinelli *et al.*, 2004) and image texture detection (Permuter *et al.*, 2006). The key points of the GMM are the following: Firstly, the GMM assumes that each class-conditional probability density is subject to Gaussian distributions with different means and covariance matrix. Secondly, under the GMM, the feature points from each specific object or class are generated from a pool of Gaussian models with different prior mixture weights.

Let the complete input data set be: $D = \{(x_1, y_1), (x_2, y_2) \ldots (x_n, y_n)\}$, which contains both vectors of hyperspectral image pixels $x_i \in R^N$ and its corresponding class label $y_i \in \{1, 2, \ldots c\}$, where R^N refers to the N-dimensional space of the observations, and c stands for the total number of classes, the jth class-conditional probability density can be written as $p(x|y_j, \theta_j)$, which is subject to multivariate Gaussian distribution with the parameter $\theta_j = \{u_j, \Sigma_j\}$, where u_j is the mean vector, and Σ_j is the covariance matrix. Assume the input data were obtained by selecting a state of nature (class) y_j with prior probability $P(y_j)$, the probability density function of the input data x is given by

$$p(x|\theta) = \sum_{j=1}^{c} p(x|y_j, \theta_j)P(y_j) \tag{3.25}$$

Equation (3.25) is called mixture density and $p(x|y_j, \theta_j)$ is the component density. The multivariate Gaussian probability density function in the N-dimensional space can be written as:

$$p(x|y_j, \theta_j) = \frac{1}{(2\pi)^{N/2}|\Sigma_j|^{1/2}} \exp\left[-\frac{1}{2}(x - \mu_j)^T \Sigma^{-1}(x - \mu_j)\right] \tag{3.26}$$

In the GMM, both θ_j and $P(y_j)$ are unknowns and need to be estimated. A maximum-likelihood estimate approach can be used to determine the above-mentioned parameters. Assume the input data are sampled from random variables that are independent and identically distributed, the likelihood function, which is the joint density of input data, can be expressed as:

$$p(D|\theta) \equiv \prod_{i=1}^{n} p(x_i|\theta) \tag{3.27}$$

Taking the log transform on both sides of Equation (3.27), the log-likelihood can be written as:

$$l = \sum_{i=1}^{n} \ln\ p(x_i|\theta) \tag{3.28}$$

The maximum-likelihood estimates of θ and $P(y_j)$, which are $\widehat{\theta}$ and $\widehat{P}(y_j)$ respectively, can be defined as:

$$\widehat{\theta} = \underset{\theta \in \Theta}{\arg\max}\, l = \underset{\theta \in \Theta}{\arg\max} \sum_{i=1}^{n} \ln\ p(x_i|\theta)$$

$$\text{Subject to} : \widehat{P}(y_i) \geq 0 \text{ and} \sum_{i=1}^{c} \widehat{P}(y_i) = 1 \tag{3.29}$$

Given an appropriate data model, a classifier is then needed to discriminate among classes. The Bayesian minimum risk classifier (Duda *et al.*, 2001; Fukunaga, 1990; Langley *et al.*, 1992), which deals with the problem in making optimal decisions in pattern recognition, was employed. The fundamental of the Bayesian classifier is to categorize testing data into given classes such that the total expected risk is minimized. In the GMM, once the maximum-likelihood estimate is used, both the prior probabilities $P(y_j)$ and the class-conditional probability density $p(x|y_j)$ are known. According to the Bayesian rule, the posterior probability $p(y_i|x)$ is given by:

$$p(y_i|x) = \frac{p(x\,|\,y_i)P(y_i)}{\sum_{j=1}^{c} p(x|y_j)P(y_j)} \tag{3.30}$$

The expected loss (i.e. the risk) associated with taking action α_k is defined as:

$$\Re(\alpha_k|x) = \sum_{i=1}^{c} \Gamma(\alpha_k|y_i)P(y_i|x) \tag{3.31}$$

where $\Gamma(\alpha_k|y_i)$ is the loss function, which stands for the loss incurred for taking action α_k when the state of nature is y_i. So the overall expected risk is written as:

$$\Re = \int \Re(\alpha(x)|x)p(x)dx \tag{3.32}$$

It is easy to show that the minimum overall risk, also called Bayes risk, is:

$$\Re^* = \min_{\alpha k} \Re\,(\alpha_k|x) \tag{3.33}$$

The 0–1 loss function can be defined:

$$\Gamma(\alpha_k|y_i) = \begin{cases} 0 & k = i \\ 1 & k \neq i \end{cases} \qquad i, k = 1, \ldots\ldots c \tag{3.34}$$

Then, the Bayesian risk can be given by:

$$\Re(\alpha_k|x) = 1 - P(y_i|x) \tag{3.35}$$

So the final minimum-risk Bayesian decision rule becomes:

$$d(x) = \underset{y_i \in \{1,2,\ldots c\}}{\arg\max}\; p(y_i|x) \tag{3.36}$$

where $d(x)$ refers to the predicted class label of sample x.

3.4. HYPERSPECTRAL IMAGE CLASSIFICATION USING NEURAL NETWORKS

An important and unique class of pattern recognition methods used in hyperspectral image processing is artificial neural networks (Bochereau *et al.*, 1992; Chen *et al.*, 1998; Das & Evans, 1992), which itself has evolved to a well-established discipline. Artificial neural networks can be further categorized as feed-forward networks, feedback networks, and self-organization networks. Compared with the conventional pattern recognition methods, artificial neural networks have several advantages. Firstly, neural networks can learn the intrinsic relationship by example. Secondly, neural networks are more fault-tolerant than conventional computational methods; and finally, in some applications, artificial neural networks are preferred over statistical pattern recognition because they require less domain-related knowledge of a specific application.

Neural networks are designed to have the ability to learn complex nonlinear input–output relationships using sequential training procedures and adapt themselves to the input data. A typical multi-layer neural network can be designed as in Figure 3.2, which includes input layer, hidden layer, and output layer. A relationship between input data and output data can be

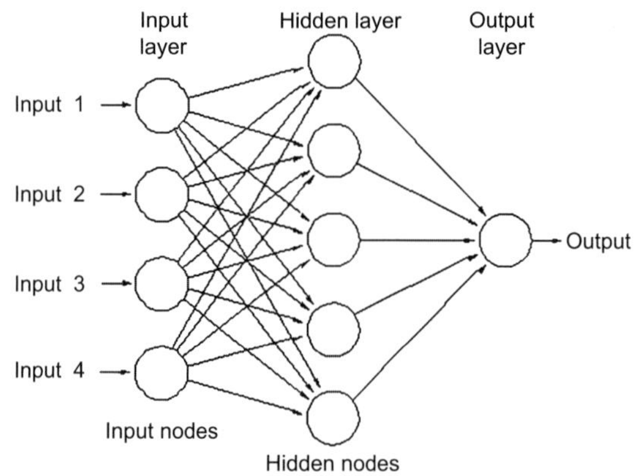

FIGURE 3.2 *A multi-layer feed-forward artificial neural network*

described by this neural network. Difference nodes in the layer have different functions and weights in the neural networks. In supervised learning, a cost function, i.e., mean-squared error, is used to minimize the average squared error between the network's output, $f(x)$, and the target value y over all the training data, here x is the input of the network. Gradient descent method is a popular way to minimize this cost function, and in this case we also called this method Multi-Layer Perceptrons. A well-known backpropagation algorithm can be applied to train neural networks. More details about the neural network can be found in Duda *et al.* (2001).

3.5. KERNEL METHOD FOR HYPERSPECTRAL IMAGE CLASSIFICATION

As a statistical learning method in data mining (Duda *et al.*, 2001; Fukunaga, 1990), Support Vector Machine (SVM) (Burges, 1998) has been used in applications such as object recognition (Guo *et al.*, 2000) and face detection (Osuna *et al.*, 1997). The basic idea of SVM is to find the optimal hyperplane as a decision surface that correctly separates the largest fraction of data points while maximizing the margins from the hyperplane to each class. The simplest support vector machine classifier is also called a maximal margin classifier. The optimal hyperplane, h, that is searched in the input space can be defined by the following equation:

$$h = w^T x + b \tag{3.37}$$

where x is the input hyperspectral image pixel vector, w is the adaptable weight vector, b is the bias, and T is the transverse operator.

Another advantage of SVM is that the above-mentioned maximization problem can be solved in any high-dimensional space other than the original input space by introducing a kernel function. The principle of the kernel method was addressed by Cover's theorem on separability of patterns (Cortes & Vapnik, 1995). The probability that the feature space is linear separable becomes higher when the low-dimensional input space is nonlinearly transformed into a high-dimensional feature space. Theoretically, the kernel function is able to implicitly and not explicitly map the input space, which may not be linearly separable, into an arbitrary high-dimensional feature space that can be linearly separable. In other words, the computation of the kernel method becomes possible in high-dimensional space, since it computes the inner product as a direct function of the input space without explicitly computing the mapping.

Suppose the input space vector $x_i \in \Re^n$ $(i = 1,\ldots, l)$ with its corresponding class label $y_i \in \{-1, 1\}$ in the two-class case, l is the number of total input data. Cortes & Vapnik (1995) showed that the above maximization problem was equal to solving the following primal convex problem:

$$\min_{\Omega,b,\xi} \quad \frac{1}{2} w^T w + C\sum_{i=1}^{l} \xi_i$$

$$\text{subject to} \quad y_i(w^T\phi(x_i) + b) \geq 1 - \xi_i \qquad \xi_i \geq 0, \quad i = 1,\ldots,l. \tag{3.38}$$

where ξ_i is the slack variable, C is a user-specified positive parameter, and w is the weighted vector. By mapping function ϕ, the input vector x_i is mapped from the input space $\Re^n$ into a higher dimensional feature space F. Thus, its corresponding dual problem is:

$$\min_{\alpha} \frac{1}{2}\alpha^T Q\alpha - e^T\alpha \qquad 0 \leq \alpha \leq C, \; i = 1,\ldots,l,$$

$$\text{subject to } y^T\alpha = 0 \tag{3.39}$$

where e is the vector of all ones, Q is an l by l positive semi-definite matrix and can be defined as:

$$Q_{ij} = y_i y_j K(x_i, x_j) \tag{3.40}$$

where $K(x_i, x_j) \equiv \phi(x_i)^T\phi(x_j)$is the kernel matrix calculated by a specified kernel function $k(x, y)$.

In general, three common kernel functions (Table 3.1), which allow one to compute the value of the inner product in F without having to carry out the

Table 3.1 Three common kernel functions

Kernel name	Kernel equations
Polynomial kernel	$k(x, y) = <x, y>^d,\ d \in R$
Gaussian kernel	$k(x, y) = \exp\left(-\frac{\lVert x - y \rVert^2}{2\sigma^2}\right),\ \sigma > 0$
Sigmoid kernel	$k(x, y) = \tanh(\kappa < x, y > + \vartheta),\ \kappa > 0,\ \vartheta > 0$

mapping ϕ, are widely used in SVM. In Table 3.1, d is the degree of freedom of polynomial kernel, σ is a parameter related with the width of the Gaussian kernel, and κ is the inner product coefficient in hyperbolic tangent function.

Assuming the training vectors x_i are projected into a higher dimensional space by the mapping ϕ, the discriminant function of SVM is (Cortes & Vapnik, 1995):

$$f(x) = \mathrm{sgn}\left(\sum_{i=1}^{l} y_i \alpha_i K(x_i, x) + b\right) \tag{3.41}$$

Besides SVM, some other kernel-based methods, such as kernel-PCA, kernel-FDA, have also been investigated in hyperspectral image classification. Details of a kernel-based method used in pattern classification can be found in the literature (Duda *et al.*, 2000).

3.7. CONCLUSIONS

In this chapter several feature selection and pattern recognition methods that are often used in hyperspectral imagery are introduced. Distance metrics and feature search strategies are two main aspects in the feature selection. The goal of linear projection-based feature selection methods is to transform the image data from original space into another space of a lower dimension. A second-order statistics-based classification method needs the assumption of a probability density model of the data, and such an assumption itself is a challenging problem. Neural networks are non-linear statistical data modeling tools which can be used to model complex relationships between inputs and outputs in order to find patterns in the image data. The kernel method appears to be especially advantageous in the analysis of hyperspectral data. For example, SVM implements a maximum margin-based geological classification strategy, which shows the robustness of high dimensionality of the hyperspectral data and low sensitivity of the number of training data.

NOMENCLATURE

Symbols

x	an N-dimensional hyperspectral grayscale vector
μ	mean of all samples
$p(x)$	probability density functions
m_i	mean of ith class samples
Σ_i	covariance of the ith class samples
D	divergence measure
S_T	covariance matrix
S_B	between-class scattering matrix
S_W	within-class scattering matrix
W	projection or weight matrix
trA	trace of matrix A
A^{-1}	the inverse of A
A^T	the transpose of A
A^H	the complex conjugate transpose of matrix A
$p(x\|y_j, \theta_j)$	the jth class-conditional probability density
θ_j	parameter set of the jth class
$P(y)$	the prior probability
y	class label of sample x
$d(x)$	predicted class label of sample x
$\Re$	overall expected risk
h	hyperplane
ϕ	mapping function
K	kernel matrix
$\Re^n$	input space
F	higher dimensional feature space

Abbreviations

FDA	Fisher's discriminant analysis
GA-SPCA	genetic-algorithm-based selective principal component analysis
GMM	Gaussian Mixture Model
ICA	independent component analysis
JM	Jefferies–Matusita distance
PCA	principal component analysis
SBS	sequential backward selection
SBFS	sequential backward floating selection

SFS	sequential forward selection
SFFS	sequential forward floating selection
SOBIICA	second-order blind identification ICA
SVM	support vector machine

REFERENCES

Belouchrani, A., Abed-Meraim, K., Cardoso, J. F., & Moulines, E. (1997). A blind source separation technique using second order statistics. *IEEE Transaction on Signal Processing, 45*(2), 434–444.

Bhattacharyya, A. (1943). On a measure of divergence between two statistical populations defined by their probability distributions. *Bulletin of the Calcutta Mathematical Society, 35*, 99–109.

Bochereau, L., Bourgine, P., & Palagos, B. (1992). A method for prediction by combining data analysis and neural networks: application to prediction of apple quality using near infra-red spectra. *Journal of Agricultural Engineering Research, 51*(3), 207–216.

Bryant, V. (1985). *Metric spaces: iteration and application*. Cambridge, UK: Cambridge University Press.

Burges, C. J. C. (1998). A tutorial on support vector machines for pattern recognition. *Data Mining and Knowledge Discovery, 2*(2), 121–167.

Campbell, J. B. (2002). *Introduction to remote sensing* (3rd ed.). Oxford, UK: Taylor & Francis.

Casasent, D., & Chen, X.-W. (2003). Waveband selection for hyperspectral data: optimal feature selection. In Optical Pattern Recognition XIV. *Proceedings of SPIE*, Vol. 5106, 256–270.

Casasent, D., & Chen, X.-W. (2004). Feature selection from high-dimensional hyperspectral and polarimetric data for target detection. In Optical Pattern Recognition XV. *Proceedings of SPIE*, Vol. 5437, 171–178.

Chen, Y. R., Park, B., Huffman, R. W., & Nguyen, M. (1998). Classification of on-line poultry carcasses with backpropagation neural networks. *Journal of Food Processing Engineering, 21*, 33–48.

Cheng, X., Chen, Y., Tao, Y., Wang, C., Kim, M., & Lefcourt, A. (2004). A novel integrated PCA and FLD method on hyperspectral image feature extraction for cucumber chilling damage inspection. *Transactions of the ASAE, 47*(4), 1313–1320.

Comon, P. (1994). Independent component analysis, a new concept? *Signal Processing, 36*(3), 287–314.

Cortes, C., & Vapnik, V. (1995). Support vector networks. *Machine Learning, 20*, 273–297.

Das, K., & Evans, M. D. (1992). Detecting fertility of hatching eggs using machine vision. II. Neural network classifiers. *Transactions of the ASAE, 35*(6), 2035–2041.

Du, H., Qi, H., Wang, X., Ramanath, R., & Snyder, W. E. (2003). Band selection using independent component analysis for hyperspectral image processing. *Proceedings of the 32nd Applied Imagery Pattern Recognition Workshop (AIPR '03). 9398.* Washington, DC, USA, October 2003.

Duda, R., Hart, P., & Stork, D. (2001). *Pattern classification* (2nd ed.). Indianapolis, IN: Wiley–Interscience.

Fukunaga, K. (1990). *Introduction to statistical pattern recognition* (2nd ed.). New York, NY: Academic Press.

Goutte, C. (1997). Note on free lunches and cross-validation. *Neural Computation, 9,* 1211–1215.

Guo, G., Li, S. Z., & Chan, K. (2000). Face recognition by support vector machines. *Proceedings of the 4th IEEE International Conference on Automatic Face and Gesture Recognition* (pp. 196–201). Grenoble, France.

Herault, J., & Jutten, C. (1986). Space or time adaptive signal processing by neural network models. AIP Conference Proceedings, *Neural Networks for Computing.* Snowbird, UT, USA.

Jain, A. K., Duin, R. P. W., & Mao, J. (2000). Statistical pattern recognition: a review. *IEEE Transactions on Pattern Analysis and Machine Intelligence, 22*(1), 4–37.

Jeffreys, H., (1946). An invariant form for the prior probability in estimation problems. *Proceedings of the Royal Society of London, Series A,* 186, 453–461.

Jiang, L., Zhu, B., Jing, H., Chen, X., Rao, X., & Tao, Y. (2007a). Gaussian Mixture Model based walnut shell and meat classification in hyperspectral fluorescence imagery. *Transactions of the ASABE, 50*(1), 153–160.

Jiang, L., Zhu, B., Rao, X., Berney, G., & Tao, Y. (2007b). Discrimination of black walnut shell and pulp in hyperspectra1 fluorescence imagery using Gaussian kernel function approach. *Journal of Food Engineering, 81*(1), 108–117.

Kim, M., Chen, Y., & Mehl, P. (2001). Hyperspectral reflectance and fluorescence imaging system for food quality and safety. *Transactions of the ASAE, 44*(3), 721–729.

Krause, E. F. (1987). *Taxicab geometry.* New York, NY: Dover.

Kudo, M., & Sklansky, J. (2000). Comparison of algorithms that select features for pattern classifiers. *Pattern Recognition, 33*(1), 25–41.

Langley, P., Iba, W., & Thompson, K. (1992). An analysis of Bayesian classifiers. *Proceedings of the 10th National Conference on Artificial Intelligence* (pp. 223–228). San Jose, CA: AAAI Press.

Lu, R. (2003). Detection of bruises on apples using near-infrared hyperspectral imaging. *Transactions of the ASAE, 46*(2), 523–530.

Marill, T., & Green, D. M. (1963). On the effectiveness of receptors in recognition systems. *IEEE Transactions on Information Theory, 9*, 11–17.

Osuna, E., Freund, R., & Girosi., F. (1997). Training support vector machines: an application to face detection. *Proceedings of CVPR'97, Puerto Rico.*

Park, B., Lawrence, K., Windham, W., & Buhr, R. (2001). Hyperspectral imaging for detecting fecal and ingesta contamination on poultry carcasses. *ASAE Paper No. 013130*. St Joseph, MI: ASAE.

Pearson, T., & Young, R. (2002). Automated sorting of almonds with embedded shell by laser transmittance imaging. *Applied Engineering in Agriculture, 18*(5), 637–641.

Pearson, T. C., Wicklow, D. T., Maghirang, E. B., Xie, F., & Dowell, F. E. (2001). Detecting aflatoxin in single corn kernels by transmittance and reflectance spectroscopy. *Transactions of the ASAE, 44*(5), 1247–1254.

Permuter, H., Francos, J., & Jermyn, I. (2006). A study of Gaussian mixture models of color and texture features for image classification and segmentation. *Pattern Recognition, 39*(4), 695–706.

Povinelli, R. J., Johnson, M. T., Lindgren, A. C., & Ye, J. (2004). Time series classification using Gaussian mixture models of reconstructed phase spaces. *IEEE Transactions on Knowledge and Data Engineering, 16*(6), 779–783.

Pudil, P., Novovicova, P. J., & Kittler, J. (1994). Floating search methods in feature selection. *Pattern Recognition Letters, 15*, 1119–1125.

Richards, J. A. (1986). *Remote sensing digital image analysis: an introduction*. Berlin: Springer-Verlag.

Searcóid, Ó. M. (2006). *Metric spaces*. Berlin: Springer. Undergraduate Mathematics Series.

Stearns, S. D. (1976). On selecting features for pattern classifiers. *Third International Joint Conference on Pattern Recognition* (pp. 71–75). *Los Alamitos CA*. IEEE Computer Society Press.

Turk, M., & Pentland, A. (1991). Eigenfaces for recognition. *Journal of Cognitive Neuroscience, 3*, 72–86.

Whitney, A. W. (1971). A direct method of nonparametric measurement selection. *IEEE Transactions on Computers, 20*, 1100–1103.

Yao, H., & Tian, L. (2003). A genetic-algorithm-based selective principal component analysis (GA-SPCA) method for high-dimensional data feature extraction. *IEEE Transactions on Geoscience and Remote Sensing, 41*(6), 1469–1478.

Ziehe, A., & Miller, K.-R. (1998). TDSEP—an efficient algorithm for blind separation using time structure. *ICANN'98, Skovde*, 675–680.

CHAPTER 4

Hyperspectral Image Processing Techniques

Michael O. Ngadi, Li Liu

Department of Bioresource Engineering, McGill University, Macdonald Campus, Quebec, Canada

CONTENTS

4.1. INTRODUCTION

Hyperspectral imaging is the combination of two mature technologies: spectroscopy and imaging. In this technology, an image is acquired over the visible and near-infrared (or infrared) wavelengths to specify the complete wavelength spectrum of a sample at each point in the imaging plane. Hyperspectral images are composed of spectral pixels, corresponding to a spectral signature (or spectrum) of the corresponding spatial region. A spectral pixel is a pixel that records the entire measured spectrum of the imaged spatial point. Here, the measured spectrum is characteristic of a sample's ability to absorb or scatter the exciting light.

The big advantage of hyperspectral imaging is the ability to characterize the inherent chemical properties of a sample. This is achieved by measuring the spectral response of the sample, i.e., the spectral pixels collected from the sample. Usually, a hyperspectral image contains thousands of spectral pixels. The image files generated are large and multidimensional, which makes visual interpretation difficult at best. Many digital image processing techniques are capable of analyzing multidimensional images. Generally, these are adequate and relevant for hyperspectral image processing. In some specific applications, the design of image analysis algorithms is required for the use of both spectral and image features. In this chapter, classic image processing techniques and methods, many of which have been widely used in hyperspectral imaging, will be discussed, as well as some basic algorithms that are special for hyperspectral image analysis.

4.2. IMAGE ENHANCEMENT

The noise inherent in hyperspectral imaging and the limited capacity of hyperspectral imaging instruments make image enhancement necessary for many hyperspectral image processing applications. The goal of image enhancement is to improve the visibility of certain image features for subsequent analysis or for image display. The enhancement process does not increase the inherent information content, but simply emphasizes certain specified image characteristics. The design of a good image enhancement algorithm should consider the specific features of interest in the hyperspectral image and the imaging process itself.

Image enhancement techniques include contrast and edge enhancement, noise filtering, pseudocoloring, sharpening, and magnifying. Normally these techniques can be classified into two categories: spatial domain methods and transform domain methods. The spatial domain techniques include methods operated on a whole image or on a local region. Examples of spatial domain methods are the histogram equalization method and the local neighborhood operations based on convolution. The transform domain techniques manipulate image information in transform domains, such as discrete Fourier and wavelet transforms. In the following sub-sections, the classic enhancement methods used for hyperspectral images will be discussed.

4.2.1. Histogram Equalization

Image histogram gives primarily the global description of the image. The histogram of a graylevel image is the relative frequency of occurrence of each graylevel in the image. Histogram equalization (Stark & Fitzgerald, 1996), or histogram linearization, accomplishes the redistribution of the image graylevels by reassigning the brightness values of pixels based on the image histogram. This method has been found to be a powerful method of enhancement of low contrast images.

Mathematically, the histogram of a digital image is a discrete function $h(k) = n_k/n$, where $k = 0,1, \ldots, L-1$ and is the kth graylevel, n_k is the number of pixels in the image having graylevel k, and n is the total number of pixels in the image. In the histogram equalization method, each original graylevel k is mapped into new graylevel i by:

$$i = \sum_{j=0}^{k} h(j) = \sum_{j=0}^{k} n_j/n \tag{4.1}$$

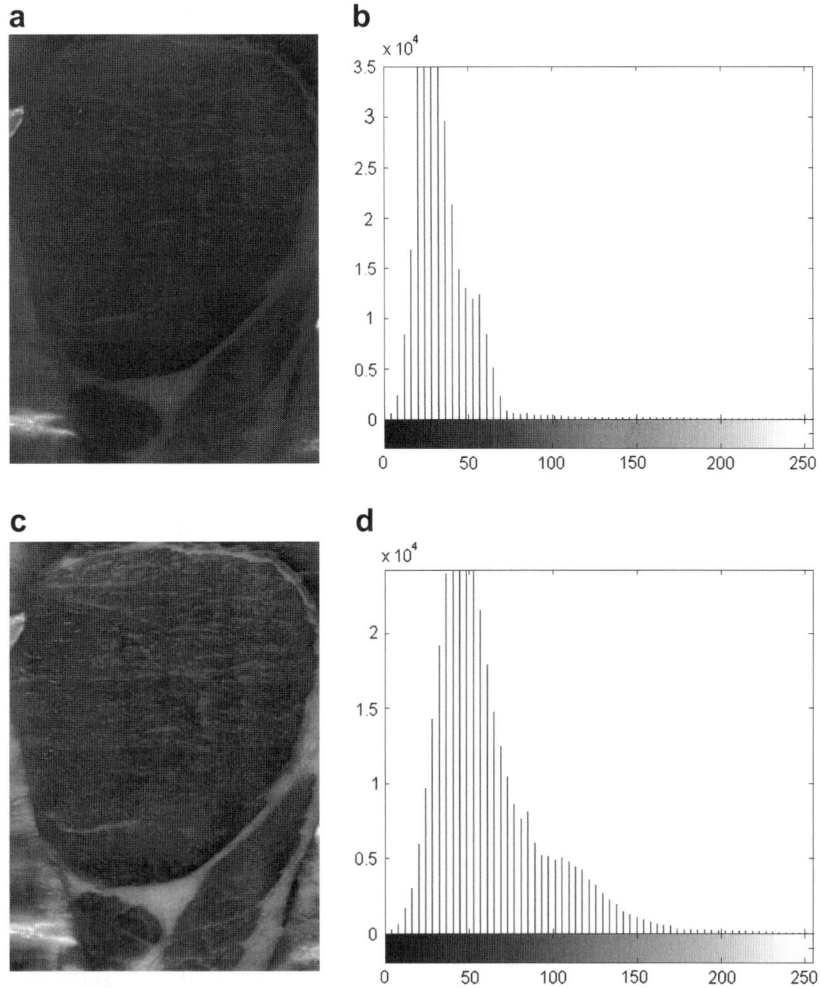

FIGURE 4.1 *Image quality enhancement using histogram equalization: (a) spectral image of a pork sample; (b) histogram of the image in (a); (c) resulting image obtained from image (a) by histogram equalization; (d) histogram of the image in (c). (Full color version available on http://www.elsevierdirect.com/companions/9780123747532/)*

where the sum counts the number of pixels in the image with graylevel equal to or less than k. Thus, the new graylevel is the cumulative distribution function of the original graylevels, which is always monotonically increasing. The resulting image will have a histogram that is "flat" in a local sense, since the operation of histogram equalization spreads out the peaks of the histogram while compressing other parts of the histogram (see Figure 4.1).

Histogram equalization is just one example of histogram shaping. Other predetermined shapes are also used (Jain, 1989). Any of these histogram-based methods need not be performed on an entire image. Enhancing a portion of the original image, rather than the entire area, is also useful in many applications. This nonlinear operation can significantly increase the visibility of local details in the image. However, it is computationally intensive and the complexity increases with the size of the local area used in the operation.

4.2.2. Convolution and Spatial Filtering

Spatial filtering refers to the convolution (Castleman, 1996) of an image with a specific filter mask. The process consists simply of moving the filter mask from point to point in an image. At each point, the response of the filter is the weighted average of neighboring pixels which fall within the window of the mask. In the continuous form, the output image $g(x, y)$ is obtained as the convolution of the image $f(x, y)$ with the filter mask $w(x, y)$ as follows:

$$g(x,y) = f(x,y) * w(x,y) \tag{4.2}$$

where the convolution is performed over all values of (x, y) in the defined region of the image. In the discrete form, convolution denotes $g_{i,j} = f_{i,j} * w_{i,j}$, and the spatial filter $w_{i,j}$ takes the form of a weight mask. Table 4.1 shows several commonly used discrete filters.

4.2.2.1. Smoothing linear filtering

A smoothing linear filter, also called a low-pass filter, is symmetric about the filter center and has only positive weight values. The response of a smoothing linear spatial filter is the weighted average of the pixels contained in the neighborhood of the filter mask. In image processing, smoothing filters are widely used for noise reduction and blurring. Normally, blurring is used in pre-processing to remove small details from an image before feature/object extraction and to bridge small gaps in lines or

Table 4.1 Examples of discrete filter masks for spatial filtering

Spatial filter	Low-pass	High-pass	Laplacian
$w_{(i,j)}$	$\frac{1}{9}\times\begin{bmatrix}1 & 1 & 1\\ 1 & 1 & 1\\ 1 & 1 & 1\end{bmatrix}$	$\begin{bmatrix}-1 & -1 & -1\\ -1 & 9 & -1\\ -1 & -1 & -1\end{bmatrix}$	$\begin{bmatrix} & -1 & \\ -1 & 4 & -1\\ & -1 & \end{bmatrix}$

curves. Noise reduction can be achieved by blurring with a linear filter or by nonlinear filtering such as a median filter.

4.2.2.2. Median filtering

A widely used nonlinear spatial filter is the median filter that replaces the value of a pixel by the median of the graylevels in a specified neighborhood of that pixel. The median filter is a type of order-statistics filter, because its response is based on ranking the pixels contained in the image area covered by the filter. This filter is often useful because it can provide excellent noise-reduction with considerably fewer blurring edges in the image (Jain, 1989). The noise-reducing effect of the median filter depends on two factors: (1) the number of noise pixels involved in the median calculation and (2) the spatial extent of its neighborhood. Figure 4.2 shows an example of impulse noise (also called salt-and-pepper noise) removal using median filtering.

4.2.2.3. Derivative filtering

There is often the need in many applications of image processing to highlight fine detail (for example, edges and lines) in an image or to enhance detail that has been blurred. Generally, an image can be enhanced by the following sharpening operation:

$$z(x, y) = f(x, y) + \lambda e(x, y) \tag{4.3}$$

where $\lambda > 0$ and $e(x, y)$ is a high-pass filtered version of the image, which usually corresponds to some form of the derivative of an image. One way to accomplish the operation is by adding gradient information to the image. An example of this is the *Sobel* filter pair that can be used to estimate the gradient in both the x and the y directions. The Laplacian

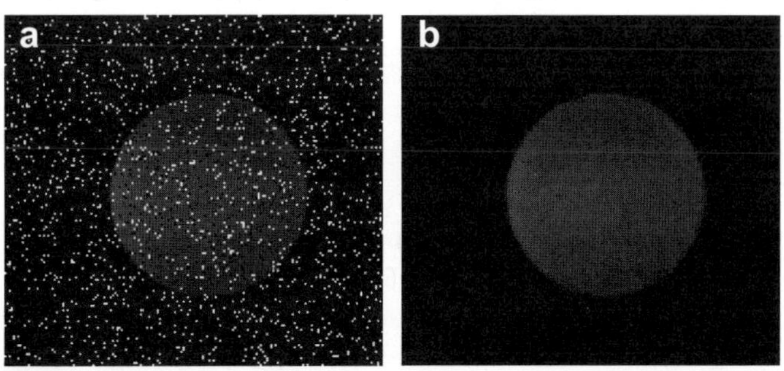

FIGURE 4.2 *Impulse noise removal by median filtering: (a) spectral image of an egg sample with salt-and-pepper noise (0.1 variance); (b) filtered image of image (a) as smoothed by a 3 × 3 median filter*

filter (Jain, 1989) is another commonly used derivative filter, which is defined as:

$$\nabla^2 f(x,y) = \left(\frac{\partial^2}{\partial x^2} + \frac{\partial^2}{\partial y^2}\right) f(x,y) \tag{4.4}$$

The discrete form of the operation can be implemented as:

$$\nabla^2 f_{i,j} = \left[f_{i+1,j} - 2f_{i,j} + f_{i-1,j}\right] + \left[f_{i,j+1} - 2f_{i,j} + f_{i,j-1}\right] \tag{4.5}$$

The kernel mask used in the discrete Laplacian filtering is shown in Table 4.1.

A Laplacian of Gaussian (LoG) filter is often used to sharpen noisy images. The LoG filter first smoothes the image with a Gaussian low-pass filtering, followed by the high-pass Laplacian filtering. The LoG filter is defined as:

$$\nabla^2 g(x,y) = \left(\frac{\partial^2}{\partial x^2} + \frac{\partial^2}{\partial y^2}\right) g_\sigma(x,y) \tag{4.6}$$

where:

$$g_\sigma(x,y) = \frac{1}{\sqrt{2\pi}\sigma} \exp\left(-\frac{x^2+y^2}{2\sigma^2}\right)$$

is the Gaussian function with variance σ, which determined the size of the filter. Using a larger filter will improve the smoothing of noise. Figure 4.3 shows the result of sharpening an image using a LoG operation.

Image filtering operations are most commonly done over the entire image. However, because image properties may vary throughout the image, it is often useful to perform spatial filtering operations in local neighborhoods.

4.2.3. Fourier Transform

In many cases smoothing and sharpening techniques in frequency domain are more effective than their spatial domain counterparts because noise can be more easily separated from the objects in the frequency domain. When an image is transformed into the frequency domain, low-frequency components describe smooth regions or main structures in the image; medium-frequency components correspond to image features; and high-frequency components are dominated by edges and other sharp transitions such as noise. Hence filters can be designed to sharpen the image while

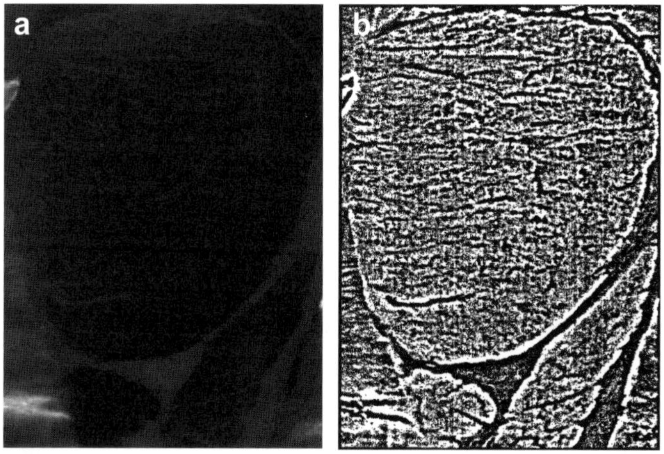

FIGURE 4.3 *Sharpening images using a Laplacian of Gaussian operation: (a) spectral image of a pork sample; (b) filtered image of image (a) as sharpened by a LoG operation*

suppressing noise by using the knowledge of the frequency components (Beghdadi & Negrate, 1989).

4.2.3.1. Low-pass filtering

Since edge and noise of an image are associated with high-frequency components, a low-pass filtering in the Fourier domain can be used to suppress noise by attenuating high-frequency components in the Fourier transform of a given image. To accomplish this, a 2-D low-pass filter transfer function $H(u, v)$ is multiplied by the Fourier transform $F(u,v)$ of the image:

$$Z(u,v) = H(u,v)F(u,v) \tag{4.7}$$

where $Z(u, v)$ is the Fourier transform of the smoothed image $z(x, y)$ which can be obtained by taking the inverse Fourier transform.

The simplest low-pass filter is called a 2-D ideal low-pass filter that cuts off all high-frequency components of the Fourier transform and has the transfer function:

$$H(u,v) = \begin{cases} 1 & \text{if } D(u,v) \leq D_0 \\ 0 & \text{otherwise} \end{cases} \tag{4.8}$$

where $D(u, v)$ is the distance of a point from the origin in the Fourier domain and D_0 is a specified non-negative value. However, the ideal low-pass filter is seldom used in real applications since its rectangular pass-band causes ringing artifacts in the spatial domain. Usually, filters with smoother roll-off characteristics are used instead. For example, a 2-D Gaussian low-pass filter is often used for this purpose:

$$H(u,v) = e^{-D^2(u,v)/2\sigma^2} = e^{-D^2(u,v)/2D_0^2} \tag{4.9}$$

where σ is the spread of the Gaussian curve, $D_0 = \sigma$ and is the cutoff frequency. The inverse Fourier transform of the Gaussian low-pass filter is also Gaussian in the spatial domain. Hence a Gaussian low-pass filter provides no ringing artifacts in the smoothed image.

4.2.3.2. High-pass filtering

While an image can be smoothed by a low-pass filter, image sharpening can be achieved in the frequency domain by a high-pass filtering process which attenuates the low-frequency components without disturbing high-frequency information in the Fourier transform. An ideal high-pass filter with cutoff frequency D_0 is given by:

$$H(u,v) = \begin{cases} 1 & \text{if } D(u,v) \geq D_0 \\ 0 & \text{otherwise} \end{cases} \tag{4.10}$$

As in the case of the ideal low-pass filter, the same ringing artifacts induced by the ideal high-pass filter can be found in the filtered image due to the sharp cutoff characteristics of a rectangular window function in the frequency domain. Therefore, one can also make use of a filter with smoother roll-off characteristics, such as:

$$H(u,v) = 1 - e^{-D^2(u,v)/2D_0^2} \tag{4.11}$$

which represents a Gaussian high-pass filter with cutoff frequency D_0. Similar to the Gaussian low-pass filter, a Gaussian high-pass filter has no ringing property and produces smoother results. Figure 4.4 shows an example of high-pass filtering using the Fourier transform.

4.2.4. Wavelet Thresholding

Human visual perception is known to function on multiple scales. Wavelet transform was developed for the analysis of multiscale image structures (Knutsson *et al.*, 1983). Rather than traditional transform domain methods such as the Fourier transform that only dissect signals into their component frequencies, wavelet-based methods also enable the analysis of the component frequencies across different scales. This makes them more suitable for such applications as noise reduction and edge detection.

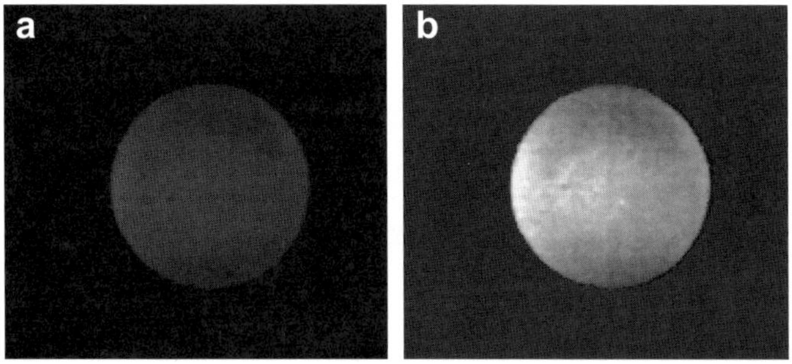

FIGURE 4.4 *High-pass filtering using the Fourier transform: (a) spectral image of an egg sample; (b) high-pass filtered image of image (a)*

Wavelet thresholding is a widely used wavelet-based technique for image enhancement which performs enhancement through the operation on wavelet transform coefficients. A nonlinear mapping such as hard-thresholding and soft-thresholding functions (Freeman & Adelson, 1991) is used to modify wavelet transform coefficients. For example, the soft-thresholding function can be defined as:

$$\theta(x) = \begin{cases} x - T & \text{if } x > T \\ x + T & \text{if } x < -T \\ 0 & \text{if } |x| \leq T \end{cases} \tag{4.12}$$

Coefficients with small absolute values (below threshold T or above −T) normally correspond to noise and thereby are reduced to a value near zero. The thresholding operation is usually performed in the orthogonal or biothorgonal wavelet transform domain. A translation-invariant wavelet transform may be a better choice in some cases (Lee, 1980). Enhancement schemes based on nonorthogonal wavelet transforms are also used (Coifman & Donoho, 1995; Sadler & Swami, 1999).

4.2.5. Pseudo-coloring

Color is a powerful descriptor that often simplifies object identification and extraction from an image. The most commonly used color space in computer vision technology is the *RGB* color space because it deals directly with the red, green, and blue channels that are closely associated with the human visual system. Another popularly employed color space is the *HSI* (hue, saturation, intensity) color space which is based on human color perception and can be described by a color cone. The hue of a color refers to the spectral wavelength that it most closely matches. The saturation is the radius of the point from the origin of the bottom circle of the cone and represents the purity of the color. The RGB and HSI color spaces can be easily converted from one to the other (Koschan & Abidi, 2008). An example of three bands from a hyperspectral image and a corresponding color image are depicted in Figure 4.5.

A pseudo-color image transformation refers to mapping a single-channel (monochrome) image to a three-channel (color) image by assigning different colors to different features. The principal use of pseudo-color is to aid human visualization and interpretation of grayscale images, since the combinations

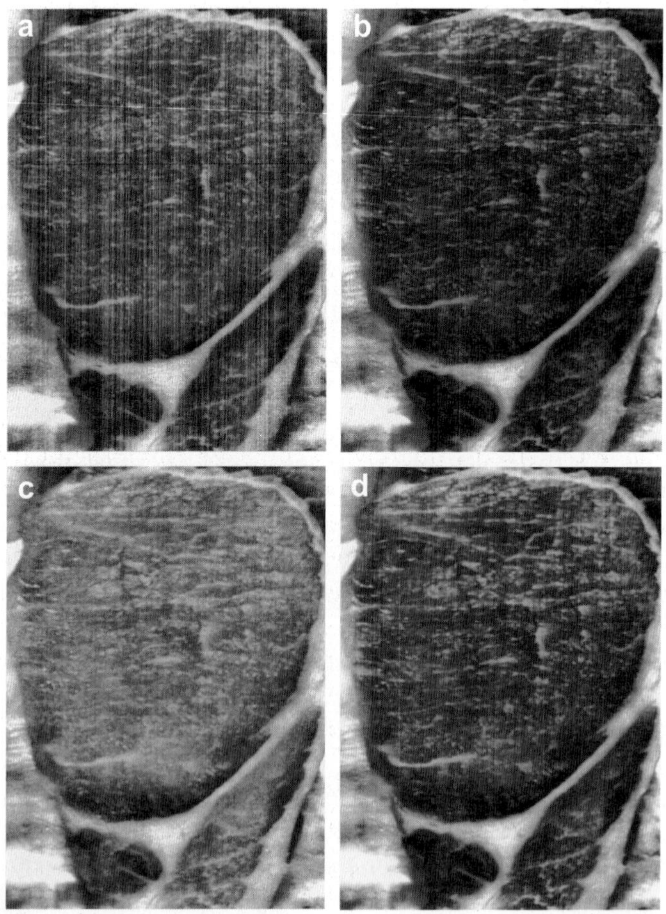

FIGURE 4.5 *RGB color image obtained from a hyperspectral image. Spectral images of a pork sample at (a) 460 nm, (b) 580 nm, and (c) 720 nm. The color image (d) in RGB was obtained by superposition of images in (a), (b), and (c). (Full color version available on http://www.elsevierdirect.com/companions/9780123747532/)*

of hue, saturation, and intensity can be discerned by humans much better than the shades of gray alone. The technique of intensity (sometimes called density) slicing and color coding is a simple example of pseudo-color image processing. If an image is interpreted as a 3-D function, this method can be viewed as one of painting each elevation with a different color. Pseudo-color techniques are useful for projecting hyperspectral image data down to three channels for display purposes.

4.2.6. Arithmetic Operations

When more than one image of the same object is available, arithmetic operations can be performed for image enhancement. For instance, averaging over N images will improve the signal-to-noise ratio by $\sqrt{N}$, and subtraction will highlight differences between images. In hyperspectral imaging, arithmetic operations are frequently used to provide even greater contrast between distinct regions of a sample (Pohl, 1998). One example is the band ratio method, in which an image at one waveband is divided by that at another wavelength (Liu *et al.*, 2007; Park *et al.*, 2006).

4.3. IMAGE SEGMENTATION

Segmentation is the process that divides an image into disjoint regions or objects. In image processing, segmentation is a major step and nontrivial image segmentation is one of the most difficult tasks. Accuracy of image segmentation determines the eventual success or failure of processing and analysis procedures. Generally, segmentation algorithms are based on one of two different but complementary perspectives, by seeking to identify either the similarity of regions or the discontinuity of object boundaries in an image (Castleman, 1996). The first approach is based on partitioning a digital image into regions that are similar according to predefined criteria, such as thresholding. The second approach is to partition a digital image based on abrupt changes in intensity, such as edges in an image. Segmentations resulting from the two approaches may not be exactly the same, but both approaches are useful for understanding and solving image segmentation problems, and their combined use can lead to improved performance (Castleman, 1996; Jain, 1989).

In this section, some classic techniques for locating and isolating regions/objects of interest in a 2-D graylevel image will be described. Most of the techniques can be extended to hyperspectral images.

4.3.1. Thresholding

Thresholding is widely used for image segmentation due to its intuitive properties and simplicity of implementation. It is particularly useful for images containing objects against a contrasting background. Assume we are interested in high graylevel regions/objects on a low graylevel background, then a thresholded image $J(x,y)$ can be defined as:

$$J(x) = \begin{cases} 1, & \text{if } I(x,y) \geq T \\ 0, & \text{otherwise} \end{cases} \tag{4.13}$$

where $I(x, y)$ is the original image, T is the threshold. Thus, all pixels at or above the threshold set to 1 correspond to objects/regions of interest (ROI) whereas all pixels set to 0 correspond to the background.

Thresholding works well if the ROI has uniform graylevel and lays on a background of unequal but uniform graylevel. If the regions differ from the background by some property other than graylevel, such as texture, one can first use an operation that converts that property to graylevel. Then graylevel thresholding can segment the processed image.

4.3.1.1. Global thresholding

The simplest thresholding technique involving partitioning the image histogram with a single global threshold is widely used in hyperspectral image processing (Liu *et al.*, 2007; Mehl *et al.*, 2004; Qin *et al.*, 2009). The success of the fixed global threshold method depends on two factors: (1) the graylevel histogram is bimodal; and (2) the threshold, T, is properly selected. A bimodal graylevel histogram indicates that the background graylevel is reasonably constant over the image and the objects have approximately equal contrast above the background. In general, the choice of the threshold, T, has considerable effect on the boundary position and overall size of segmented objects. For this reason, the value of the threshold must be determined carefully.

4.3.1.2. Adaptive thresholding

In practice, the background graylevel and the contrast between the ROI and the background often vary within an image due to uneven illumination and other factors. This indicates that a threshold working well in one area of an image might work poorly in other areas. Thus, global thresholding is unlikely to provide satisfactory segmentation results. In such cases, an adaptive threshold can be used, which is a slowly varying function of position in the image (Liu *et al.*, 2002).

One approach to adaptive thresholding is to partition an original $N \times N$ image into subimages of $n \times n$ pixels each ($n < N$), analyze graylevel histograms of each subimage, and then utilize a different threshold to segment each subimage. The subimage should be of proper size so that the number of background pixels in each block is sufficient enough to allow reliable estimation of the histogram and setting of a threshold.

4.3.2. Morphological Processing

A set of morphological operations may be utilized if the initial segmentation by thresholding is not satisfactory. The binary morphological operations are neighborhood operations by sliding a structuring element over the image. The structuring element can be of any size, and it can contain any complement of 1s and 0s. There are two primitive operations to morphological processing: dilation and erosion. Dilation is the process of incorporating into an object all the background points which connect to the object, while erosion is the process of eliminating all the boundary points from the object. By definition, a boundary point is a pixel that is located inside the object but has at least one neighbor pixel outside the object. Dilation can be used to bridge gaps between two separated objects. Erosion is useful for removing from a thresholded image the irrelevant detail that is too small to be of interest.

The techniques of morphological processing provide versatile and powerful tools for image segmentation. For example, the boundary of an object can be obtained by first eroding the object by a suitable structuring element and then performing the difference between the object and its erosion; and dilation-based propagation can be used to fill interior holes of segmented objects in a thresholded image (Qiao *et al.*, 2007b). However, the best-known morphological processing technique for image segmentation is the watershed algorithm (Beucher & Meyer, 1993; Vincent & Soille, 1991), which often produces stable segmentation results with continuous segmentation boundaries.

A one-dimensional illustration of the watershed algorithm is shown in Figure 4.6. Here the objects are assumed to have a low graylevel against a high graylevel background. Figure 4.6 shows the graylevels along one scan line that passes through two objects in close proximity. Initially, a lower threshold is used to segment the image into the proper number of objects. The threshold is then slowly raised, one graylevel at a time. This makes the boundaries of the objects expand accordingly. The final boundaries are determined at the moment that the two objects touch each other. In any case, the procedure ends before the threshold reaches the background's graylevel.

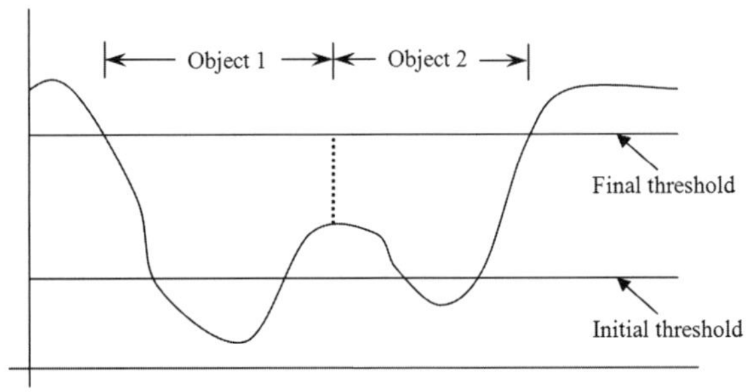

FIGURE 4.6 *Illustration of the watershed algorithm*

Unlike the global thresholding, which tries to segment the image at the optimum graylevel, the watershed algorithm begins the segmentation with a low enough threshold to properly isolate the objects. Then the threshold is raised slowly to the optimum level without merging the objects. This is useful to segment objects that are either touching or in too close a proximity for global thresholding to function. The initial and final threshold graylevels must be well chosen. If the initial threshold is too low, objects might be over-segmented and objects with low contrast might be missed at first and then merged with objects in a close proximity as the threshold increases. If the initial threshold is too high, objects might be merged at the start. The final threshold value influences how well the final boundaries fit the objects.

4.3.3. Edge-based Segmentation

In an image, edge pixels correspond to those points at which graylevel changes dramatically. Such discontinuities normally occur at the boundaries of objects. Thus, image segmentation can be implemented by identifying the edge pixels located at the boundaries.

4.3.3.1. Edge detection

Edges in an image can be detected by computing the first- and second-order digital derivatives, as illustrated in Figure 4.7. There are many derivative operators for 2-D edge detection and most of them can be classified as gradient-based or Laplacian-based methods. The first method locates the edges by looking for the maximum in the first derivative of the image, while the second method detects edges by searching for zero-crossings in the second derivative of the image.

For both edge detection methods, there are two parameters of interest: slope and direction of the transition. Edge detection operators examine each

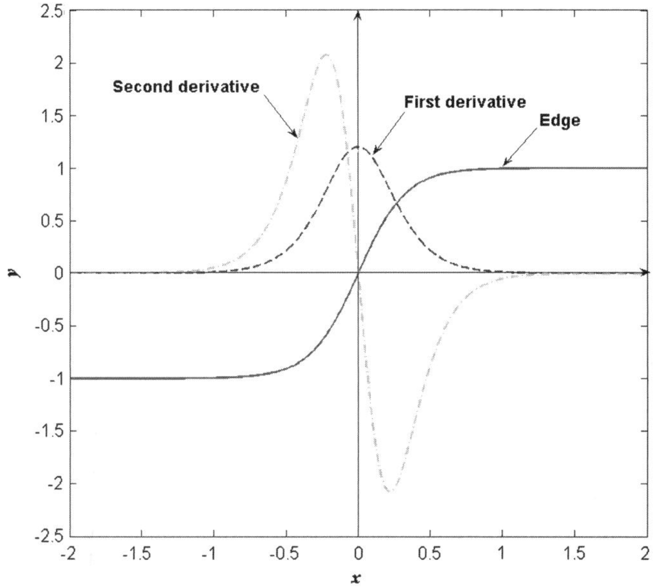

FIGURE 4.7 *An edge and its first and second derivatives. (Full color version available on http://www.elsevierdirect.com/companions/9780123747532/)*

pixel neighborhood and quantify the slope and the direction of the graylevel transition. Most of these operators perform a 2-D spatial gradient measurement on an image $I(x, y)$ using convolution with a pair of horizontal and vertical derivative kernels, g_x and g_y, which are designed to respond maximally to edges running in the x- and y-direction, respectively. Each pixel in the image is convolved with the two orthogonal kernels. The absolute magnitude of the gradient $|G|$ and its orientation α at each pixel can be estimated by combining the outputs from both kernels as:

$$|G| = \left[G_x^2 + G_y^2\right]^{1/2} \quad (4.14)$$

$$\alpha = \arctan\left(\frac{G_y}{G_x}\right) \quad (4.15)$$

where:

$$G_x = I(x,y) * g_x, \; G_y = I(x,y) * g_y \quad (4.16)$$

Table 4.2 lists the classic derivative-based edge detector.

Table 4.2 Derivative-based kernels for edge detection

Derivative kernels	Roberts	Prewitt	Sobel
g_x	$\begin{bmatrix} 1 & 0 \\ 0 & -1 \end{bmatrix}$	$\begin{bmatrix} -1 & 0 & 1 \\ -1 & 0 & 1 \\ -1 & 0 & 1 \end{bmatrix}$	$\begin{bmatrix} -1 & 0 & 1 \\ -2 & 0 & 2 \\ -1 & 0 & 1 \end{bmatrix}$
g_y	$\begin{bmatrix} 0 & 1 \\ -1 & 0 \end{bmatrix}$	$\begin{bmatrix} -1 & -1 & -1 \\ 0 & 0 & 0 \\ 1 & 1 & 1 \end{bmatrix}$	$\begin{bmatrix} -1 & -2 & -1 \\ 0 & 0 & 0 \\ 1 & 2 & 1 \end{bmatrix}$

4.3.3.2. Edge linking and boundary finding

In practice, the edge pixels yielded by the edge detectors seldom form closed connected boundaries due to noise, breaks in the edge from nonuniform illumination, and other effects. Thus, another step is usually required to complete the delineation of object boundaries for image segmentation.

Edge linking is the process of assembling edge pixels into meaningful edges so as to create a closed connected boundary. It can be achieved by searching a neighborhood around an endpoint for other endpoints and then filling in boundary pixels to connect them. Typically this neighborhood is a square region of 5×5 pixels or larger. Classic edge linking methods include heuristic search (Nevatia, 1976), curve fitting (Dierckx, 1993), and Hough transform (Ballard, 1981).

Edge linking based techniques, however, often result in only coarsely delineated object boundaries. Hence, a boundary refinement technique is required. A widely used boundary refinement technique is the active contour, also called a snake. This model uses a set of connected points, which can move around so as to minimize an energy function formulated for the problem at hand (Kass *et al.*, 1987). The curve formed by the connected points delineates the active contour. The active contour model allows a simultaneous solution for both the segmentation and tracking problems and has been applied successfully in a number of ways.

4.3.4. Spectral image segmentation

Segmentation of the sample under study is a necessary precursor to measurement and classification of the objects in a hyperspectral image. For biological samples, this is a significant problem due to the complex nature of the samples and the inherent limitation of hyperspectral imaging. Traditionally, segmentation is viewed as a low-level operation decoupled from higher-level analysis such as measurement and classification. Each pixel has a scalar graylevel value and objects are first isolated from the background based on graylevels and then identified based on a set of measurements reflecting their morphology. With hyperspectral imaging, however, each pixel is a vector of intensity values, and the identity of an object is encoded in that vector. Thus, segmentation and classification are more closely related and can be integrated into a single operation. This approach has been used with success in chromosome analysis and in optical character recognition (Agam & Dinstein, 1997; Martin, 1993).

4.4. OBJECT MEASUREMENT

Quantitative measurement of a region of interest (ROI) extracted by image segmentation is required for further data analysis and classification. In hyperspectral imaging, object measurement is based on a function of the intensity distribution of the object, called graylevel object measures. There are two main categories of graylevel object measurements. Intensity-based measures are normally defined as first-order measures of the graylevel distribution, whereas texture-based measures quantify second- or higher-order relationships among graylevel values.

If a hyperspectral image is obtained in the reflectance mode, all spectral reflectance images are required to correct from the dark current of the camera prior to image processing and object measurement (ElMasry *et al.*, 2007; Jiang *et al.*, 2007; Mehl *et al.*, 2004; Park *et al.*, 2006). To obtain the relative reflectance, correction is performed on the original hyperspectral reflectance images by:

$$I = \frac{I_0 - B}{W - B} \tag{4.17}$$

where I is the relative reflectance, I_0 is the original image, W is the reference image obtained from a white diffuse reflectance target, B is the dark current image acquired with the light source off and a cap covering the zoom lens. Hence, under the reflectance mode, all measures introduced in this section will be based on the relative reflectance.

4.4.1. Intensity-based measures

The regions of interest extracted by segmentation methods often contain areas that have heterogeneous intensity distributions. Intensity measures can be used to quantify intensity variations across and between objects. The most widely used intensity measure is the mean spectrum (ElMasry *et al.*, 2007; Park *et al.*, 2006; Qiao *et al.*, 2007a, 2007b), which is a vector consisting of the average intensity of the ROI at each wavelength. When normalized over the selected range of the wavelengths, the mean spectrum is the probability density function of the wavelengths (Qiao *et al.*, 2007b). Thus, measures derived from the normalized mean spectrum of the range of wavelengths provide statistical descriptors characterizing the spectral distribution. The same normalization operation can also be applied on each hyperspectral pixel, since the hyperspectral pixel can be viewed as a vector containing spectral signature/intensity over the range of wavelengths (Qin *et al.*, 2009).

First-order measures calculated on the normalized mean spectrum generally include mean, standard deviation, skew, energy, and entropy, while common second-order measures are based on joint distribution functions and normally are representative of the texture.

4.4.2. Texture

In image processing and analysis, texture is an attribute representing the spatial arrangement of the graylevels of pixels in the region of interest (IEEE, 1990). Broadly speaking, texture can be defined as patterns of local variations in image intensity, which are too fine to be distinguished as separate objects at the observed resolution (Jain *et al.*, 1995). Textures can be characterized by statistical properties such as standard deviation of graylevel and autocorrelation width, and also can be measured by quantifying the nature and directionality of the pattern, if it has any.

4.4.2.1. Graylevel co-occurrence matrix

The graylevel co-occurrence matrix (GLCM) provides a number of second-order statistics which describe the graylevel relationships in a neighborhood around a pixel of interest (Haralick, 1979; Kruzinga & Petkov, 1999; Peckinpaugh, 1991). It perhaps is the most commonly used texture measure in hyperspectral imaging (ElMasry *et al.*, 2007; Qiao *et al.*, 2007a; Qin *et al.*, 2009). The GLCM, $\mathbf{P}_D$, is a square matrix with elements specifying how often two graylevels occur in pairs of pixels separated by a certain offset distance in a given direction. Each entry (i, j) in $\mathbf{P}_D$ corresponds to the number of occurrences of the graylevels, i and j, in pairs of pixels that are separated by the chosen distance and direction in the image. Hence, for a given image, the GLCM is a function of the distance and direction.

Several widely used statistical and probabilistic features can be derived from the GLCM (Haralick & Shapiro, 1992). These include contrast (also called variance), which is given as:

$$V = \sum_{i,j} (i-j)^2 \mathbf{P}_D(i,j) \tag{4.18}$$

inverse differential moment (IDM, also called homogeneity), given by:

$$IDM = \sum_{i,j} \frac{\mathbf{P}_D(i,j)}{1+(i-j)^2} \tag{4.19}$$

angular second moment, defined as:

$$ASM = \sum_{i,j} [\mathbf{P}_D(i,j)]^2 \tag{4.20}$$

entropy, given as:

$$H = -\sum_{i,j} \mathbf{P}_D(i,j)\log(\mathbf{P}_D(i,j)) \tag{4.21}$$

and correlation, denoted by:

$$C = \frac{\sum_{i,j} (ij)\mathbf{P}_D(i,j) - \mu_i \mu_j}{\sigma_i \sigma_j} \tag{4.22}$$

where μ_i, μ_j, σ_i, and σ_j are the means and standard deviations, respectively, of the sums of rows and columns in the GLCM matrix. Generally, contrast is used to express the local variations in the GLCM. Homogeneity usually measures the closeness of the distribution of elements in the GLCM to its diagonal. Correlation is a measure of image linearity among pixels and the lower that value, the less linear correlation. Angular second moment (ASM) is used to measure the energy. Entropy is a measure of the uncertainty associated with the GLCM.

4.4.2.2. Gabor filter

A texture feature quantifies some characteristic of the graylevel variation within an object and can also be extracted by image processing techniques (Tuceryan & Jain, 1999). Among the image processing methods, the 2-D Gabor filter is perhaps the most popular method for image texture extraction and analysis. Its kernel is similar to the response of the 2-D receptive field profiles of the mammalian simple cortical cell, which makes the 2-D Gabor filter have the ability to achieve certain optimal joint localization properties in the spatial domain and in the spatial frequency domain (Daugman, 1980, 1985). This ability exhibits desirable characteristics of capturing salient visual properties such as spatial localization, orientation selectivity, and spatial frequency. Such characteristics make it an effective tool for image texture extraction and analysis (Clausi & Ed Jernigan, 2000; Daugman, 1993; Manjunath & Ma, 1996).

A 2-D Gabor function is a sinusoidal plane wave of a certain frequency and orientation modulated by a Gaussian envelope (Tuceryan & Jain, 1999) and is given by:

$$G(x,y;u,\sigma,\theta) = \frac{1}{2\pi\sigma^2}\exp\left\{-\frac{x^2+y^2}{2\sigma^2}\right\}\cos[2\pi u(x\cos\theta + y\sin\theta)] \quad (4.23)$$

where (x, y) is the coordinate of point in 2-D space, u is the frequency of the sinusoidal wave, θ controls the orientation of the Gabor filter, and σ is the standard deviation of the Gaussian envelope. When the spatial frequency information accounts for the major differences among texture, a circular symmetric Gabor filter can be used (Clausi & Ed Jernigan, 2000; Ma *et al.*, 2002), which is a Gaussian function modulated by a circularly symmetric sinusoidal function and has the following form (Ma *et al.*, 2002):

$$G(x,y;u,\sigma) = \frac{1}{2\pi\sigma^2}\exp\left\{-\frac{x^2+y^2}{2\sigma^2}\right\}\cos\left[2\pi u\left(\sqrt{x^2+y^2}\right)\right] \quad (4.24)$$

Figure 4.8 clearly shows the difference between an oriented Gabor filter and a circularly symmetric Gabor filter. In order to make Gabor filters more robust against brightness difference, discrete Gabor filters can be tuned to zero DC (direct current) with the application of the following formula (Zhang *et al.*, 2003):

$$\tilde{G} = G - \frac{\sum_{i=-n}^{n}\sum_{j=-n}^{n} G[i,j]}{(2n+1)^2} \quad (4.25)$$

where $(2n+1)^2$ is the size of the filter. Figure 4.9 illustrates how the two types of discrete Gabor filters work on a spectral image.

4.5. HYPERSPECTRAL IMAGING SOFTWARE

Many software tools have been developed for hyperspectral image processing and analysis. One of the most popular, commercially available

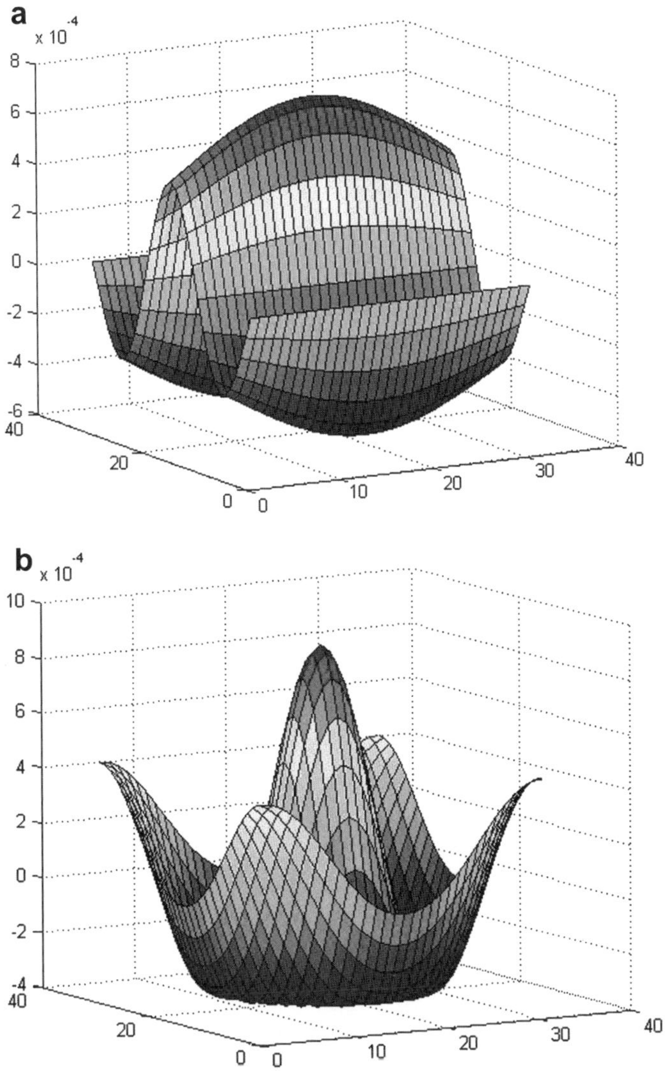

FIGURE 4.8 *Gabor filters: (a) shows example of an oriented Gabor filter defined in Equation (4.23) and (b) illustrates a circular symmetric Gabor filter defined in Equation (4.24). (Full color version available on http://www.elsevierdirect.com/companions/9780123747532/)*

analytical software tools is the Environment for Visualizing Images (ENVI) software (Research Systems Inc., Boulder, CO, USA) which is widely used in food engineering (ElMasry *et al.*, 2007; Liu *et al.*, 2007; Mehl *et al.*, 2004; Park *et al.*, 2006; Qiao *et al.*, 2007a, 2007b; Qin *et al.*, 2009). ENVI is

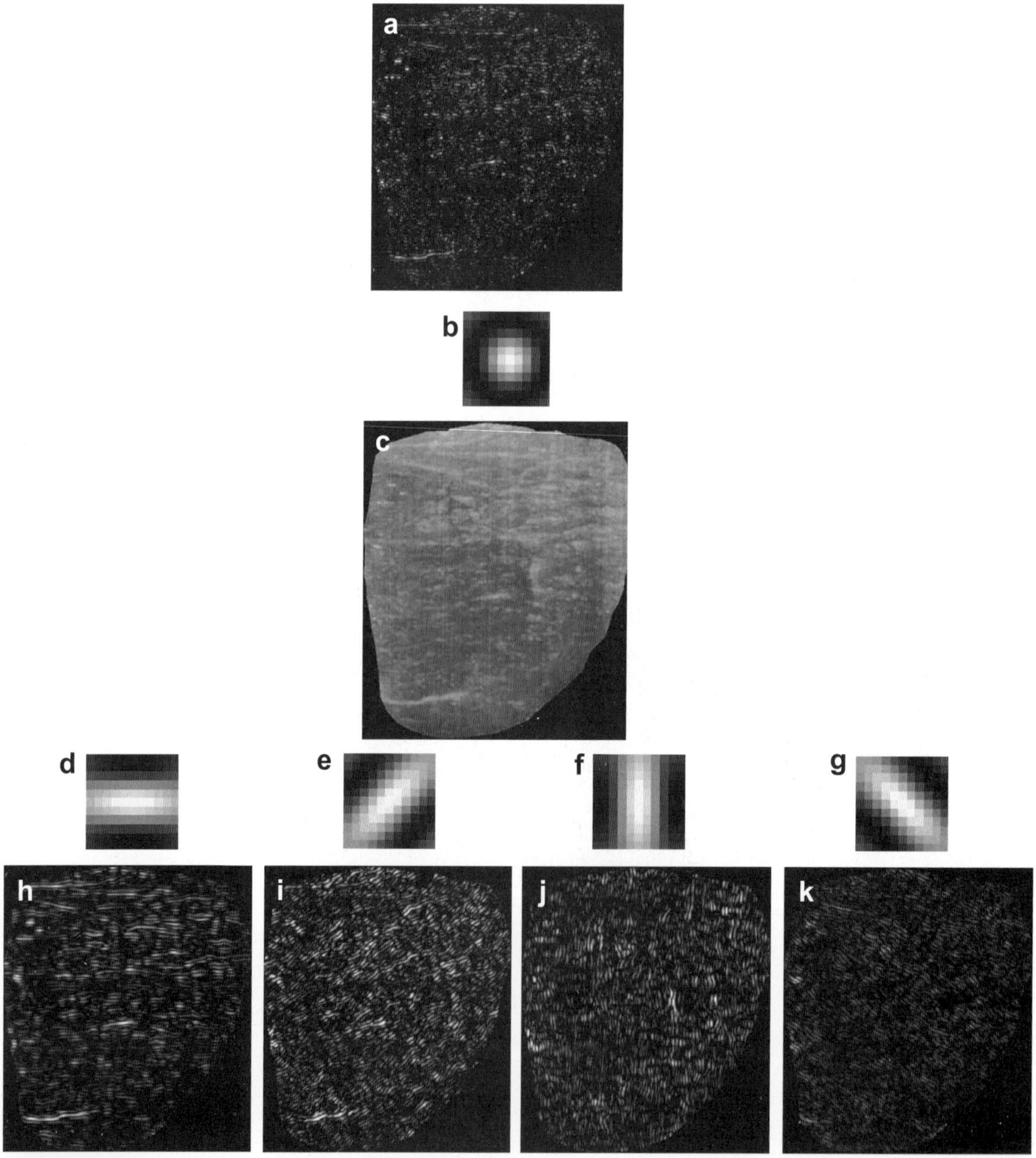

FIGURE 4.9 *A spectral image (c) is filtered by a circular Gabor filter (b) and four oriented Gabor filters in the direction of 0° (d), 45° (e), 90° (f), and 135° (g). Responses from the Gabor filters are shown in (a) and (h)–(k), respectively*

a software tool that is used for hyperspectral image data analysis and display. It is written totally in the interactive data language (IDL), which is based on array and provides integrated image processing and display capabilities. ENVI can be used to extract spectra, reference spectral libraries, and analyze high spectral resolution images from many different sensors. Figure 4.10 shows a user interface and imagery window from ENVI for a pork sample.

MATLAB (The Math-Works Inc., Natick, MA, USA) is another widely used software tool for hyperspectral image processing and analysis, which is a computer language used to develop algorithms, interactively analyze data, and view data files. MATLAB is a powerful tool for scientific computing and can solve technical computing problems more flexibly than ENVI and faster than traditional programming languages, such as C, C++, and Fortran. This makes it more and more popular in food engineering (ElMasry *et al.*, 2007;

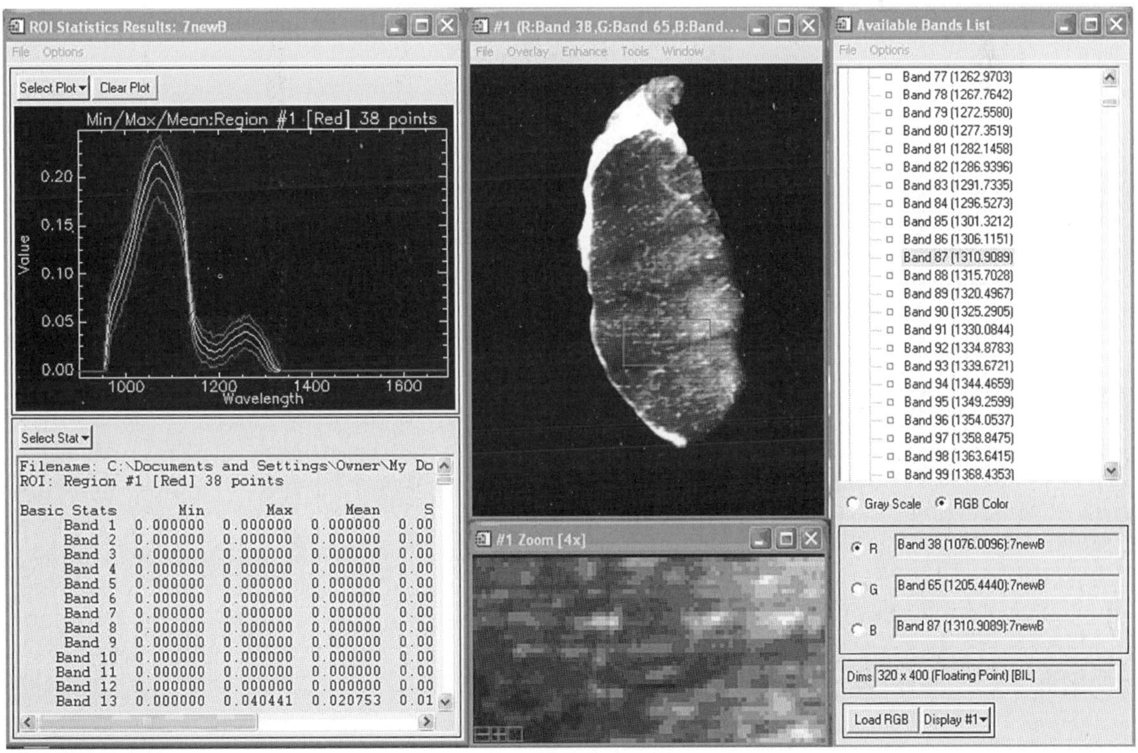

FIGURE 4.10 *ENVI user interface and a pork sample imagery. (Full color version available on http://www.elsevierdirect.com/companions/9780123747532/)*

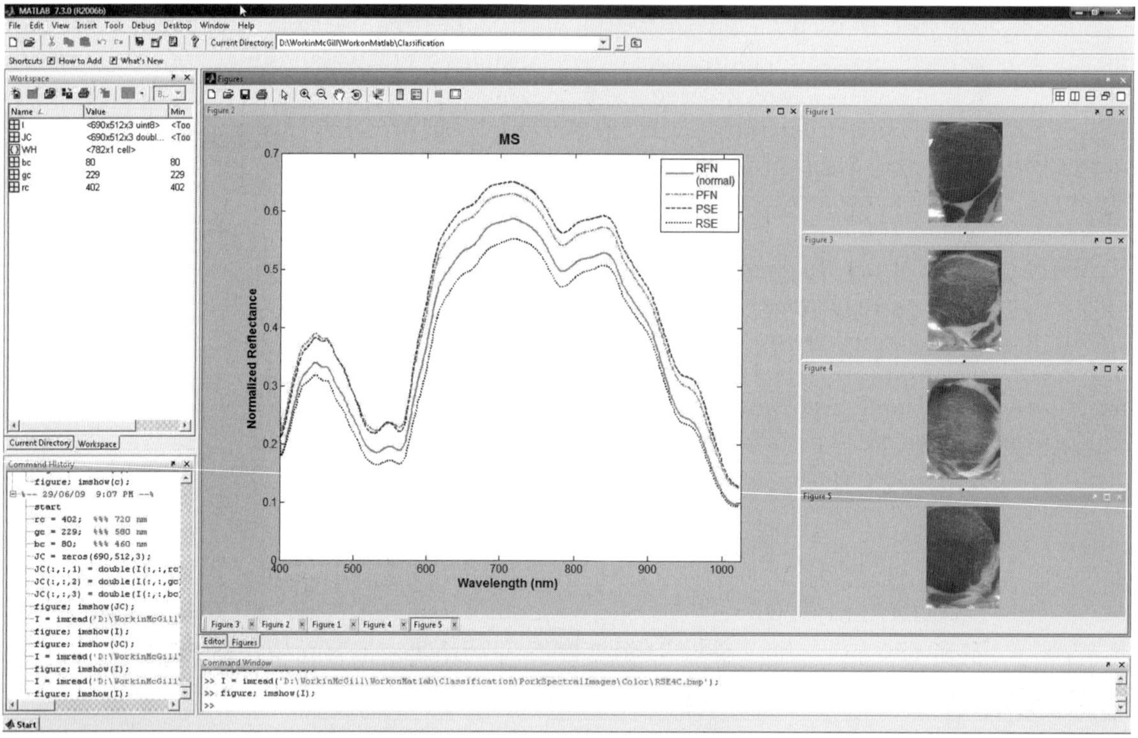

FIGURE 4.11 *A sample window in MATLAB. (Full color version available on http://www.elsevierdirect.com/companions/9780123747532/)*

Gomez-Sanchis *et al.*, 2008; Qiao *et al.*, 2007a, 2007b; Qin *et al.*, 2009; Qin & Lu, 2007). The graphics features which are required to visualize hyperspectral data are available in MATLAB. These include 2-D and 3-D plotting functions, 3-D volume visualization functions, and tolls for interactively creating plots. Figure 4.11 shows a sample window of MATLAB which collects four images of different kinds of pork samples as well as the corresponding spectral signatures.

There are also some enclosure, data acquisition, and preprocessing software tools available for simple and useful hyperspectral image processing, such as SpectraCube (Auto Vision Inc., CA, USA) and Hyperspec (Headwall Photonics, Inc., MA, USA). Figure 4.12 and Figure 4.13 illustrate the graphical user interface for a pork image acquisition and spectral profile analysis using SpectraCube and Hyperspec, respectively. In addition to these commercially available software tools, one can develop one's own software for hyperspectral image processing based on a certain computer language such as C/C++, Fortran, Java, etc.

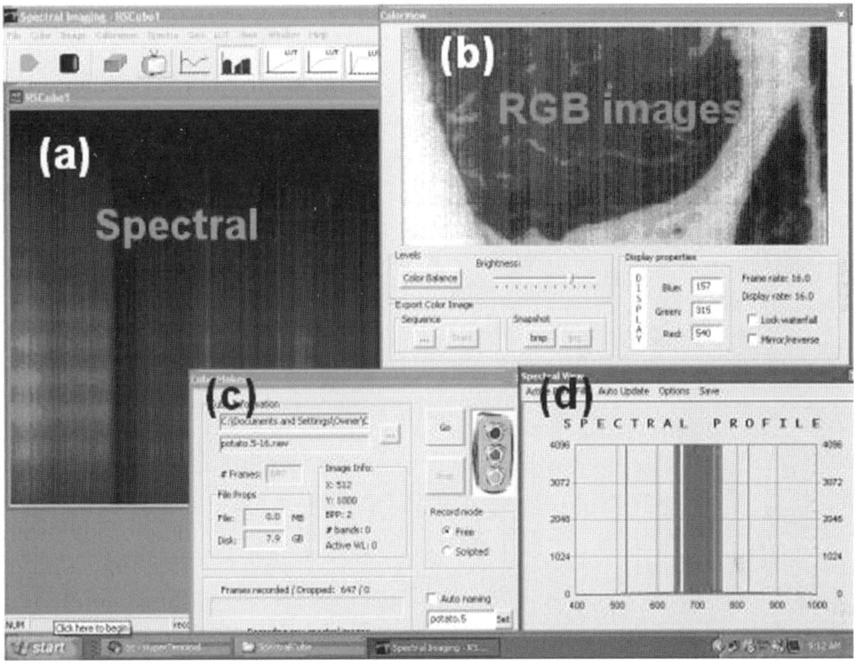

FIGURE 4.12 *The graphical user interface of the SpectraCube software for image acquisition and processing. (Full color version available on http://www.elsevierdirect.com/companions/9780123747532/)*

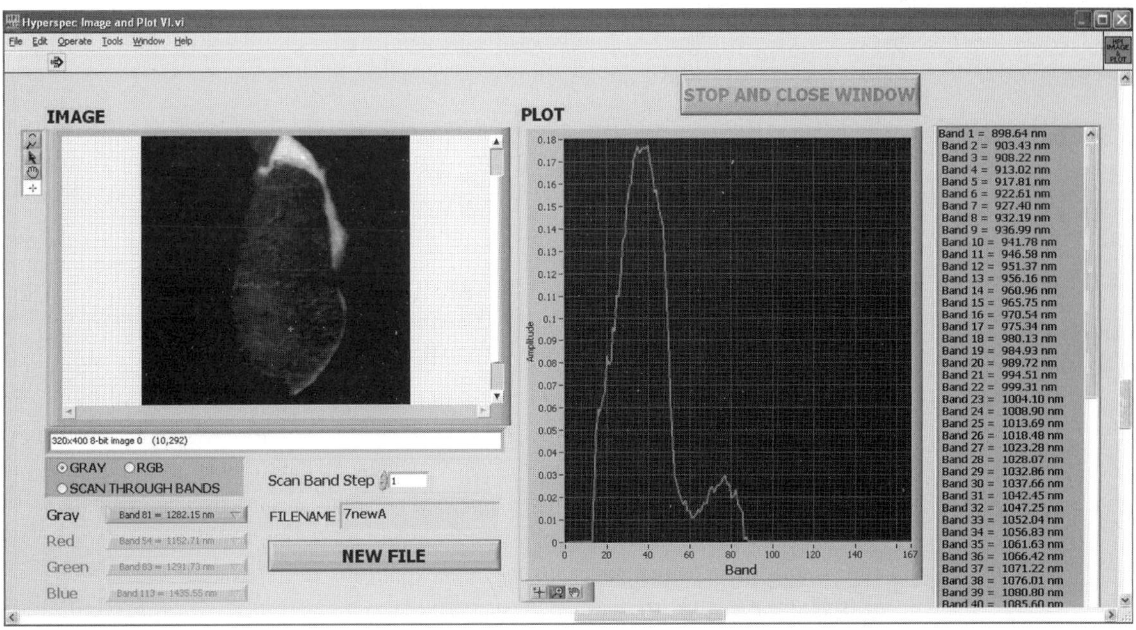

FIGURE 4.13 *The imaging user interface and sample imagery of the Hyperspec software. (Full color version available on http://www.elsevierdirect.com/companions/9780123747532/)*

4.6. CONCLUSIONS

Hyperspectral imaging is a growing research field in food engineering and has become more and more important for food quality analysis and control due to the ability of characterizing inherent chemical constituents of a sample. This technique involves the combined use of spectroscopy and imaging. This chapter focused on the image processing methods and algorithms which can be used in hyperspectral imaging. Most standard image processing techniques and methods can be generalized for hyperspectral image processing and analysis. Since hyperspectral images are normally too big and complex to be interpreted visually, image processing is often necessary in hyperspectral imaging for further data analysis. Many commercially analytical software tools such as ENVI and MATLAB are available for hyperspectral image processing and analysis. In addition, one can develop one's own hyperspectral image processing software for some specific requirement and application based on some common computer languages.

NOMENCLATURE

Symbols

n_k	number of pixels in the image having graylevel k
σ	standard deviation of the Gaussian envelope
$F(u, v)$	Fourier transform
D_0	cutoff frequency
g_x/g_y	horizontal/vertical derivative kernel
W	reference image obtained from a white diffuse reflectance target
B	dark current image
$\mathbf{P}_D$	graylevel co-occurrence matrix
μ_i/μ_j	mean of the sum of rows/columns in the GLCM matrix
σ_i/σ_j	standard deviation of the sum of rows/columns in the GLCM matrix
θ	orientation of the Gabor filter

Abbreviations

ASM	angular second moment
DC	direct current
ENVI	Environment for Visualizing Images software
GLCM	graylevel co-occurrence matrix

HSI	hue, saturation, intensity
IDM	inverse differential moment
RGB	red, green, and blue

REFERENCES

Agam, G., & Dinstein, I. (1997). Geometric separation of partially overlapping nonrigid objects applied to automatic chromosome classification. *IEEE Transactions on Pattern Analysis and Machine Intelligence, 19*(11), 1212–1222.

Ballard, D. (1981). Generalizing the Hough transform to detect arbitrary shapes. *Pattern Recognition, 13*, 111–122.

Beghdadi, A., & Negrate, A. L. (1989). Contrast enhancement technique based on local detection of edges. *Computer Vision and Graphical Image Processing, 46*, 162–174.

Beucher, S., & Meyer, F. (1993). The morphological approach to segmentation: the watershed transformation. In E. Dougherty (Ed.), *Mathematical morphology in image processing* (pp. 433–481). New York, NY: Marcel Dekker.

Castleman, K. R. (1996). *Digital image processing*. Englewood Cliffs, NJ: Prentice–Hall.

Clausi, D. A., & Ed Jernigan, M. (2000). Designing Gabor filters for optimal texture separability. *Pattern Recognition, 33*(1), 1835–1849.

Coifman, R. R., & Donoho, D. L. (1995). Translation-invariant denoising. In Anestis Antoniadis, & Georges Oppenheim (Eds.), *Wavelets and statistics*. New York, NY: Springer-Verlag.

Daugman, J. G. (1980). Two-dimensional spectral analysis of cortical receptive field profiles. *Vision Research, 20*, 847–856.

Daugman, J. G. (1985). Uncertainty relation for resolution in space, spatial frequency, and orientation optimized by two-dimensional visual cortical filters. *Journal of the Optical Society of America A, 2*(7), 1160–1169.

Daugman, J. G. (1993). High confidence visual recognition of persons by a test of statistical independence. *IEEE Transactions on Pattern Analysis and Machine Intelligence, 15*(11), 1148–1161.

Dierckx, P. (1993). *Curve and surface fitting with splines*. New York, NY: Oxford University Press.

ElMasry, G., Wang, N., Elsayed, A., & Ngadi, M. O. (2007). Hyperspectral imaging for non-destructive determination of some quality attributes for strawberry. *Journal of Food Engineering, 81*(1), 98–107.

Freeman, W. T., & Adelson, E. H. (1991). The design and use of steerable filters. *IEEE Transactions on Pattern Analysis and Machine Intelligence, 13*(9), 891–906.

Gomez-Sanchis, J., Molto, E., Camps-Valls, G., Gomez-Chova, L., Aleixos, N., & Blasco, J. (2008). Automatic correction of the effects of the light source on spherical objects: an application to the analysis of hyperspectral images of citrus fruits. *Journal of Food Engineering, 85*, 191–200.

Haralick, R. M. (1979). Statistical and structural approaches to texture. *Proceedings of IEEE, 67*(5), 786–804.

Haralick, R. M., & Shapiro, L. G. (1992). *Computer and robot vision*. Boston, MA: Addison–Wesley.

IEEE Standard 601.4-1990. (1990). *IEEE standard glossary of image processing and pattern recognition terminology*. Los Alamitos, CA: IEEE Press.

Jain, A. K. (1989). *Fundamentals of digital image processing*. Englewood Cliffs, NJ: Prentice–Hall.

Jain, R., Kasturi, R., & Schunk, B. G. (1995). *Machine vision*. New York, NY: McGraw–Hill.

Jiang, L., Zhu, B., Rao, X. Q., Berney, G., & Tao, Y. (2007). Discrimination of black walnut shell and pulp in hyperspectral fluorescence imagery using Gaussian kernel function approach. *Journal of Food Engineering, 81*(1), 108–117.

Kass, M., Witkin, A., & Terzopoulos, D. (1987). Snakes: active contour models. *Proceedings of the First International Conference on Computer Vision*, 259–269.

Knutsson, H., Wilson, R., & Granlund, G. H. (1983). Anisotropic non-stationary image estimation and its applications. Part I: Restoration of noisy images. *IEEE Transactions on Communications, 31*(3), 388–397.

Koschan, A., & Abidi, M. A. (2008). *Digital color image processing*. Hoboken, NJ: John Wiley & Sons, Inc.

Kruzinga, P., & Petkov, N. (1999). Nonlinear operator for oriented texture. *IEEE Transactions on Image Processing, 8*(10), 1395–1407.

Liu, F., Song, X. D., Luo, Y. P., & Hu, D. C. (2002). Adaptive thresholding based on variational background. *Electronics Letters, 38*(18), 1017–1018.

Liu, Y., Chen, Y. R., Kim, M. S., Chan, D. E., & Lefcourt, A. M. (2007). Development of simple algorithms for the detection of fecal contaminants on apples from visible/near infrared hyperspectral reflectance imaging. *Journal of Food Engineering, 81*(2), 412–418.

Lee, J. S. (1980). Digital image enhancement and noise filtering by local statistics. *IEEE Transactions on Pattern Analysis and Machine Intelligence, 2*, 165–168.

Ma, L., Tan, T., Wang, Y., & Zhang, D. (2002). Personal identification based on iris texture analysis. *IEEE Transactions on Pattern Recognition and Machine Intelligence, 25*(12), 1519–1533.

Manjunath, B. S., & Ma, W. Y. (1996). Texture feature for browsing and retrieval of image data. *IEEE Transactions on Pattern Analysis and Machine Intelligence, 18*(8), 837–842.

Martin, G. (1993). Centered-object integrated segmentation and recognition of overlapping handprinted characters. *Neural Computation, 5*(3), 419–429.

Mehl, P. M., Chen, Y. R., Kim, M. S., & Chan, D. E. (2004). Development of hyperspectral imaging technique for the detection of apple surface defects and contaminations. *Journal of Food Engineering, 61*(1), 67–81.

Nevatia, R. (1976). Locating object boundaries in textured environments. *IEEE Transactions on Computers, 25*, 1170–1180.

Park, B., Lawrence, K. C., Windham, W. R., & Smith, D. (2006). Performance of hyperspectral imaging system for poultry surface fecal contaminant detection. *Journal of Food Engineering, 75*(3), 340–348.

Peckinpaugh, S. H. (1991). An improved method for computing graylevel co-occurrence matrix-based texture measures. *Computer Vision, Graphics and Image Processing, 53*(6), 574–580.

Pohl, C. (1998). Multisensor image fusion in remote sensing. *International Journal of Remote Sensing, 19*(5), 823–854.

Qiao, J., Ngadi, M. O., Wang, N., Gariepy, C., & Prasher, S. O. (2007a). Pork quality and marbling level assessment using a hyperspectral imaging system. *Journal of Food Engineering, 83*, 10–16.

Qiao, J., Wang, N., Ngadi, M. O., Gunenc, A., Monroy, M., Gariepy, C., & Prasher, S. O. (2007b). Prediction of drip-loss, pH, and color for pork using a hyperspectral imaging technique. *Meat Science, 76*, 1–8.

Qin, J., Burks, T. F., Ritenour, M. A., & Gordon Bonn, W. (2009). Detection of citrus canker using hyperspectral reflectance imaging with spectral information divergence. *Journal of Food Engineering, 93*(2), 183–191.

Qin, J., & Lu, R. (2007). Measurement of the absorption and scattering properties of turbid liquid foods using hyperspectral imaging. *Applied Spectroscopy, 61*(4), 388–396.

Sadler, B. M., & Swami, A. (1999). Analysis of multiscale products for step detection and estimation. *IEEE Transactions on Information Theory, 45*(3), 1043–1051.

Stark, J. A., & Fitzgerald, W. J. (1996). An alternative algorithm for adaptive histogram equalization. *Graphical Models and Image Processing, 56*(2), 180–185.

Tuceryan, M., & Jain, A. K. (1999). Texture analysis. In C. H. Chen, L. F. Pau, & P. S. P. Wang (Eds.), *Handbook of pattern recognition and computer vision*. Singapore: World Scientific Books.

Vincent, L., & Soille, P. (1991). Watersheds in digital spaces: an efficient algorithm based on immersion simulations. *IEEE Transactions on Pattern Analysis and Machine Intelligence, 13*(6), 583–598.

Zhang, D., Kong, W. K., You, J., & Wong, M. (2003). Online palmprint identification. *IEEE Transactions on Pattern Analysis and Machine Intelligence, 25*(9), 1041–1050.

CHAPTER 5

Hyperspectral Imaging Instruments

Jianwei Qin

US Department of Agriculture, Agricultural Research Service, Henry A. Wallace Beltsville Agricultural Research Center, Beltsville, Maryland, USA

CONTENTS

5.1. INTRODUCTION

Optical sensing technologies offer great potential for nondestructive evaluation of agricultural commodities. Approaches based on imaging and spectroscopy have been intensively investigated and developed for many years. Although they have been used in various agricultural applications, conventional imaging and spectroscopy methods have limitations to obtain sufficient information from individual food items. In recent years, hyperspectral imaging has emerged as a better solution for quality and safety inspection of food and agricultural products. A comparison for the three approaches mentioned above may help better understand the merits of the hyperspectral imaging technique. General system configurations for conventional imaging, conventional spectroscopy, and hyperspectral imaging are illustrated in Figure 5.1. A conventional imaging system mainly consists of a light source and an area detector. The light source provides illumination to the sample, and the area detector captures mixed spectral contents from the sample. Spatial information of the sample is obtained in the forms of monochromatic or colorful images. Major components in a conventional spectroscopy system include a light source, a wavelength dispersion device, and a point detector. Light is dispersed into different wavelengths after interaction with the sample, and the point detector collects the dispersed light to obtain spectral information from the sample. Due to the size limitation of the point detector, the spectroscopy measurement cannot cover large areas or small areas with high spatial resolution. Hyperspectral imaging technique combines

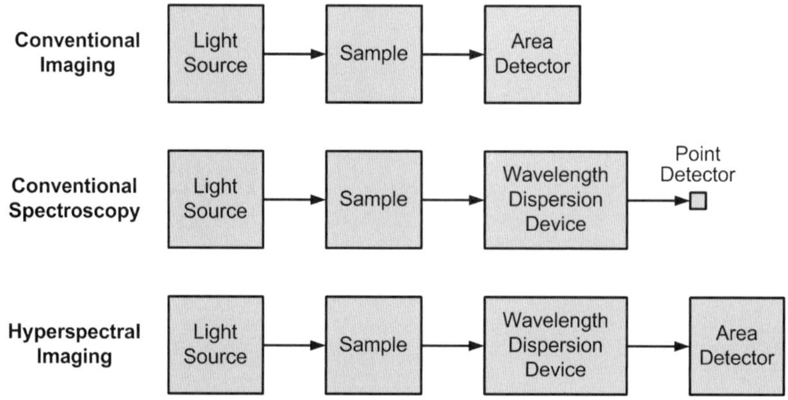

FIGURE 5.1 *General system configurations for conventional imaging, conventional spectroscopy, and hyperspectral imaging*

conventional imaging and spectroscopy techniques. A typical hyperspectral system consists of a light source, a wavelength dispersion device, and an area detector. It is capable of acquiring both spatial and spectral information from the sample in a form of spatially organized spectroscopy. If conventional imaging tries to answer the question *where* and conventional spectroscopy tries to answer the question *what*, then hyperspectral imaging tries to answer the question *where is what*.

Instrumentation is the base of any reliable measurement system. Selection of the instruments and design of their setup and calibrations are crucial for the performance of hyperspectral imaging systems. With proper arrangement, some instruments used for conventional imaging and spectroscopy can also be used for hyperspectral imaging. There are also instruments specifically designed for hyperspectral imaging. This chapter primarily focuses on instrumentation for hyperspectral imaging technique, with an emphasis on those that have found applications in food quality analysis and control. There is a brief introduction of methods for hyperspectral image acquisition (Section 5.2), with basic concepts and ground rules for the rest of the chapter. Main attention is paid to introduce a variety of essential components for constructing hyperspectral imaging systems (Section 5.3), including light sources, wavelength dispersion devices, and detectors. Instruments and methods for calibrating hyperspectral imaging systems such as spatial calibration, spectral calibration, and flat-field correction are also discussed (Section 5.4). Conclusions are given by summarizing the chapter and addressing the future development of hyperspectral imaging instruments (Section 5.5).

5.2. METHODS FOR HYPERSPECTRAL IMAGE ACQUISITION

Hyperspectral images are three-dimensional (3-D) in nature. Generally there are four approaches that can be used for acquiring 3-D hyperspectral image cubes [hypercubes (x, y, λ)]. They are point scanning, line scanning, area scanning, and the single shot method, as illustrated in Figure 5.2. In the point-scanning method (also known as the whiskbroom method), a single point is scanned along two spatial dimensions (x and y) by moving either the sample or the detector. A spectrophotometer equipped with a point detector is used to acquire a spectrum for each pixel in the scene. Hyperspectral image data are accumulated pixel by pixel in an exhaustive manner. Two-axis motorized positioning tables are usually needed to finish the image acquisition. The line-scanning method (also known as the pushbroom method)

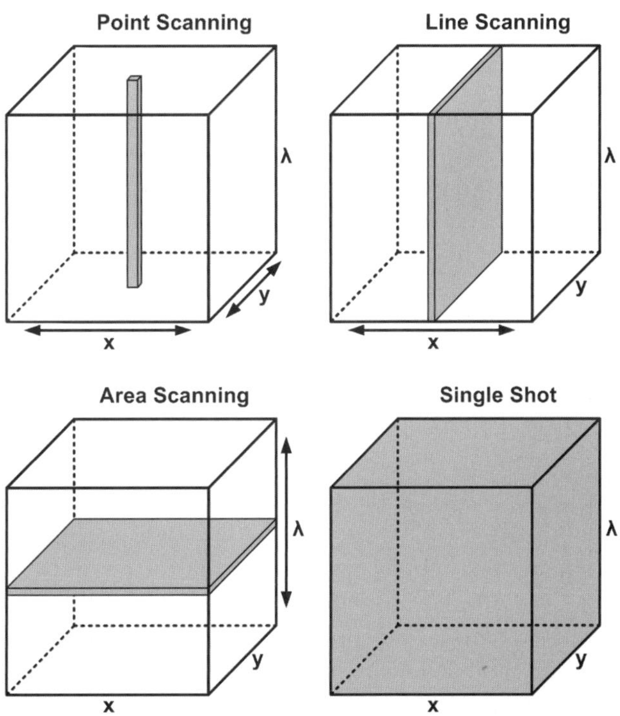

FIGURE 5.2 *Methods for acquiring three-dimensional hyperspectral image cubes containing spatial (x and y) and spectral (λ) information. Arrows represent scanning directions, and gray areas represent data acquired at a time*

can be considered as an extension of the point-scanning method. Instead of scanning one point each time, this method simultaneously acquires a slit of spatial information as well as spectral information corresponding to each spatial point in the slit. A special 2-D image (y, λ) with one spatial dimension (y) and one spectral dimension (λ) is taken at a time. A complete hypercube is obtained as the slit is scanned in the direction of motion (x). Hyperspectral systems based on imaging spectrographs with either fixed or moving slits work in the line-scanning mode.

Both point scanning and line scanning are spatial-scanning methods. The area-scanning method (also known as band sequential method), on the other hand, is a spectral-scanning method. This approach acquires a single band 2-D grayscale image (x, y) with full spatial information at once. A hypercube containing a stack of single band images is built up as the scanning is performed in the spectral domain through a number of wavelengths. No relative movement between the sample and the detector is required for this method. Imaging systems using filters (e.g., filter wheels containing fixed bandpass filters and electronically tunable filters) or Fourier transform imaging spectrometers belong to the area-scanning method. At last, the single shot method is intended to record both spatial and spectral information on an area detector with one exposure. No scanning in either spatial or spectral domains is needed for obtaining a 3-D image cube, making it attractive for applications requiring fast hyperspectral image acquisitions. This method is still in the early stage and not fully developed. Only a few implementations that rely on complicated fore-optics design and computationally intensive postprocessing for image reconstructions are currently available, with limitations for ranges and resolutions for spatial and spectral dimensions.

The 3-D hyperspectral image cubes acquired by point-scanning, line-scanning, and area-scanning methods are generally stored in the formats of Band Interleaved by Pixel (BIP), Band Interleaved by Line (BIL), and Band Sequential (BSQ), respectively. Different formats have different advantages in terms of image processing operations and interactive analysis. The BIP and BSQ formats offer optimal performances for spectral and spatial accesses of the hyperspectral image data, respectively. The BIL format gives a compromise in performance between spatial and spectral analysis. The three data storage formats can be converted to each other. The single shot method usually utilizes a large area detector to capture the images. The spatial and spectral contents from each frame can be transformed in either format mentioned above using appropriate reconstruction algorithms.

5.3. INSTRUMENTS FOR CONSTRUCTING HYPERSPECTRAL IMAGING SYSTEMS

The essential components for constructing hyperspectral imaging systems include light sources, wavelength dispersion devices, and area detectors. They are introduced in the following sections.

5.3.1. Light Sources

Light serves as an information carrier for vision-based inspection systems. Light sources generate light that illuminates or excites the target. Choice of the light sources and design of the lighting setup are critical for the performance and reliability of any imaging system. There are numerous types of light sources available for imaging or non-imaging applications. In this section, selected representative illumination and excitation light sources suitable for hyperspectral imaging applications are introduced.

5.3.1.1. Halogen lamps

Halogen lamps are the most common broadband illumination sources used in visible (VIS) and near-infrared (NIR) spectral regions. In their typical form, a lamp filament made of tungsten wire is housed in a quartz glass envelope filled with halogen gas. Light is generated through incandescent emission when a high operation temperature is on the filament. The halogen gas helps remove the deposited tungsten on the inside of the envelope and return it to the filament, maintaining the bulb is cleanliness and a long-term stable output for the lamp. The output light of quartz–tungsten–halogen (QTH) lamps forms a smooth continuous spectrum without sharp spectral peaks in the wavelength range from visible to infrared. The QTH lamps are bright light sources and are operated with low voltage, and they are the popular all-purpose illumination sources. The disadvantages of the halogen lamps include large heat generation, relatively short lifetime, output variations due to operating voltage fluctuations, spectral peak shift due to temperature change, and sensitivity to vibration.

The halogen lamps are commercially available in various forms. They can be used directly to illuminate the target (like room lighting) or be put in a lamp housing, from which light is delivered through an optical fiber. Figure 5.3 shows a DC-regulated halogen fiber-optic illuminator produced by TechniQuip (Danville, CA, USA). It generates light by a 150-watt halogen lamp inside, and offers a variable intensity control from 0 to 100%. A cold mirror is placed on the backside of the lamp to reflect the light to the fiber bundle. Coupled with proper fiber-optic light guides, the unit can deliver

FIGURE 5.3 *A halogen fiber-optic illuminator produced by TechniQuip (photo courtesy of TechniQuip, Danville, CA, USA). (Full color version available on http://www.elsevierdirect.com/companions/9780123747532/)*

broadband light for different illumination purposes (e.g., line light for hyperspectral line scanning and ring light for hyperspectral area scanning). The tungsten halogen lamps have been intensively used as light sources in hyperspectral reflectance measurements for surface inspections (Kim *et al.*, 2001; Lu, 2003; Park *et al.*, 2002). High intensity lamps have also been used in hyperspectral transmittance measurements for detecting inside agricultural commodities (Ariana & Lu, 2008; Qin & Lu, 2005; Yoon *et al.*, 2008).

5.3.1.2. Light emitting diodes

Owing to the demand for cheap, powerful, robust, and reliable light sources, light emitting diode (LED) technology has advanced rapidly during the past few years. Unlike tungsten halogen lamps, LEDs do not have a filament for incandescent emission. Instead, they are solid state sources that emit light when electricity is applied to a semiconductor. They can generate narrowband light in the VIS region at different wavelengths (or colors), depending on the materials used for the p–n junction inside the LED. Recently, LEDs that can produce high intensity broadband white light have been developed (Steigerwald *et al.*, 2002). Currently there are two major approaches to generate white light with LEDs. One approach mixes red, blue, and green monochromatic lights from three independent LEDs to generate the white light (Muthu *et al.*, 2002). The other approach utilizes a blue LED to excite a phosphor coating to form a phosphor-converted LED (pcLED) (Mueller-Mach *et al.*, 2002). The phosphor converts partial energy of the blue light into red and green light. The white light is created by mixing the generated red and green light with the rest of the blue light. This is the commonly used approach at present. Figure 5.4(a) shows a spectrum emitted by a white LED using the pcLED approach. It has a fairly good output in the VIS region. A spectral peak can be observed in the blue region around 470 nm due to the leaked blue light.

A picture of a LED line light produced by Advanced Illumination (Rochester, VT, USA) is shown in Figure 5.4(b). It is a high intensity source that can provide white light for long working distance or large area imaging

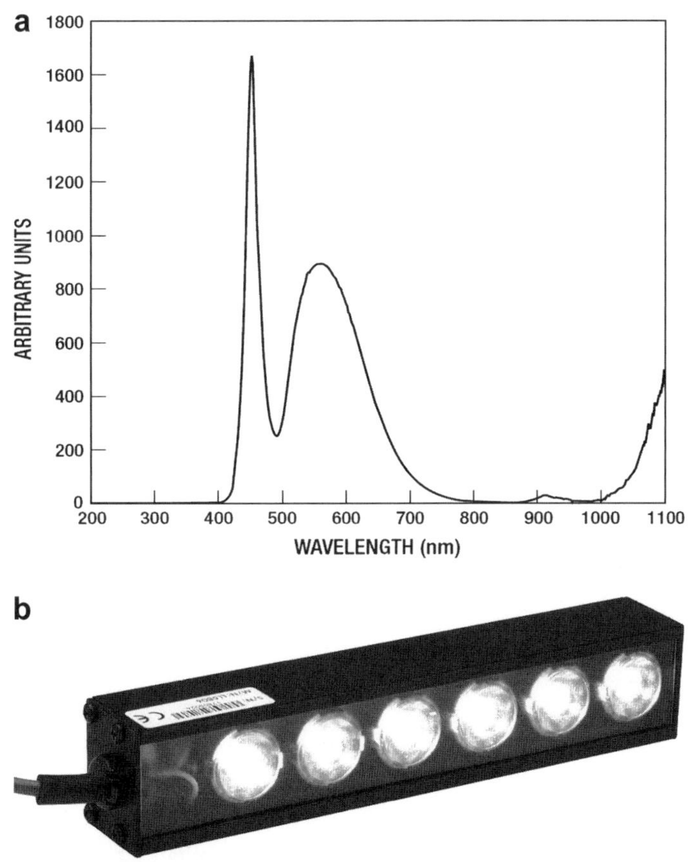

FIGURE 5.4 *Light emitting diode (LED): (a) emission spectrum of a white LED (courtesy of Newport Corporation, Irvine, CA, USA) and (b) a LED line light produced by Advanced Illumination (photo courtesy of Advanced Illumination, Inc., Rochester, VT, USA). (Full color version available on http://www.elsevierdirect.com/companions/9780123747532/)*

applications. Its operating temperature is below 60 °C, and the lamp lifetime is 50 000 hours, which is at least one order higher than that of most tungsten halogen lamps. As a new type of light source, LEDs have a lot of advantages over traditional lighting, such as long lifetime, low power consumption, low heat generation, small size, fast response, robustness, and insensitivity to vibration. They can be assembled in different arrangements (e.g., spot, line, and ring lights) to satisfy different illumination requirements. The LED technology is still ongoing with the development of new materials and electronics. LEDs have great potential to become mainstream light sources beyond their traditional uses such as small indicator lights on instrument panels. With the various benefits mentioned above, LED lights have started to find uses in the area of food quality and safety inspection (Chao *et al.*, 2007; Lawrence *et al.*, 2007). The use of LEDs as new light sources for hyperspectral imaging applications is likely to expand in the near future.

5.3.1.3. Lasers

Tungsten halogen lamps and white LEDs are illumination sources that are generally used in hyperspectral reflectance and transmittance imaging applications. The spectral constitution of the incident broadband light is not changed after light-sample interactions. The measurement is performed based on intensity changes at different wavelengths. Unlike broadband illumination sources, lasers are powerful directional monochromatic light sources. Light from lasers is generated through stimulated emission, which typically occurs inside a resonant optical cavity filled with a gain medium, such as gas, dye solution, semiconductor, and crystal. They can operate in CW (continuous wave) mode or pulse mode in terms of temporal continuity of the output. Lasers are widely used as excitation sources for fluorescence and Raman measurements owing to their unique features such as highly concentrated energy, perfect directionality, and real monochromatic emission.

Excited by a monochromatic light with a high energy, some biological materials (e.g., animal and plant tissues) emit light of a lower energy in a broad wavelength range. The energy change (or frequency shift) can cause fluorescence emission or Raman scattering that carries composition information of the target. Both fluorescence imaging and Raman imaging are sensitive optical techniques that can detect subtle changes of biological materials. Lasers have found applications in hyperspectral fluorescence imaging for inspection of agricultural commodities. For example, Kim *et al.* (2003) used a 355 nm pulsed Nd:YAG laser as an excitation source to perform fluorescence measurement for contaminant detection of apple and pork samples. Noh & Lu (2007) applied a 408 nm CW blue diode laser on apples to excite chlorophyll fluorescence. The hyperspectral laser-induced fluorescence images were analyzed for evaluating apple fruit quality. Lasers have also been utilized as excitation sources in hyperspectral Raman imaging applications (Jestel *et al.*, 1998; Wabuyele *et al.*, 2005). Besides lasers, other types of sources such as high-pressure arc lamps (e.g., xenon), low-pressure metal vapor lamps (e.g., mercury), and ultraviolet (UV) fluorescent lamps can also serve as excitation sources. In addition, LEDs that can produce narrowband pulsed or continuous light have started to be used as excitation sources, although at present their output has lower intensities and broader bandwidths than lasers.

5.3.1.4. Tunable sources

The configuration shown in Figure 5.1 is adopted by most current hyperspectral imaging systems for food quality and safety inspection. That is, the wavelength dispersion device is positioned between the sample and the detector. Light is dispersed into different wavelengths after interaction with the sample. There is another equivalent approach that puts the wavelength dispersion device in the illumination light path instead of the imaging light path (Figure 5.5a). This approach can be used by the hyperspectral systems that rely on broadband illumination (e.g., reflectance and transmittance imaging). Combined with the wavelength dispersion device, the white light source becomes a tunable light source. Incident light is dispersed before reaching the sample. There is no difference in principle between the two approaches for hyperspectral measurements. The major advantage of the tunable source approach is that the wavelength dispersion device does not need to maintain the spatial information of the target (Klein *et al.*, 2008). The detector directly performs area scanning to obtain both spatial and spectral information from the sample. The wavelength dispersion device should be synchronized with the detector to achieve automatic image acquisition. The intensity of the illumination using tunable sources is relatively weak since only narrowband light is incident on the sample at a time.

Tunable light sources are still in an early stage of development. Various wavelength dispersion methods have the potential to be adopted for making the tunable sources. Figure 5.5(b) shows an example of tunable source based on an acousto–optic tunable filter (AOTF) produced by Brimrose (Sparks, MD, USA). Its major components include a white light source and an AOTF device and its driver. Narrowband light is generated at a time when the white light interacts with the AOTF device. The source operates in the wavelength range of 360–560 nm with a spectral resolution up to 1 nm. To

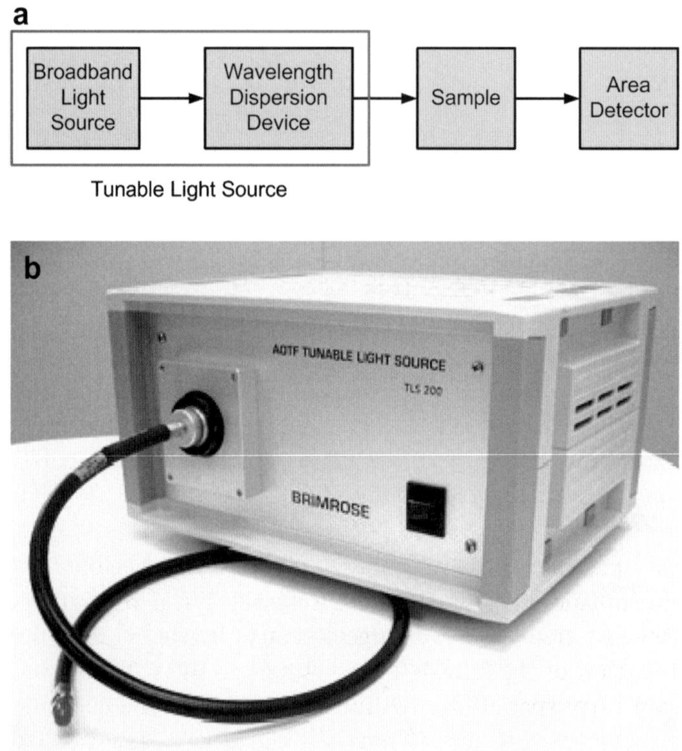

FIGURE 5.5 *Tunable light source: (a) concept and (b) a tunable light source based on acousto-optic tunable filter (AOTF) produced by Brimrose (photo courtesy of Brimrose Corporation, Sparks, MD, USA). (Full color version available on http://www.elsevierdirect.com/companions/9780123747532/)*

date, the use of tunable light sources for hyperspectral imaging applications is still limited because of the immature development of the related hardware. Efforts have been made to apply the tunable sources to hyperspectral reflectance and transmittance imaging, especially for the measurement conditions where weak illumination is desired to protect the target (e.g., document analysis and verification). Brauns & Dyer (2006) used a Michelson interferometer in front of a tungsten source to provide illumination for document samples at different wavelengths. Hyperspectral transmittance images were acquired for identification of fraudulent documents. Klein *et al.* (2008) put discrete bandpass filters between a broadband source and the target to fulfill a tunable light source. Hyperspectral reflectance measurement was performed for analyzing historical documents. Details on the operating principles of the wavelength dispersion devices mentioned above (i.e., AOTF, Michelson interferometer, and bandpass filter) can be found in

Section 5.3.2. The introduction of tunable sources opens a new avenue for implementation of hyperspectral image acquisition. Their feasibility for agricultural applications needs to be explored.

5.3.2. Wavelength Dispersion Devices

Wavelength dispersion devices are the heart of any hyperspectral imaging system. Various optical and electro–optical instruments can be used in hyperspectral imaging systems for dispersing broadband light into different wavelengths. The commonly used wavelength dispersion devices as well as some newly developed instruments are presented in this section. Their advantages and disadvantages for hyperspectral imaging applications are also discussed.

5.3.2.1. Imaging spectrographs

An imaging spectrograph is an optical device that is capable of dispersing incident broadband light into different wavelengths instantaneously for different spatial regions from a target surface. It can be considered as an enhanced version of the traditional spectrograph in that the imaging spectrograph can carry spatial information in addition to the spectral information. The imaging spectrograph generally operates in line-scanning mode, and it is the core component for the pushbroom hyperspectral imaging systems. Most contemporary imaging spectrographs are built based on diffraction gratings. A diffraction grating is a collection of transmitting or reflecting elements separated by a distance comparable to the wavelength of the light under investigation. The fundamental physical characteristic of the diffraction grating is the spatial modulation of the refractive index, by which the incident electromagnetic wave has its amplitude and/or phase modified in a predictable manner (Palmer, 2005). There are two main approaches in constructing imaging spectrographs, and they are transmission gratings (i.e., a grating laid on a transparent surface) and reflection gratings (i.e., a grating laid on a reflective surface).

Figure 5.6(a) illustrates the general configuration of an imaging spectrograph utilizing a transmission grating. Specifically, the operating principle of this imaging spectrograph is based on a prism–grating–prism (PGP) construction. An incoming light from the entrance slit of the spectrograph is collimated by the front lens. The collimated beam is dispersed at the PGP component so that the direction of the light propagation depends on its wavelength. The central wavelength passes symmetrically through the prisms and grating and stays at the optical axis. The shorter and longer

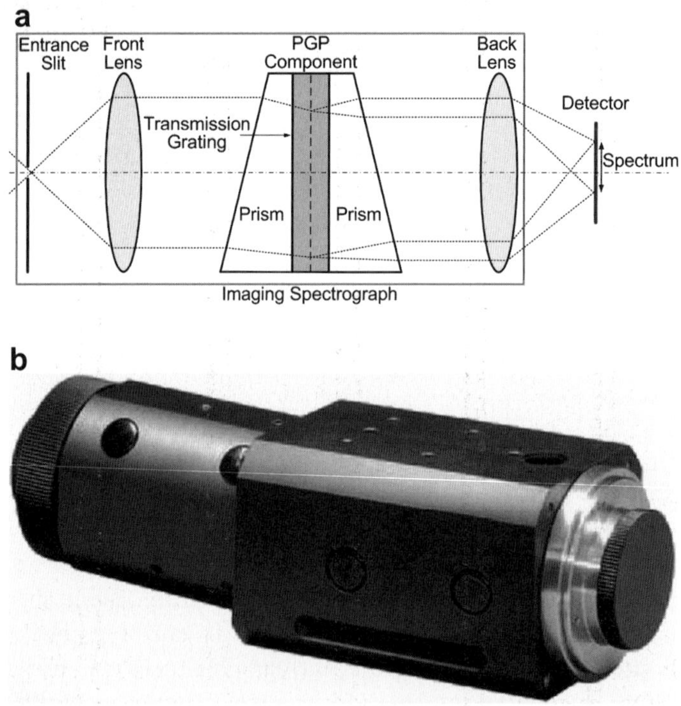

FIGURE 5.6 *Prism–grating–prism (PGP) imaging spectrograph: (a) operating principle and (b) an ImSpector imaging spectrograph produced by Spectral Imaging Ltd. (photo courtesy of Spectral Imaging Ltd., Oulu, Finland). (Full color version available on http://www.elsevierdirect.com/companions/9780123747532/)*

wavelengths are dispersed up and down relative to the central wavelength. This design results in a minimum deviation from the ideal on-axis condition and minimizes geometrical aberrations in both spatial and spectral axes (Spectral Imaging, 2003). As a result, the light from the scanning line is dispersed into different wavelengths, and they are projected onto an area detector through the back lens, creating a special two-dimensional image: one dimension represents spatial information and the other dimension spectral. As shown in Figure 5.6(a), each vertical line along the spectral axis of the 2-D area detector forms a continuous spectrum from a fixed spatial point of the object surface.

Figure 5.6(b) shows a commercialized PGP-based imaging spectrograph (ImSpector series) produced by Spectral Imaging Ltd. (Oulu, Finland). The ImSpector series includes several versions of imaging spectrographs covering different wavelength ranges e.g., UV (200–400 nm), VIS (380–780 nm), and NIR (900–1700 nm). Besides the standard series, enhanced and fast versions of the ImSpectors are also available to meet the requirements of high spectral and spatial resolutions as well as high speed spectral image acquisitions. The one shown in Figure 5.6(b), for example, is an ImSpector V10E imaging spectrograph. It is designed for the VIS and short-wavelength NIR region. The spectral range covered by this imaging spectrograph is 400–1000 nm. The slit length is 14.2 mm, and the spectral resolution under the default silt width (30 μm) is 2.8 nm. The slit width is customizable to realize different spectral resolutions. The ImSpectors have the merits of small size, ease of mounting, and common straight optical axis. They can be easily attached to a lens and a monochrome area detector to form a line-scanning spectral camera system. For the past decade, the ImSpector imaging spectrographs have been widely used throughout the world in standard or customized forms for developing many hyperspectral imaging systems. The ImSpector-based measurement systems have been applied for analyzing physical and/or chemical properties of a broad range of food and agricultural products. Examples include detecting contaminants on apples (Kim *et al.*, 2001) and poultry carcasses (Park *et al.*, 2002), tumors on chicken skin (Chao *et al.*, 2002), bruises on apples (Lu, 2003), pits in tart cherries (Qin & Lu, 2005), internal defects in cucumbers (Ariana & Lu, 2008), canker lesions on citrus (Qin *et al.*, 2008), and cracks in the shell of eggs (Lawrence *et al.*, 2008).

Reflection gratings are intensively used for making various conventional monochromators and spectrographs. Depending on the surface geometry of the diffraction gratings, plane gratings and curved gratings (i.e., concave and convex) are two basic types of the reflection gratings that are used in practice. Many optical layouts exist for constructing different types of monochromators and spectrographs. Examples include Czerny–Turner, Ebert–Fastie, Monk–Gillieson, Littrow, Rowland, Wadsworth, and flat-field configurations (Palmer, 2005). Reflection gratings have recently been used to build imaging spectrographs. For example, Headwall Photonics (Fitchburg, MA, USA) developed hyperspectral imaging spectrographs (Hyperspec series, Figure 5.7b) based on the Offner configuration (Figure 5.7a). The unit is constructed entirely from reflective optics. The basic structure of the design involves a pair of concentric spherical mirrors coupled with an aberration-corrected convex reflection grating. As shown in Figure 5.7(a), the lower mirror is used to guide the incoming light from the entrance slit to the reflection grating, where the incident beam is dispersed into different wavelengths in a reflection manner. The upper mirror then reflects the dispersed light to the detector, where a continuous spectrum is formed. This configuration offers the advantage of high image quality, free of

a

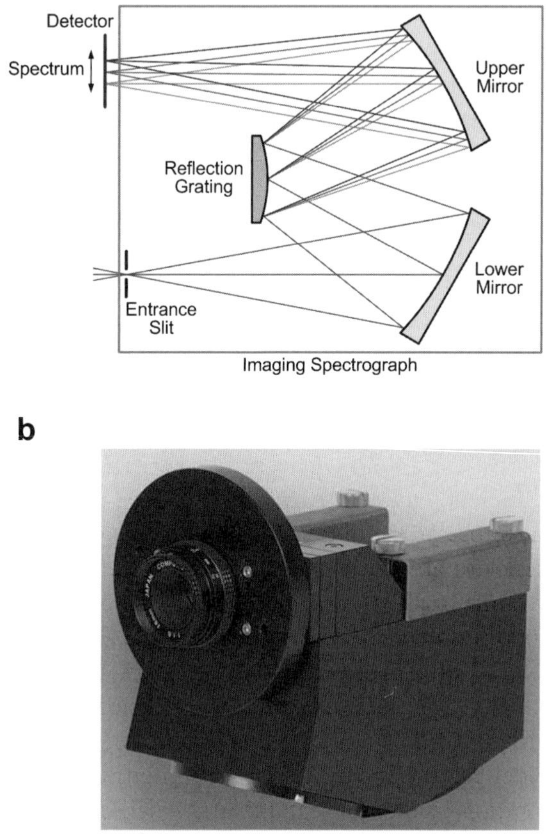

b

FIGURE 5.7 *Offner imaging spectrograph: (a) operating principle and (b) a Hyperspec imaging spectrograph produced by Headwall Photonics (photo courtesy of Headwall Photonics, Fitchburg, MA, USA). (Full color version available on http://www.elsevierdirect.com/companions/9780123747532/)*

higher-order aberrations, low distortion, low f-number, and large field size (Bannon & Thomas, 2005). The reflection gratings are not limited by the transmission properties of the grating substrate. Additionally, the reflective optical components (e.g., mirrors) generally have higher efficiencies than the transmission components (e.g., prisms). Thus the reflection grating-based imaging spectrographs are ideal for the situations where high signal-to-noise ratio (SNR) is crucial for measurement. The imaging spectrograph utilizing the reflection grating approach represents an increasingly accepted instrument for line-scanning hyperspectral imaging systems, especially for the low light measuring conditions such as fluorescence imaging and Raman imaging.

5.3.2.2. Filter wheels

The most basic implementation of spectral imaging is the use of a rotatable disk called a filter wheel carrying a set of discrete bandpass filters (Figure 5.8b). The main characteristic of the bandpass filters is that they transmit a particular wavelength with high efficiency while rejecting light energy out of the passband (Figure 5.8a). As the filter wheel employs mechanical rotation, the light perpendicularly transmits across different filters, generating a series of narrow band images at different predetermined wavelengths. Interference filters are commonly used as optical bandpass filters. Modern interference filters are constructed with a series of thin films (usually a few nanometers thick) between two glass plates. Each film layer is made from a dielectric material with a specified refractive index. The incident light to the filter is affected by interferences due to different refractive indices of the films. High

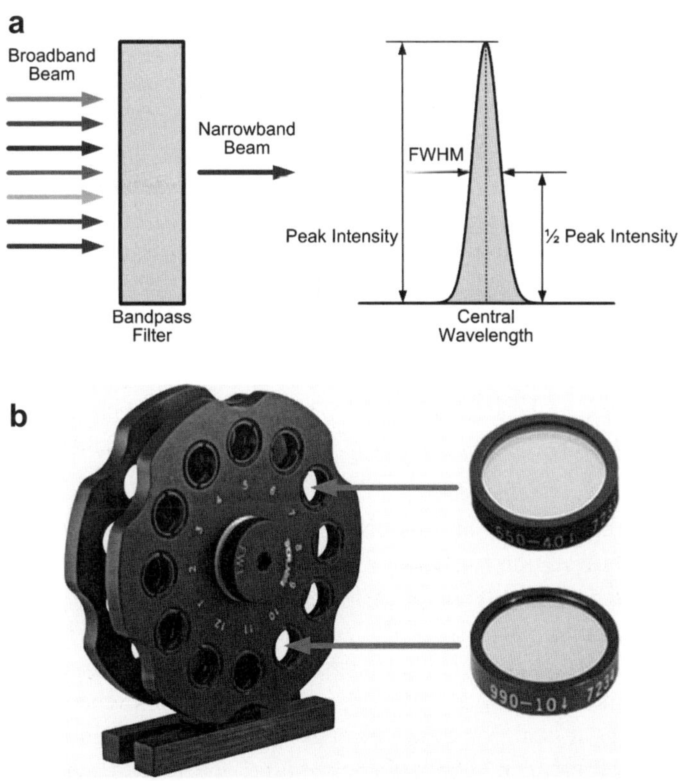

FIGURE 5.8 *Optical bandpass filter: (a) concept and (b) filter wheel and interference bandpass filters (photo courtesy of Thorlabs, Newton, NJ, USA). (Full color version available on http://www.elsevierdirect.com/companions/9780123747532/)*

reflectance will occur for the wavelengths undergoing destructive interferences, whereas high transmittance will occur for other wavelengths undergoing constructive interferences. The interference bandpass filters are generally designed for collimated light that is normally incident on the filter surface. The light with incident angles other than normal will cause an undesired output such as shift of the central wavelength and change of the transmission region. Large incident angles will cause a significant decrease for the transmittance of the passband.

Central wavelength (i.e., wavelength corresponding to peak transmission) and spectral bandwidth that is defined as full width at half maximum (FWHM) (Figure 5.8a) are two key parameters for the bandpass filters. A broad range of filters with various specifications are commercially available to meet the requirements of different applications. Different mechanical configurations of the filter wheels (e.g., single-wheel and dual-wheel) can hold different numbers of filters. Beside manual control, filter wheels that are electronically controlled are also available. They can be synchronized with the camera system to fulfill automatic filter switching and image acquisition. The filter wheels are easy to use and relatively inexpensive. However, they have some limitations for hyperspectral imaging applications, such as narrow spectral range and low resolution, slow wavelength switching, mechanical vibration from moving parts, and image misregistration due to the filter movement. The spectral range and the resolution are determined by the number and the bandwidth of the filters that can be housed in the wheels. The one with double filter holders shown in Figure 5.8(b) can carry up to 24 filters. If the filters with 10 nm FWHM are used, the wavelength range covered by the filter wheel system is 240 nm.

5.3.2.3. Acousto–optic tunable filters

An acousto–optic tunable filter (AOTF) is a solid state device that works as an electronically tunable bandpass filter based on light–sound interactions in a crystal. The major function of the AOTF is to isolate a single wavelength of light from a broadband source in response to an applied acoustic field. The operating principle of the AOTF is illustrated in Figure 5.9(a). It mainly consists of a crystal, an acoustic transducer, an acoustic absorber, a variable source working at radio frequencies (RF), and a beam stop. The most common crystal for constructing the AOTF is Tellurium Dioxide (TeO_2). The transducer, which is bonded to one side of the crystal and controlled by the RF source, generates high frequency acoustic waves through the crystal. The acoustic waves change the refractive index of the crystal by compressing and relaxing the crystal lattice. The variations of

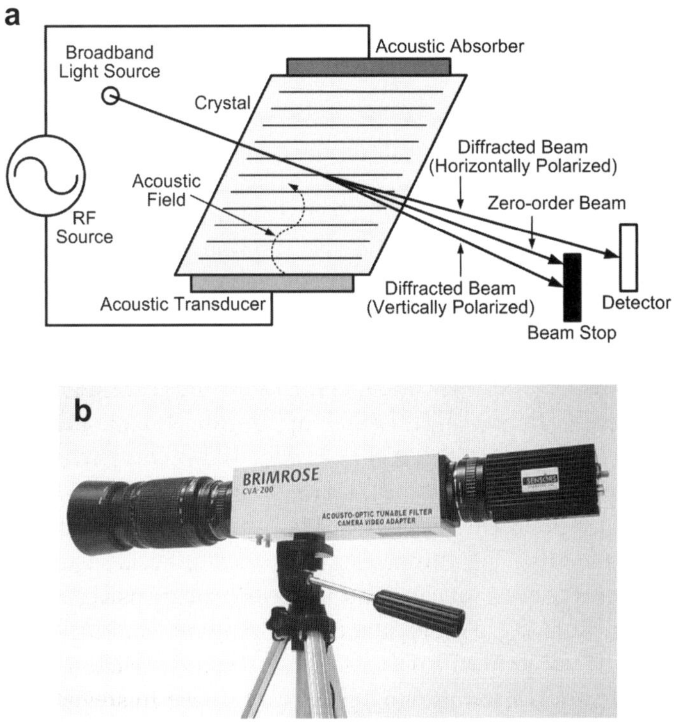

FIGURE 5.9 *Acousto—optic tunable filter (AOTF): (a) operating principle and (b) an AOTF camera video adapter produced by Brimrose (photo courtesy of Brimrose Corporation, Sparks, MD, USA). (Full color version available on http://www.elsevierdirect.com/companions/9780123747532/)*

the refractive index make the crystal like a transmission diffraction grating. The incident light is diffracted after going through the AOTF. As shown in Figure 5.9(a), the diffracted light is divided into two first-order beams with orthogonal polarizations (i.e., horizontally polarized and vertically polarized). Both diffracted beams can be used in certain applications. The undiffracted zero-order beam and the undesired diffracted beam (e.g., vertically polarized beam in Figure 5.9a) are blocked by the beam stop. Similar to a bandpass filter with a narrow bandwidth, the AOTF only diffracts light at one particular wavelength at a time. The wavelength of the isolated light is a function of the frequency of the acoustic waves that are applied to the crystal. Therefore, the wavelength of the transmitted light can be controlled by varying the frequency of the RF source. Wavelength switching for the AOTF is very fast (typically in tens of microseconds) owing to the fact that the tuning speed is only limited by the speed of the sound propagation in the crystal. In addition to the wavelength separation, the bandwidth and the intensity of the filtered light can also be adjusted through the control of the RF source.

Important features of the AOTF include high optical throughput, moderate spectral resolution, broad spectral range, fast wavelength switching, accessibility of random wavelength, and flexible controllability and programmability (Morris *et al.*, 1994). The AOTFs have the ability to transmit single-point signals and 2-D images in the VIS and NIR spectral regions. They can be used to build spectrophotometers as well as hyperspectral imaging systems. Figure 5.9(b) shows a commercial AOTF camera video adapter produced by Brimrose (Sparks, MD, USA). The adapter is designed for acquiring hyperspectral images in the VIS and NIR spectral regions. The aperture size of the adapter is 10 × 10 mm. It is available in three wavelength ranges (i.e., 400–650 nm, 550–1000 nm, and 900–1700 nm) by using different AOTF devices. The corresponding spectral resolutions are in the range of 2 to 20 nm. A zoom lens and a CCD (charge-coupled device) camera are mounted at the front and back ends of the AOTF adapter, respectively. The AOTF hyperspectral imaging system provides narrow bandwidth, fast wavelength selection, and intensity control of the output light. The AOTF-based hyperspectral imaging systems have been used in agricultural applications, such as estimation of leaf nitrogen and chlorophyll concentrations (Inoue & Penuelas, 2001) and detection of green apples in the field (Safren *et al.*, 2007).

5.3.2.4. Liquid crystal tunable filters

A liquid crystal tunable filter (LCTF) is a solid state instrument that utilizes electronically controlled liquid crystal cells to transmit light with a specific wavelength with the elimination of all other wavelengths. The LCTF is constructed from a series of optical stacks, each consisting of a combination of a birefringent retarder and a liquid crystal layer inserted between two parallel polarizers. A single filter stage including the essential optical components is shown in Figure 5.10(a). The incident light is linearly polarized through the polarizer. It is then separated into two rays (i.e., ordinary and extraordinary) by the fixed retarder. The ordinary and the extraordinary rays have different optical paths through the retarder, and they emerge with a phase delay that is dependent upon the wavelength of the light. The polarizer behind the retarder only transmits those wavelengths of light in phase to the next filter stage. Each stage transmits the light as a sinusoidal function of the wavelength, with the frequency determined by the thickness of the retarder and the difference of the refractive index between the ordinary and the extraordinary rays at the wavelength of the light. The transmitted light adds constructively in the desired passband and destructively in the

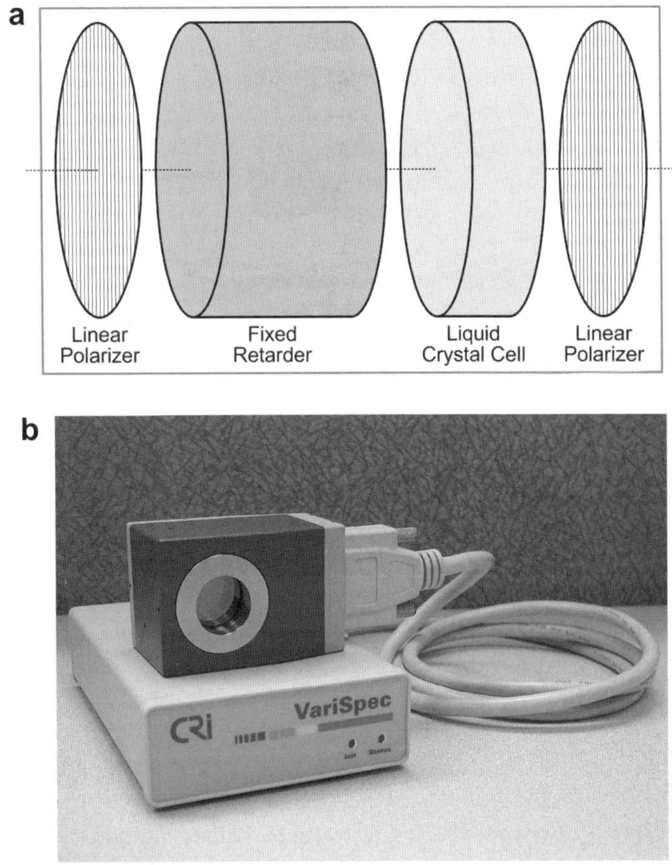

FIGURE 5.10 *Liquid crystal tunable filter (LCTF): (a) single filter stage and (b) a VariSpec LCTF and its controller produced by Cambridge Research and Instrumentation (CRi) (photo courtesy of Cri, Inc., Woburn, MA, USA). (Full color version available on http://www.elsevierdirect.com/companions/9780123747532/)*

other spectral regions. All the individual filter stages are connected in series, and they function together to transmit a single narrow band. A liquid crystal cell is used in each filter stage to realize electronic tunability. An electric field is applied between the two polarizers which causes small retardance changes to the liquid crystal layer. The electronic controller of the LCTF is able to shift the narrow passband region throughout the entire wavelength range covered by the filter unit. A single LCTF unit generally covers a specific wavelength range because the components for constructing the filter have different characteristics that can only accommodate a particular spectral region. The wavelength switching speed depends on the relaxation time of the liquid crystal as well as the number of stages in the filter. Typically, it takes tens of milliseconds to switch from one wavelength to another, which is far longer than the response time of the AOTFs.

A picture of a commercial LCTF unit and its controller produced by Cambridge Research and Instrumentation (Woburn, MA, USA) is shown in Figure 5.10(b). The VariSpec series LCTFs can cover the VIS to NIR range from 400 to 2450 nm, with the use of four different LCTF units [i.e., VIS (400–720 nm), SNIR (650–1100 nm), LNIR (850–1800 nm), and XNIR (1200–2450 nm)]. The VariSpec devices have relatively large apertures (20–35 mm), and the bandwidths of the filters are in the range of 7–20 nm, making them suitable for imaging and non-imaging applications requiring moderate spectral resolutions. The LCTF approach for hyperspectral imaging has found a number of applications in the research of food quality and safety inspection, such as estimation of apple fruit firmness (Peng & Lu, 2006), fungal detection in wheat (Zhang *et al.*, 2007), and early inspection of rottenness on citrus (Gómez-Sanchis *et al.*, 2008a). Compared to the fixed interference filters used in the filter wheels, the electronically tunable filters including AOTFs and LCTFs can be flexibly controlled through the computer. Also, they do not have moving parts and therefore do not suffer the problems associated with the rotating filter wheels, such as speed limitation, mechanical vibration, and image misregistration.

5.3.2.5. Fourier transform imaging spectrometers

Self interference of a broadband light can generate an interferogram that carries its spectral information. An inverse Fourier transform to the generated interferogram can reveal the constitution of the frequencies (or wavelengths) of the broadband light. That is the fundamental principle of Fourier transform interference spectroscopy. The simplest form of the two-beam interferometers is the Michelson interferometer, which is widely used in commercial Fourier transform spectrometers working in the infrared region. It consists of a beamsplitter and two flat mirrors (fixed mirror and moving mirror) that are perpendicular each other (Figure 5.11a). Light from the source is divided into two beams at a beamsplitter that has a semi-reflecting coating on the surface. The light is partially reflected to the fixed mirror, and the remaining energy is transmitted through the beamsplitter to the moving mirror, which moves in a parallel direction with the incident light. The beams reflected back from the two mirrors are recombined by the same beamsplitter. The moving mirror introduces optical path difference (OPD) between the two beams. Interferograms are then generated and collected by the detector.

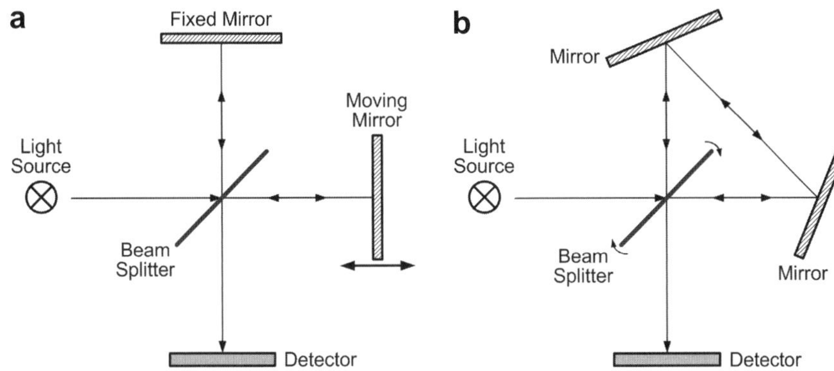

FIGURE 5.11 *Principles of interferometers: (a) Michelson interferometer and (b) Sagnac interferometer*

Different from the Michelson interferometer, the Sagnac interferometer is a common-path two-beam interferometer. The major components of the Sagnac interferometer include two fixed mirrors arranged in a specified angle and a beamsplitter that can be slightly rotated (Figure 5.11b). Two separated beams from the beamsplitter travel the same path in opposite directions. They are recombined at the beamsplitter after traversing the triangular loops. The OPD between the two beams is a function of the angular position of the beamsplitter. Interference fringes can be created by tuning the beamsplitter at very small angles. Hyperspectral images can be acquired by rotating the beamsplitter in a stepwise manner. An interferogram is generated for each spatial point on the sample surface. The spectral information is obtained by Fourier analysis of the interferograms. Although most interferometers are susceptible to vibrations, especially for light with short wavelengths, the Sagnac interferometers are stable and easy to align owing to the fact that they rely on the beamsplitter's rotation other than the mirror's translation to generate interference patterns (Hariharan, 2007). This advantage extends the working range of the traditional Fourier transform interference spectroscopy from the infrared to the visible and short-wavelength near-infrared regions. The wavelength dispersion devices based on Fourier transform techniques have the advantages of high optical throughput, high spectral resolution, and flexible selection of the wavelength range. Varying sensitivity in the entire spectral region and intense computation for data transform are two shortcomings for practical applications.

Applied Spectral Imaging (Vista, CA, USA) developed hyperspectral imaging systems (SpectraCube series) based on the rotating Sagnac interferometer (Malik *et al.*, 1996). Several settings can be adjusted depending on the field of view, spatial resolution, spectral region and resolution, and signal-to-noise ratio. Spectral resolution is uneven across the whole wavelength range. Shorter wavelengths have higher resolutions than longer wavelengths. Image acquisition speed is moderate. According to Pham *et al.* (2000), it takes 40 s to acquire a hyperspectral image cube with a dimension of 170 × 170 × 24 (24 bands) covering the spectral region of 550–850 nm. SpectraCube imaging systems have been used in biomedical research, such as examination of human skin lesions (Orenstein *et al.*, 1998), quantification of absorption and scattering properties of turbid materials (Pham *et al.*, 2000), and spectral karyotyping for prenatal diagnostics (Mergenthaler-Gatfield *et al.*, 2008).

5.3.2.6. Single shot imagers

One example of single shot hyperspectral imagers is the computed tomography imaging spectrometer (CTIS), which can be considered as an application of computed tomography (CT) in imaging spectrometry (Descour & Dereniak, 1995; Okamoto & Yamaguchi, 1991). In this method, multiplexed spatial and spectral data are collected simultaneously to fulfill the acquisition of a complete hyperspectral image cube using one exposure of an area detector. Implementation of a CTIS generally involves a computer-generated hologram (CGH) disperser, a large 2-D area detector, and other optical components for light collimation and image formation (Descour *et al.*, 1997). The CGH element is the central component of the CTIS, and its function is to disperse the field of view into multiple diffraction orders. The dispersed images form a mosaic on the large area detector. Each subimage is not a single band image, but the result of both spectral and spatial multiplexing. The spectral information of the original scene is encoded in the positions and intensities of the subimages in the mosaic. Reconstruction algorithms similar to those used in tomographic imaging techniques are utilized to rebuild the 3-D hypercubes from the original 2-D image data.

More recently, Bodkin Design and Engineering (Wellesley Hills, MA, USA) developed a hyperspectral imager with the capacity to acquire a hypercube in one snapshot (Figure 5.12). The design is based on the company's so called HyperPixel Array technology (Bodkin, 2007). The imaging system includes two stages for optical signal processing. A 2-D lenslet array or a 2-D pinhole array is used to resample an image from the fore-optics (i.e., the first stage) of the imager. The field of view is divided into multiple spatial channels. Each channel is then dispersed into multiple spectral signatures, and they are collected by a 2-D focal plane array. The detector can obtain spectral content of all the pixels (so called HyperPixels) in

FIGURE 5.12 *A single shot hyperspectral imager produced by Bodkin Design and Engineering (photo courtesy of Bodkin Design and Engineering, Wellesley Hills, MA, USA). (Full color version available on http://www.elsevierdirect.com/companions/9780123747532/)*

real time. Generation of the hypercubes purely relies on the parallel optical signal processing performed in the second stage, making it not dependent on computations for image reconstructions. Details for the optical system design of the Bodkin hyperspectral imagers can be found in Bodkin *et al.* (2008). The one shown in Figure 5.12 (VNIR-20) is able to capture hyperspectral images with a dimension of 100 × 180 × 20 (20 bands) at a speed of 20 cubes/s. It works in the VIS range (425–675 nm) with a low spectral resolution (12.5 nm/pixel on average). Another model (VNIR-90) works in the spectral region of 490–925 nm with a higher spectral resolution (3.9 nm/pixel on average). The spatial resolution of this imager is relatively low, and it can acquire hyperspectral images with a dimension of 55 × 44 × 90 (90 bands) at a speed of 15 cubes/s.

The major advantage of single shot hyperspectral imagers is their speed for capturing 3-D images. The line-scanning and area-scanning methods are time-consuming for building hypercubes. It is difficult to perform hyperspectral image acquisitions for fast-moving samples using scanning imagers. The single shot systems can obtain all the spatial and spectral data from a sample at video frame rates, making it possible to generate a hypercube in tens of milliseconds. This feature is especially useful for real-time hyperspectral imaging applications, such as on-line quality and safety inspection of food and agricultural products. The current single shot imagers can work in a broad wavelength range with high spectral resolution at a cost of scarifying spatial resolution. Improvements are needed to address the issue of low spatial resolution, which could limit their applications for circumstances requiring high-resolution spatial data. Single shot devices that can capture 3-D hypercubes without any scanning represent a new trend in instrument development for hyperspectral imaging techniques.

5.3.2.7. Other instruments

Besides the wavelength dispersion devices described above, there are also other types of imaging spectrometers that can be used in hyperspectral imaging systems. Examples include circular variable filters (CVF) and linear variable filters (LVF) (Min *et al.*, 2008), Hadamard transform imaging spectrometer (Hanley *et al.*, 1999), digital array scanned interferometer (DASI) (Smith & Hammer, 1996), volume holographic imaging spectrometer (VHIS) (Liu *et al.*, 2004), tunable etalon imaging spectrometer (Marinelli *et al.*, 1999), etc. Details for the operating principles of these designs are omitted for brevity, and they can be found in the literature provided. Wavelength dispersion instruments are the core of hyperspectral imaging systems. New technologies are being introduced to create new devices in this area. For example, a new type of electronically tunable filter has recently been developed based on microelectromechanical systems (MEMS) technology (Abbaspour-Tamijani *et al.*, 2003; Goldsmith *et al.*, 1999). Such filters are constructed using MEMS variable capacitors, and they have similar functions with AOTFs and LCTFs. Owing to their merits such as extremely small size and low power consumption, the MEMS-based tunable filters have the potential to be used to build miniature hyperspectral imaging systems (e.g., hand-held instruments). Meanwhile, current instruments can also be modified or improved to satisfy specific requirements of different applications. For example, moving slit design can be introduced to imaging spectrographs so that line scanning can be performed with both sample and detector remaining stationary (Lawrence *et al.*, 2003). Introduction of new design concepts and improvement of current instruments are main drivers for the future development of hyperspectral imaging technology.

5.3.3. Area Detectors

After interacting with the target and going through the wavelength dispersion device, light carrying the useful information will eventually be acquired by a detector. The function of the detector is to measure the intensity of the collected light by converting radiation energy into electrical signals. The performance of the detector directly determines the quality of the images. Two major types of solid state area detectors including CCD (charge-coupled device) and CMOS (complementary metal-oxide-semiconductor) cameras are introduced in this section.

5.3.3.1. CCD cameras

The CCD sensor is composed of many (usually millions) small photodiodes (called pixels) that are made of light sensitive materials such as silicon (Si) or indium gallium arsenide (InGaAs). Each photodiode acts like an individual spot detector that converts incident photons to electrons, generating an electrical signal that is proportional to total light exposure. All the electrical signals are shifted out of the detector in a predefined manner and then are digitalized to form the images. The pixels in the CCD sensor can be arranged in one-dimensional or two-dimensional arrays, resulting in line detector and area detector, respectively. Hyperspectral imaging systems usually use area detectors to obtain the image data. Thus emphasis is put on the introduction to the CCD area detectors.

Generally there are four types of CCD architectures that are used for reading out the data from the area sensors, and they are full frame, frame transfer, interline transfer, and frame interline transfer (Figure 5.13). The full

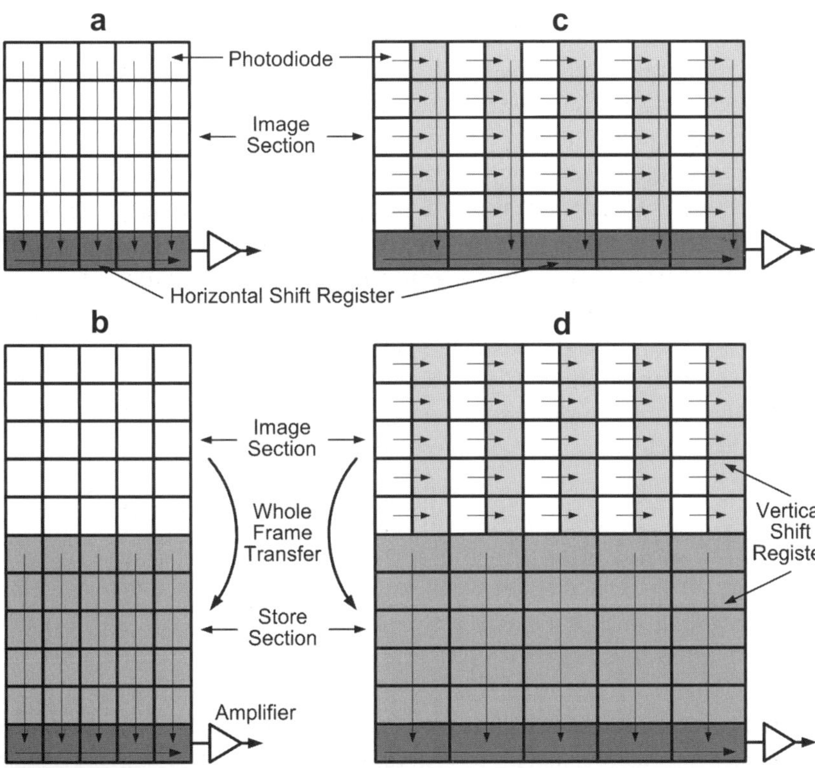

FIGURE 5.13 *Typical CCD architectures for different data transfer methods: (a) full frame; (b) frame transfer; (c) interline transfer; and (d) frame interline transfer*

frame structure is the simplest form for constructing the CCD. Electric charges are accumulated in the photosensitive section (image section) during the light integration period. Then they are vertically shifted row by row into a horizontal shift register, where each row is exported to form an array of pixels (known as a progressive scan). A mechanical shutter is usually used to cover the sensor during the process of data transfer to avoid interference of newly generated charges, making this CCD architecture relatively slow for image acquisition. The frame transfer approach extends the full frame structure by adding a new store section (normally with identical size of the image section) that is covered by a mask all the time. Accumulated charges from the image section are rapidly transferred to the store section for each whole frame. While the next light signal is integrated in the image section, the charges in the store section are shifted vertically into the horizontal register. This structure has faster frame rates than the full frame structure, at a cost of larger size of the image sensor. The interline structure, on the other hand, transfers the charge from each pixel into a corresponding vertical shift register (called interline mask), which is immediately adjacent to each photodiode and shielded from the incident light. The subsequent process is the same with the frame transfer structure. This structure is also quick at shifting the data. A disadvantage of this approach is that the interline mask on the sensor decreases the effective area for collecting the light signal. Lastly, the frame interline transfer is a combination of the frame transfer and the interline transfer. Charges in the interline mask are transferred to the store section as a whole frame, which further accelerates the data shift speed. However, it bears the disadvantages of high cost for the large sensor and reduced sensitive area. The architectures of full frame and frame transfer are adopted by most scientific cameras for quantitative measurement applications, while the two architectures using interline transfer are commonly used in various video cameras.

Many factors (e.g., sensor size, pixel size, dynamic range, readout speed, dark noise, readout noise, spectral response, cooling method, image output form, computer interface, and synchronization option) need to be considered when choosing a CCD camera for a specific application. Spectral response of the CCD sensor is an important characteristic that determines the working wavelength range of the camera. A measure of this feature, quantum efficiency (QE) quantifies the relationship between the wavelength of the incident light and the sensitivity of the camera. The QE of the CCD is primarily governed by the substrate materials used to make the photodiodes. Owing to its natural sensitivity to visible light, silicon is intensively used as sensor material for the CCD cameras working in the VIS and short-wavelength NIR regions. The spectral response of the silicon image sensors is

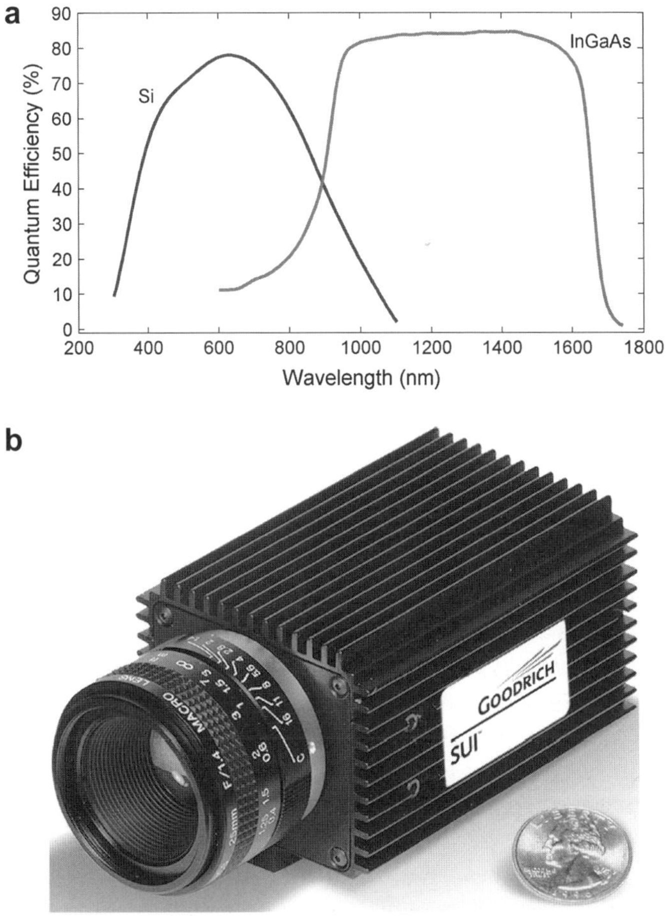

FIGURE 5.14 *Indium gallium arsenide (InGaAs) image sensor: (a) typical quantum efficiencies of silicon (Si) and InGaAs image sensors and (b) an InGaAs camera produced by Sensors Unlimited (data and photo courtesy of Sensors Unlimited, Inc., Princeton, NJ, USA). (Full color version available on http://www.elsevierdirect.com/companions/9780123747532/)*

a bell-shaped curve with QE values declined towards both UV and NIR regions (Figure 5.14a). The silicon-based CCD cameras have been widely used in hyperspectral reflectance and transmittance imaging systems for inspection of agricultural commodities using spectral information in the VIS and short-wavelength NIR regions (Kim *et al.*, 2001; Park *et al.*, 2002; Qin & Lu, 2005).

The NIR spectral region also carries plenty of useful information for food quality and safety inspection. The InGaAs image sensor, which is made of an alloy of indium arsenide (InAs) and gallium arsenide (GaAs), has fairly flat and high quantum efficiency in the NIR region (Figure 5.14a). Standard InGaAs (53% InAs and 47% GaAs) image sensors cover the wavelength range of 900–1700 nm. Extended wavelength range (e.g., 1100–2200 nm and 1100–2600 nm) can be achieved by changing the percentages of InAs and GaAs for making the InGaAs sensors (Sensors Unlimited, 2006). In terms of quantum efficiency, the InGaAs camera starts from where the silicon camera declines, making the InGaAs camera a good choice for hyperspectral imaging systems working in the NIR region for agricultural applications (Lu, 2003; Nicolai *et al.*, 2006; Zhang *et al.*, 2007). The InGaAs camera produced by Sensors Unlimited (Princeton, NJ, USA) is shown in Figure 5.14(b). It utilizes a standard InGaAs image sensor with a sensitivity range from 900 to 1700 nm. The QE of the camera is greater than 65% in the wavelength range of 1000–1600 nm. It can work at room temperature and the frame rate is up to 60 Hz. Detectors for the mid-infrared region are also available, such as lead selenide (PbSe), indium antimonide (InSb), and mercury cadmium telluride (MCT).

The CCD camera can deliver high quality images when there is sufficient light reaching the image sensor and no short exposure is required, which is a typical condition for hyperspectral reflectance and transmittance measurements. However, for low light applications such as fluorescence imaging and Raman imaging, the regular CCD camera may not be able to obtain the data that satisfy the application requirements. High performance cameras such as Electron Multiplying CCD (EMCCD) and Intensified CCD (ICCD) cameras are usually used to acquire the images with high signal-to-noise ratio. EMCCD is a quantitative digital camera technology that is capable of detecting single photon events whilst maintaining high quantum efficiency (Andor, 2006). An EMCCD differs from a traditional CCD by adding a unique solid state electron multiplication register to the end of the normal readout register (Figure 5.15a). This built-in multiplication register multiplies the weak charge signals before any readout noise is imposed by the output amplifier, achieving real gain for the useful signals. Figure 5.15(b) shows an EMCCD camera (iXon series) produced by Andor (South Windsor, CT, USA). The electron multiplier gain of this camera can be adjusted in the range of 1–1000 through the camera software control. When there is plenty of light, the gain function can also be switched off to change the EMCCD camera to a conventional CCD camera. EMCCD cameras have started to find their applications for inspection of food and agricultural products. Kim *et al.* (2007) developed an EMCCD-based hyperspectral system to perform both reflectance and fluorescence measurements for on-line defect and fecal

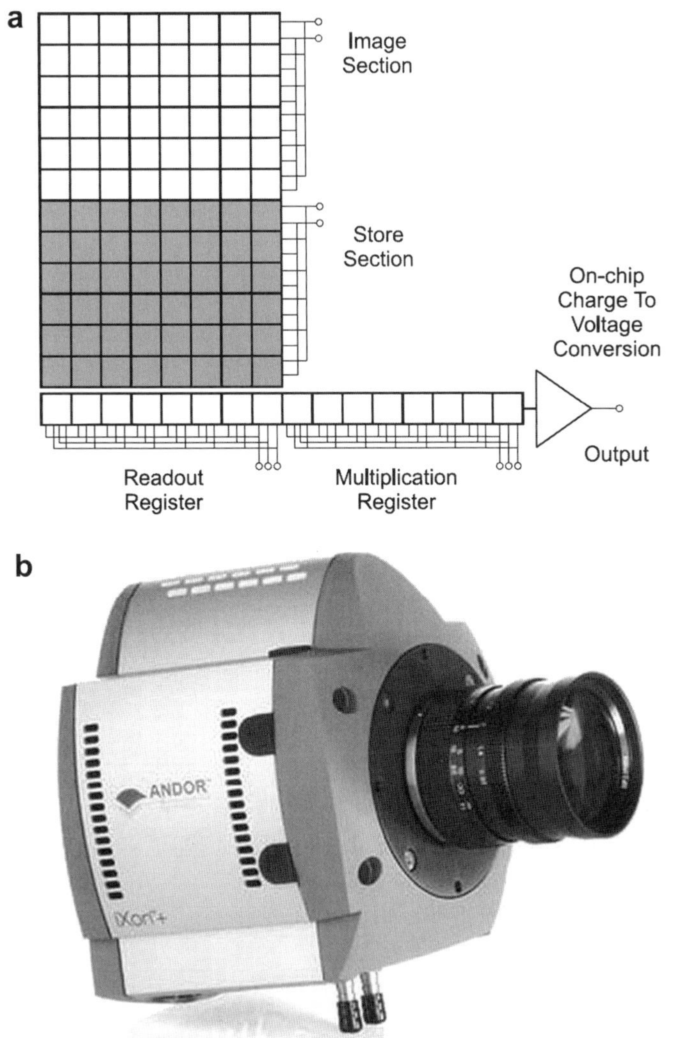

FIGURE 5.15 *Electron Multiplying CCD (EMCCD): (a) architecture and (b) an EMCCD camera produced by Andor (illustration and photo courtesy of Andor Technology PLC, South Windsor, CT, USA). (Full color version available on http://www.elsevierdirect.com/companions/9780123747532/)*

contamination detection of apples. ICCD is another type of high performance image sensor that can detect weak optical signals. Instead of adding a multiplication register after photon to electron conversion (EMCCD's approach), the ICCD utilizes an image intensifier tube to apply the gain to the incident light before it reaches the image sensor. The amplified light signals are then coupled to the CCD. Hence the EMCCD is based on electronic amplification, while the ICCD is based on optical amplification. Besides the gain function, ICCD cameras have another important feature of being able to realize very fast gate times (in nanoseconds or picoseconds). This feature makes them suitable for detecting time-resolved signals with very short duration, such as time-dependent fluorescence emissions induced by pulsed lasers (Kim *et al.*, 2003).

5.3.3.2. CMOS cameras

Currently CCD cameras are the dominant devices in the area of image acquisition, especially for technical applications. The CMOS image sensor is another major type of solid state area detector that has the potential to compete with CCD. The major difference between these two types of detectors is that the CMOS image sensor includes both photodetector and readout amplifier in each pixel (called active pixel) (Litwiller, 2005). A typical architecture of the CMOS image sensor is shown in Figure 5.16. A photodiode is still used to sense the incident light, as it does in the CCD. After the photon to electron conversion, a set of optically insensitive transistors adjacent to the photodiode will convert the integrated charge to

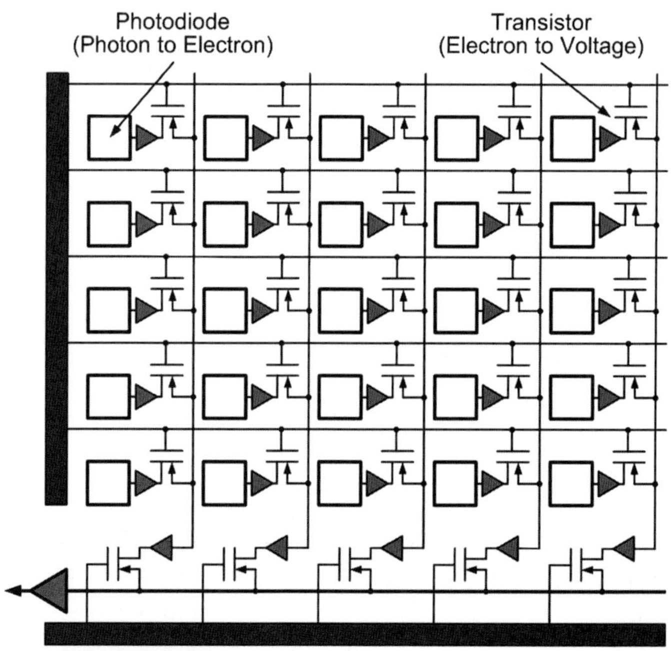

FIGURE 5.16 *Architecture of the CMOS image sensor*

a voltage signal immediately. The electron to voltage conversion occurs inside each pixel, and the generated voltage signals are then read out over the wires. Compared to the vertical and horizontal registers used by the CCD to shift the charges (see Figure 5.13), the wires used in the CMOS image sensor are much faster for signal transfer, making the CMOS cameras especially suitable for high speed imaging applications such as on-line industrial inspection. Owing to the addressability of the wires arranged in rows and columns, it is possible to extract a region of interest (ROI) from the sensor rather than the whole image, which can be utilized for on-chip image manipulations (e.g., zoom and pan). Besides the features of high speed and random addressing, the CMOS cameras have other advantages such as low cost, low power consumption, single power supply, and small size for system integration, which makes them prevail in the consumer electronics market (e.g., low-end camcorders and cell phones). The main reason that limits their applications in quantitative measurements is that the current CMOS image sensors have higher noise and higher dark current than the CCDs due to the on-chip circuits used for signal amplification and transfer. Consequently the dynamic range and the sensitivity are lower than those of CCDs. Hyperspectral imaging systems generally have higher requirements for cameras than conventional imaging systems since they also need to acquire spectral information. The CMOS cameras still need substantial performance improvement to challenge the CCD cameras in hyperspectral imaging as well as other scientific applications.

5.4. INSTRUMENTS FOR CALIBRATING HYPERSPECTRAL IMAGING SYSTEMS

Before proper measurements can be achieved, appropriate calibrations for hyperspectral imaging systems are needed. The commonly used calibration methods and instruments are introduced in the following sections.

5.4.1. Spatial Calibration

Spatial calibration for hyperspectral imaging systems is intended to determine the range and the resolution for the spatial information contained in the hypercubes. The calibration results are useful for adjusting the field of view and estimating the spatial detection limit. Different spatial calibration methods can be used for the imaging systems utilizing different image acquisition modes. The hyperspectral systems working in the area-scanning mode generate a series of single band images at different wavelengths. Each single band image is a regular 2-D grayscale image with full spatial information. Hence the spatial calibration can be performed at a selected wavelength using printed targets with square grids or standard test charts such as US Air Force 1951 test chart. The area-scanning systems generally have the same resolution for both spatial dimensions if the same binning is used for the horizontal and vertical axis of the camera. For the line-scanning imaging systems, the resolution for the two spatial dimensions could be different. The *x* direction is for the stepwise movement of the samples (see Figure 5.2), and the resolution depends on the step size of the movement. The *y* direction is parallel to the slit of the imaging spectrograph, and the resolution is determined by the combination of the working distance, lens, imaging spectrograph, and camera.

An example of spatial calibration for a line-scanning hyperspectral imaging system is shown in Figure 5.17. The system is developed based on an imaging spectrograph (ImSpector V10, Spectral Imaging Ltd., Oulu, Finland), and it works in line-scanning mode to collect hyperspectral reflectance images from fruit and vegetable samples carried by a precision motor-controlled stage (Qin & Lu, 2008). The step size of the stage used for image acquisition is 1.0 mm. Thus the spatial resolution for the *x* direction (see Figure 5.2) of the hypercubes is 1.0 mm/pixel. The spatial range for the

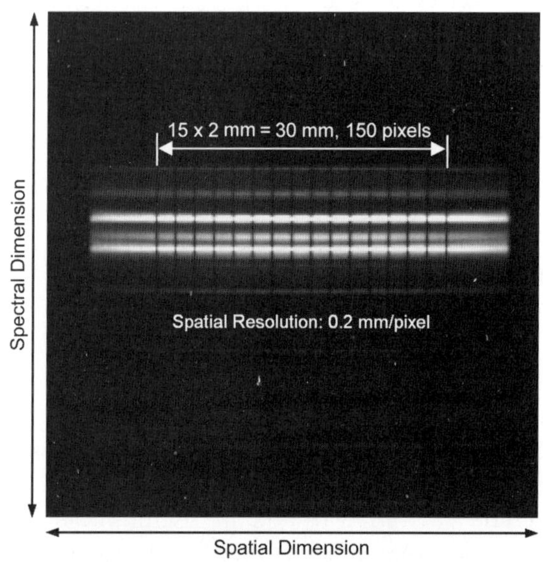

FIGURE 5.17 *Spatial calibration for a line-scanning hyperspectral imaging system using a white paper printed with thin parallel lines 2 mm apart*

x direction is determined by the number of scans. The spatial axis of the imaging spectrograph is aligned to the horizontal dimension of the CCD detector. Thus the horizontal dimension of the line-scanning images represents spatial information and the vertical dimension spectral. The one shown in Figure 5.17 is a line-scanning image with a dimension of 256 × 256. It is obtained from a white paper printed with thin parallel lines 2 mm apart, which is illuminated by a fluorescent lamp. The spatial resolution for the y direction (see Figure 5.2) of the hypercubes can be determined by dividing the real spatial distance by the number of image pixels in this range. Specifically, there are 150 pixels within 30 mm spatial distance (15 intervals with 2 mm apart for adjacent lines), thus the spatial resolution for the *y* direction can be calculated as 30 mm/150 pixels = 0.2 mm/pixel. The spatial range for the *y* direction covered by the imaging system is 0.2 mm/pixel × 256 pixels = 51.2 mm.

5.4.2. Spectral Calibration

Spectral calibration for hyperspectral imaging systems is intended to define the wavelengths for the pixels along the spectral dimension of the hypercubes. The calibration results can be used for determining the range and the resolution for the spectral information contained in the hypercubes. The area-scanning systems using fixed or tunable filters can generate single band images at a series of known wavelengths. Therefore the spectral calibration is usually not necessary. The central wavelengths of the interference bandpass filters housed in the filter wheel are generally used as the wavelengths for the corresponding single band images. The wavelengths through the tunable filters (e.g., AOTFs and LCTFs) are determined by their electronic controllers. On the other hand, imaging spectrograph-based line-scanning systems generate hypercubes with unknown wavelengths. Hence spectral calibration is needed to map the pixel indices along the spectral dimension to the exact wavelengths. The calibration can be performed utilizing spectrally well-known light sources, such as spectral calibration lamps, lasers (e.g., 632.8 nm by helium–neon [HeNe] lasers), fluorescent lamps, and broadband lamps equipped with interference bandpass filters. Spectral calibration lamps are the most commonly used calibration sources. They generate narrow, intense spectral lines from the excitation of various rare gases and metal vapors. Because a given chemical element only emits radiation at specific wavelengths, the wavelengths produced by the calibration lamps are considered to be absolute, and they are used as standards for spectral calibration. Various calibration lamps are available for the wavelength range from UV to NIR. Choices include lamps

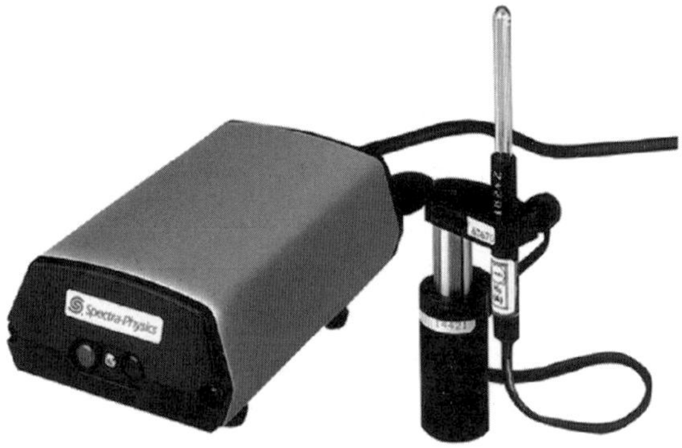

FIGURE 5.18 *A pencil style spectral calibration lamp and its power supply produced by Newport (photo courtesy of Newport Corporation, Irvine, CA, USA)*

using argon, krypton, neon, xenon, mercury, mercury–argon, mercury–neon, mercury–xenon, etc. Such calibration lamps are commercially available for use under different circumstances (e.g., pencil style lamps, battery powered lamps, and high power lamps). Figure 5.18 shows a pencil style spectral calibration lamp and its power supply produced by Newport (Irvine, CA, USA).

An example of spectral calibration for a line-scanning hyperspectral imaging system is illustrated in Figure 5.19. The imaging system is the same as that used for demonstration of spatial calibration (Figure 5.17). Details for the hyperspectral system can be found in (Qin & Lu, 2008). The spectral calibration is performed using two pencil style spectral calibration lamps (i.e., a xenon lamp [model 6033] and a mercury–argon lamp [model 6035], Newport, Irvine, CA, USA), which have several good peaks in the wavelength range of 400–1000 nm. Two images on the top are original line-scanning images from xenon and mercury–argon lamps. Two spectral profiles are extracted along the vertical axis (spectral dimension) of the line-scanning images. The spectral peaks from each lamp and their corresponding pixel positions in the vertical axis are identified. The relationship between the vertical pixel indices and the known wavelengths from the two lamps is established using a linear regression function. The resulting linear model then can be used to determine all the wavelengths along the spectral dimension. Nonlinear regression models are also used for the spectral calibration (Chao *et al.*, 2007; Park *et al.*, 2002). Nominal spectral resolution of the imaging spectrograph is linearly dependent on the slit width (Spectral Imaging, 2003). The spectrograph used in this calibration (ImSpector V10, Spectral Imaging Ltd., Oulu, Finland) has a 25 μm slit width, and its nominal resolution is 3 nm. The calculated spectral resolution

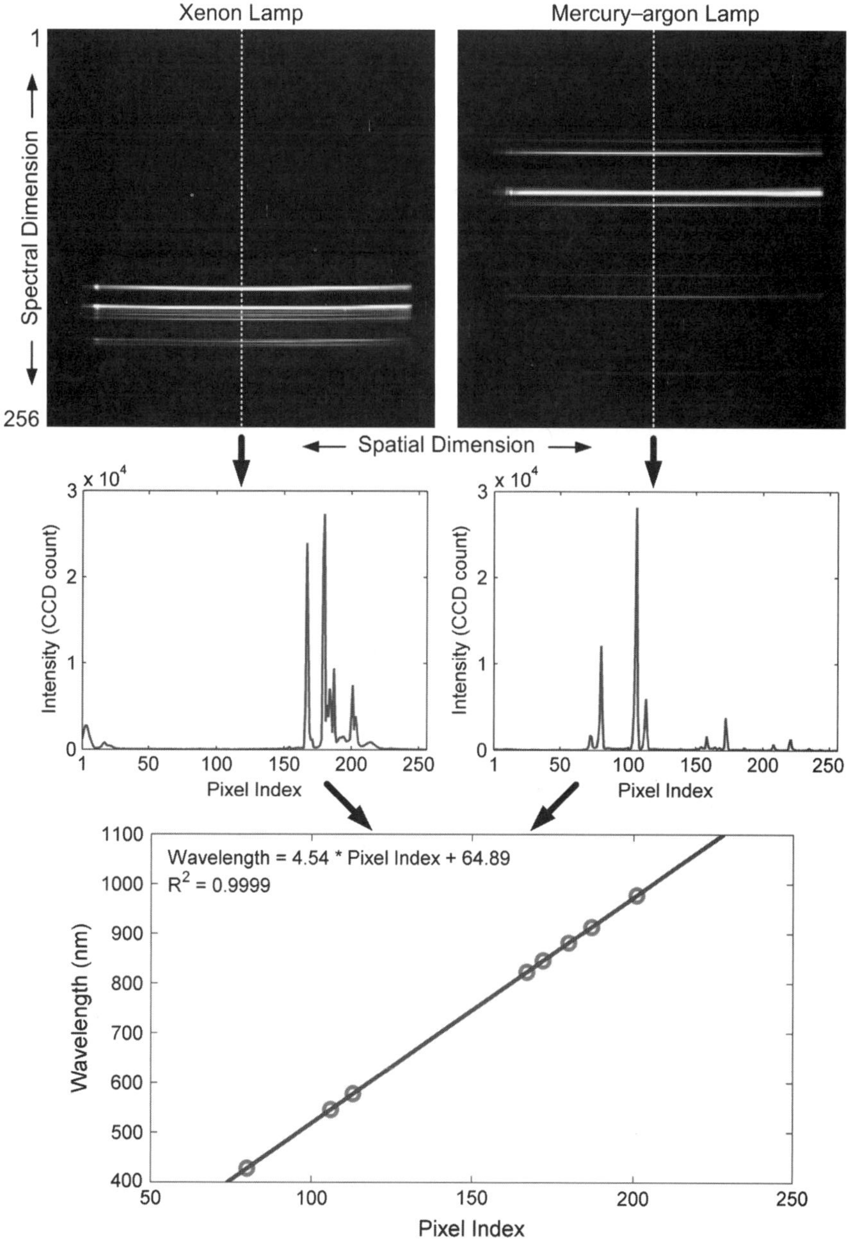

FIGURE 5.19 *Spectral calibration for a line-scanning hyperspectral imaging system using xenon and mercury–argon calibration lamps. (Full color version available on http://www.elsevierdirect.com/companions/9780123747532/)*

from the linear model shown in Figure 5.19 is 4.54 nm, which is slightly lower than the one of the spectrograph. It should be noted that it is the nominal spectral resolution of the imaging spectrograph that determines the accuracies for the spectral measurements. The camera merely collects dispersed light signals passing through the spectrograph. The calculated resolution based on the image pixel measurements is determined by the nominal resolution of the imaging spectrograph as well as the binning for the vertical axis of the detector.

5.4.3. Flat-field Correction

Raw hyperspectral images contain noises and artifacts due to measurement environments and imperfections of each component (e.g., source, lens, filter, spectrograph, and camera) in the optical path of the imaging system. During the image acquisition, the noise counts accumulated on the detector may increase the pixel values beyond the true intensities. Various image artifacts can be generated by factors such as nonuniform illumination, dust on the lens surface, and pixel-to-pixel sensitivity variations of the detector, making the original images unsuitable for quantitative analysis. Flat-field correction is intended to remove the effects of the noises and artifacts. The resulting relative (or percent) reflectance instead of the absolute intensity data is usually used for further data analysis.

White diffuse reflectance panels (Figure 5.20), which have high and flat reflectance over a broad wavelength range (e.g., 250–2500 nm), are usually used as standards for the flat-field correction for hyperspectral reflectance measurement. The flat-field correction can be performed using the following equation:

$$Rs(\lambda) = \frac{Is(\lambda) - Id(\lambda)}{Ir(\lambda) - Id(\lambda)} \times Rr(\lambda) \tag{5.1}$$

where *Rs* is the relative reflectance image of the sample, *Is* is the intensity image of the sample, *Ir* is the reference image obtained from the white panel, *Id* is the dark current image acquired with the light source off and the lens covered, *Rr* is the reflectance factor of the white panel, and λ is the wavelength. All the variables in Equation 5.1 are wavelength dependent, and corrections should be conducted for all the wavelengths covered by the imaging system. A constant reflectance factor (*Rr*) of 100% can be used for simplification, although the actual reflectance values of the white panel are slightly lower and they also have small variations over a certain spectral region. Since most samples have lower reflectance than the white panel, the relative reflectance values obtained by Equation 5.1 are in the

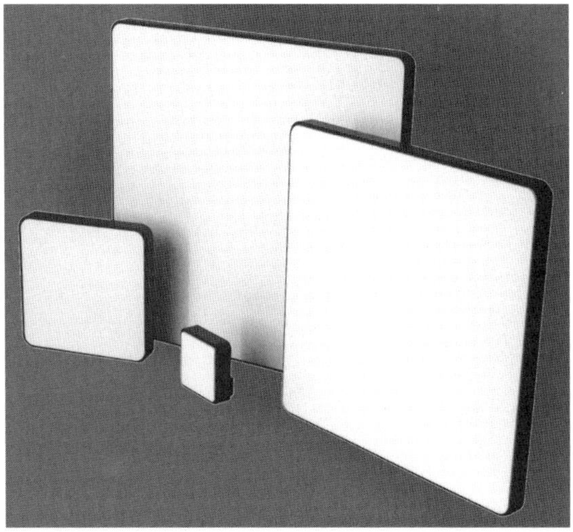

FIGURE 5.20 *White diffuse reflectance panels that can be used for flat-field corrections (photo courtesy of Labsphere, Inc., North Sutton, NH, USA)*

range of 0–100%. They can be multiplied by a constant factor (e.g., 10 000) to have a large dynamic data range and to reduce rounding errors for further data analysis.

Figure 5.21 shows an example of flat-field correction for hyperspectral reflectance measurement of a leaf sample. The plots shown in Figure 5.21(a) are original reflectance spectra extracted from three hypercubes (i.e., leaf sample, white panel [Spectralon SRT-99-100, Labsphere Inc., North Sutton, NH, USA], and dark current). Ideally, the reflectance profile of the white panel should be flat. However, the measured spectrum is bell shaped with a peak around 700 nm due to the combined spectral response of the imaging system. The white panel has the highest reflectance, and the values of the dark current are relatively low and flat over the entire wavelength range. The reflectance intensities of the leaf sample are in between. After the flat-field correction using Equation 5.1, a relative reflectance spectrum of the leaf sample is obtained (Figure 5.21b). Light absorption due to chlorophyll in the leaf can be observed around 670 nm.

5.4.4. Other Calibrations

Besides the common calibration methods described above, there are also other types of calibrations that can be performed for hyperspectral imaging systems to satisfy different measurement requirements. For example,

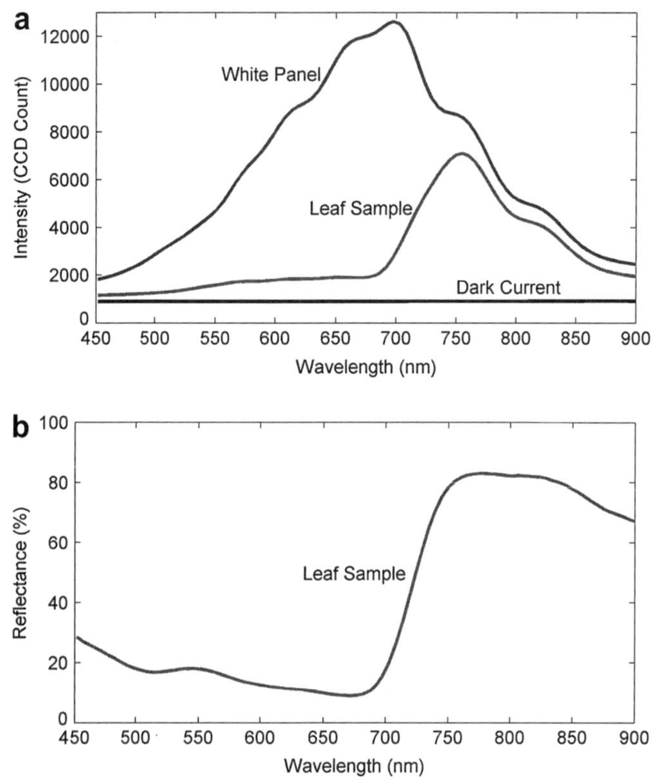

FIGURE 5.21 *Flat-field correction for hyperspectral reflectance measurement: (a) original reflectance spectra and (b) relative reflectance spectrum after flat-field correction. (Full color version available on http://www.elsevierdirect.com/companions/9780123747532/)*

radiometric calibration is required when the absolute spectral radiance of the sample is to be determined. An integrating sphere typically serves as a radiance standard for the radiometric calibration. Particular agricultural applications utilizing hyperspectral imaging can also generate particular calibration needs. For example, hyperspectral reflectance measurements for spherical fruit can not be successfully corrected by the flat-field correction due to the curvature effects. To tackle this problem, Qin & Lu (2008) developed a method for correcting spatial profiles extracted from line-scanning images of apple samples using an imaging spectrograph-based hyperspectral system. Gómez-Sanchis *et al.* (2008b) also developed a method for correcting area-scanning images from citrus samples using a LCTF-based hyperspectral system. Efforts have been made to develop various elaborate calibration methods and procedures (Burger and Geladi, 2005; Lu & Chen

1998; Lawrence *et al.*, 2003; Polder *et al.*, 2003; Qin & Lu, 2007). New effective and efficient calibration and correction approaches are expected in the future to better utilize the hyperspectral imaging techniques.

5.5. CONCLUSIONS

This chapter has presented methods for hyperspectral image acquisition and instruments for constructing and calibrating hyperspectral imaging systems. Point scanning, line scanning, area scanning, and single shot are four major methods for acquiring hyperspectral images. Various line-scanning and area-scanning hyperspectral measurement systems have been developed and used successfully for the quality and safety inspection of food and agricultural products. Related instruments for constructing and calibrating such scanning imaging systems, such as light sources, wavelength dispersion devices, detectors, standard test charts, calibration sources, and standard reflectance panels, are already commercially available. Single shot hyperspectral imagers can capture 3-D hypercubes at high speed without any scanning. They represent a new direction in hyperspectral instrument development, while such devices are still in the early development stage. New instrument design concepts will be continuously introduced, and current instruments and systems can also be improved to achieve better performance. The advances in hyperspectral imaging instruments along with the progress in hyperspectral image processing techniques will inspire the future development of hyperspectral imaging technology.

NOMENCLATURE

Abbreviations

AOTF	acousto–optic tunable filter
BIL	band interleaved by line
BIP	band interleaved by pixel
BSQ	band sequential
CCD	charge-coupled device
CGH	computer-generated hologram
CMOS	complementary metal-oxide-semiconductor
CTIS	computed tomography imaging spectrometer
CVF	circular variable filter
CW	continuous wave
DASI	digital array scanned interferometer

EMCCD	electron multiplying CCD
FWHM	full width at half maximum
ICCD	intensified CCD
InGaAs	indium gallium arsenide
InSb	indium antimonide
LCTF	liquid crystal tunable filter
LED	light emitting diode
LVF	linear variable filter
MCT	mercury cadmium telluride
MEMS	microelectromechanical systems
NIR	near-infrared
OPD	optical path difference
PbSe	lead selenide
pcLED	phosphor-converted LED
PGP	prism–grating–prism
QE	quantum efficiency
QTH	quartz–tungsten–halogen
RF	radio frequency
ROI	region of interest
SNR	signal-to-noise ratio
UV	ultraviolet
VHIS	volume holographic imaging spectrometer
VIS	visible

REFERENCES

Abbaspour-Tamijani, A., Dussopt, L., & Rebeiz, G. M. (2003). Miniature and tunable filters using MEMS capacitors. *IEEE Transactions on Microwave Theory and Techniques, 51*(7), 1878–1885.

Andor. (2006). *Scientific Digital Camera Solutions: 2006 Catalog*. South Windsor, CT, USA: Andor Technology PLC.

Ariana, D. P., & Lu, R. (2008). Detection of internal defect in pickling cucumbers using hyperspectral transmittance imaging. *Transactions of the ASABE, 51*(2), 705–713.

Bannon, D., & Thomas, R. (2005). Harsh environments dictate design of imaging spectrometer. *Laser Focus World, 41*(8), 93–95.

Bodkin, A. (2007). Hyperspectral imaging at the speed of light. SPIE Newsroom, December 11, 2007; doi: 10.1117/2.1200712.0845.

Bodkin, A., Sheinis, A. I., & Norton, A. (2008). Hyperspectral imaging systems. *US Patent Application Publication*. Pub. No.: US 2008/0088840 A1.

Brauns, E. B., & Dyer, R. B. (2006). Fourier transform hyperspectral visible imaging and the nondestructive analysis of potentially fraudulent documents. *Applied Spectroscopy, 60*(8), 833–840.

Burger, J., & Geladi, P. (2005). Hyperspectral NIR image regression. Part I: Calibration and correction. *Journal of Chemometrics, 19*(5), 355–363.

Chao, K., Mehl, P. M., & Chen, Y. R. (2002). Use of hyper- and multi-spectral imaging for detection of chicken skin tumors. *Applied Engineering in Agriculture, 18*(1), 113–119.

Chao, K., Yang, C. C., Chen, Y. R., Kim, M. S., & Chan, D. E. (2007). Fast line-scan imaging system for broiler carcass inspection. *Sensing and Instrumentation for Food Quality and Safety, 1*(2), 62–71.

Descour, M. R., & Dereniak, E. L. (1995). Computed-tomography imaging spectrometer—experimental calibration and reconstruction results. *Applied Optics, 34*(22), 4817–4826.

Descour, M. R., Volin, C. E., Dereniak, E. L., Gleeson, T. M., Hopkins, M. F., Wilson, D. W., & Maker, P. D. (1997). Demonstration of a computed-tomography imaging spectrometer using a computer-generated hologram disperser. *Applied Optics, 36*(16), 3694–3698.

Goldsmith, C. L., Malczewski, A., Yao, Z. J., Chen, S., Ehmke, J., & Hinzel, D. H. (1999). RF MEMs variable capacitors for tunable filters. *International Journal of RF and Microwave Computer-Aided Engineering, 9*(4), 362–374.

Gómez-Sanchis, J., Gómez-Chova, L., Aleixos, N., Camps-Valls, G., Montesinos-Herrero, C., Moltó, E., & Blasco, J. (2008a). Hyperspectral system for early detection of rottenness caused by *Penicillium digitatum* in mandarins. *Journal of Food Engineering, 89*(1), 80–86.

Gómez-Sanchis, J., Moltó, E., Camps-Valls, G., Gómez-Chova, L., Aleixos, N., & Blasco, J. (2008b). Automatic correction of the effects of the light source on spherical objects: an application to the analysis of hyperspectral images of citrus fruits. *Journal of Food Engineering, 85*(2), 191–200.

Hanley, Q. S., Verveer, P. J., & Jovin, T. M. (1999). Spectral imaging in a programmable array microscope by hadamard transform fluorescence spectroscopy. *Applied Spectroscopy, 53*(1), 1–10.

Hariharan, P. (2007). *Basics of interferometry* (2nd ed., pp. 13–22). San Diego, CA: Elsevier.

Inoue, Y., & Penuelas, J. (2001). An AOTF-based hyperspectral imaging system for field use in ecophysiological and agricultural applications. *International Journal of Remote Sensing, 22*(18), 3883–3888.

Jestel, N. L., Shaver, J. M., & Morris, M. D. (1998). Hyperspectral Raman line imaging of an aluminosilicate glass. *Applied Spectroscopy, 52*(1), 64–69.

Kim, M. S., Chen, Y. R., Cho, B. K., Chao, K., Yang, C. C., Lefcourt, A. M., & Chan, D. (2007). Hyperspectral reflectance and fluorescence line-scan imaging for online defect and fecal contamination inspection of apples. *Sensing and Instrumentation for Food Quality and Safety, 1*(3), 151–159.

Kim, M. S., Chen, Y. R., & Mehl, P. M. (2001). Hyperspectral reflectance and fluorescence imaging system for food quality and safety. *Transactions of the ASAE, 44*(3), 721–729.

Kim, M. S., Lefcourt, A. M., & Chen, Y. R. (2003). Multispectral laser-induced fluorescence imaging system for large biological samples. *Applied Optics, 42*(19), 3927–3934.

Klein, M. E., Aalderink, B. J., Padoan, R., de Bruin, G., & Steemers, T. A. G. (2008). Quantitative hyperspectral reflectance imaging. *Sensors, 8*(9), 5576–5618.

Lawrence, K. C., Park, B., Heitschmidt, G. W., Windham, W. R., & Thai, C. N. (2007). Evaluation of LED and tungsten–halogen lighting for fecal contaminant detection. *Applied Engineering in Agriculture, 23*(6), 811–818.

Lawrence, K. C., Park, B., Windham, W. R., & Mao, C. (2003). Calibration of a pushbroom hyperspectral imaging system for agricultural inspection. *Transactions of the ASAE, 46*(2), 513–521.

Lawrence, K. C., Yoon, S. C., Heitschmidt, G. W., Jones, D. R., & Park, B. (2008). Imaging system with modified-pressure chamber for crack detection in shell eggs. *Sensing and Instrumentation for Food Quality and Safety, 2*(3), 116–122.

Litwiller, D. (2005). CMOS vs. CCD: Maturing technologies, maturing markets. *Photonics Spectra, 39*(8), 54–58.

Liu, W., Barbastathis, G., & Psaltis, D. (2004). Volume holographic hyperspectral imaging. *Applied Optics, 43*(18), 3581–3599.

Lu, R. (2003). Detection of bruises on apples using near-infrared hyperspectral imaging. *Transactions of the ASAE, 46*(2), 523–530.

Lu, R., & Chen, Y. R. (1998). Hyperspectral imaging for safety inspection of food and agricultural products. In Pathogen Detection and Remediation for Safe Eating. *Proceedings of SPIE, Vol. 3544*, 121–133.

Malik, Z., Cabib, D., Buckwald, R. A., Talmi, A., Garini, Y., & Lipson, S. G. (1996). Fourier transform multipixel spectroscopy for quantitative cytology. *Journal of Microscopy, 182*(2), 133–140.

Marinelli, W. J., Gittins, C. M., Gelb, A. H., & Green, B. D. (1999). Tunable Fabry–Perot etalon-based long-wavelength infrared imaging spectroradiometer. *Applied Optics, 38*(12), 2594–2604.

Mergenthaler-Gatfield, S., Holzgreve, W., & Hahn, S. (2008). Spectral karyotyping (SKY): Applications in prenatal diagnostics. In S. Hahn, & L. G. Jackson (Eds.), *Prenatal diagnosis: methods in molecular biology, vol. 444* (pp. 3–26). Totowa, NJ: Humana Press Inc.

Min, M., Lee, W. S., Burks, T. F., Jordan, J. D., Schumann, A. W., Schueller, J. K., & Xie, H. K. (2008). Design of a hyperspectral nitrogen sensing system for orange leaves. *Computers and Electronics in Agriculture, 63*(2), 215–226.

Morris, H. R., Hoyt, C. C., & Treado, P. J. (1994). Imaging spectrometers for fluorescence and Raman microscopy: acousto–optic and liquid-crystal tunable filters. *Applied Spectroscopy, 48*(7), 857–866.

Mueller-Mach, R., Mueller, G. O., Krames, M. R., & Trottier, T. (2002). High-power phosphor-converted light-emitting diodes based on III-nitrides. *IEEE Journal of Selected Topics in Quantum Electronics, 8*(2), 339–345.

Muthu, S., Schuurmans, F. J. P., & Pashley, M. D. (2002). Red, green, and blue LEDs for white light illumination. *IEEE Journal of Selected Topics in Quantum Electronics, 8*(2), 333–338.

Nicolai, B. M., Lotze, E., Peirs, A., Scheerlinck, N., & Theron, K. I. (2006). Non-destructive measurement of bitter pit in apple fruit using NIR hyperspectral imaging. *Postharvest Biology and Technology, 40*(1), 1–6.

Noh, H. K., & Lu, R. (2007). Hyperspectral laser-induced fluorescence imaging for assessing apple fruit quality. *Postharvest Biology and Technology, 43*(2), 193–201.

Okamoto, T., & Yamaguchi, I. (1991). Simultaneous acquisition of spectral image-information. *Optics Letters, 16*(16), 1277–1279.

Orenstein, A., Kostenich, G., Rothmann, C., Barshack, I., & Malik, Z. (1998). Imaging of human skin lesions using multipixel Fourier transform spectroscopy. *Lasers in Medical Science, 13*(2), 112–118.

Palmer, C. (2005). *Diffraction grating handbook* (6th ed., pp. 14–42). Rochester, NY: Newport Corporation.

Park, B., Lawrence, K. C., Windham, W. R., & Buhr, R. J. (2002). Hyperspectral imaging for detecting fecal and ingesta contaminants on poultry carcasses. *Transactions of the ASAE, 45*(6), 2017–2026.

Peng, Y., & Lu, R. (2006). An LCTF-based multispectral imaging system for estimation of apple fruit firmness. Part I: Acquisition and characterization of scattering images. *Transactions of the ASAE, 49*(1), 259–267.

Pham, T. H., Bevilacqua, F., Spott, T., Dam, J. S., Tromberg, B. J., & Andersson-Engels, S. (2000). Quantifying the absorption and reduced scattering coefficients of tissuelike turbid media over a broad spectral range with noncontact Fourier-transform hyperspectral imaging. *Applied Optics, 39*(34), 6487–6497.

Polder, G., van der Heijden, G. W. A. M., Keizer, L. C. P., & Young, I. T. (2003). Calibration and characterisation of imaging spectrographs. *Journal of Near Infrared Spectroscopy, 11*(3), 193–210.

Qin, J., Burks, T. F., Kim, M. S., Chao, K., & Ritenour, M. A. (2008). Citrus canker detection using hyperspectral reflectance imaging and PCA-based image classification method. *Sensing and Instrumentation for Food Quality and Safety, 2*(3), 168–177.

Qin, J., & Lu, R. (2005). Detection of pits in tart cherries by hyperspectral transmission imaging. *Transactions of the ASAE, 48*(5), 1963–1970.

Qin, J., & Lu, R. (2007). Measurement of the absorption and scattering properties of turbid liquid foods using hyperspectral imaging. *Applied Spectroscopy, 61*(4), 388–396.

Qin, J., & Lu, R. (2008). Measurement of the optical properties of fruits and vegetables using spatially resolved hyperspectral diffuse reflectance imaging technique. *Postharvest Biology and Technology, 49*(3), 355–365.

Safren, O., Alchanatis, V., Ostrovsky, V., & Levi, O. (2007). Detection of green apples in hyperspectral images of apple-tree foliage using machine vision. *Transactions of the ASABE, 50*(6), 2303–2313.

Sensors Unlimited. (2006). *What is InGaAs?* Princeton, NJ: Sensors Unlimited, Inc. Application Note.

Smith, W. H., & Hammer, P. D. (1996). Digital array scanned interferometer: Sensors and results. *Applied Optics, 35*(16), 2902–2909.

Spectral Imaging. (2003). *ImSpector imaging spectrograph user manual* Ver. 2.21. Oulu, Finland: Spectral Imaging, Ltd.

Steigerwald, D. A., Bhat, J. C., Collins, D., Fletcher, R. M., Holcomb, M. O., Ludowise, M. J., & Rudaz, S. L. (2002). Illumination with solid state lighting technology. *IEEE Journal of Selected Topics in Quantum Electronics, 8*(2), 310–320.

Wabuyele, M. B., Yan, F., Griffin, G. D., & Vo-Dinh, T. (2005). Hyperspectral surface-enhanced Raman imaging of labeled silver nanoparticles in single cells. *Review of Scientific Instruments, 76*(6), 063710.

Yoon, S. C., Lawrence, K. C., Smith, D. P., Park, B., & Windham, W. R. (2008). Bone fragment detection in chicken breast fillets using transmittance image enhancement. *Transactions of the ASABE, 51*(1), 331–339.

Zhang, H., Paliwal, J., Jayas, D. S., & White, N. D. G. (2007). Classification of fungal infected wheat kernels using near-infrared reflectance hyperspectral imaging and support vector machine. *Transactions of the ASABE, 50*(5), 1779–1785.

PART 2

Applications

CHAPTER 6

Meat Quality Assessment Using a Hyperspectral Imaging System

Gamal ElMasry [1,2], Da-Wen Sun [1]
[1] *University College Dublin, Agriculture and Food Science Centre, Belfield, Dublin, Ireland*
[2] *Agricultural Engineering Department, Suez Canal University, Ismailia, Egypt*

CONTENTS

6.1. INTRODUCTION

Assessment of meat quality parameters has always been a big concern in all processes of the food industry because consumers are always demanding superior quality of meat and meat products. Interest in meat quality is driven by the need to supply the consumer with a consistent high quality product at an affordable price. Indeed, high quality is a key factor for the modern meat industry because the high quality of the product is the basis for success in today's highly competitive market. To meet the consumers' needs, it is a crucial element within the meat industry to correctly assess meat quality parameters by improving modern techniques for quality evaluation of meat and meat products (Herrero, 2008). Therefore, the meat industry should exert cooperative efforts to improve the overall quality and safety of meat and meat products to gain a share in both local and international markets. Maintaining and increasing demand for meat, in both local and international markets, depends heavily on such factors as assurances of food safety, animal welfare, and the final quality of the product. Animal welfare is a major concern in meat production due to the fact that consumers are increasingly demanding that animals are produced, transported, and slaughtered in a humane way. Therefore meat production continues to be reformed by the rapidly growing demands of customers. Although health concerns may influence the decision of whether or not to eat meat, or how often and how much to eat, economic factors such as meat prices and consumers' incomes also influence the choice of consuming meat. The great variability in raw meat leads to highly variable products being marketed without a controlled level of quality. This problem is aggravated when the industry is unable to satisfactorily characterize this level of quality and cannot therefore market products with a certified quality level (Damez & Clerjon, 2008).

Generally, meat quality can be defined in terms of consumer appreciation of texture and flavour, and food safety, which includes the health implications of both compositional and microbiological properties. The ultimate quality of meat is a direct integration of parameters and conditions such as feeding and management of animals during their growth, pre-slaughter stress, stunning method, electrical stimulation, cooling method and rate, maturing time, freezing and thawing, and cooking conditions as well as handling and processing techniques and composition of meat products (Liu *et al.*, 2003a).

The visual appearance, textural patterns, geometrical features, and color of fresh meat products are the main criteria used by consumers for choosing and purchasing high quality meat. These parameters are linked to some chemical properties such as water holding capacity, intramuscular fat (marbling), and protein contents. The conventional methods for determining such parameters rely on subjective visual judgment and then laboratory chemical tests. In addition, traditional meat grading routines and quality evaluation methods are time consuming, destructive, and are associated with inconsistency and variability due to human inspection. Therefore, evaluation of meat quality in recent meat processing lines requires instrumentation that is fast, specific, robust, and durable enough for the harsh environments of processing plants in order to overcome all disadvantages of traditional methodology (Herrero, 2008). These instrumentations also have to be cost-effective to reflect the competitive nature of the food and agriculture markets.

The meat industry is currently undergoing dramatic changes in applying the most advanced technological inventions that have gained acceptance and respect in handling, quality control and assurance, packaging, and distribution (Shackelford *et al.*, 2004). The changes are noticed in many fields because there is an increasing demand from the consumers and the media for optimal quality, consistency, safety, animal welfare, and environmental issues. Many different methods for measuring meat quality traits are available which are based on different principles, procedures, and/or instruments. Over the past few years, a number of methods have been developed to objectively measure meat quality traits (Abouelkaram *et al.*, 1997, 2006; Liu *et al.*, 2003a; Shackelford *et al.*, 2005; Vote *et al.*, 2003). One of these methods is the imaging technique that has been applied for visual evaluation of meat quality. On the other hand, the spectroscopic technique is finding increasing use owing to its rapidity, simplicity, and safety, as well as its ability to measure multiple attributes simultaneously without monotonous sample preparation. However, the spectroscopic technique alone is not able to provide some fundamental information where demonstration of the spatial distribution of quality parameters is essential. Hyperspectral imaging has thus emerged to integrate both the spectroscopic and imaging techniques for providing spectral and spatial information simultaneously to cope with the increasing demand for safe foods.

Hyperspectral imaging technique is an upcoming and promising field of research for non-destructive quality assessment of agricultural and food products including meat (Cluff *et al.*, 2008; Naganathan *et al.*, 2008a, 2008b). In recent years, there has been growing interest in this technology from researchers around the world. The main impetus for developing hyperspectral imaging system is to integrate spectroscopy and imaging techniques to make direct identification of different components and their spatial distribution in the tested sample. The commercial growth of hyperspectral imaging lies in its ability to solve some application problems, such as those associated with industrial process monitoring and control, diagnosis, inspection, and quality-related assessments. Although this technology has not yet been sufficiently exploited in meat processing lines and quality assessment, its potential is promising. In contrast to conventional methods for the determination of meat quality parameters, the hyperspectral imaging technique is a sensitive, fast, and non-destructive analytical technique with simplicity in sample preparation allowing simultaneous assessment of numerous meat properties. For instance, hyperspectral imaging can be used to identify a particular type of meat (Qiao *et al.*, 2007a, 2007b), as some meat (species, cuts or grades) are more valuable for the consumers than others (Alomar *et al.*, 2003). Some other key potential applications include overall inspection and disease detection in different meat products (Chau *et al.*, 2009; Kim *et al.*, 2004; Wold *et al.*, 2006). In addition, hyperspectral imaging can be used as an authentication tool in order to prevent fraud as well as to estimate chemical composition with acceptable accuracy and even to detect handling aspects of the product. Therefore, developing a quality evaluation system based on hyperspectral imaging technology to assess meat quality parameters and to ensure its authentication would bring economic benefits to the meat industry by increasing consumer confidence in the quality of the meat products. In this chapter an overview of the current meat quality assessment techniques is provided with an emphasis on hyperspectral imaging method. In particular, latest research results on using hyperspectral imaging technology for assessing quality of red meat (beef, lamb, and pork) and white meat (poultry and fish) will be highlighted and described in more detail.

6.2. MEAT QUALITY EVALUATION TECHNIQUES

Meat is a perishable, nutritious, and expensive food commodity, and its quality concept is related to individual experience and preference. To facilitate marketing, grading standards have been developed to classify carcasses into quality and yield grades. Although these standards are not universal, they basically include kind of meat, sex classification, maturity evaluation, and color and texture of muscles. *Quality* is the general term to express the compositional quality, relative desirability or expected palatability of the meat in a carcass or cut. It refers to a combination of traits, which result in an edible product that is attractive in appearance, and is nutritious and palatable after cooking. In general, quality of food products covers many aspects, such as functional, technological, sensory, nutritional, toxicological, regulatory, and ethical aspects (Herrero, 2008). *Meat quality* is always defined by the compositional quality (lean to fat ratio, meat percentage, intramuscular fat, marbling, protein, and muscle area), functional quality (water holding capacity, isometric tension, muscle fiber shortening, pH, and cooking loss), and eating quality or *palatability* (appearance, juiciness, tenderness, and flavour) (AMSA, 2001). Therefore, the term "meat quality" covers many different properties that must be considered from the perspective of producers, packers, retailers, and consumers.

Many countries such as Canada, Japan, United States, and Australia initiated their own quality standard charts for meat (AMSA, 2001), which are slightly different but always based on visual comparison of the primary lean quality traits such as color, wetness, firmness, texture, and marbling content of the exposed loin eye. This grading is normally applicable for beef, lamb, veal, and pork. Poultry and fish are not in this classification because of the difference in lean and fat content and color patterns and because it is different from meat in having a negligible fat content. However, from growth to slaughter, there are many parameters affecting meat quality, such as genetic factors, pre-slaughter stress, aging, pH and other factors during handling, and loading and transport of meat. Also, the characteristics of raw meats are greatly influenced by animal (breed, sex, age), environment (feeding, transporting, and slaughtering condition), and processing (storing time/temperature condition) (Liu *et al.*, 2003a).

The purpose of evaluating meat quality is to identify physical attractiveness and to predict the palatability of the cooked lean meat. It is impossible to develop models for predicting the quality of meat throughout the meat production chain without assessing essential quality parameters. Visible quality traits are not precise palatability predictors, but are reasonably useful to identify cuts that will be tender/tough, juicy/dry, and flavourful/off-flavour cooked products. The main quality features to be evaluated visually include color and texture of the lean meat, degree of marbling, and color of fat for beef, veal, pork and lamb. To measure the other quality traits related to compositional, functional, hygiene, and sensory parameters, it is normally necessary to apply destructive tests to find the actual values of these traits. Fortunately, there are a lot of non-destructive methods that can substitute the destructive measurements.

6.2.1. Destructive Measurements of Major Meat Quality Parameters

Since *color* is related to the level of the protein pigment, myoglobin, present in the muscle, it can be estimated chemically by analyzing the pigments present in the meat by extracting these pigments from meat followed by spectrophotometric determination of pigment concentration. For objective measurements of color, it is usually performed by using the Commission International de l'Eclairage (CIE) color system (Yam & Papadakis, 2004; Valkova *et al.*, 2007). In this system, color is usually measured in the L*a*b* scale, where L* denotes the brightness, a* the red–blue color and b* the green–yellow color. Based on color measurements, meat can be broadly classified as "red" or "white" depending on the concentration of myoglobin in muscle fiber.

The *water content* of meat is another important criterion for two reasons. First, meat is sold by weight, so water loss is an important economic factor. Secondly, the water content of meat determines to a large extent the juiciness of meat and thereby the eating quality. Indeed, the reduction of pH post mortem normally results in a reduction in water holding, so that exudates leak out of cut muscle surfaces during post-mortem storage. Since water holding capacity (WHC) is the ability of meat to hold all or part of its own water during application of external forces like cutting, heating, grinding or pressing, it is considered one of the most important quality factors that need to be determined. The most acceptable ways for WHC determination are through destructive measurements, either mechanically by applying mechanical force through positive or negative pressure like centrifugation and suction, or thermally by applying thermal force by heating and measuring the cooking loss (Honikel, 1998). Figure 6.1 depicts the traditional methods of measuring color, water holding capacity (WHC) and pH.

Another important quality parameter is the tenderness. *Tenderness* as a general term for meat texture is a crucial sensory quality attribute

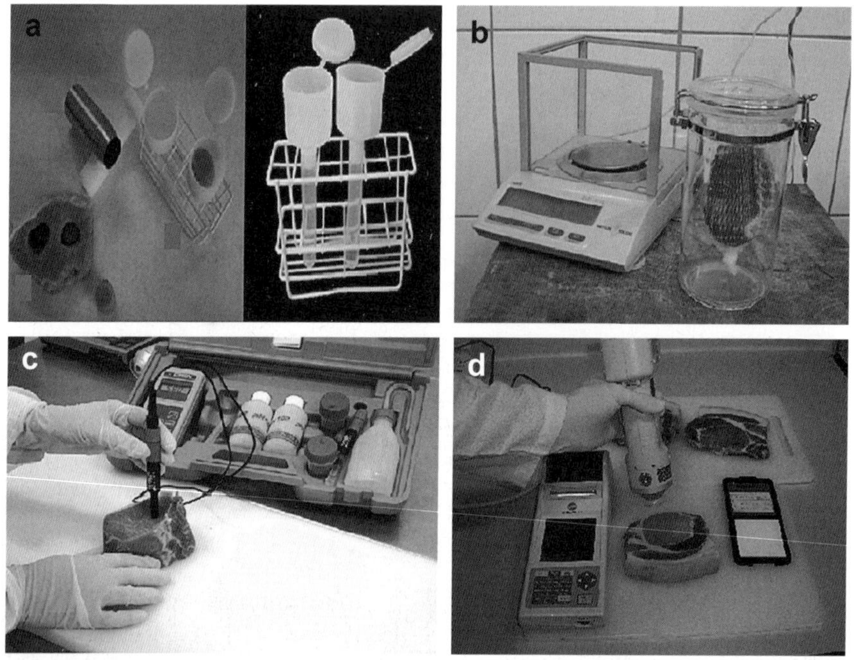

FIGURE 6.1 *Traditional methods for measuring meat quality parameters. (a) Measuring water holding capacity (WHC) by using EZ-Drip loss method (Rasmussen & Andersson, 1996); (b) measuring WHC by using bag method (Honikel, 1998); (c) measuring pH by using pH meter; and (d) measuring color by using a portable Minolta colorimeter. (Full color version available on http://www.elsevierdirect.com/companions/9780123747532/)*

associated with consumer satisfaction as consumers consider tenderness as the primary factor in eating satisfaction, and they are willing to pay more for tender meat (Lusk *et al.*, 2001) as tenderness is positively related to juiciness and flavour (Winger & Hagyard, 1994). Meat tenderness is related to muscle structure and biochemical activity in the period between slaughtering and meat consumption. As proposed by Dransfield (1994), the tenderness issue is separated into three components: tenderization, ageing, and tenderness. The tenderization is the enzymatic proteolysis, which cannot be measured early post mortem because of muscle contraction up to rigor mortis. The meat aging is the maturing of the meat, which is the traditional method of enhancing the meat tenderness by storage for up to 3 weeks. The last component of tenderness is the tenderness of the end product (the cooked meat). This component is related to an integration of components such as connective tissue, muscle shortening, sarcomere length, and fat and water content.

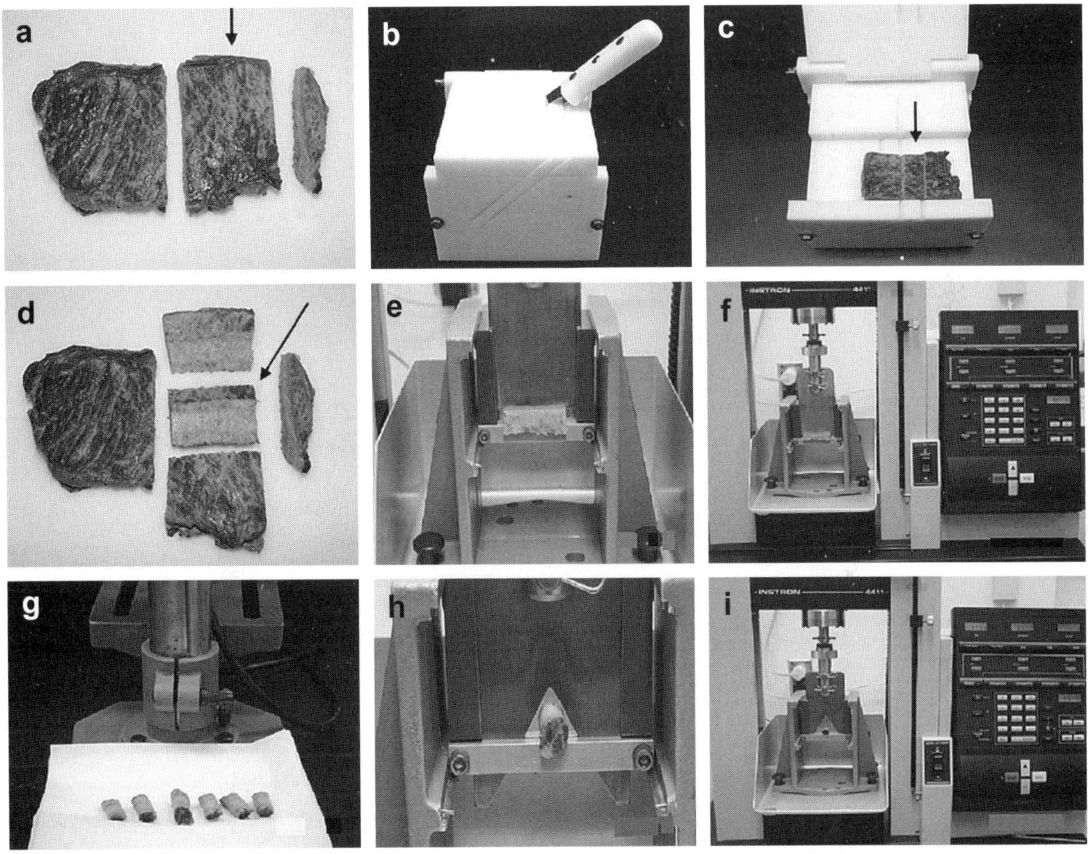

FIGURE 6.2 *Destructive determination methods of meat tenderness. (a–f) using slice shear force "SSF", (g–i) using Warner–Bratzler shear force "WBSF". [*Slice shear force "SSF" method*: a single slice of 5 cm long from the center of a cooked steak is removed parallel to the long dimension (a); using a double-blade knife, two parallel cuts are simultaneously made through the length of the 5 cm long steak portion at a 45° angle to the long axis and parallel to the muscle fibers (b–c), this results in a slice of 5 cm long and 1 cm thick parallel to the muscle fibers (d), and the slice is then sheared once perpendicular to the muscle fibers using universal testing machine equipped with a flat, blunt-end blade (e–f);* Warner–Bratzler shear force "WBSF" method*: six core samples of 12.7 mm in diameter are taken from a cooked steak parallel to the longitudinal orientation of the muscle fibers (g), each core is then sheared using a universal testing machine equipped with a triangular slotted blade (h–i). In both methods the maximum shear force (meat tenderness) is the highest peak of the force–deformation curve.] (Full color version available on http://www.elsevierdirect.com/companions/9780123747532/)*

The most common approach to assess meat texture is by measuring the mechanical properties of the sample using Warner–Bratzler shear force "WBSF" or slice shear force "SSF" methods as shown in Figure 6.2 (Shackelford *et al.*, 1997). For Warner–Bratzler shear force (WBSF) determination, six cylindrical, 1.27 cm diameter cores are typically removed from each steak; while for SSF determination, a single slice 1 cm thick, 5 cm long is removed from the lateral end of each *longissimus* steak. For both techniques, samples should be removed parallel to the muscle fiber orientation and sheared across the fibers. WBSF uses a V-shaped blade, while SSF uses a flat blade with the same thickness and degree of bevel on the shearing edge. However, both methods are not suitable for the commercial and fast-paced production environment. In the meat marketing system, meat products leave the packing plant at about three days post mortem, and reach the consumer after approximately 14 days. The meat industry needs an instrument that can scan fresh meat at 2–3 days post mortem and ultimately predict its tenderness when the consumer cooks it about two weeks later.

6.2.2. Necessity of Objective Methods for Meat Quality Evaluation

In practice, the quality of meat is normally assessed subjectively by an experienced grader. This method relies greatly on human skills and is subject to non-objective results. Hence, the outcome of subjective grading may vary between different analysts. Presently, all meat quality evaluation systems are unable to incorporate a direct measurement of some quality parameters such as tenderness because there is no accurate, rapid, and non-destructive method for predicting tenderness available to the meat industry. Thus, meat cuts are not priced on the basis of actual tenderness, creating a lack of incentive for producers to supply a tender product. Moreover, traditional quality evaluation methods such as the Warner–Bratzler method for tenderness and impedance measurements for detecting frozen meats and fat content are time-consuming, demand high labor costs, and require lengthy sample preparation associated with inconsistency and variability (Damez *et al.*, 2008; Shackelford *et al.*, 1995). Furthermore, these methods are destructive and only able to predict the global characteristics of a meat sample without considering the spatial distribution of these characteristics. Therefore, these methods are not practical when fast analysis and early detection of quality parameters in industrial and commercial processing lines are required (Damez & Clerjon 2008). As a result objective and fast assessment of meat quality has been desirable for a long time in the industry and there have been many research efforts in the direction of developing the required instrumentation.

Indeed, recent advances in computer technology have led to the development of imaging systems capable of rapidly identifying quality parameters on the processing line, with the minimum of human intervention (Brosnan & Sun, 2004; Du & Sun, 2004; Yang *et al.*, 2009). On the other hand, as one of the major optical applications, spectroscopy has been widely used to detect the chemical attributes of meat and meat products. Near infrared spectroscopy (NIRS) is always one of the most promising techniques for large-scale meat quality evaluation as it offers a number of important advantages over conventional quality evaluation methods such as rapid and frequent measurements, no sample preparation required, suitability for on-line use, and simultaneous determination of different attributes. The main disadvantages of the method are its dependence on reference method, weak sensitivity to minor constituents, limited transfer of calibration between different instruments, complicated spectral data interpretation, and particularly, the low spatial resolution for analysis of food samples with non-homogeneous composition as they are found in meats and meat products (Prevolnik *et al.*, 2004).

As an extension of traditional imaging and spectroscopic techniques, hyperspectral imaging technology, known also as imaging spectroscopy or imaging spectrometry, is developed to combine the advantages of both techniques to perform so many tasks in quality evaluation purposes such as identification, classification, mapping, and target detection. This technology is based on the utilization of an integrated hardware and software platform that combines conventional imaging and spectroscopy to attain both spatial and spectral information from each pixel. In recent years there has been growing interest in this technology from researchers around the world for non-destructive analysis in many research and industrial sectors (Cluff *et al.*, 2008; ElMasry *et al.*, 2007, 2009; Naganathan *et al.*, 2008a, 2008b; Noh & Lu, 2007).

6.2.3. Non-destructive Techniques for Measuring Meat Quality

6.2.3.1. Computer vision

The design of artificial vision systems that attempt to emulate the human sense of sight is a very attractive field of research because it is considered an expeditious, safe, hygienic, and versatile technique. Building a machine that can sense its environment visually and perform some useful functions has been the subject of investigations for many years. Computer vision utilizing imaging technique has been developed as an inspection tool for quality and safety assessment of a variety of meat products (Sun, 2008a). The flexibility and the non-destructive nature of this technique help to maintain its attractiveness for applications in the food industry (ElMasry *et al.*, 2008). Computer vision has long been seen as a potential solution for various automated visual quality evaluation processes. It is recognized as the integrated use of devices for non-contact optical sensing and computing and decision processes to receive and interpret an image of a real scene automatically, in order to detect defects, to evaluate quality, and to improve operating efficiency and the safety of both products and processes.

Application of computer vision depends on many disciplines, such as image processing, image analysis, mathematics, computer science, and software programming. As automated visual inspection is the most common and rapid way for the quality assessment of meat products applied to the production chain, computer vision has been recognized as a promising approach for the objective assessment of meat quality, and computer vision systems have found widespread usages in quality evaluation of different meat products and in analysis of surface defects and color classification. Detecting visible characteristics of the tested samples is the basis for computer vision in the quality assessment of meat. Based on this technique, some commercial technologies utilizing computer vision systems have been introduced to evaluate the overall quality and for grading purposes. Belk *et al.* (2000) reported that a prototype video imaging system (BeefCam) could identify carcasses that would yield steaks that would be "tender" after aging and cooking. However, this prototype BeefCam has limitations that prevent its use in a commercial setting. Vote *et al.* (2003) carried out four independent experiments in two commercial packing plants that utilize electrical stimulation, to determine the effectiveness of the computer vision system equipped with a BeefCam module (CVS BeefCam) for predicting the Warner–Bratzler shear force (WBSF) values of *longissimus* muscle steaks from carcasses and classifying these carcasses according to beef tenderness differences, in a commercial setting. The system captured and segmented video images at commercial packing-plant chain speeds to produce information useful in explaining observed variation in Warner–Bratzler shear force values of steaks, even when there is a narrow range of marbling scores. This information could be used to sort carcasses according to expected palatability differences of their steaks. However, a conventional imaging technique is not suitable for certain industrial applications especially when the tested samples have similar colors and when chemical compositions of these samples are required to be quantitatively assessed as well as when some invisible potentially harmful concentrations of hazardous residues on foods need to be detected (Park *et al.*, 2006a).

6.2.3.2. Spectroscopy

For many years, spectroscopy has been used intensively as an analytical technique for meat and meat products. The basic principle of spectroscopy is to radiate the sample with a controlled wavelength and measure the response from the sample (Sun, 2008b). In optic spectroscopy the sample is excited by illumination from a light source, then the light is transmitted, absorbed, and reflected by the sample and this response can be measured with a detector. As explained in Chapter 1, spectroscopic methods provide detailed fingerprints of the biological sample to be analyzed using physical characteristics of the interaction between electromagnetic radiation and the sample material such as reflectance, transmittance, absorbance, phosphorescence, fluorescence, and radioactive decay. Recently, the near-infrared spectroscopy (NIRS) technique has received considerable attention as a means for the non-destructive sensing of meat quality (Sun, 2008b). More vitally, NIRS has the potential for simultaneously measuring multiple quality attributes. Applications of near-infrared spectroscopy (NIRS) have increased in food product quality analysis. Specifically, NIRS has been widely used to predict the quality of fresh meat and has been shown to be a rapid and effective tool for meat quality assessment with most attention being focused on the prediction of beef tenderness (Liu *et al.*, 2003a; Park *et al.*, 1998; Ripoll *et al.*, 2008; Rust *et al.*, 2007; Shackelford *et al.*, 2005) in order to substitute other commonly used destructive methods.

Unfortunately, NIRS is unable to provide constituent gradients because the analysis focuses on only a relatively small part of the material analysed. In other words, NIRS techniques rely on measuring the aggregate amount of light reflected or transmitted from only a specific area of a sample (point measurement where the sensor is located), and does not contain information on the spatial distribution of quality traits on the sample (Ariana *et al.*, 2006; Prevolnik *et al.*, 2004). Thus, it may lead to inconsistency between predicted and measured values of a certain constituent simply because it produces an average value of this constituent in the whole sample using only the data extracted from a small portion of the sample. Generally speaking, by using an imaging technique alone it is easy to know the location of certain features, but it is not easy to discover the quantitative information of these features.

6.2.3.3. Hyperspectral imaging

As previously mentioned, hyperspectral imaging combines the major advantages of imaging and spectroscopy for acquiring both contiguous spectral and spatial information from an object simultaneously, which otherwise cannot be achieved with either conventional imaging or spectroscopy. Hyperspectral imaging sensors measure the radiance of the materials within each pixel area at a very large number of contiguous spectral wavelength bands (Manolakis *et al.*, 2003). Therefore, hyperspectral imaging refers to the imaging of a scene over a large number of discrete, contiguous spectral bands such that a complete reflectance spectrum can be obtained for the region being imaged. The spectra on the surface of food materials contain characteristic or diagnostic absorption features to identify a number of important inherent characteristics. Moreover, hyperspectral imaging can provide spectral measurements at the entire surface area of the product while conventional spectrometers only give point measurements. By combining the chemical selectivity of spectroscopy with the power of image visualization, hyperspectral imaging is particularly useful in situations where multiple quality attributes must be considered and when either machine vision or spectroscopy is not suitable. This is due to the fact that hyperspectral imaging enables a more complete description of ingredient concentration and distribution in any kind of heterogeneous sample (Gowen *et al.*, 2008).

In classification or grading of meat products, multiple extrinsic and intrinsic factors are often needed to judge the overall quality. Hyperspectral imaging could be an effective technique to grade meat based on both extrinsic, like appearance (e.g. size, intramuscular fat, color), and intrinsic (tenderness and chemical composition) properties, which are all important in determining the overall quality of meat. The non-destructive nature of hyperspectral imaging is an attractive characteristic for application on raw materials and final product quality (Folkestad *et al.*, 2008; Wold *et al.*, 2006). Because the scope of this chapter is about the hyperspectral imaging system and its potential in meat quality evaluation, more technical details will be given in the next sections.

6.3. HYPERSPECTRAL IMAGING SYSTEM

Nowadays, the hyperspectral imaging technique has entered a new era of industrial applications for real-time inspection of food and agricultural products. One of the major advantages of hyperspectral imaging comes from the possibility of using intact samples presented directly to the system without any pretreatment and supplying qualitative and quantitative assessments simultaneously. The main configuration, design, image acquisition modes as well as the fundamentals, characteristics, terminologies, advantages, disadvantages, and constraints of hyperspectral imaging systems are described in detail in Chapter 1.

Optical measurements through hyperspectral imaging techniques are commonly implemented in one of the major three sensing modes: reflectance, transmittance or interactance. In reflectance mode, the light reflected by the illuminated sample is captured by the detector in a specific conformation to avoid specular reflection. This technique is commonly used to detect external quality characteristics such as color, size, shape, and external features and defects. In transmittance mode the image is acquired with the light source positioned opposite to the detector and the sample in between; this method is commonly used to detect internal defects of fish, fruits, and vegetables. In interactance mode the light source and the detector are positioned parallel to each other; this arrangement must be specially set up in order to prevent specular reflection entering the detector (ElMasry & Wold, 2008; Nicolai *et al.*, 2007).

6.3.1. Chemical Imaging

Recent developments in hyperspectral imaging allow the method to be applied to assess the spatial distribution of food composition (Millar *et al.*, 2008), and to make measurements of selected regions of food samples. The hyperspectral imaging technique has been extensively applied to visualize the chemical composition of various food materials in a methodology known as *chemical imaging*. For detailed food analysis, concentration gradients of certain chemical components are often more interesting than average concentrations, no matter how accurately the latter are determined. It is advantageous to know and understand the heterogeneity of samples in understandable images known as *chemical images*, *spectrally classified images* or *spectral maps*. It is sometimes necessary to analyze and establish the local distribution of properties of interest in a sample that is spatially non-homogeneous. With conventional spectroscopy one can either tediously scan the entire sample with a focused optical probe point by point or obtain average properties over the entire sample using a single measurement. This is where hyperspectral imaging provides huge potential. The value of spectral imaging lies in the ability to resolve spatial heterogeneities in solid-state samples like meat samples. The combination of spectral data and spatial details together enables the high-speed analysis of chemical content, uniformity, quality, and a host of other product characteristics and attributes. For any point (pixel) in the image, the chemical spectra or *spectral signature* of this particular point can be determined while maintaining the integrity of spatial information obtained. The spectrum of any point in the sample can be used for calculating concentrations of some chemical compositions, e.g. fat, protein, water, carbohydrates etc., because each pixel has a corresponding spectrum. Hence, the hyperspectral images consist of a spectrum for each pixel allowing, in theory, the prediction of component concentrations at each pixel, leading to the creation of concentration images or maps, i.e., the *chemical images* (Burger & Geladi, 2006).

Recently, there is an increasing need for the identification, quantification, and distribution of minor and major components of biological materials especially food products. The interaction of the light beam with the sample causes the generation of many signals carrying varied information that could be used simultaneously to create an image and derive data from the specimen's chemical composition. The main advantage of this technique is that it is a chemical-free assessment method where sample preparation is eliminated and thus reduces time for analysis and eliminates all types of artifacts. Chemical imaging is the final goal of hyperspectral imaging to produce such images to show the gradients and spatial distributions of chemical compositions of the samples based on their spectral signatures by applying one or more chemometric tools, such as principal component regression (PCR) or partial least squares regression (PLSR). Based on collected spectra for regions (pixels) having different levels of components of interest (e.g. moisture content and fat content), calibrations are then derived. By applying these calibrations to unknown pixels, images of distribution of the relevant components are then generated. This, therefore, can allow extraction and visualization of extra information that the human fails to capture. The chemical imaging trend of hyperspectral imaging is a relatively young technique that has gained popularity and acceptance for the analysis of manufactured products. Generally, chemical imaging is the procedure of creating visual images to produce quantitative spatial distribution of sample components by using simultaneous measurement of spectra to represent chemical characterizations of these components. The contrast in the images is based on the chemical differences between the various components of heterogeneous samples. The power of chemical imaging resides in the quick access to the spatial distribution of chemical compositions and their relative concentrations. Recently, this technique has found widespread applications in many fields such as chemistry, medicine, pharmacy, food science, biotechnology, agriculture, and industry (Bonifazi & Serranti, 2008; de Juan *et al.*, 2004; ElMasry & Wold, 2008; Leitner *et al.*, 2003; Rutlidge & Reedy, 2009; Šašic, 2007; Sugiyama, 1999).

Chemical imaging is usually used to answer three kinds of question: what, how much, and where. Therefore it will be effectively used first to identify sample components (what), then to determine the quantity or concentration of these components (how much), and finally to visually demonstrate the spatial distribution of these components in the samples (where). Consequently, any samples having chemical gradients are suitable to be investigated by this technique, which couples spatial and chemical characterization in chemical imaging. Chemical imaging not only allows visualization of the chemical information on the tested sample, but it is also a non-destructive technique so that samples are preserved for further testing. Chemical imaging is particularly useful for performing rapid, reproducible, reliable, non-contact, and non-destructive analyses of samples. The abundant information characterizing both chemical and morphological features opens the door to chemical imaging techniques to be implemented in several applications, not only in laboratory and research contexts but also in the food industry.

Because chemical imaging combines the digital imaging with the attributes of spectroscopic measurements, the configuration of chemical imaging instrumentation is the same as any hyperspectral imaging system that is composed of an illumination source, a spectrograph, a detector array (the camera) to collect the images, and a computer supported with image acquisition software. The resulting hypercube can be visually presented as a series of spectrally resolved images where each image plane corresponds to the image at one wavelength. However, the spectrum measured at a particular spatial location can be easily viewed which is useful for chemical identification and building the final chemical images. In some circumstances, selecting one image plane at a particular wavelength can highlight the spatial distribution of sample components, provided that their spectral signatures are different at the selected wavelength. However, having only one image at a single wavelength is sometimes not enough to view all spatial differences in chemical composition of the sample under investigation simply because each component has its own spectral features at different wavelengths compared with the other components. In addition, some components have unique spectral features at more than one wavelength. Consequently, manipulating the hyperspectral datacube by one of the calibrated multivariate approaches such as PLS1, PLS2 or PCR to separate spectral signatures of sample components and to relate spectral data with the real content (concentration) of the these components is essential when the spatial distribution of one or more chemical components is required to be viewed precisely. However, detecting certain components in the sample is strongly influenced by particle size, the chemical and spatial heterogeneity of the sample, and the spatial resolution of the image (Burger & Geladi, 2006).

6.3.2. Data Exploitation

A full-size hyperspectral image is very large. For instance, a hypercube of 256×256 pixel in the spatial dimension and 100 bands (in the spectral dimension) has a size of 6.55 Mega pixels, and when digitized to 10 or 12 bits, the file size becomes 13.1 Mega bytes. Handling, displaying, visualizing, and processing such files requires efficient analysis tools (Bro *et al.*, 2002; Hruschka, 2001). Analysing hyperspectral images and treatments of the vast data have been concerns for all applications of this technique in identification, detection, classification, and mapping purposes. Classification enables the recognition of regions with similar spectral characteristics without conducting chemical background determination of these regions. For quantitative assessment, it is necessary to extract chemical information from hyperspectral images by carrying out correlation between spectral information and real chemical concentrations obtained by established conventional chemical determination methods for attaining physical and chemical properties. This step is called the calibration process, which needs to be tested and validated with different meat samples. Chemical validation is necessary in order to estimate if a calibration model based on spectroscopic data is suitable for the practical purpose it was designed for, for example as a quality control tool in the meat industry. In this respect, hyperspectral imaging is considered as an indirect method by using obvious correlations between spectral measurements and meat component properties. Taking these calculations and modeling into consideration, the major spatial and spectral features involved can help to improve our understanding of meat properties and thus of eating quality.

As multivariate data, hyperspectral imaging data are usually analyzed by applying the same mathematical approaches as those applied in spectroscopic data. This is due to the fact that the spectrum retained in each pixel in the hyperspectral image is equivalent to a single point spectrum extracted from spectroscopy; therefore all pre-processing, chemometric, and pattern recognition techniques could be used with the same aim to perform a qualitative or quantitative characterization of the sample components. The most efficient tool for exploratory multivariate data analysis is chemometrics, which provides practical solution of spectral data problems by efficient utilization of experimental data. Chemometric methods are mathematical and statistical methods that decompose complex multivariate data into simple and easier interpretable structures that can improve the understanding of chemical and biological information of the tested samples (Bro *et al.*, 2002; Geladi, 2003). For instance, principal component analysis (PCA) is considered a powerful and robust tool for obtaining an overview of complex data, such as spectral hypercubes of meat samples, in order to discover groupings and trends in the data. Chemometric methods have been developed to account for the limitations of traditional statistics, which suffer from two drawbacks when related to the multivariate data. First, multivariate data, such as hyperspectral data, suffer from co-linearity problems of adjacent wavelengths. Second, the usual statistical assumption of normal distribution is rarely fulfilled in chemical data series. The combined spectral/spatial analysis for hyperspectral image cubes takes advantage of tools borrowed from spatial image processing, chemometrics, and specifically spectroscopy, resulting in new custom exploitation tools being developed specifically for these applications. In the spectral domain, a hyperspectral image is characterized by its high dimensionality which needs to be reduced to the most meaningful dimension without losing the informative power of the original image. A dimensionality reduction technique is performed to remove redundant information from the hyperspectral image, thus creating simplified data. Therefore, various data analysis methodologies comprised of computer programs and algorithms are required for that task to analyze hyperspectral images and then to generate data that describe material properties of the tested samples. Reducing the dimensionality of hyperspectral data may include, for example, removing redundant information by performing a principal component analysis (PCA) or a partial least squares regression (PLSR). As with conventional spectroscopy, chemometrics can be applied, not only for dimensionality reduction, but also to extract relevant information relating to the spectral content, allowing sample classifications or quantitative determinations. When additional quantitative information is available for calibrating hypercubes, partial least squares and other regression models can be created for predicting future test set hypercubes. Readers who are interested about these issues are advised to refer to the relevant chapters of this book if they need more details about data exploitation using these analytical techniques.

On the other hand, in the spatial domain each hyperspectral image at one wavelength is equivalent to a digital image and standard image analysis can be used for feature extraction. For instance, the analysis may include extracting image-textural features from a hyperspectral image and relating this feature with a real meat trait such as tenderness (Naganathan *et al.*, 2008a, 2008b). Extracting image-textural features could be done by performing a co-occurrence matrix analysis, a wavelet analysis, or an analysis that utilizes Gabor filters. Additionally, the analysis may also include pattern recognition algorithms such as regression, discriminant analysis, neural networks, and fuzzy modeling to relate image features to properties associated with the object. In general, Gat (1999) stated that the typical objectives of exploitation techniques are to:

1. classify and segment the image into areas exhibiting similar spectral properties;
2. search for areas that exhibit a particular spectral signature of interest;
3. locate signatures of unresolved objects (those that are spatially smaller than a single pixel); and
4. determine the composition of a mixture of material within a spatial resolution (an image pixel).

Table 6.1 Comparison between spatial processing and spectral processing of an image.

Spatial processing	Spectral processing
■ Information is embedded in the spatial arrangement of pixels in every spectral band (two-dimensional image)	■ Each pixel has an associated spectrum that can be used to identify different chemical components in the sample
■ Image processing exploits geometrical shape information	■ Processing can be done in a pixel-wise manner at a time
■ High spatial resolution is required to identify objects by shape, color and or texture (by using many pixels on the sample)	■ No need for high spatial resolution (since the spectral information is residing in the pixel itself)
■ Data volume grows with the square of the spatial resolution	■ Data volume increases linearly with the number of spectral bands
■ Limited success in developing fully automated spatial-feature exploitation algorithms in complex applications	■ Fully automated algorithms for spectral-feature exploitation have been successfully developed for various applications

Data exploitation in both spatial and spectral domains needs much contemplation from the researchers by applying the relevant spatial or spectral processing routines to fulfill the main goal of the experiment. The main differences between spatial and spectral analyses of the hyperspectral data are summarized in Table 6.1.

6.4. HYPERSPECTRAL IMAGING FOR MEAT QUALITY EVALUATION

As a hyperspectral imaging system is a very useful tool in several distinct applications, such as classification of food into different groups, either by separating different types of food items or by sorting a single food source into a quality stack, process control to exclude contaminated, sub-standard or by-product food stuffs from the food chain with the minimum of additional cost, and food uniformity monitoring where the quality of the product can be affected by some variable in the process which results in improved food quality (Driver, 2009), several studies have accentuated the possible applications of hyperspectral imaging for quality evaluation of meat and meat products. As a non-destructive and promising inspection method, hyperspectral imaging techniques have been widely studied for determining properties of meat products, but less for meat cuts as

Table 6.2 Various applications of hyperspectral imaging technique for evaluating different quality parameters of meat (beef, pork, fish and chicken).

Product	Imaging mode	λ (nm)	Quality attributes	Author(s) / year
Beef	Reflectance	496–1036	Tenderness	Cluff *et al.*, 2008
	Reflectance	400–1000	Tenderness	Naganathan *et al.*, 2008a
	Reflectance	900–1700	Tenderness	Naganathan *et al.*, 2008b
	Reflectance	400–1100	Tenderness	Peng & Wu, 2008
Pork	Reflectance	430–1000	Quality classification and marbling	Qiao *et al.*, 2007a
	Reflectance	430–980	Quality classification, color, texture, and exudation	Qiao *et al.*, 2007b
	Reflectance	400–1000	Drip loss, pH, and color	Qiao *et al.*, 2007c
Fish	Transflection	400–1000	Ridge detection and automatic fish fillet inspection	Sivertsen *et al.*, 2009
	Interactance	760–1040	High-speed assessment of water and fat contents in fish fillets	ElMasry & Wold, 2008
	Reflectance	892–2495	Determination of fish freshness	Chau *et al.*, 2009
	Transmittance	400–1000	Detection of nematodes and parasites in fish fillets	Wold *et al.*, 2001; Heia *et al.*, 2007
Chicken	Reflectance	400–1000	Faecal contaminants detection	Heitschmidt *et al.*, 2007
	Fluorescence	425–711	Skin tumor detection	Kong *et al.*, 2004
	Reflectance	400–900	Surface contaminants detection	Lawrence *et al.*, 2004
	Reflectance	447–733	Skin tumors detection	Nakariyakul & Casasent, 2004
	Reflectance	430–900	Detection of fecal contaminants	Park *et al.*, 2006a
	Reflectance	400–900	Feces and ingesta detection on the surface of poultry carcasses	Park *et al.*, 2002
	Reflectance	400–900	Contaminants classification	Park *et al.*, 2007
	Reflectance/ Transmittance	400–1000	Bone fragment detection in breast fillets	Yoon *et al.*, 2008

compared to horticultural products. To our knowledge there is no publication available for its application for predicting quality traits in lamb. However, many studies have confirmed the ability of the hyperspectral imaging technique to predict the quality traits such as color, tenderness, marbling, pH, moisture and water holding capacity in beef, pork, poultry, and fish (ElMasry & Wold, 2008; Naganathan *et al.*, 2008a, 2008b; Park *et al.*, 2007; Qiao *et al.*, 2007a, 2007b, 2007c; Sivertsen *et al.*, 2009;). Table 6.2 presents the main papers published in the area of meat quality and composition assessment during last decade (2000–2009).

6.4.1. Beef

The use of hyperspectral imaging for the assessment of beef quality criteria has been studied by many researchers. In particular, the hyperspectral imaging technique has been used to develop models of various accuracies for predicting beef tenderness (Cluff *et al.*, 2008; Naganathan *et al.*, 2008a, 2008b; Peng & Wu, 2008). Like other spectroscopic methods, hyperspectral imaging techniques offer some solutions of chemical assessment problems in terms of accuracy and reproducibility. Also, the technique offers outstanding solutions in some cases where the samples are not homogenous, which is of great importance with respect to the spatial distribution of chemical constituents in every spot in the sample. Moreover, the hyperspectral data residing in each image contains abundant physical and chemical information about the sample being analyzed. If this information is properly analyzed, it can be used to characterize the sample itself. However, because beef is a variable product with respect to muscle fiber arrangement, pH, and connective tissue content, it is extremely difficult to standardize a way of interpreting the spectral data (Swatland, 1989). On the other hand, to replace the ordinary and time-consuming chemical method with a more precise and faster hyperspectral imaging technique, it is important to relate spectral data with those determined by the reference method through the calibration step. An estimation of the uncertainty of the chemical reference methods can be of a great value in order to judge whether the hyperspectral imaging method is suited as a practical replacement for the chemical method. Therefore, multivariate analyses could be a useful tool for qualitative and quantitative assays based on the extracted hyperspectral data to allow classification without using laborious chemical determination.

In beef, tenderness is the most relevant and most widely discussed quality characteristic. This is because tenderness is the most important factor in the consumer perception of beef palatability or quality (Savell *et al.*, 1989). Tenderness is a property of a cooked product and predicting this property from a fresh steak poses considerable challenges. Direct evaluation of tenderness is absent because there is currently no accepted method available for predicting tenderness on-line. One of the most common ways for predicting tenderness non-destructively is through using a video imaging technique as an objective technique instead of the non-destructive methods such as Warner–Bratzler shearing force (WBSF) or slice shearing force (SSF) methods. Research on computer vision-based beef quality evaluation has shown that texture features computed from muscle images are useful indicators of beef tenderness (Du *et al.*, 2008; Jackman *et al.*, 2009a). The addition of image texture features to color and marbling parameters significantly improves the accuracy of tenderness prediction (Jackman *et al.*, 2010). While a rigorous definition of image texture is not yet available, it generally refers to image characteristics like coarseness, graininess, uniformity, and consistency. Textural features represent the spatial distribution of tonal variations in an image at any wavelength in the visible and/or infrared region of the spectrum (Kavdır & Guyer, 2004). A number of methods have been suggested in the literature for texture analysis, but the graylevel of co-occurrence matrix (GLCM) method is the most reported one. Among the available techniques, the wavelet transform technique is a key approach to decompose beef muscle images into textural primitives or elements of different sizes (Jackman *et al.*, 2008, 2009b, 2009c, 2009d). Image texture features computed from the textural primitives have been used to classify beef samples into different tenderness categories. While image texture features alone may not be sufficient to classify beef into multiple levels of tenderness, they certainly appear to be useful contributors to beef tenderness prediction and deserve inclusion in the pool of tenderness indicators (Li *et al.*, 2001).

On the other hand, several studies have shown that near-infrared reflectance spectroscopy can be used to predict beef tenderness with various success (Andrés *et al.*, 2008; Leroy *et al.*, 2003; Park *et al.*, 1998). Some attempts have been made by Shackelford *et al.* (2005) to develop a high-speed spectroscopic system for on-line determination of beef tenderness. In the research by Shackelford *et al.* (2005), spectroscopic measurement was performed on-line at two large-scale commercial fed-beef processing facilities, with data extracted on the beef grading bloom chain approximately 2 min after the carcasses were ribbed. The field of view was restricted to 50 mm in diameter to sample a large area of the cross-section of the *longissimus* muscle. To build prediction models, the data extracted from spectroscopy were calibrated against a destructive measurement of tenderness using the slice shear force (SSF) method. A regression model was calibrated using 146 carcasses and tested against an additional 146 carcasses. Their experiment indicates that US "Select" carcasses can be non-invasively classified for *longissimus* tenderness using visible and near-infrared spectroscopy. Also, Rust *et al.* (2007) developed an on-line near infrared (NIR) spectral reflectance system to predict 14-day aged cooked beef tenderness in a real-world processing plant environment. Near-infrared (NIR) analyses were performed in reflectance mode with a VIS/NIR spectrophotometer. The spectrometer used in this study was capable of collecting light in the visible and NIR regions (400–2500 nm). A fiber-optic contact probe was used to transmit light reflected from the beef surface to three internal detectors. Light was supplied by a 20-W halogen light source and a diffuse reflection probe with

35° geometry with an effective measuring area of 1 mm^2. The halogen lamp was powered by a feedback controller to stabilize illumination level. The detectors consisted of a silicon photodiode array, a thermoelectrically (TE) cooled indium gallium arsenide (InGaAs) detector, and a TE-cooled extended InGaAs detector to measure the 350–1000 nm, 1001–1670 nm, and 1671–2500 nm wavelength domains, respectively. The results indicated that the tested spectrometer appears to perform with a similar level of accuracy as the system described by Shackelford *et al.* (2005), but it is unclear if it would perform as well with a higher percentage of tough carcasses.

The imaging system alone is able to capture images in three distinct wavelengths or bands (RGB: red, green, blue). These images usually have a high spatial resolution, but have a limited spectral resolution. Also, spectroscopy alone provides high spectral resolution information over both VIS and NIR spectral regions but with virtually no spatial information. Therefore, the most crucial stage for tenderness prediction is to integrate the powers of both imaging and spectroscopy techniques in one approach utilizing hyperspectral imaging as performed by Naganathan *et al.* (2008a) and Grimes *et al.* (2007, 2008), who developed a pushbroom hyperspectral imaging system in the wavelength range of 400–1000 nm with a diffuse-flood lighting system (Figure 6.3). Hyperspectral images of beef-ribeye steaks

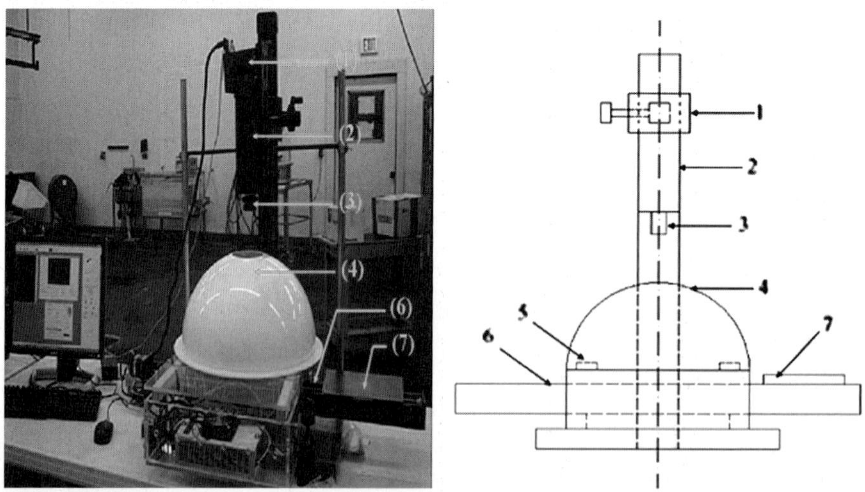

FIGURE 6.3 *Visible and NIR hyperspectral imaging system for beef tenderness prediction (adapted with permission from Naganathan* et al., *2008a (© 2008 Elsevier) and from Grimes* et al., *2008). (1) CCD camera; (2) spectrograph; (3) lens; (4) diffuse lighting chamber; (5) tungsten halogen lamps; (6) linear slide; (7) sample plate. (Full color version available on http://www.elsevierdirect.com/companions/9780123747532/)*

(*longissimus dorsi*) between the 12th and 13th ribs (n = 111) at 14 days post mortem were acquired. After imaging, steaks were cooked and slice shear force (SSF) values were collected as a tenderness reference. All images were corrected for reflectance. After reflectance calibration, a region of interest (ROI) of 200 × 600 pixels at the center of each steak was selected and principal component analysis (PCA) was carried out on the ROI images to reduce the dimension along the spectral axis. The first five principal components explained over 90% of the variance of all spectral bands in the image. The principal component analysis was conducted for each hyperspectral image steak by steak instead of considering the overall hyperspectral images for all steaks. This method is considered as a non-traditional PCA approach where each hyperspectral image is considered as a separate data set and PCA is conducted for each image separately to retain spatial variability of samples. The loading vectors or Eigenvectors are different among images. Actually, this approach explains "within sample" variation. Graylevel textural co-occurrence matrix (GLCM) analysis was conducted to extract second-order statistical textural features from the principal component images. The second-order textural feature extraction routine produced textural tonal images, mean, variance, homogeneity, contrast, dissimilarity, entropy, second moment, and correlation. The average value of each textural band was then calculated and used in developing a canonical discriminant model to classify steaks into three tenderness categories, namely tender (SSF ≤205.80 N), intermediate (205.80N <SSF <254.80 N), and tough (SSF ≥254.80 N). Figure 6.4 depicts the distribution of the tested beef samples in the canonical discriminant model. With a leave-one-out cross-validation procedure, the model predicted the three tenderness categories with an accuracy of 96.4%. The result indicated that hyperspectral imaging was able to identify all tough samples and has considerable promise for predicting beef tenderness. However, before suggesting this method for industrial implementation, the model must be validated with new and larger sets of samples. Also, this method has a big drawback because the selected ROIs were always in the middle of each steak. In addition, using this method is very time-consuming since it depends on calculating PCs for each steak and then calculating a huge number of textural features from certain PC images selected by another algorithm. Any instrumentation that is meant for beef tenderness evaluation should be fast enough to keep up with a speed at which a beef carcass moves in a production line and should have the ability to be implemented on-line. The developed hyperspectral imaging system is an off-line system which needs 10 seconds to acquire an image of a beef sample, and 10 minutes to assign a tenderness category by the previously mentioned algorithm. This time could be reduced significantly by reducing the high

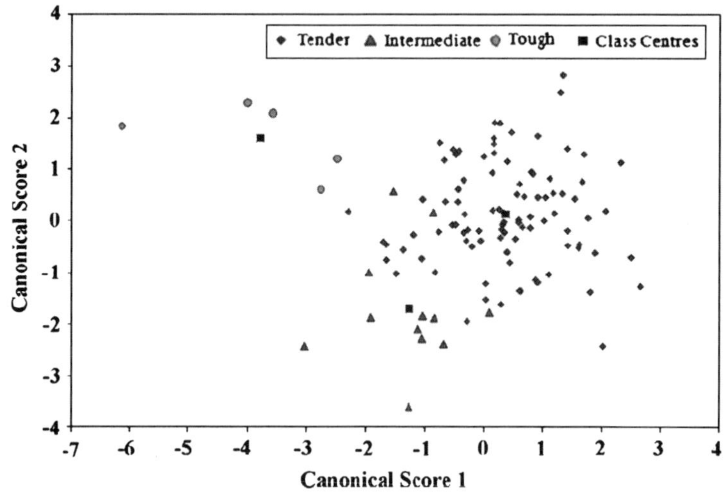

FIGURE 6.4 *Distribution of samples in the canonical space (reproduced from Naganathan* et al., *2008a. © 2008 with permission from Elsevier). (Full color version available on http://www.elsevierdirect.com/companions/9780123747532/)*

dimensionality of the hyperspectral images to form a multispectral imaging system consisting of a few important spectral wavebands for definite applications. The hyperspectral analyses should be conducted once in off-line mode, and then the outcomes of these analyses are applicable for on-line implementations. Implementing image processing routines at these selected bands would decrease the processing time significantly. Another use of these selected wavebands is to reduce image acquisition time by acquiring images at those selected wavelengths and such an approach is called multispectral imaging. The resulting multispectral imaging system could be established on-line similar to current video image analysis systems used for yield grade predictions in the beef industry.

In another attempt to enhance the performance of a hyperspectral imaging system for classifying beef steaks based on their tenderness, Naganathan *et al.* (2008b) repeated the same protocol explained in Naganathan *et al.* (2008a) but used a hyperspectral imaging system in the spectral range of 900–1700 nm to forecast 14-day aged, cooked beef tenderness from the hyperspectral images of fresh ribeye steaks (n = 319) acquired at 3–5 days post mortem. They used PLSR as a dimensionality reduction technique instead of PCA by considering slice shear force (SSF) as a dependent variable, and then PLSR loading vectors were obtained. A graylevel co-occurrence matrix (GLCM) with two graylevel quantization levels (64 and 256) was used to extract image-textural features from the PLSR bands. In total, 48 features

(6 PLSR bands × 8 textural features per PLSR band) were extracted from each beef steak image. These features were then used in a canonical discriminant model to predict three beef tenderness categories. The model with a quantization level of 256 performed better than the one with a quantization level of 64. This model correctly classified 242 out of 314 samples with an overall accuracy of 77.0%. Also, some key wavelengths (1074, 1091, 1142, 1176, 1219, 1365, 1395, 1408, and 1462 nm) corresponding to fat, protein, and water absorptions were identified. Further work is needed to relate the spectral response at these key wavelengths with the biochemical properties of beef muscle. The results show that NIR hyperspectral imaging holds promise as an instrument for forecasting beef tenderness.

In general, the automatic system of high performance in tenderness measurement should be more expeditious, reliable, and flexible in addition to its ability for mapping tenderness values in all points of the sample. The beef industry is interested in an instrument that can assess the average tenderness of the whole steak. However, mapping of tenderness is very important in the case of beef steaks having different tenderness values due to their own structures. Moreover, mapping tenderness is essential to identify different steaks of various tenderness values or in the case of beef cuts containing several muscles of different tenderness. This identification is really important for trimming the beef cuts to a certain tenderness level based on consumer requirements. Categorizing meat cuts by tenderness would enhance economic opportunities for cattle producers and processors by improving assessment of beef product quality to meet consumer expectations. Also, increasing consistency in tenderness a major challenge faced by the beef industry, will lead to improved consumer satisfaction and hence promote frequent purchases. Besides, labeling accurate quality factors on the packaging of retail cuts would add another value to the products and benefit consumers. Yet some consumers are willing to pay higher prices for beef protected by quality labels guaranteeing the homogeneity of their products (Dransfield, 1994; Leroy *et al.*, 2003). In fact, this kind of hyperspectral imaging system offering this vital advantage in beef products is still missing. Research is still needed to develop such a system by applying different algorithms and spectral image processing routines to generate a clear overview of the real tenderness of any muscles and/or beef cuts.

Diffuse reflectance implemented in most hyperspectral imaging systems is not the only way to characterize the meat properties. Beef absorption coefficients are related to the sample chemical compositions such as the concentration of myoglobin and its derivatives; while scattering coefficients depend on meat structural properties such as sarcomere length and collagen concentration. Structural properties are also key factors in determining beef tenderness. The reflectance measured at the sample surface is the result of both scattering and absorption processes involved in light–muscle interactions. The measured diffuse reflectance reflects those photons that have survived absorption and have been scattered diffusely in meat and have eventually escaped from the meat surface. Hence, the conventional absorbance is the combined result of the absorbing and scattering effects and is different from the derived absorption coefficient, which is independent of scattering. The absorbance calculated from reflectance depends on the measurement position; while the absorption coefficients represent the samples' absorbing characteristics and are solely determined by the sample itself (i.e. its chemical compositions). Absorbance cannot provide an accurate absorption spectrum because the scattering effect is not excluded. Similarly, the scattering coefficients are independent of sample chemical compositions and are solely determined by sample structure properties (Xia *et al.*, 2007). Based on that, a hyperspectral imaging system can be used to collect the scattering profile with very high spatial and spectral resolutions with short acquisition times, the scattered light can thus be captured with high spatial resolution at many wavelengths simultaneously in the hyperspectral image. Cluff *et al.* (2008) developed a non-destructive method using a hyperspectral imaging system (496–1036 nm) for predicting cooked beef (44 strip loin and 17 tenderloin cuts) tenderness by optical scattering of light on fresh beef muscle tissue. The hyperspectral image consisted of 120 bands with spectral intervals of 4.54 nm. In total, 40 hyperspectral images representing scattering profiles at 40 different locations in each steak were acquired, and then these images were averaged to produce a representative hyperspectral image with a high signal-to-noise ratio. Figure 6.5 presents the averaged hyperspectral image of the optical scattering within the beef steak. The optical scattering profiles were derived from the hyperspectral images and fitted to the modified Lorentzian function (Cluff *et al.*, 2008). As a photon of light enters the steak and hits one of the scattering centers, the light scatters and comes back out of the steak toward the camera. Tissue scattering is related to the morphology and refractive index distributions of the tissue composition. It is well known that connective tissue and myofibrillar proteins are the most important factors influencing meat tenderness. These tissue structures are also the primary component of meat texture associated with the light scattering properties of meat. Therefore, light scattering could potentially be used as an indicator of beef tenderness and the changes in scattering profiles are believed to represent the changes in tenderness. Figure 6.6 illustrates the change in scattering profiles between the strip loin and tenderloin portions of the porterhouse steaks. Parameters, such as the peak height, full scattering width at half maximum (FWHM), and the slope around the FWHM were

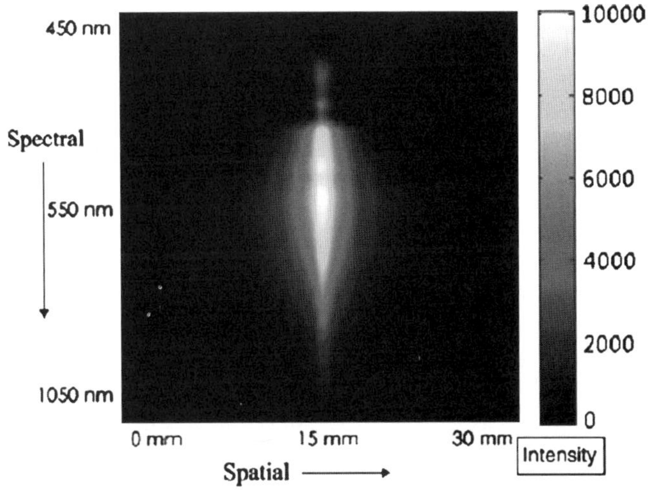

FIGURE 6.5 *Hyperspectral image of optical scattering in beef steak. Y-axis represents spectral information with intervals of 4.54 nm and X-axis represents spatial distance with a spatial resolution of 0.2 mm. The optical scattering can be seen to vary with wavelength (reproduced with permission from Cluff* et al., *2008. © with permission from Springer Science+Business Media 2008). (Full color version available on http://www.elsevierdirect.com/companions/9780123747532/)*

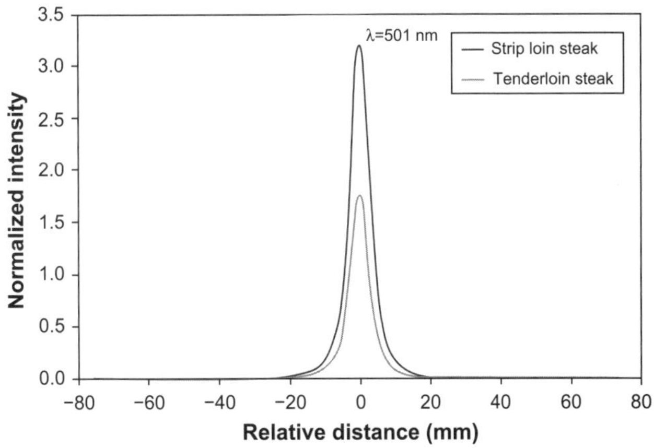

FIGURE 6.6 *Difference in the averaged optical scattering profiles of the porterhouse strip steak (WBSF = 28.9 N) and tenderloin (WBS = 24.7 N) (reproduced with permission from Cluff* et al., *2008. © with permission from Springer Science+Business Media 2008). (Full color version available on http://www.elsevierdirect.com/companions/9780123747532/)*

determined at each wavelength. The stepwise regression was able to identify seven parameters and wavelengths from the scattering profiles that could be used to predict the WBSF scores. The results indicated that the model was able to predict WBSF scores with an $R = 0.67$, indicating that the optical scattering implemented with hyperspectral imaging has not proved a remarkable success for predicting the current status of tenderness in beef steak. If the predicted WBSF values were used to classify the samples into categories "tender" and "intermediate" (there were no "tough" samples) as described by Naganathan *et al.* (2008a), the accuracy would be 98.4%.

6.4.2. Pork

The desirable high quality of pork is usually associated with the factors that influence the processing of lean and fat tissues and the consumer acceptability and palatability of both fresh and processed products. The preferred method of assessing pork quality is via the direct evaluation of the exposed loin eye at the 10th/11th rib interface of the *longissimus* muscle. Quality of fresh pork varies greatly and is traditionally classified into different categories based on color, texture (firmness) and exudation (drip loss) (Qiao *et al.*, 2007a; Warner *et al.*, 1997). Good quality pork meat is typically red, firm, and nonexudative (RFN). Pork meats that are classified as RFN have desirable color, firmness, normal water holding capacity, minimal drip loss, and moderate decline rate of pH. Various combinations of color, texture, and driploss determine other quality grades of pork meat, such as RSE (red, soft, and exudative), PFN (pale, firm, and non-exudative) and PSE (pale, soft, and exudative). Based on spectroscopic studies, Xing *et al.* (2007) showed that it was possible to separate pale meat from red meat with an accuracy of about 85%, and to distinguish PFN meat from PSE meat using visible spectroscopy (400–700 nm). However, the visible spectral information was not sufficient to separate all the four quality groups proposed (RFN, RSE, PFN, PSE). A hyperspectral imaging system is able to fill this gap with more promising results in pork quality assessment. Researchers have tested hyperspectral imaging for indirect determination of pork quality parameters such as drip loss, pH, marbling, water holding capacity, texture and exudation (Qiao *et al.*, 2007a, 2007b, 2007c). Results for pork meat classification based on color, texture, and exudation with artificial neural network predictions resulted in 87.5% correct classification (Qiao *et al.*, 2007b). The same group of researchers tested different methods for pork classification including principal component analysis, a feed-forward neural network, and PLSR.

In the first attempt for determining pork quality parameters, Qiao *et al.* (2007a) used a hyperspectral imaging system in the spectral range of

400–1000 nm to extract spectral characteristics of 40 pork samples of different quality grades (RFN, PSE, PFN, and RSE) for classification purposes and marbling assessment. The hyperspectral imaging system was a push-broom type which utilized a complementary metal-oxide-semiconductor (CMOS) camera, a spectrograph, a fiber-optic line light and a convey or belt controlled by a computer. The appropriate convey or speed was selected to avoid distortion on image size and spatial resolution and fit the predetermined camera exposure time. After image acquisition and routine calibration for each image, the main spectrum was extracted from a ROI of 10 000 pixels to represent each sample. Each spectrum was then smoothed by a 10-points mean filter, and their second derivatives were calculated to correct multiplicative scatter, avoid the overlapping peaks, and also correct the baseline. The extracted spectral data were then liable to dimensionality reduction by using principal component analysis (PCA). The authors (Qiao *et al.*, 2007a) then used a different number of principal components (PCs) to build their prediction models for sample classification either by cluster analysis or an artificial neural network (ANN). In fact, there is no noteworthy difference between the methodology applied in this study and the spectroscopic technique because this study does not consider the full strength of hyperspectral imaging in terms of spatial and spectral dimensions. The only advantage was the flexibility of using different ROIs from representative pixels in the image and using bigger ROIs (10 000 pixels) compared with spectroscopy which only considers a small portion (point) of the sample where the sensor is located. However, this study could be considered as a preliminary investigation of using hyperspectral imaging in pork quality assessment.

Figure 6.7 shows the difference in spectral characteristics of the tested four quality levels (Qiao *et al.*, 2007a). The PFN and PSE showed a higher reflectance than that of RFN and RSE. The differences in spectral data suggested a possibility of classifying the quality levels of pork samples using their spectral features. Different numbers of principal components (PCs) of 5, 10, and 20 with explained variance of 62%, 76%, and 90% were used as the input for cluster analysis and artificial neural network (ANN) modelling. By using the above-mentioned protocol, the obtained overall accuracy for pork sample classification using cluster analysis was reported to be 75–80%. By using an ANN model, the corrected classification was 69% and 85% when using 5 and 10 PCs, respectively.

In their later study, Qiao *et al.* (2007b) increased the number of pork samples to 80 steaks and then extracted average spectral features from the whole pork steak instead of a small ROI. Also, they reduced the spectral range to 430–980 nm instead of 400–1000 nm by removing high noisy spectra. In addition they used PCA and stepwise regression to pick up the

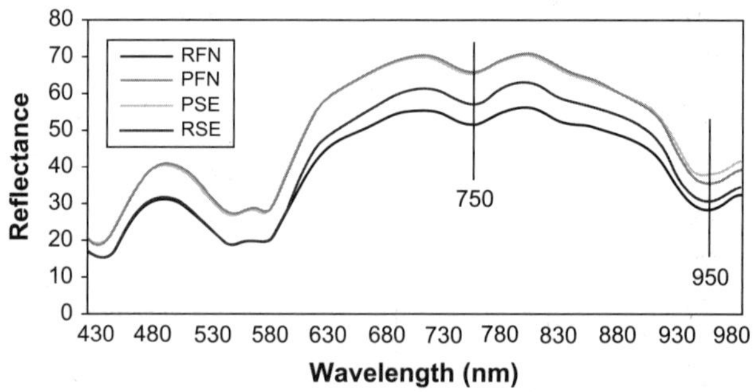

FIGURE 6.7 *Spectral characteristics of different quality levels of pork samples with water absorbing bands at 750 and 950 nm indicated (reproduced from Qiao* et al., *2007a. © 2007 with permission from Elsevier). (Full color version available on http://www.elsevierdirect.com/companions/9780123747532/)*

most important wavebands instead of using the whole spectral range. PCA was conducted for the mean spectra, and the important wavelengths were selected based on the greater weight of the first three PCs.

Moreover, as the stepwise regression is always used to develop a subset of data that is useful to predict the target variable, and to eliminate those data that do not provide additional prediction in the regression equation, the stepwise was performed on the average spectra with their corresponding pork quality class indicated by a qualified expert in their study.

As they found in their previous study (Qiao *et al.*, 2007a), Qiao *et al.* (2007b) emphasized that there were spectral differences among the four classes, indicating that there were some differences in their physicochemical attributes. The differences in spectral data suggested a possibility of classifying the quality classes of pork samples using their spectral features. The important wavelengths at which the main difference between the pork classes occurred were selected by PCA and stepwise regression as indicated in Table 6.3. Classification results using these selected wavelengths showed

Table 6.3 Selected wavelengths for pork classification by using principal component analysis (PCA) and stepwise regression

Methods	Selected wavelengths (nm)	No.
PCA for the main spectra	481, 530, 567, 701, 833, 859, 881, 918, 980	9
Stepwise for the main spectra	615, 627, 934, 961	4
PCA for the first derivative spectra	496, 583, 622, 657, 690, 737, 783, 833, 851, 927, 961	11
Stepwise for the first derivative spectra	430, 458, 527, 560, 571, 600, 690, 707, 737, 851, 872, 896, 975	13

Source: *Qiao* et al., *2007b*

a performance of 67.5 to 87.5% with the best result of 87.5% using the wavelengths selected by PCA experienced in the first derivative spectra. As seen from the results, the classification accuracy was enhanced compared with the first study.

The most important step in non-destructive assessment of pork quality is to use hyperspectral imaging for predicting pork quality attributes like color, firmness, water holding capacity, drip loss, and pH. In this trend, Qiao *et al.* (2007c) continued their studies using the same hyperspectral imaging system for predicting drip-loss, pH, and color of pork meat. Simple correlation analysis was conducted between the spectral response at each wavelength from 430 to 980 nm and corresponding drip loss, pH, and color, respectively. The wavelengths at which the highest correlation coefficient (r) was found were selected. Simple correlation analyses showed that high correlation coefficients (r) were found at 459, 618, 655, 685, 755, and 953 nm for drip loss, 494, 571, 637, 669, 703, and 978 nm for pH, and 434, 494, 561, 637, 669, and 703 for color. The results using only spectral data at these wavelengths instead of the whole spectral range showed that the drip loss, pH, and color of pork meat could be predicted with correlation coefficients of 0.77, 0.55, and 0.86, respectively. Such findings represent an obvious advantage for promising non-contact pork quality determination as pork traits and the softness of the lean meat are more difficult to appreciate from a distance, particularly in the case of the RSE class.

6.4.3. Fish

Quality assessment and documentation of the chemical composition of fish and seafood products is extremely important for both producers and consumers. The need for implementing reliable, accurate, expeditious, and non-destructive on-line techniques for quality assessment and monitoring of fish and seafood products is essential in todays growing markets. The term "quality" in the case of fish refers to the aesthetic appearance and freshness or degree of spoilage which the fish has undergone (Huidobro *et al.*, 2001). It may also involve safety aspects such as being free from harmful bacteria, parasites or chemicals. It is important to remember that quality implies different things to different people and is a term that must be defined in association with an individual product type. On-line quality monitoring systems may provide better means for sorting fish as well as provide better documentation of product quality for price differentiation and promotion. Applications of hyperspectral imaging for the quality assessment of fish and seafood products are mainly focused on qualitative and quantitative aspects in terms of overall fish freshness and chemical composition determination.

The method enables both spatial and spectral identification of irregularities in the muscle, and makes it possible to extract information regarding the chemical composition of fish or areas in the image, such as blood, water or fat. Determination of chemical composition of fish or fishery products have been reported by various researchers using spectroscopic techniques (Herrero, 2008; Khodabux *et al.*, 2007; Nortvedt *et al.*, 1998; Wold & Isaksson, 1997), however how these chemical compounds of different concentration gradients are visually distributed has not been reported in many research articles. Compared with NIR spectroscopy, applications of hyperspectral imaging in the fish industry has a limited number of publications in this field, although there is a prevailing tendency of promising success. That is probably because hyperspectral imaging is a relatively new technique, and its full potential has yet to be exploited.

If applied in an on-line inspection system, hyperspectral imaging would offer big advantages in the fish industry in quality assurance and quality control programs. This requires fast algorithms that can handle large amount of data and make reliable decisions in fractions of a second (Sivertsen *et al.*, 2009). On the one hand, today's chemical methods for chemical determination of quality attributes are highly destructive and time-consuming and require use of hazardous chemicals that may be harmful to analysts and the environment. Also, screening of every single fish requires an on-line method for chemical composition determination to fulfill speed requirements for mass production. On the other hand, hyperspectral imaging could enable optimized processing of the raw fish, correct pricing and labeling, and gives the opportunity to sort fish with different chemical composition content according to market requirements and product specifications.

6.4.3.1. Quantitative measurement of fish quality parameters

The first application of spectral imaging technique in a high-speed implementation was reported by Wold *et al.* (2006) for inspecting dried salted coalfish (bacalao) using a non-contact transflectance near infrared spectral imaging system in the visible and near infrared regions. The same system was later used by ElMasry & Wold (2008) in interactance mode to determine water and fat content distribution in the fillets of six fish species (Atlantic halibut, catfish, cod, mackerel, herring, and saithe) in real time using PLSR as the calibration method. The spectral imaging system used in these studies, called the Qmonitor scanner, is designed and manufactured by Qvision Inc. (AS, Oslo, Norway) to work in a Matlab graphical user interface. This industrial on-line scanner records spectral images in the VIS and NIR range of 460–1040 nm with a spectral resolution of approximately 20 nm at the speed of 10 000 spectra per second. The pixel absorption spectra in the

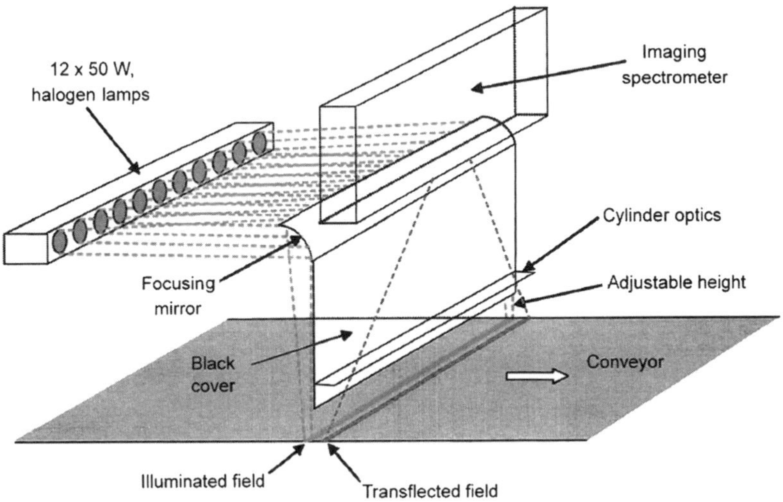

FIGURE 6.8 *Layout of the main configuration of the NIR spectral imaging system (reproduced with permission from ElMasry & Wold, 2008. © 2008 with permission from American Chemical Society). (Full color version available on http://www.elsevierdirect.com/companions/9780123747532/)*

NIR range (760–1040 nm) consists of 15 wavelengths. Figure 6.8 shows how the system is designed to measure in the interactance mode. The light is focused along a line across the conveyor belt. A metal plate is used to shield the detector from unwanted reflected light from the fish surface. In that way, it is assured that the detected light has been transmitted into the fish and then back scattered to the surface, i.e. only the light that has traversed the interior of the fish is analyzed. Interactance thus probes deeper into the fillet compared to reflectance and suppresses surface effects. Interactance has a practical advantage over transmission in that both illumination and detection are on the same side of the sample. This reduces the influence of thickness considerably and provides a fairly clear measurement situation. Fish was put on the conveyor belt and moved at a speed of approximately 0.1 ms^{-1}, which resulted in a spatial resolution of 0.3 mm along the conveyor belt. Image size varied according to the length of the fish. The fish was scanned line-by-line to collect the entire spectral image.

The ultimate goal of this study was to build chemical images to demonstrate how fat and water contents were distributed with different concentration gradients in the fish fillet by using the flow chart shown in Figure 6.9. Spectral data were extracted from different spots of all species of fish fillets, and then a PLSR model was applied to relate these data to the real chemical measurements in the same spots. The PLSR models were used to

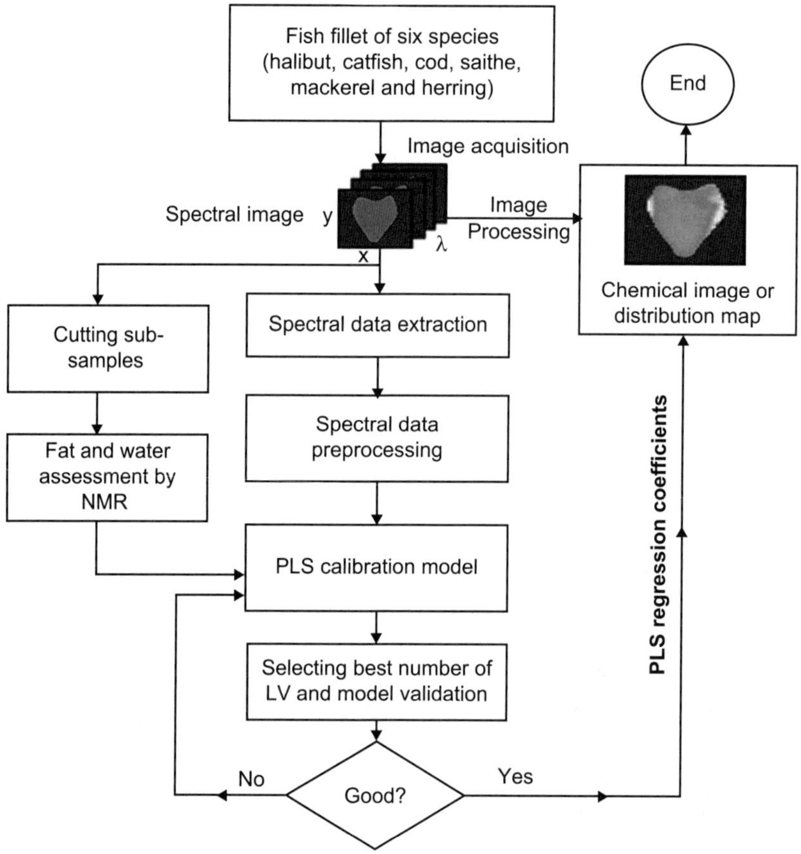

FIGURE 6.9 *Key steps for building chemical images (distribution maps) (reproduced with permission from ElMasry & Wold, 2008. © 2008 with permission from American Chemical Society). (Full color version available on http://www.elsevierdirect.com/companions/9780123747532/)*

predict water and fat concentrations in each pixel of the spectral image. This was done by calculating the dot product between each pixel spectrum and the coefficient vector obtained from the PLSR model. The resulting chemical images were displayed in colors, where the colors represented different concentrations. Figure 6.10 shows some chemical images of fat and water content distribution of the tested fish species. The changes in fat and water contents were assigned with a linear color scale. Although it is impossible to differentiate the fat and water distribution in the fillet by the naked eye, the spatial distribution of water and fat could be visualized by the NIR interactance imaging system. It is also clear how the concentrations of fat and water vary drastically between different parts of the same fillet. The fish

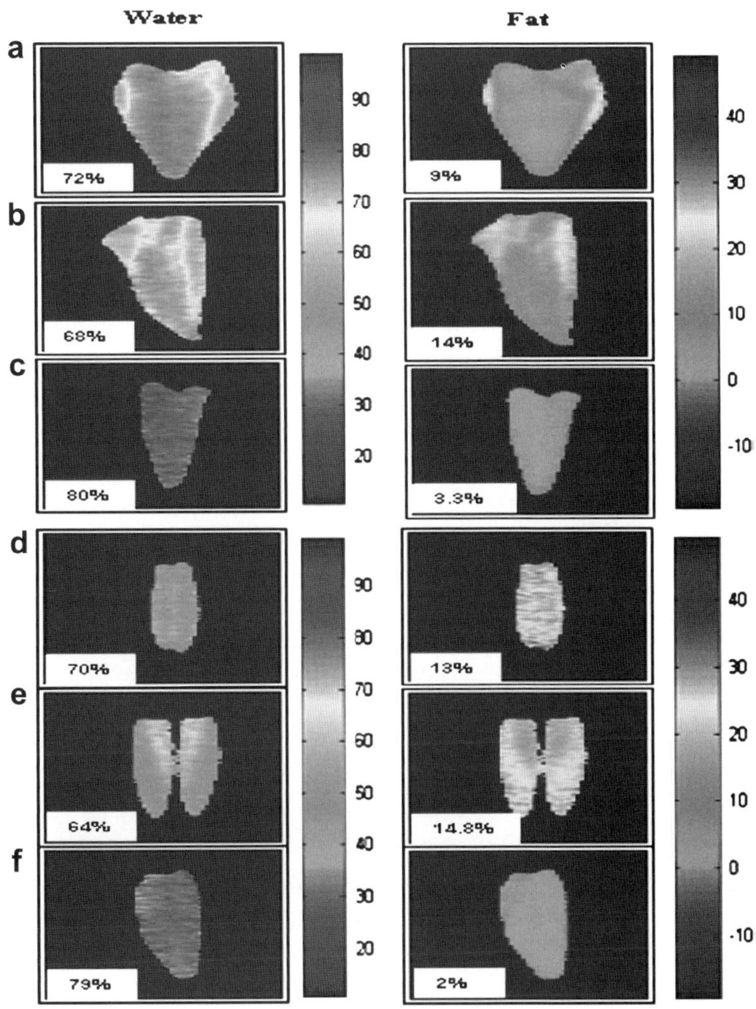

FIGURE 6.10 *Water and fat distribution maps in fillets, the values in the left bottom corner of the figure represent the average concentrations of water and fat in the whole fillet: (a) Atlantic halibut; (b) catfish; (c) cod; (d) herring; (e) mackerel; and (f) saithe (reproduced with permission from ElMasry & Wold, 2008. © 2008 with permission from American Chemical Society). (Full color version available on http://www.elsevierdirect.com/companions/9780123747532/)*

industry can benefit from the possibility of performing this non-destructive technique at an early stage of processing without additional laborious chemical analysis. This enables early sorting of products and thereby improves quality management. Also, fish manufacturers who wish to cut away fillets with certain threshold concentrations could perform this task easily with limited modification in their production lines. The technique can be implemented as a key component of computer-integrated manufacturing and provide smart opportunities for various applications, not only in the fish fillet industry but also in various food quality monitoring processes. The wide application of this automated system would seem to offer a number of potential advantages, including reduced labor costs, elimination of human error and/or subjective judgment, and the creation of product data in real time for documentation, traceability, and labeling (ElMasry & Wold, 2008).

6.4.3.2. Qualitative measurement of fish

In a qualitative study carried out by Chau *et al.* (2009) for determining the freshness of cod fish (*Gadus morhua*) as a main element of fish quality, a hyperspectral imaging system with a wavelength range of 892–2495 nm range was used. It was believed that a suitable system for the objective analysis of fish freshness would improve the ability to market fish on value and to monitor and manage the freshness of fish in the supply chain to reduce waste. Instead of evaluating the fish freshness at specific regions on the fish surface as in NIR spectroscopy, hyperspectral imaging evaluated the freshness in all spots (pixels) of the fish. The system included a shortwave near infrared (SWIR) spectral camera (Specim Ltd, Oulu, Finland) containing a cooled 320×256 pixel HgCdTe detector and a spectrograph. This was mounted above a motorized translation stage, operating in a "pushbroom" configuration as shown in Figure 6.11. Whole fish were presented on a metal tray and fish fillets were presented on a black painted tray. Before acquiring images, whole fish and fillets were gently patted with paper towels before scanning to remove excess water, and any adhering ice was also removed. Samples were scanned immediately after presentation to minimize any heating of them by the lamps.

The results showed that there was a difference among the mean spectra of a whole fish, the fillet flesh and belly flap regions of a fillet at several locations of the spectra as indicated in their corresponding spectral curves. These dissimilarities between these spectra are attributed basically to the significant differences between chemical compositions of these objects. Based on this spectral difference, the fillets part could be identified as the red region in Figure 6.12 after excluding the regions in which the spectral data were saturated, typically corresponding to specular reflections as indicated in green in Figure 6.12 and the regions of belly flap indicated in blue in Figure 6.12.

Mean spectra for the whole cod fish showed evidence of an increase in reflectance (decrease in log [1/R]) with storage time. The best correlation ($R^2 = 0.59$) of the average value for whole cod at a single waveband against storage days on ice was at 1164 nm. NIR images of the whole cod fish at this

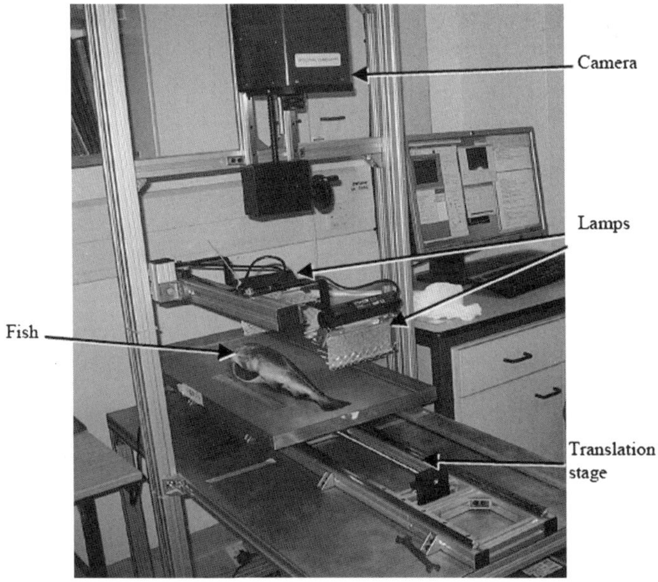

FIGURE 6.11 *Hyperspectral imaging system in the shortwave infrared (SWIR) region for qualitative freshness evaluation of whole fish and fillets (Chau* et al., *2009, reproduced with permission from the Seafish Industry Authority). (Full color version available on http://www.elsevierdirect.com/companions/9780123747532/)*

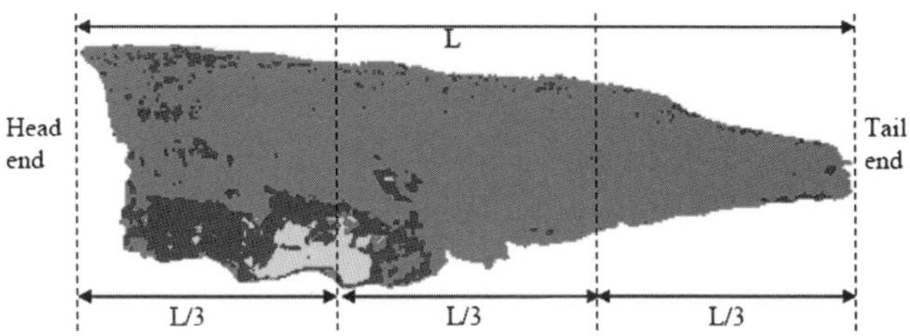

FIGURE 6.12 *Identifying fillet flesh region of interest (red) after excluding the oversaturated specular area (green) and belly area (blue) based on their spectral data (Chau* et al., *2009, reproduced with permission from the Seafish Industry Authority). (Full color version available on http://www.elsevierdirect.com/companions/9780123747532/)*

particular wavelength are shown in Figure 6.13 with log (1/R) values shown on a false color scale. The average absorbance of the whole cod with days on ice ranged from 0.7 to 0.95, which corresponds to the region between cyan and green on the chosen color scale for Figure 6.8. Although many variations in reflectance can be seen across the body of each fish, the general trend of an increase in overall reflectance with storage time is clear, signified by a shift

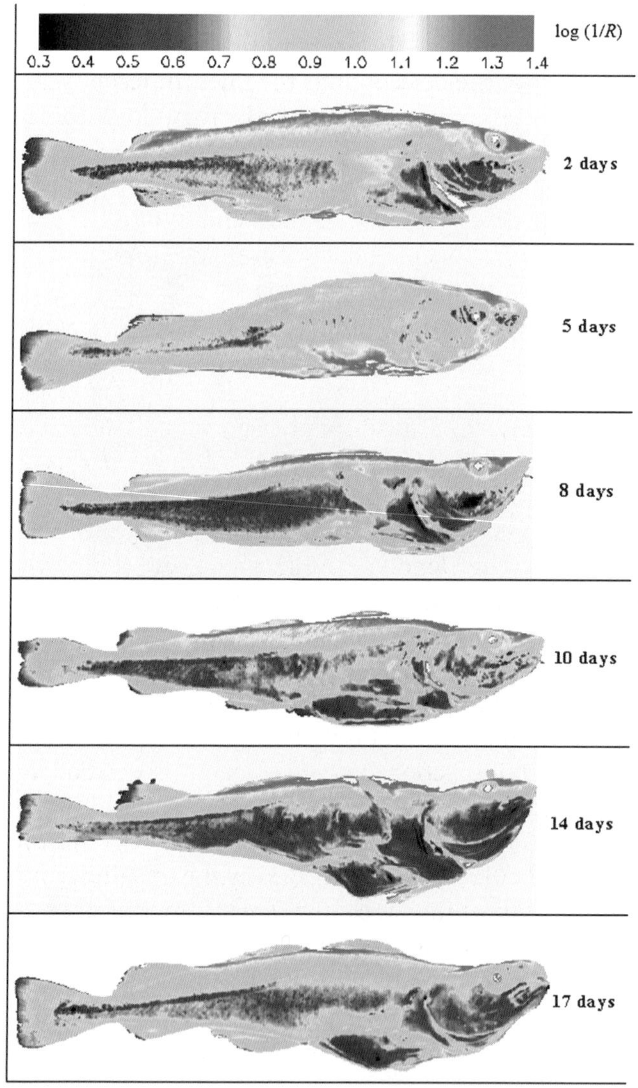

FIGURE 6.13 *False color images of the whole cod fish at 1164 nm (Chau* et al., *2009, reproduced with permission from the Seafish Industry Authority). (Full color version available on http://www.elsevierdirect.com/companions/9780123747532/)*

towards the blue end of the false color scale. This of course was a clear indication of fish freshness status.

In fish fillets in general and in whitefish fillet in particular, quality inspection is currently carried out manually on candling tables. The candling table consists of a bright diffuse light source that is directed to shine through the translucent flesh of the fillet (Valdimarsson *et al.*, 1985). This is a very labor-intensive method, where the operator first has to inspect the fillet and then manually remove the defects from the fillet. In recent years, there has been an increased focus on research and development for automatic fish fillet inspection systems. The reasons for this are the need for reliable and objective documentation of quality and more cost-effective production. The first difficulty facing developing this recommended automatic system is the difficulty of identifying different regions of similar colors on the fillet surface which makes distinguishing these regions by the naked eye or even by machine vision systems a very tough task. Another difficulty is that some defects like bruising or presence of nematodes are invisible to the inspectors. Common for all fillet inspection systems, the main complements are an imaging system, a rejection mechanism, and a computer. The computer automatically analyzes the image of the fillet by a set of different processing steps. One of the most important steps in hyperspectral image processing algorithms is segmentation, where different regions in the image are labeled. For fish fillets, this can be done to identify which part of the fillet belongs to the loin, belly flap, center cut, and tail. Segmentation requires a robust spatial reference system that is invariant to rotation and warping of the fillet. In cod fillets, the centerline, consisting of veins and arteries cut off during filleting, is always visible on cod fillets and hence could be used as a good reference for segmentation. Sivertsen *et al.* (2009) attempted to detect cod fish ridge in automatic fish fillet inspection by centerline segmentation by using the absorption characteristics of hemoglobin.

In their study (Sivertsen *et al.*, 2009), the fillets were placed on a slow moving conveyer belt (3.5 cm/s) and imaged by the imaging spectrometer. All data analysis was done at line, meaning that the hyperspectral images were collected and stored to disc for later analysis. The images were acquired in the transflection (interactance) mode in which illumination and measurements were performed on the same side of the sample. However, the illumination was focused on an area adjacent and parallel to the detector's field of view. Keeping the angle of illumination parallel to measurement is important in order to have a constant optical path length, with varying sample thickness. Transflection can eliminate the effect of specular reflection, and increase the signal received from inside the sample. The fillet was scanned line by line, neck first, as the fillet passed on the conveyer belt and the resulting hyper-cube data recorded were saved for later analysis. The centerline is marked with the green dashed line in Figure 6.14. It consists mainly of blood remnants in arteries and veins that have been cut off during filleting, which gives the centerline its red/ brown color. Hemoglobin in the arteries is mainly bound to oxygen and is referred to as oxyhemoglobin (O_2Hb). The

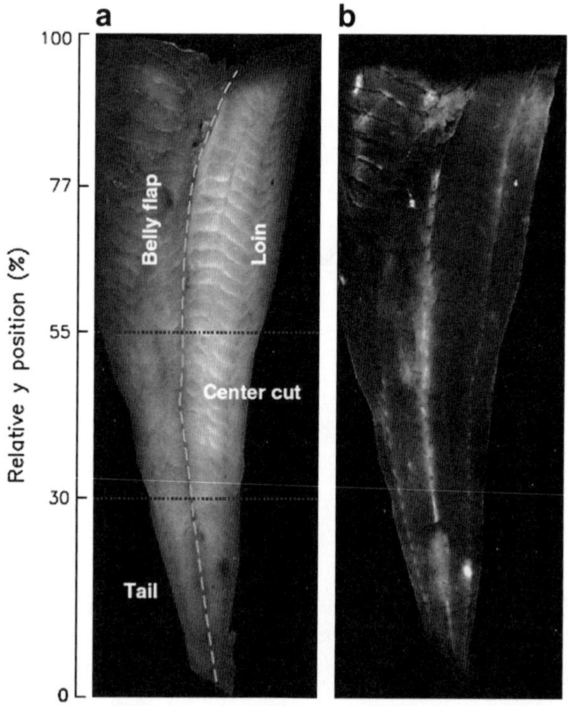

FIGURE 6.14 *Fish fillet images: (a) color image where the red, green, and blue channel is represented by spectral image at 640, 550 and 460 nm, respectively; the green dashed line indicates the manually detected centerline and the blue dotted lines indicate the transition between tail to center cut and center cut to loin/belly-flap respectively; (b) centerline enhanced image, the axis on the left-hand side indicates the position along the fillet in percent relative to fillet length (reproduced from Sivertsen* et al., *2009. © 2009 with permission from Elsevier). (Full color version available on http://www.elsevierdirect.com/companions/9780123747532/)*

hemoglobin in the veins has released most of its oxygen to the muscle and the main part of the blood in the veins is deoxyhemoglobin (HHb).

To better discriminate the centerline from the surrounding cod muscle, it is necessary to increase the signal of the centerline and at the same time lower the signal from the surrounding muscle. This can be done by performing numerical division between two different wavelength bands: one band is at where the centerline and the surrounding muscle is similar with respect to absorbance wavelength, and the other wavelength band is at where the centerline absorbs significantly more than the surrounding muscle. By using the 715 nm band for the first task, and 525 nm band for the second task, the centerline could be represented as pixels with high intensities where the surrounding muscle has a lower intensity (Figure 6.14b). A drawback with this method is that it does not only enhance the centerline, but all areas of high blood content. By applying a method named directional average ridge follower (DARF) to follow the ridge in the direction of the maximum directional average, the centerline can be detected with an average accuracy of 1 mm from the tail and 77% into the fillet relative to its total length. The error increases rapidly in the neck region, and typical errors of 4 mm are reported. This method is ready for industrial use with respect to both accuracy and computational requirements (Sivertsen *et al.*, 2009).

In one of the most interesting applications of hyperspectral imaging, Wold *et al.* (2001) and Heia *et al.* (2007) used a hyperspectral system to detect nematodes in cod fish fillets. Existence of parasites in fish muscle will preliminarily cause immediate consumer rejection of the product, and it will also lead to distrust of fish as a healthy and wholesome product. To fulfill market requirements and avoid complaints, the fish industry must be able to deliver parasite-free products. Several approaches have been tried in the effort to develop an efficient method to detect the parasites, but so far, the only reasonable solution has been manual inspection and trimming of each fillet on a candling table. Detection by *candling* is based on human vision and the ability to register differences in color and structure. As the fillets are placed onto a white light table, parasites embedded to a depth of 6 mm into the fillet can be spotted and removed manually; however, the detection efficiency is reported to be low. With the combination of color and morphological features, it is quite easy to distinguish visible worms from the fish flesh. A computer-based detection system using only morphological image features will probably have limited performance, because the parasites can appear in any shape and are often very similar to features in the fish fillet. Spectral properties are independent of the parasite's shape. Therefore, it is important to develop a reliable and non-destructive technique for detecting parasites and other illness-causing agents. An instrumental detection method based on the optical properties of the fish muscle and the parasites is therefore considered of interest. Although they are sometimes invisible to human eyes, nematodes could be easily detected by hyperspectral imaging technology due to the fact that presence of nematodes in fish flesh presents distinctive spectral fingerprints compared with the normal fish muscles. In their work Wold *et al.* (2001) and Heia *et al.* (2007) proposed that by applying a white light transmission setup and imaging spectroscopy to cod fillets, it is possible to make spectral images containing information to differentiate between fish muscle and parasites. The aim of the analysis was to evaluate the capability of identifying nematodes embedded in fish muscle based on spectral information. The method has proven to be an effective method for automatic detection of parasites even at 6 mm

(Wold *et al.*, 2001) and 8 mm (Heia *et al.*, 2007) below the fillet surface, which is 2–3 mm deeper than what can be found by manual inspection of fish fillets.

The first step in parasite detection protocol is to manually measure the actual depth of the parasite calculated from the surface of the fillet and then find out at which depth the spectral imaging system is able to detect these parasites. To make this protocol more reliable in industrial applications, the investigations should also address the varying colors of nematodes from dark to light brown as well as the red ones. In their experiments for detecting parasites in cod fish fillets, Wold *et al.* (2001) used a multispectral imaging system in the transmittance mode to investigate whether parasites could be distinguished in cod fillets purely on the basis of spectral characteristics in the VIS and NIR region. Spectral images of 8–14 channels as characterized in Table 6.4 were created by using interference filters in front of the lens. The filters were placed in drawers and exchanged manually. Exposure time at each channel was set to provide high signal-to-noise ratio but prevent saturation. The use of different exposure times for different wavelengths was mainly a consequence of different transmission properties in the fish and filters. The imaged area of the samples was adjusted to be about 35 × 35 mm. The length of the worms spanned from 15 to 40 mm, and the color varied from dark

Table 6.4 Interference filter properties and exposure time for collecting spectral images

Channel	Wavelength (nm)	Bandwidth* (nm)	Exposure time (s)
1	400	40	1.5
2	450	40	1.5
3	500	40	0.5
4	550	40	0.5
5	600	40	0.5
6	650	40	0.5
7	700	40	0.3
8	800	10	1.0
9	850	10	1.0
10	900	10	0.3
11	975	10	1.0
12	1000	50	0.8
13	1000	10	1.0
14	1050	10	5.0
15	1100	10	5.0

* *Bandwidth is FWHM (full width at half-maximum).*

Source: *Wold*, et al., *2001*

brown via yellow/reddish to almost white. Most of the detected parasites were curled in a characteristic spiral shape, appearing like vague elliptical dark areas in the fish fillets.

As expected, there was a significant difference in the spectral properties between parasites at different depth and those of normal flesh, as shown in Figure 6.15. It is clear that the averaged spectra from selected areas of a fish sample are thoroughly dissimilar. The spectra are from two parasites (one 1 mm and one 4 mm below the fillet surface), white muscle, dark muscle, and blood, as well as skin remnant. Spectra vary in both intensity and shape, depending on factors such as the color and depth of the parasite, the concentration and depth of the blood spot, the thickness of the dark muscle, and so on. It can be seen that transmission in white muscle is the highest at all wavelengths. At 400 nm, transmission is in general low for all components; while dark muscle and blood are relatively low at 500 and 550 nm (dark muscle contains much blood). The spectral signatures of the two parasites are quite different since they are measured at different depths. The deep parasite has higher transmittance, resulting in lower contrast compared to white muscle. As shown in Figure 6.16, parasitic nematodes in cod fillets can be automatically detected on the basis of spectral characteristics by use of

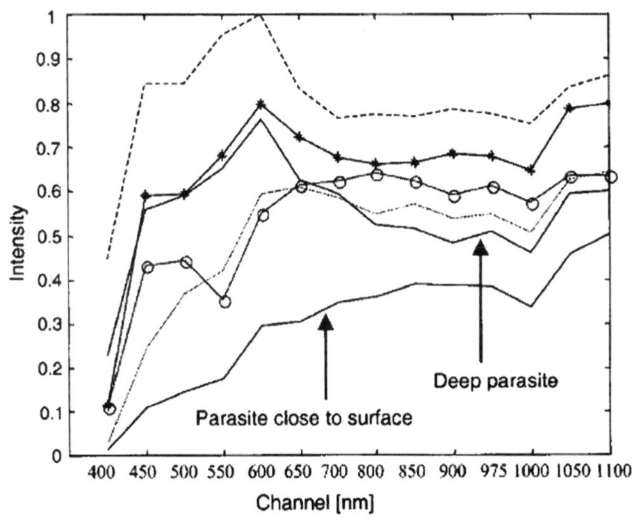

FIGURE 6.15 *Spectra features of different components in fish fillet as measured in a spectral image. Parasites (—); white muscle (- - - -); dark muscle (··········); skin remnant (*); blood (-o-). The deeper embedded parasite lies at about 4 mm (reproduced from Wold* et al.*, 2001. © with permission from Optical Society of America 2001)*

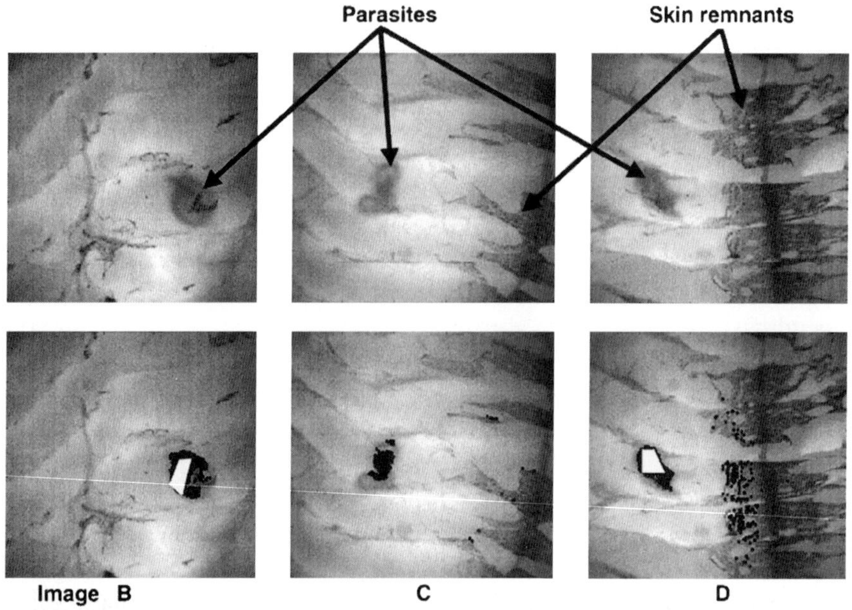

FIGURE 6.16 *Fish fillet images with nematodes. Upper row shows three samples from different fillets (B, C, and D), the nematodes are naturally at 1–4 mm depth. Lower row shows the same images with classification results, the SIMCA model was based on the pixels within both the white areas in images B and D. The model was then tested on all pixels in the three images, and the black points indicate pixels classified as parasite (reproduced from Wold* et al., *2001. © with permission from Optical Society of America 2001)*

spectral imaging and SIMCA (soft independent modeling of class analogies) classification. The method is sensitive to the depth and color of the parasites, as well as the surrounding fish tissue. Image channels in the near-infrared have the potential to "see" deeper than those in the visible area, but the best classification is obtained by combining channels from both regions. The fact that the parasites are visible in the NIR area is interesting for two reasons. First, scattering in the NIR is lower than in the VIS area. This consideration could enable detection of parasites embedded deeper than 6 mm. Second, if detection can be based on NIR channels, the method could be independent of the parasites' color. The method has potential for on-line implementation, but further studies are required to verify feasibility for the fish industry (Wold *et al.*, 2001).

In another experiment using spectral imaging in transmission mode where the light source is located below the fillet, Heia *et al.* (2007) acquired spectral images of cod fish fillets infested by parasites of different colors and at different depths. The samples are imaged with full spectral information in the range from 350 to 950 nm with spectral resolution of approximately 2–3 nm and spatial resolution of 0.5 × 0.5 mm. One spectral image of a fillet sample contained 290 × 290 pixels, where each pixel was represented with a spectrum ranging from 350 to 950 nm. For training, the X-matrix was constructed using the selected subset of pixels representing nematodes and normal muscle. Each row in the X-matrix consisted of the mean centered full spectrum from 1 pixel. The elements of the Y-vector used for training were designed with ones and zeros representing nematode and fish muscle, respectively. For classification of new samples the trained discriminant partial least squares (DPLS) regression model was applied to each pixel in the spectral image. The classifier output was numbers of one and zero representing nematode and no nematode, respectively. Hence the multispectral image was reduced to a single-band image, representing the classification result.

Due to the distinctive spectral characteristics of nematodes which differ sufficiently from those of fish flesh, fairly good classifications are expected to be obtained. Figure 6.17 demonstrates example results of the experiments. Figure 6.17(a) shows an image of the sample captured with a standard digital camera. In Figure 6.17(b) the image at 540 nm is shown with marked nematodes, blood spots, and black lining. The result from applying the DPLS regression is shown in Figure 6.17(c), where the classification before applying the threshold is positioned in green on top of the original image. As can be

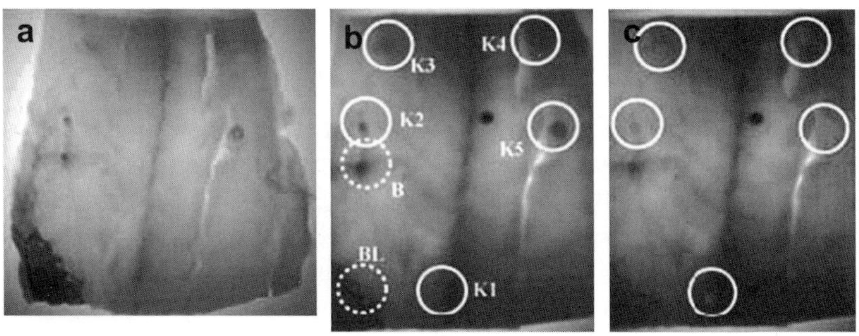

FIGURE 6.17 *Section of cod fillet. (a) Image of the cod sample captured with RGB digital camera. (b) Spectral image at 540 nm where the nematodes, K1 to K5, are indicated with white circles, a blood spot, B, marked with a dotted circle and black lining, BL, marked with a dotted circle. (c) Classification result, in green, on top of the 540 nm image before thresholding. For the naked eye the bloodspot may appear as a parasite. Bloodspots were not identified as parasites by the classification (reproduced by permission from Heia* et al., *2007. © with permission from Institute of Food Technologists 2007). (Full color version available on http://www.elsevierdirect.com/companions/9780123747532/)*

seen from Figure 6.17(c), some nematodes are clearly visible while those deeply embedded appear more diffuse. The five nematodes marked in Figure 6.17(b) were initially different in color and were located at different depths. In this respect these results are very promising, indicating that instrumental detection may perform better than today's manual procedure. The spectral imaging system as shown here is proved to be feasible in view of the on-line requirements of the fish-processing industry. Further examples of using the hyperspectral imaging technique can be found in Chapter 8.

6.4.4. Chicken

The use of hyperspectral imaging for quality evaluation and monitoring of chicken and poultry products in off-line and on-line applications has been demonstrated and intensively studied in many research endeavors (Chao *et al.*, 2008; Lawrence *et al.*, 2004; Park, Lawrence *et al.*, 2006; Park *et al.*, 2007; Windham *et al.*, 2003; Yang *et al.*, 2009). Most of these studies focused on either differentiation between wholesome and unwholesome freshly slaughtered chickens or on detection of various contaminations on the surface of the poultry carcasses. The core idea behind this technique is to identify the spectral difference among different components in the sample. Some of these systems have already been installed in a real-time inspection line where a spectral image is captured for each bird, and the image is then processed by the system's computer to determine whether or not the bird has a disease, a contaminant or a systemic defect. In addition, the system could also provide some information to detect small birds, broken parts, bruising, tumors, and air sacs. The limit on production throughput due to the limitation of manual inspection, combined with increases in chicken consumption and demand over the past two decades, places additional pressure to develop a reliable, non-invasive, quick inspection system. Unlike visual inspection by the naked eye of workers, the spectral imaging technique is able to provide a constant and reliable tool to accurately monitor chicken overall quality and therefore it has the potential to augment and enhance the human inspection process. The spectral diagnostic system could be used as a non-invasive tool to monitor a production line of chicken carcasses by developing spectral profiles from hyperspectral images taken during all stages of production. In the following sections, some application examples are given. Further developments in the area can be found in Chapter 7.

6.4.4.1. Detection of contamination

Risks from microbiological and chemical hazards are of a serious concern to human health. Chicken meat is essentially sterile at the time of slaughter.

However, the necessary skinning, evisceration, and cutting exposes the carcass to large numbers of naturally occurring microorganisms. The level of contamination differs with the processing and handling procedures employed. In the area of contamination detection that frequently occurred on the surfaces of poultry carcasses, researchers have developed hyperspectral imaging systems of different designs and sensitivities for the identification of fecal matter and ingesta (Park, Lawrence *et al.*, 2006; Windham *et al.*, 2003, 2005a, 2005b). Poultry carcasses with pathological problems must be identified and removed from food processing lines to meet the requirement of high standards of food safety. Without proper inspection protocols during slaughter and processing, the edible portions of the poultry carcasses can become contaminated with bacteria capable of causing illness in humans. Therefore, regulation emphasized that a carcass with visible fecal contamination has to be removed in order to prevent cross-contamination among carcasses. For safety purposes, the identification and separation of poultry carcasses contaminated by feces and/or ingesta are very important to protect the public health from a potential source of food poisoning.

Currently, inspection of fecal contamination in poultry carcasses is through human visual observation where the criteria of color, consistency, and composition are used for identification. Trained human inspectors carry out the inspection and examine a small number of representative samples from a large production run. In addition to being a very tedious task, manual inspection is both labor-intensive and prone to human error and inspector-to-inspector variation (Liu *et al.*, 2003b). Human inspectors are often required to examine 30–35 poultry samples per minute. Such working conditions can lead to repetitive motion injuries, distracted attention and fatigue problems, and result in inconsistent quality (Du *et al.*, 2007). Spectral properties of normal and contaminated surfaces should be identified first and algorithms should be developed to enhance the identification of contamination. These features could be transferred to an imaging system that can scan carcasses at line speeds. In general, the hyperspectral imaging technique has demonstrated the ability to recognize spectral signatures associated with contaminated poultry carcasses even in shadowy regions of the carcasses. In conjunction with an appropriate image processing algorithm, the hyperspectral imaging system is proven to be an effective tool for the identification of different contaminants on poultry carcasses.

The USDA's Agricultural Research Service is the pioneer research institution for developing hyperspectral and multispectral imaging techniques to detect different contaminants on poultry carcasses. Intensive research has been exerted by this research group for calibrating hyperspectral imaging systems, identifying spectral signatures of different contaminants in the VIS and NIR regions, developing algorithms for fecal detection and spectral image processing and finally exploiting the system in on-line multispectral application (Lawrence *et al.*, 2003; Liu *et al.*, 2003b; Park *et al.*, 2002).

Detecting contaminants in poultry carcasses could be performed either by inspecting contaminated birds or typically by applying certain amounts of possible contaminant, e.g. of feces (from the duodenum, ceca, and colon) and ingesta, varying the type of contaminants, contaminated spot size, and location on the carcass. The threshold at which hyperspectral imaging is able to identify contaminant should then be determined. For instance, Windham *et al.* (2005b) contaminated one-half of their experimental units of carcasses by homogenized multiple cecal contaminants of different amounts (10, 50, or 100 mg) and kept the other half uncontaminated as a negative control. Typically, fecal or cecal materials detected (Figure 6.18) have spectra that increase in reflectance from 420 nm to 780 nm whereas most other spectra (skin, meat, bones, blood, etc.) decreased from 500 to 560 nm. Therefore, dividing an image at 565 nm by an image at 517 nm would result in contaminants with values greater than one while

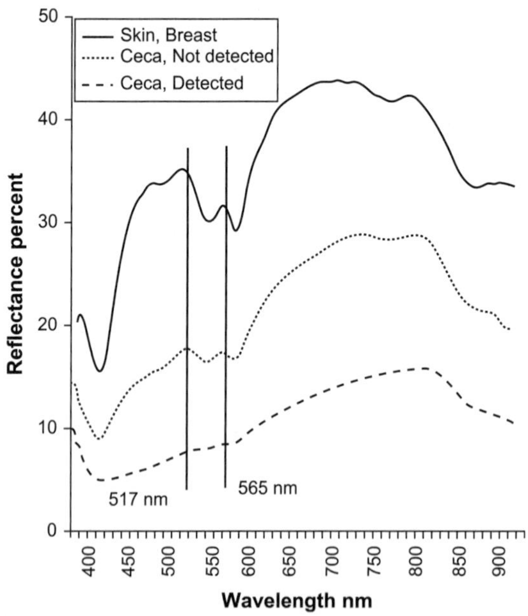

FIGURE 6.18 *Typical mean spectra of poultry breast skin and cecal contaminant. Mean spectra were obtained by spatially averaging over cecal detected and cecal edge pixels not detected (reproduced from Windham* et al., *2005b. © with permission from Asian Network for Scientific Information 2005).*

non-contaminants would have values less than one. To test the effectiveness of the hyperspectral imaging system for detecting cecal contaminants varying in mass, thresholds of 1.0, 1.05, and 1.10 were applied to the masked-ratio image (565/517) to delineate the cecal contaminants from the remainder of the image. Their results showed that the hyperspectral imaging system correctly detected 100% of the cecal spots applied to the carcass halves at a threshold of 1.00 and 1.05. Spectra from pixels on the boundary of the cecal contaminant not detected (Figure 6.19) are a mixture of the cecal and skin as indicated by the reflectance peaks for myoglobin in the skin. Mixed pixels are problematic to detect with a 565/517 nm ratio regardless of the contaminant type (i.e., ingesta, duodenum, colon, cecal) because the wavelength values are too close to each other. Moreover, the hyperspectral imaging system appeared to be more effective than the traditional microbiological method for detecting 10 mg contaminants.

After the necessary spectral and spatial calibration of the hyperspectral imaging system by using a procedure similar to the one explained in Chapter 1,

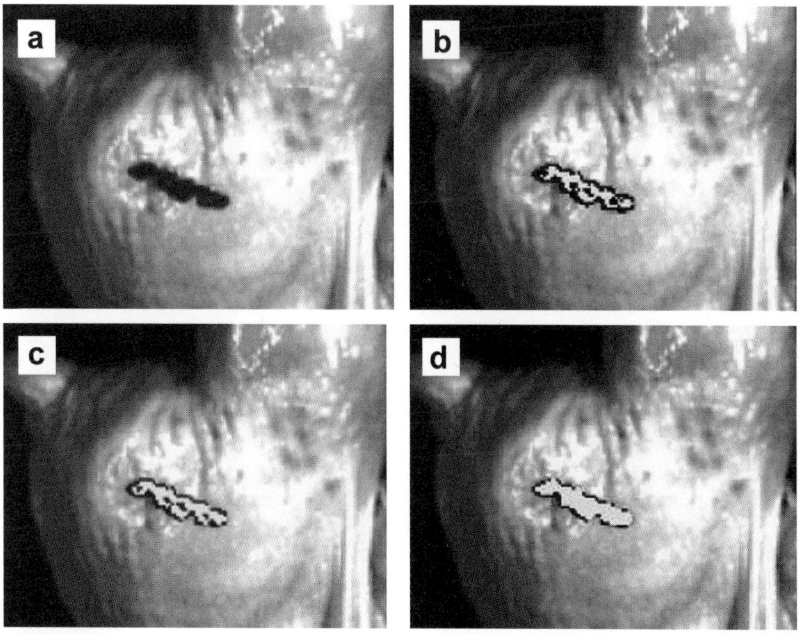

FIGURE 6.19 *Color composite image of a 90 mg cecal mass contaminant: (a) pixels not detected (black); (b) pixels detected (gray or yellow) with a 1.10 threshold; (c) pixels detected with a 1.05 threshold; (d) pixels detected with a 1.10 threshold (reproduced from Windham* et al., *2005b. © with permission from Asian Network for Scientific Information 2005). (Full color version available on http://www.elsevierdirect.com/companions/9780123747532/)*

the next step in detecting contaminant is to identify key wavelengths by spectral analysis of spectra extracted from either visible/NIR spectroscopy (400–2498 nm) or from hypercubes (430–900 nm) of hyperspectral images (Liu *et al.*, 2003b; Park *et al.*, 2002). The selected key wavelengths should be validated to ensure correct detection of fecal and ingesta. The method depends on processing and analyzing hyperspectral images with different preprocessing methods considering calibration and spectral smoothing approaches. Figure 6.20 shows visual results of a poultry carcass with the image-processing algorithm applied to a calibrated smoothed preprocessed hyperspectral image. In the ratio images (I_{565}/I_{517}) as shown in Figure 6.20(c), there was a notable difference in the contrast between the

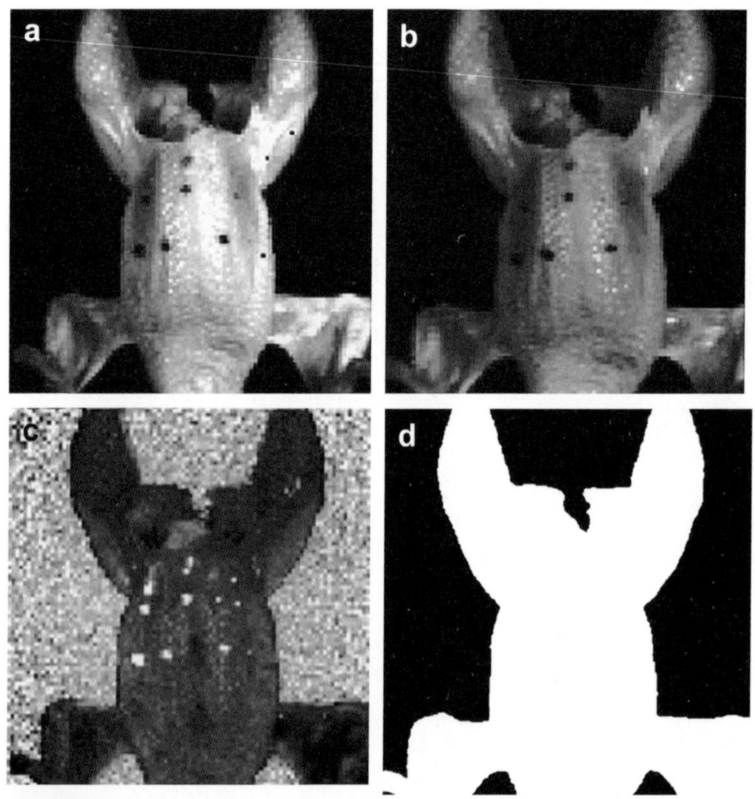

FIGURE 6.20 *Hyperspectral image processing of a poultry carcass: (a) color composite image (pseudo-RGB image); (b) calibrated color image; (c) ratio image (I_{565}/I_{517}); (d) background mask (reproduced from Park* et al., *2006a. © 2006 with permission from Elsevier). (Full color version available on http://www.elsevierdirect.com/companions/9780123747532/)*

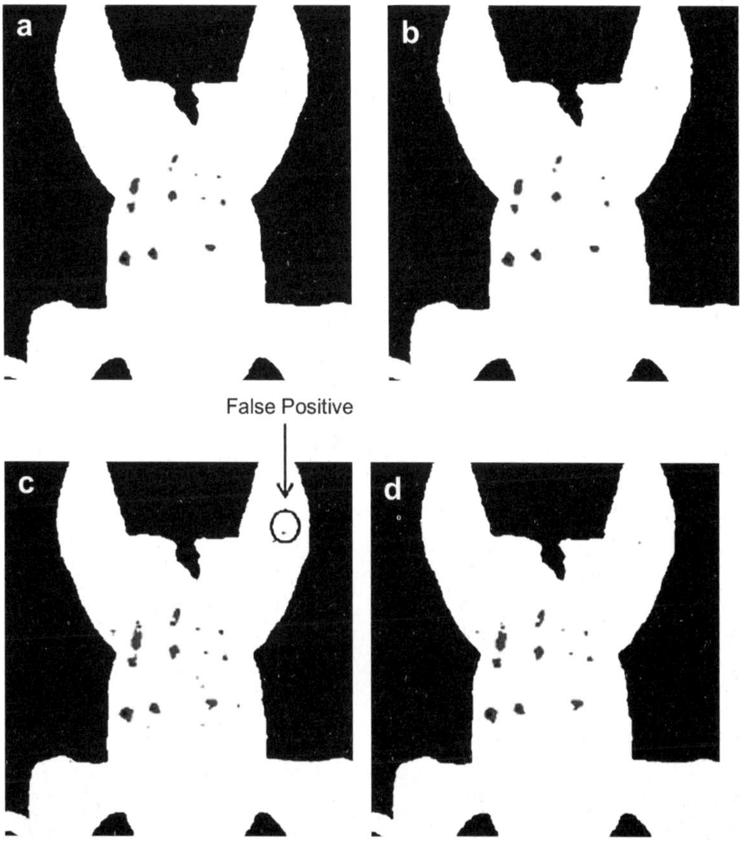

FIGURE 6.21 *Band ratio of two wavelengths (517 and 565 nm) selected by regression model and scanning monochromator: (a) threshold = 1.0; (b) threshold = 1.0 with filter; (c) threshold = 0.95; (d) threshold = 0.95 with filter (reproduced from Park* et al.*, 2006a. © 2006 with permission from Elsevier). (Full color version available on http://www.elsevierdirect.com/companions/9780123747532/)*

carcass, the background and the contaminant which could be easily detected by employing various threshold values as shown in Figure 6.21. Although a band-ratio of three wavelengths $(I_{576}–I_{616})/(I_{529}–I_{616})$ had some success in contaminant detection as well, a band-ratio image-processing algorithm with two bands (I_{565}/I_{517}) performed very well with 96.4% accuracy for detecting both feces (duodenum, ceca, colon) and ingesta contaminants (Park *et al.*, 2006a).

Figure 6.22 shows the detailed steps for detecting contaminants in poultry carcasses (Park *et al.*, 2002), which includes selecting the dominant wavelengths by PCA loadings and calibration regression coefficients;

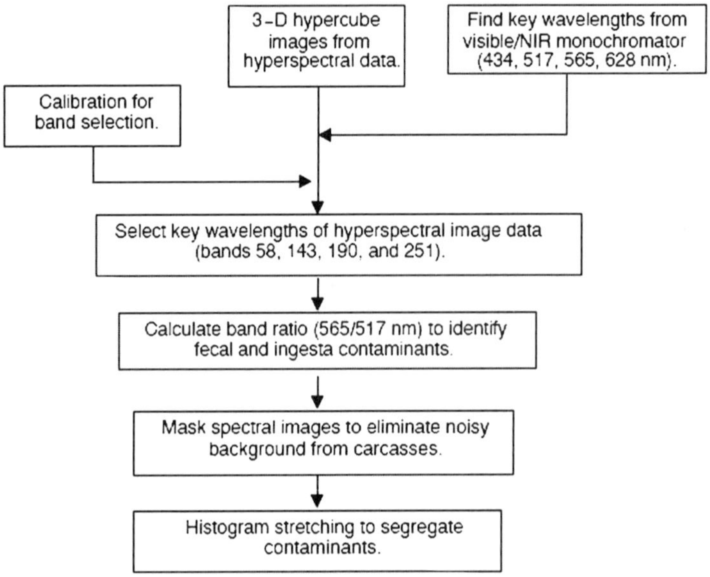

FIGURE 6.22 *Key steps for the algorithms used to detect feces and ingesta on poultry carcasses using visible/near-infrared spectroscopy and hyperspectral imaging (reproduced from Park* et al., *2002. © with permission from American Society of Agricultural and Biological Engineers 2002)*

calculating the band ratios among the selected spectral images; removing the background noise from the carcasses' masked image; and finally applying histogram stretching to all masked images to visually segregate individual fecal and ingesta contaminants. Four dominant wavelengths (434, 517, 565, and 628 nm) were selected by principal component analysis from VIS/NIR spectroscopy for wavelength selection of hyperspectral images. Hyperspectral image processing algorithms, specifically band ratio of dual-wavelength (565/517) images and histogram stretching, were effective in the identification of fecal and ingesta contamination of poultry carcasses (Figure 6.23). Test results indicated that the detection accuracy was 97.3% for linear and 100% for non-linear histogram stretching.

Generally, detection of contaminants depends on the largest difference in spectral difference between contaminants and normal skin. Also, the wavelengths at which the contaminants gave the highest contrast with the normal skin would be deemed key wavelengths. In their earlier work, Windham *et al.* (2003) selected four key wavelengths (434, 517, 565, and 628 nm) to detect feces and ingesta on poultry carcasses. The method developed was able to detect 100% of the fecal contaminants in a limited population of broilers

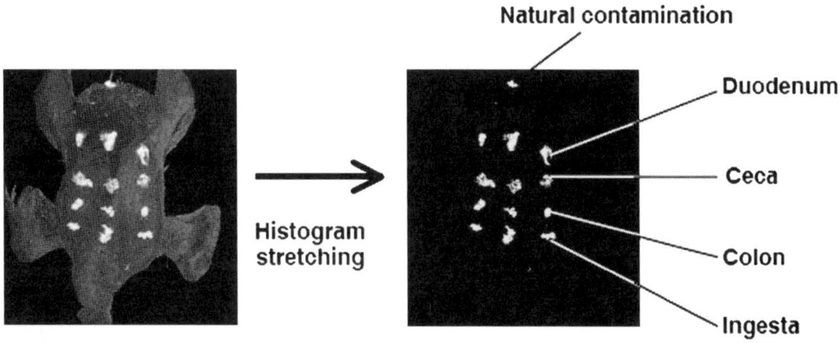

FIGURE 6.23 *Band-ratio (565/517) image presented high contrast compared to the normal skin (*left*) and applying histogram stretching could separate feces and ingesta from the carcass (*right*) (reproduced from Park* et al., *2002. © with permission from American Society of Agricultural and Biological Engineers 2002)*

especially with a ratio of 565/517 nm. Then they used 565/517 nm to classify uncontaminated skin from feces/ingesta using a single-term linear regression. By using another approach, Windham *et al.* (2005a) determined the effectiveness of hyperspectral imaging for detecting ingesta contamination spots varying in mass from the crop and gizzard from the upper digestive tract. They applied a decision tree classifier algorithm to selected images at wavelengths of 517, 565, and 802 nm, producing a Boolean output image with gizzard and crop contaminates identified. The spectral imaging system correctly detected 100% of the crop and gizzard contents regardless of the mass or spot size. However, not every pixel associated with a given spot was detected.

6.4.4.2. Tumor and diseased chicken detection

Skin tumor in chickens is an ulcerous lesion region surrounded by a region of thickened-skin. Skin cancer causes skin cells to lose the ability to divide and grow normally, and induces abnormal cells to grow out of control to form tumors. Tumorous carcasses often demonstrate swollen or enlarged tissue caused by the uncontrolled growth of new tissue. Tumor is not as visually obvious as other pathological diseases such as septicemia, air sacculitis, and bruise since its spatial signature appears as shape distortion rather than a discoloration. Therefore, conventional vision-based inspection systems operating in the visual spectrum may reveal limitations in detecting skin tumors on poultry carcasses (Du *et al.*, 2007). Detection of chicken tumors is a difficult issue because the tumors vary in size and shape and sometimes tumors appear on both sides of the chicken. In addition, tumors have different spectral responses and even different areas of normal chicken skin have a similar spectral response to that of tumors making their identification a difficult task (Nakariyakul & Casasent, 2004). Thus, proper classification and detection routines are needed to alleviate this problem. Recently, several researchers have utilized hyperspectral imaging techniques for the detection of chicken skin tumor with good results (Kim *et al.*, 2004; Kong *et al.*, 2004; Nakariyakul & Casasent, 2004).

In detecting chicken skin tumors, the classification algorithm must be trained before performing a real classification, chicken carcasses should be examined first using spectral information, and the results can then be used to determine candidate regions for skin tumors. For training, pixels from tumor areas as well as normal skin areas should be carefully selected by visual inspection of hyperspectral images based on the spectral characteristics of the spectra of both normal skin and tumors. Furthermore, based on wavelength-dependent responses between the normal skins and tumors, relationships between some wavebands can further amplify the differences between the two classes. A potential tumor is a region that consists of pixels identified as a tumor in spectral classification. The spectral map defined from spectral analysis is then used as an input to a spatial classification depending on structural properties of the tumors such as size, filling ratio, and ratio of major to minor axes. The spatial classification algorithm selects the real tumor spots from the candidate regions. The final output of the spatial classifier shows the locations of tumors detected. By this method, it will be possible to detect chicken carcasses with tumors, but sometimes the method could fail to detect some tumors that are very small in size. In their experiment for detecting tumors in chicken carcasses, Kim *et al.* (2004) presented a method for detecting skin tumors on chicken carcasses using a hyperspectral fluorescence imaging system; however they failed to detect some tumors that were smaller than 3 mm in diameter. Their resultant detection rate, false positive rate, and missing rate of the proposed method were 76%, 28%, and 24%, respectively.

To overcome the computational restrictions in real-time processing, the speed of a tumor detection procedure can be hastened by selecting only a few wavebands from hyperspectral data and employing classification algorithms such as fuzzy logic and support vector machine classifiers, which have been illustrated by investigations conducted by Chao *et al.* (2008). Waveband selection methods are intended to identify a subset of significant spectral bands in terms of information content and to remove the bands of less importance. Among widely used dimensionality reduction methods, the principal component analysis (PCA) rearranges the data in terms of the significance measured by the eigenvalues of the data covariance matrix. For selecting key wavelengths in tumour detection dilemmas, PCA of some ROIs representing normal and tumor areas provided an efficient mechanism for selection of some narrow-band wavelength regions for use in a multispectral imaging system.

6.4.4.3. On-line inspection of chicken

The USDA's Agricultural Research Service and Instrumentation and Sensing Laboratory have an ongoing program of developing real-time on-line systems for poultry inspection. Park, Kise *et al.* (2006) developed a prototype real-time multispectral imaging system for fecal and ingesta contaminant detection on broiler carcasses. The prototype system includes a common aperture camera with three optical trim filters (517, 565, and 802 nm wavelengths) selected by VIS/NIR spectroscopy and validated by a hyperspectral imaging system with a decision tree algorithm. Figure 6.24 shows the diagram of poultry processing for a real-time fecal inspection imaging system at a pilot-scale plant in Russell Research Center, Athens, Georgia, USA. The on-line testing results showed that the multispectral imaging technique can be used effectively for detecting feces (from duodenum, ceca, and colon) and ingesta on the surface of poultry carcasses with a processing speed of 140 birds per minute. The real-time imaging software for on-line inspection was developed using the object-oriented Unified Modeling Language (UML), which is a useful language for specifying, visualizing, constructing, and documenting software systems. Both hardware and software for real-time fecal detection were tested at the pilot-scale poultry processing plant. The runtime of the software including on-line calibration was fast enough to inspect carcasses on-line to industry requirements. Based on the preliminary test at the pilot-scale processing line, the system was able to acquire and process poultry images in real-time. According to the test results, the imaging system is reliable in harsh environments and the Unified Modeling Language- (UML-) based image processing software is flexible and easy to be updated when additional parameters are needed for in-plant trials.

Because the ideal inspection regulations require zero tolerance for unwholesome chickens exhibiting symptoms of septicemia or toxemia, these unwholesome chickens must be removed from the processing line. Septicemia is caused by the presence of pathogenic microorganisms or their toxins in the bloodstream, and toxemia results from toxins produced by cells at a localized infection or from the growth of microorganisms. Therefore, an on-line line-scan imaging system (Figure 6.25) was developed and tested on an

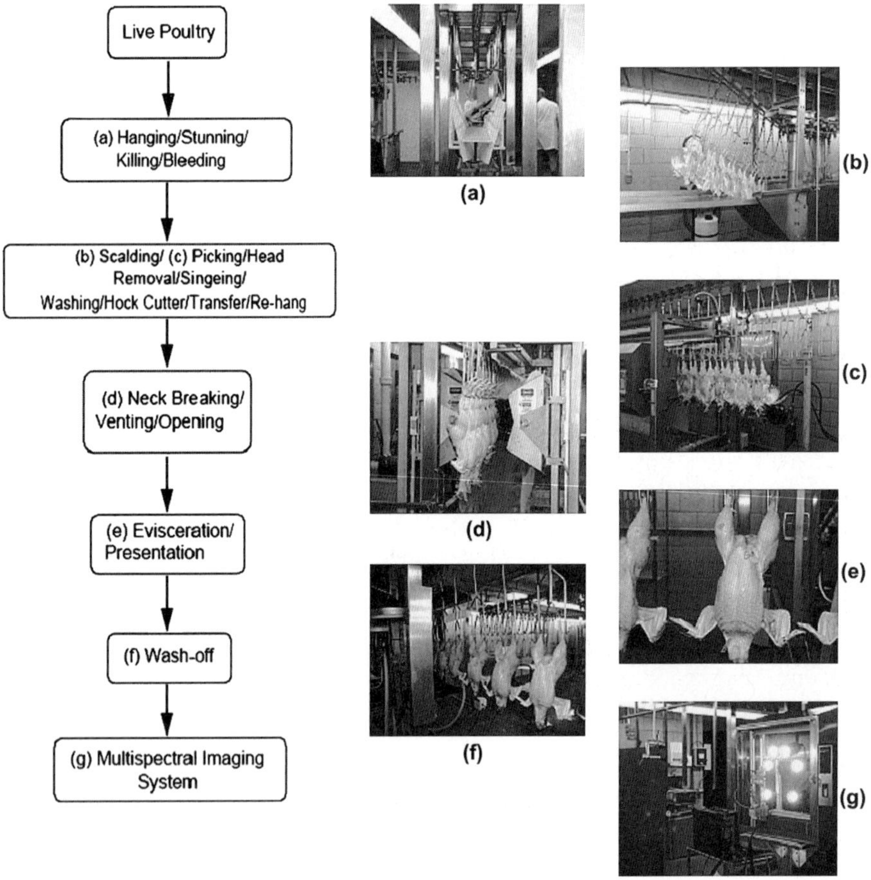

FIGURE 6.24
Flowchart of a poultry processing line in a real-time fecal inspection imaging system at pilot-scale plant in Russell Research Center (reproduced from Park et al., *2006b. © with permission from SPIE 2006). (Full color version available on http://www.elsevierdirect.com/companions/9780123747532/)*

eviscerating line at a poultry processing plant with 140 birds per minute (bpm) for differentiation of wholesome and systemically diseased chickens (Chao *et al.*, 2008). Further details can be found in Chapter 7.

6.5. CONCLUSIONS

Hyperspectral imaging has passed the stage of scientific curiosity and it is now under dynamic evaluation by researchers in dozens of fields. Deployment of this technology in various food science sectors has become one of the researchers' big responsibilities for more widespread utilization of this new-emerging technology. Whilst some of the applications explored require

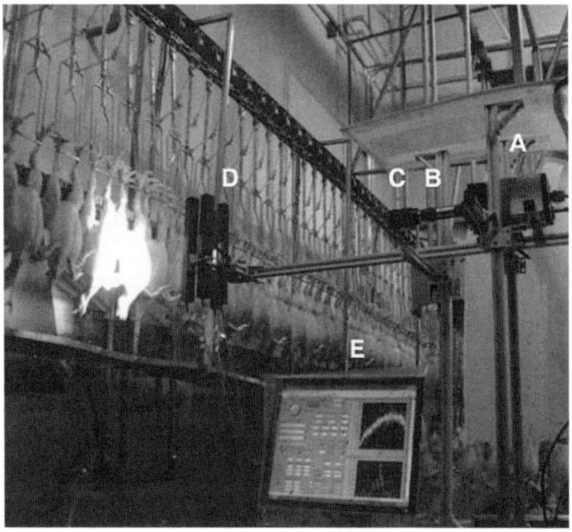

FIGURE 6.25 *A hyperspectral/multispectral imaging inspection system on a commercial chicken processing line. A, Electron-multiplying charge-coupled-device (EMCCD) camera; B, line-scan spectrograph; C, lens assembly; D, LED lighting system; E, data processing unit (reproduced from Chao* et al., *2008. © with permission from American Society of Agricultural and Biological Engineers 2008)*

further development, the potential for wider exploitation of this non-destructive method for the assessment of numerous food products is anticipated in the coming few years. Throughout this chapter, it is confirmed that hyperspectral imaging techniques provide an attractive solution for the analysis of meat and meat products. The various applications outlined show the benefits of this technique for sample characterization, defect and disease detection, spatial visualization of chemical attributes (chemical images), and evaluations of overall quality parameters of beef, pork, fish, and chicken. By combining spatial and spectral details together in one system, hyperspectral imaging has been proved to be a promising technology for objective meat quality evaluation. In addition to its ability for effectively quantifying and characterizing quality attributes of some important visual features of meat such as color, marbling, maturity and texture, it is able to measure multiple chemical constituents simultaneously without monotonous sample preparation. After developing, calibrating, testing, and validating the hyperspectral imaging system, a multispectral imaging system employing only few effective wavebands can then be used for certain applications in a real-time implementation. Despite these achievements, there are still many challenges facing the full exploitation of the technique in terms of processing speed, hardware limitations, and calibrations. To establish this technology in meat quality assurance and quality control programs to solve certain manufacturing problems, researchers must devote more effort to enhancing the technique by developing effective methodologies for more consistent and expeditious regimes adapted for meat quality evaluation.

NOMENCLATURE

ANN	artificial neural network
bpm	birds per minute
CCD	charge-coupled device
CIE	Commission International De I'Eclairage
CMOS	complementary metal-oxide-semiconductor
r	correlation coefficient
DPLS	discriminant partial least squares
FWHM	full width at half maximum
GLCM	graylevel of co-occurrence matrix
NIR	near infrared
NIRS	near infrared spectroscopy
PCA	principal component analysis
PCR	principal component regression
PFN	pale, firm and non-exudative
PLSR	partial least squares regression
PSE	pale, soft and exudative
RFN	red, firm, and nonexudative
RGB	red, green, and blue (components of a color image)
ROI	region of interest
RSE	red, soft, and exudative
SSF	slice shear force
SWIR	shortwave infrared
UML	unified modeling language
VIS	visible light
WBSF	Warner–Bratzler shear force
WHC	water holding capacity

REFERENCES

Abouelkaram, S., Berge, P., & Culioli, J. (1997). Application of ultrasonic data to classify bovine muscles. *Proceedings of IEEE Ultrasonics Symposium, 2*, 1197–1200.

Abouelkaram, S., Chauvet, S., Strydom, P., Bertrand, D., & Damez, J. L. (2006). Muscle study with multispectral image analysis. In Declan Troy (Ed.), *Proceedings of the 52nd International Congress of Meat Science and Technology* (pp. 669–670). Wagening: Wageningen Academic Publishers.

Alomar, D., Gallo, C., Castañeda, M., & Fuchslocher, R. (2003). Chemical and discriminant analysis of bovine meat by near infrared reflectance spectroscopy (NIRS). *Meat Science, 63*, 441–450.

AMSA. (2001). *Meat evaluation handbook*. Savoy, IL: American Meat Science Association.

Andrés, S., Silva, A., Soares-Pereira, A. L., Martins, C., Bruno-Soares, A. M., & Murray, I. (2008). The use of visible and near infrared reflectance spectroscopy to predict beef *M. longissimus thoracis et lumborum* quality attributes. *Meat Science, 78*, 217–224.

Ariana, D. P., Lu, R., & Guyer, D. E. (2006). Near-infrared hyperspectral reflectance imaging for detection of bruises on pickling cucumbers. *Computers and Electronics in Agriculture, 53*(1), 60–70.

Belk, K. E., Scanga, J. A., Wyle, A. M., Wulf, D. M., Tatum, J. D., Reagan, J. O., & Smith, G. C. (2000). The use of video image analysis and instrumentation to predict beef palatability. *Proceeding Reciprocal Meat Conference, 53*, 10–15.

Bonifazi, G., Serranti, S., (2008). Hyperspectral imaging applied to complex particulate solids systems. In F. Berghmans, A. G. Mignani, A. Cutolo, P. P. Meyrueis, & T. P. Pearsall (Eds), *Optical Sensors 2008. Proceedings of SPIE*, 70030F.

Bro, R., van den Berg, F., Thybo, A., Anderseny, C. M., Jøgensenx, B. M., & Andersen, H. (2002). Multivariate data analysis as a tool in advanced quality monitoring in the food production chain. *Trends in Food Science and Technology, 13*, 235–244.

Brosnan, T., & Sun, D.-W. (2004). Improving quality inspection of food products by computer vision–a review. *Journal of Food Engineering, 61*(1), 3–16.

Burger, J., & Geladi, P. (2006). Hyperspectral NIR imaging for calibration and prediction: a comparison between image and spectrometer data for studying organic and biological samples. *The Analyst, 131*, 1152–1160.

Chao, K., Yang, C.-C., Kim, M. S., & Chan, D. E. (2008). High throughput spectral imaging system for wholesomeness inspection of chicken. *Applied Engineering in Agriculture, 24*(4), 475–485.

Chau, A., Whitworth, M., Leadley, C., & Millar, S. (2009). Innovative sensors to rapidly and non-destructively determine fish freshness. *Seafish Industry Authority*. Report No. CMS/REP/110284/1.

Cluff, K., Naganathan, G. K., Subbiah, J., Lu, R., Calkins, C. R., & Samal, A. (2008). Optical scattering in beef steak to predict tenderness using hyperspectral imaging in the VIS-NIR region. *Sensing and Instrumentation for Food Quality and Safety, 2*, 189–196.

Damez, J. L., & Clerjon, S. (2008). Meat quality assessment using biophysical methods related to meat structure. *Meat Science, 80*(1), 132–149.

Damez, J. L., Clerjon, S., Abouelkaram, S., & Lepetit, J. (2008). Beef meat electrical impedance spectroscopy and anisotropy sensing for non-invasive early assessment of meat ageing. *Journal of Food Engineering, 85*(1), 116–122.

de Juan, A., Tauler, R., Dyson, R., Marcolli, C., Rault, M., & Maeder, M. (2004). Spectroscopic imaging and chemometrics: a powerful combination for global and local sample analysis. *Trends in Analytical Chemistry, 23*(1), 70–79.

Dransfield, E. (1994). Optimisation of tenderisation, ageing and tenderness. *Meat Science, 36*, 105–121.

Driver, R. D., (2009). Quantification and threshold detection in real-time hyperspectral imaging. In S. Moon S. Kim, Shu-I Tu, Kaunglin Chao (Eds.), *Sensing for Agriculture and Food Quality and Safety. Proceeding of the SPIE*, 73150L.

Du, C.-J., & Sun, D.-W. (2004). Recent developments in the applications of image processing techniques for food quality evaluation. *Trends in Food Science & Technology, 15*, 230–249.

Du, C.-J., Sun, D.-W., Jackman, P., & Allen, P. (2008). Development of a hybrid image processing algorithm for automatic evaluation of intramuscular fat content in beef. *M. longissimus dorsi. Meat Science, 80*(4), 1231–1237.

Du, Z., Jeong, M. K., & Kong, S. G. (2007). Band selection of hyperspectral images for automatic detection of poultry skin tumors. *IEEE Transactions on Automation Science and Engineering, 4*(3), 332–339.

ElMasry, G., Nassar, A., Wang, N., & Vigneault, C. (2008). Spectral methods for measuring quality changes of fresh fruits and vegetables. *Stewart Postharvest Review, 4*(4), 1–13.

ElMasry, G., Wang, N., ElSayed, A., & Ngadi, M. (2007). Hyperspectral imaging for nondestructive determination of some quality attributes for strawberry. *Journal of Food Engineering, 81*, 98–107.

ElMasry, G., Wang, N., & Vigneault, C. (2009). Detecting chilling injury in Red Delicious apple using hyperspectral imaging and neural networks. *Postharvest Biology and Technology, 52*(1), 1–8.

ElMasry, G., & Wold, J. P. (2008). High-speed assessment of fat and water content distribution in fish fillets using online imaging spectroscopy. *Journal of Agricultural and Food Chemistry, 56*(17), 7672–7677.

Folkestad, A., Wold, J. P., Rorvik, K.-A., Tschudi, J., Haugholt, K. H., Kolstad, K., & Morkore, T. (2008). Rapid and non-invasive measurements of fat and pigment concentrations in live and slaughtered Atlantic salmon (*Salmo salar* L.). *Aquaculture, 280*(1–4), 129–135.

Gat, N. (1999). Directions in environmental spectroscopy. *Spectroscopy Showcase*. March issue.

Geladi, P. (2003). Chemometrics in spectroscopy. Part 1: Classical chemometrics. *Spectrochimica Acta Part B, 58*, 767–782.

Gowen, A. A., O'Donell, C. P., Cullen, P. J., & Bell, S. E. J. (2008). Recent applications of chemical imaging to pharmaceutical process monitoring and quality control. *European Journal of Pharmaceutics and Biopharmaceutics, 69*, 10–22.

Grimes, L. M., Naganathan, G. K., Subbiah, J., & Calkins, C. R., (2007). *Hyperspectral imaging: a non-invasive technique to predict beef tenderness*. Animal Science Department, Nebraska Beef Cattle Reports, University of Nebraska–Lincoln, USA, pp. 97–99.

Grimes, L. M., Naganathan, G. K., Subbiah, J., & Calkins, C. R., (2008). *Predicting aged beef tenderness with a hyperspectral imaging system*. Animal Science Department, Nebraska Beef Cattle Reports, University of Nebraska–Lincoln, USA, pp. 138–139.

Heia, K., Sivertsen, A. H., Stormo, S. K., Elvevoll, E., Wold, J. P., & Nilsen, H. (2007). Detection of nematodes in cod (*Gadus morhua*) fillets by imaging spectroscopy. *Journal of Food Science, 72*(1), E011–E015.

Heitschmidt, G. W., Park, B., Lawrence, K. C., Windham, W. R., & Smith, D. P. (2007). Improved hyperspectral imaging system for fecal detection on poultry carcasses. *Transactions of the ASABE, 50*(4), 1427–1432.

Herrero, A. N. (2008). Raman spectroscopy a promising technique for quality assessment of meat and fish: A review. *Food Chemistry, 107*(4), 1642–1651.

Honikel, K. O. (1998). Reference methods for the assessment of physical characteristics of meat. *Meat Science, 49*(4), 447–457l.

Hruschka, W. R., (2001). Data analysis: wavelength selection methods. In P. Williams, & K. Norris (Eds.), *Near infrared technology in the agricultural and food industries* (2nd ed., pp. 30–58). St Paul, MN: American Association of Cereal Chemists.

Huidobro, A., Pastor, A., Lopez-Caballero, M. E., & Tejada, M. (2001). Washing effect on the quality index method (QIM) developed for raw gilthead seabream (*Sparus aurata*). *European Food Research and Technology, 212*, 408–412.

Jackman, P., Sun, D.-W., & Allen, P. (2009a). Automatic segmentation of beef *longissimus dorsi* muscle and marbling by an adaptable algorithm. *Meat Science, 83*(2), 187–194.

Jackman, P., Sun, D.-W., & Allen, P. (2009b). Comparison of various wavelet texture features to predict beef palatability. *Meat Science, 83*(1), 82–87.

Jackman, P., Sun, D.-W., & Allen, P. (2009d). Comparison of the predictive power of beef surface wavelet texture features at high and low magnification. *Meat Science, 82*(3), 353–356.

Jackman, P., Sun, D.-W., & Allen, P. (2010). Prediction of beef palatability from color, marbling and surface texture features of *longissimus dorsi*. *Journal of Food Engineering, 96*(1), 151–165.

Jackman, P., Sun, D.-W., Du, C.-J., & Allen, P. (2009c). Prediction of beef eating qualities from color, marbling and wavelet surface texture features using homogenous carcass treatment. *Pattern Recognition, 42*(5), 751–763.

Jackman, P., Sun, D.-W., Du, C.-J., Allen, P., & Downey, G. (2008). Prediction of beef eating quality from color, marbling and wavelet texture features. *Meat Science, 80*(4), 1273–1281.

Kavdır, I., & Guyer, D. E. (2004). Comparison of artificial neural networks and statistical classifiers in apple sorting using textural features. *Biosystems Engineering, 89*(3), 331–344.

Khodabux, K., L'Omelette, M. S. S., Jhaumeer-Laulloo, S., Ramasami, P., & Rondeau, P. (2007). Chemical and near-infrared determination of moisture, fat and protein in tuna fishes. *Food Chemistry, 102*(3), 669–675.

Kim, I., Kim, M. S., Chen, Y. R., & Kong, S. G. (2004). Detection of skin tumors on chicken carcasses using hyperspectral fluorescence imaging. *Transactions of the ASAE, 47*(5), 1785–1792.

Kong, S. G., Chen, Y.-R., Kim, I., & Kim, M. S. (2004). Analysis of hyperspectral fluorescence images for poultry skin tumor inspection. *Applied Optics, 43*(4), 824–833.

Lawrence, K. C., Windham, W. R., Park, B., & Buhr, R. J. (2003). Hyperspectral imaging system for identification of faecal and ingesta contamination on poultry carcasses. *Journal of Near Infrared Spectroscopy, 11*, 261–281.

Lawrence, K. C., Windham, W. R., Park, B., Smith, D. P., & Poole, G. H. (2004). Comparison between visible/NIR spectroscopy and hyperspectral imaging for detecting surface contaminants on poultry carcasses. In B.S. Bennedsen, Y.-R. Chen, G. E. Meyer, A. G. Senecal, & S.-I. Tu (Eds.),. *Monitoring Food Safety, Agriculture, and Plant Health. Proceedings of SPIE, Vol. 5271*, 35–42.

Leitner, R., Mairer, H., & Kercek, A. (2003). Real-time classification of polymers with NIR spectral imaging and blob analysis. *Real-Time Imaging, 9*(4), 245–251.

Leroy, B., Lambotte, S., Dotreppe, O., Lecocq, H., Istasse, L., & Clinquart, A. (2003). Prediction of technological and organoleptic properties of beef *longissimus thoracis* from near-infrared reflectance and transmission spectra. *Meat Science, 66*, 45–54.

Li, J., Tan, J., & Shatadal, P. (2001). Classification of tough and tender beef by image texture analysis. *Meat Science, 57*, 341–346.

Liu, Y., Lyon, B. G., Windham, W. R., Realini, C. B., Pringle, T. D. D., & Duckett, S. (2003a). Prediction of color, texture, and sensory characteristics of beef steaks by visible and near infrared reflectance spectroscopy: a feasibility study. *Meat Science, 65*, 1107–1115.

Liu, Y., Windham, W. R., Lawrence, K. C., & Park, B. (2003b). Simple algorithms for the classification of visible/NIR and hyperspectral imaging spectra of chicken skins, feces, and fecal contaminated skins. *Journal of Applied Spectroscopy, 57*, 1609–1612.

Lusk, J. L., Fox, J. A., Schroeder, T. C., Mintert, J., & Koohmaraie, M. (2001). In-store evaluation of steak tenderness. *American Journal of Agricultural Economics, 83*(3), 539–550.

Manolakis, D., Marden, D., & Shaw, G. A. (2003). Hyperspectral image processing for automatic target detection applications. *Lincoln Laboratory Journal, 14*(1), 78–116.

Millar, S. J., Whitworth, M. B., Chau, A., & Gilchrist, J. R. (2008). Mapping food composition using NIR hyperspectral imaging. *New Food, 11*(3), 34–39.

Naganathan, G. K., Grimes, L. M., Subbiah, J., Calkins, C. R., Samal, A., & Meyer, G. E. (2008a). Visible/near-infrared hyperspectral imaging for beef tenderness prediction. *Computers and Electronics in Agriculture, 64*(2), 225–233.

Naganathan, G. K., Grimes, L. M., Subbiah, J., Calkins, C. R., Samal, A., & Meyer, G. E. (2008b). Partial least squares analysis of near-infrared hyperspectral images for beef tenderness prediction. *Sensing and Instrumentation for Food Quality and Safety, 2*, 178–188.

Nakariyakul, S., & Casasent, D. P. (2004). Hyperspectral feature selection and fusion for detection of chicken skin tumors. In B. S. Bennedsen, Y.-R. Chen, G. E. Meyer, A. G. Senecal, & S.-I. Tu (Eds.), Monitoring Food Safety, Agriculture, and Plant Health. *Proceedings of SPIE, Vol. 5271*, 128–139.

Nicolai, B. M., Beullens, K., Bobelyn, E., Peirs, A., Saeys, W., Theron, K. I., & Lammertyn, J. (2007). Nondestructive measurement of fruit and vegetable quality by means of NIR spectroscopy: a review. *Postharvest Biology and Technology, 46*(1), 99–118.

Noh, H. K., & Lu, R. (2007). Hyperspectral laser-induced fluorescence imaging for assessing apple fruit quality. *Postharvest Biology and Technology, 43*, 193–201.

Nortvedt, R., Torrissen, O. J., & Tuene, S. (1998). Application of near-infrared transmittance spectroscopy in the determination of fat, protein and dry matter in Atlantic halibut fillet. *Chemometrics and Intelligent Laboratory Systems, 42*, 199–207.

Park, B., Chen, Y. R., Hruschka, W. R., Shackelford, S. D., & Koohmaraie, M. (1998). Near-infrared reflectance analysis for predicting beef *longissimus* tenderness. *Journal of Animal Science, 76*, 2115–2120.

Park, B., Kise, M., Lawrence, K. C., Windham, W. R., Smith, D. P., & Thai, C. N. (2006b). Real-time multispectral imaging system for online poultry fecal inspection using UML. In Y.-R. Chen, G. E. Meyer, & S.-I. Tu (Eds.), *Optics for Natural Resources, Agriculture, and Foods. Proceeding of SPIE, 63810W-1-63810W-12.*

Park, B., Lawrence, K. C., Windham, W. R., & Buhr, R. J. (2002). Hyperspectral imaging for detecting fecal and ingesta contamination on poultry carcasses. *Transactions of the ASAE, 45*(6), 2017–2026.

Park, B., Lawrence, K. C., Windham, W. R., & Smith, D. P. (2006a). Performance of hyperspectral imaging system for poultry surface fecal contaminant detection. *Journal of Food Engineering, 75*(3), 340–348.

Park, B., Windham, W. R., Lawrence, K. C., & Smith, D. (2007). Contaminant classification of poultry hyperspectral imagery using a spectral angle mapper algorithm. *Biosystems Engineering, 96*(3), 323–333.

Peng, Y., Wu, J., (2008). Hyperspectral scattering profiles for prediction of beef tenderness. *ASABE Annual International Meeting*, Rhode Island, USA, Paper No. 080004.

Prevolnik, M., Candek-Potokar, M., & Skorjanc, D. (2004). Ability of NIR spectroscopy to predict meat chemical composition and quality – a review. *Czech Journal of Animal Science, 49*(11), 500–510.

Qiao, J., Ngadi, M. O., Wang, N., Gariepy, C., & Prasher, S. O. (2007a). Pork quality and marbling level assessment using a hyperspectral imaging system. *Journal of Food Engineering, 83*(1), 10–16.

Qiao, J., Ngadi, M., Wang, N., Gunenc, A., Monroy, M., Gariepy, C., & Prasher, S. (2007b). Pork quality classification using a hyperspectral imaging system and neural network. *International Journal of Food Engineering, 3*(1). Article No. 6.

Qiao, J., Wang, N., Ngadi, M., Gunenc, A., Monroy, M., Gariepy, C., & Prasher, S. (2007c). Prediction of drip-loss, pH, and color for pork using a hyperspectral imaging technique. *Meat Science, 76*(1), 1–8.

Rasmussen, A. J., & Andersson, M., (1996). New method for determination of drip loss in pork muscles. In *Proceedings 42nd International Congress of Meat Science and Technology*, Lillehammer, Norway, pp. 286–287.

Ripoll, G., Albertí, P., Panea, B., Olleta, J. L., & Sañudo, C. (2008). Near-infrared reflectance spectroscopy for predicting chemical, instrumental and sensory quality of beef. *Meat Science, 80*, 697–702.

Rust, S. R., Price, D. M., Subbiah, J., Kranzler, G., Hilton, G. G., Vanoverbeke, D. L., & Morgan, J. B. (2007). Predicting beef tenderness using near-infrared spectroscopy. *Journal of Animal Science, 86*, 211–219.

Rutlidge, H. T., & Reedy, B. J. (2009). Classification of heterogeneous solids using infrared hyperspectral imaging. *Applied Spectroscopy, 63*(2), 172–179.

Šašic, S. (2007). An in-depth analysis of Raman and near-infrared chemical images of common pharmaceutical tablets. *Applied Spectroscopy, 61*, 239–250.

Savell, J. W., Cross, H. R., Francis, J. J., Wise, J. W., Hale, D. S., Wilkes, D. L., & Smith, G. C. (1989). National consumer retail beef study: interaction of trim level price and grade on consumer acceptance of beef steaks and roasts. *Journal of Food Quality, 12*, 251.

Shackelford, S. D., Wheeler, T. L., & Koohmaraie, M. (1995). Relationship between shear force and trained sensory panel tenderness ratings of 10 major muscles from Bos-Indicus and Bos-Taurus cattle. *Journal of Animal Science, 73*(11), 3333–3340.

Shackelford, S. D., Wheeler, T. L., & Koohmaraie, M. (1997). Tenderness classification of beef: I. Evaluation of beef *longissimus* shear force at 1 or 2 days postmortem as a predictor of aged beef tenderness. *Journal of Animal Science, 75*, 2417–2422.

Shackelford, S. D., Wheeler, T. L., & Koohmaraie, M. (2004). Development of optimal protocol for visible and near-infrared reflectance spectroscopic evaluation of meat quality. *Meat Science, 68*, 371–381.

Shackelford, S. D., Wheeler, T. L., & Koohmaraie, M. (2005). On-line classification of US select beef carcasses for *longissimus* tenderness using visible and near-infrared reflectance spectroscopy. *Meat Science, 69*(3), 409–415.

Sivertsen, A. H., Chu, C.-K., Wang, L.-C., Godtliebsen, F., Heia, K., & Nilsen, H. (2009). Ridge detection with application to automatic fish fillet inspection. *Journal of Food Engineering, 90*, 317–324.

Sugiyama, J. (1999). Visualization of sugar content in the flesh of a melon by near-infrared imaging. *Journal of Agricultural and Food Chemistry, 47*, 2715–2718.

Sun, D.-W. (Ed.). (2008a). *Computer vision technology for food quality evaluation*. San Diego, CA: Academic Press/Elsevier.

Sun, D.-W. (Ed.). (2008b). *Infrared spectroscopy for food quality analysis and control*. San Diego, CA: Academic Press/Elsevier.

Swatland, H. J. (1989). A review of meat spectrophotometry (300–800 nm). *Canadian Institute of Food Science and Technology, 22*, 390–402.

Valdimarsson, G., Einarsson, H., & King, F. J. (1985). Detection of parasites in fish muscle by candling technique. *Journal of Association of Official Analytical Chemists, 68*, 549.

Valkova, V., Salakova, A., Buchtova, H., & Tremlova, B. (2007). Chemical, instrumental and sensory characteristics of cooked pork ham. *Meat Science, 77*(4), 608–615.

Vote, D. J., Belk, K. E., Tatum, J. D., Scanga, J. A., & Smith, G. C. (2003). Online prediction of beef tenderness using a computer vision system equipped with a BeefCam module. *Journal of Animal Science, 81*, 457–465.

Warner, R. D., Kauffman, R. G., & Greaser, M. L. (1997). Muscle protein changes post mortem in relation to pork quality traits. *Meat Science, 45*(3), 339–352.

Windham, W. R., Heitschmidt, G. W., Smith, D. P., & Berrang, M. E. (2005a). Detection of ingesta on pre-chilled broiler carcasses by hyperspectral imaging. *International Journal of Poultry Science, 4*(12), 959–964.

Windham, W. R., Smith, D. P., Berrang, M. E., Lawrence, K. C., & Feldner, P. W. (2005b). Effectiveness of hyperspectral imaging system for detecting cecal contaminated broiler carcasses. *International Journal of Poultry Science, 4*(9), 657–662.

Windham, W. R., Smith, D. P., Park, B., & Lawrence, K. C. (2003). Algorithm development with visible/near infrared spectra for detection of poultry feces and ingesta. *Transactions of the ASAE, 46*, 1733–1738.

Winger, R. C., & Hagyard, C. J. (1994). Juiciness—its importance and some contribution factors. In A. M. Pearson A.M., & T. R. Dustson (Eds.), *Quality Attributes and Their Measurement in Meat, Poultry and Fish Products* (pp. 94–124). London, UK: Blackie Academic & Professional, Advances in Meat Research Series 9.

Wold, J. P., & Isaksson, T. (1997). Non-destructive determination of fat and moisture in whole Atlantic salmon by near-infrared diffuse spectroscopy. *Journal of Food Science, 62*(4), 734–736.

Wold, J. P., Johansen, T., Haugholt, K. H., Tschudi, J., Thielemann, J., Segtnan, V. H., Narum, B., & Wold, E. (2006). Non-contact transflectance near infrared imaging for representative on-line sampling of dried salted coalfish (bacalao). *Journal of Near Infrared Spectroscopy, 14*(1), 59–66.

Wold, J. P., Westad, F., & Heia, K. (2001). Detection of parasites in cod fillets by using SIMCA classification in multispectral images in the visible and NIR region. *Applied Spectroscopy, 55*(8), 1025–1034.

Xia, J. J., Berg, E. P., Lee, J. W., & Yao, G. (2007). Characterizing beef muscles with optical scattering and absorption coefficients in VIS–NIR region. *Meat Science, 75*(1), 78–83.

Xing, J., Ngadi, M., Gunenc, A., Prasher, S., & Gariepy, C. (2007). Use of visible spectroscopy for quality classification of intact pork meat. *Journal of Food Engineering, 82*, 135–141.

Yam, K. L., & Papadakis, S. E. (2004). A simple digital imaging method for measuring and analyzing color of food surfaces. *Journal of Food Engineering, 61*(1), 137–142.

Yang, C.-C., Chao, K., & Kim, M. S. (2009). Machine vision system for online inspection of freshly slaughtered chickens. *Sensing and Instrumentation for Food Quality and Safety, 3*(1), 70–80.

Yoon, S. C., Lawrence, K. C., Smith, D. P., Park, B., & Windham, W. R. (2008). Embedded bone fragment detection in chicken fillets using transmittance image enhancement and hyperspectral reflectance imaging. *Sensing and Instrumentation for Food Quality and Safety, 2*, 197–207.

CHAPTER 7

Automated Poultry Carcass Inspection by a Hyperspectral–Multispectral Line-Scan Imaging System

Kuanglin Chao

US Department of Agriculture, Agricultural Research Service, Henry A. Wallace Beltsville Agricultural Research Center, Environmental Microbial and Food Safety Laboratory, Beltsville, Maryland, USA

CONTENTS

7.1. INTRODUCTION

Hyperspectral imaging is one of the latest technologies to be developed for effective and non-destructive quality and safety inspection in the area of food processing. The technology takes the most useful characteristics of both machine vision and spectroscopy, two technologies already widely used in the food and agricultural industries. *Machine vision* imaging is commonly used to detect surface features (color, size/shape, surface texture, or defects) in food inspection, but cannot identify or detect chemical, biological, or material properties or characteristics from the product. In contrast, *spectroscopy* can evaluate these properties and characteristics, but does not provide the spatial information that is often critical in food inspection. *Hyperspectral imaging* integrates the main features of imaging and spectroscopy to simultaneously acquire both spectral and spatial information, which is key to evaluating food safety and quality attributes. As a result, the technology provides us with unprecedented detection capabilities, which otherwise cannot be achieved with either imaging or spectroscopy alone. Hyperspectral imaging technology has been a proven tool for developing methods of

CONTENTS

automated multispectral inspection (Kim *et al.*, 2001; Lu & Chen, 1998). However, one major limiting factor that initially hindered direct commercial application of hyperspectral technology for on-line use was the speed needed for rapid acquisition and processing of large-volume hyperspectral image data (Chen *et al.*, 2002; Gowen *et al.*, 2007). More recently, advanced computer and optical sensing technologies are gradually overcoming this problem, as demonstrated by the development of on-line hyperspectral detection systems for inspection of poultry carcasses for wholesomeness and fecal contamination (Chao, Chen *et al.*, 2002; Lawrence, Park *et al.*, 2003; Lawrence, Windham *et al.*, 2003; Park *et al.*, 2002), inspection of apples for fecal contamination (Kim *et al.*, 2002, 2004; Mehl *et al.*, 2002), and sorting and grading of fruits for internal quality (Lu & Peng, 2006; Noh & Lu, 2007; Qin & Lu, 2006). In particular, the line-scan imaging technology has demonstrated significant advantages for the direct implementation of hyperspectral imaging for rapid automated food quality and safety inspection such as on-line poultry inspection (Chao *et al.*, 2007; Yang *et al.*, 2009).

The Agricultural Research Service (ARS), an agency of the United States Department of Agriculture (USDA), has had a long-term interest in developing automated inspection methods for food and agricultural products. Beginning in the 1960s and continuing to this day, ARS research on spectroscopy and spectral imaging methods for non-destructive food quality and safety measurement has included the development of visible/near-infrared (VIS/NIR) techniques for grain and oilseed quality, for fruit and vegetable quality, for food quality and safety of dairy, meat, and poultry. The first computerized NIR spectrophotometer was developed by ARS researchers and ultimately led to the now widespread use of the technology in the grain industry. Current ongoing research includes the development of automated spectral imaging for the detection of surface contaminants on fresh produce and for wholesomeness inspection and contamination detection on poultry carcasses, all on high-speed processing lines.

This chapter describes the development of automated chicken inspection techniques by ARS researchers that has led to the latest hyperspectral line-scan imaging system for wholesomeness inspection on commercial high-speed processing lines, which is now under commercial development for industrial use. In addition to the usual problems inherent to the fundamental research for developing feasible spectral methods to assess poultry characteristics, researchers also had to address significant challenges in adapting the scientific findings to implement them in the current ARS imaging system for practical real-world use in automated high-speed processing environments. The current USDA poultry inspection program and the progression of

ARS VIS/NIR spectroscopy methods, target-imaging, and hyperspectral/multispectral line-scan imaging for poultry inspection are discussed in the following sections.

7.2. CURRENT UNITED STATES POULTRY INSPECTION PROGRAM

The 1957 Poultry Product Inspection Act (PPIA) mandates post-mortem inspection of every bird carcass processed by a commercial facility for human consumption. Since then, the USDA has employed inspectors from its Food Safety and Inspection Service (FSIS) agency to conduct on-site organoleptic inspection of all chickens processed in poultry plants in the United States for indications of disease or defect conditions. At inspection stations on commercial evisceration lines, FSIS inspectors examine by sight and by touching the body, the inner body cavity surfaces, and the internal organs of every chicken carcass on the evisceration line (USDA, 2005). Most chicken plants in the United States operate evisceration lines at speeds of 70 birds per minute (bpm) or 91 bpm, using either a Streamlined Inspection System (SIS) or a New Line Speed (NELS) inspection system. Additionally, some newer high-speed chicken processing systems operate evisceration lines at even higher speeds, as high as 140 bpm. By law, the human inspectors may work at a maximum speed of 35 bpm, which results in multiple inspection stations along a single line. In this way, for example, one inspector examines every fourth chicken on a 140 bpm line and the line is equipped with four inspection stations in the USDA inspection zone (Figure 7.1).

Poultry processing plants in the United States process over 8.9 billion broilers (young chickens, 4–6 weeks old) annually, more than any other country and valued at over $31 billion (USDA, 2008). Broiler production has increased dramatically over the years to meet rising market demand. The domestic per capita consumption of broilers increased from 27 kg in 1990 to 34.9 kg in 2000, and reached 39.5 kg in 2006. Currently, about 2200 FSIS poultry inspectors are employed to inspect roughly 8.9 billion broilers per year. With ever-increasing consumer demand for poultry products, human inspection capability is becoming the limiting factor to increased production throughput.

The inspection program implemented by FSIS as a result of the 1957 PPIA addresses mandatory inspection of poultry products for food safety only, not for quality attributes. Products that pass this safety inspection are labeled as having been USDA Inspected and Passed. The human inspectors are trained to visually and manually inspect poultry carcasses and viscera

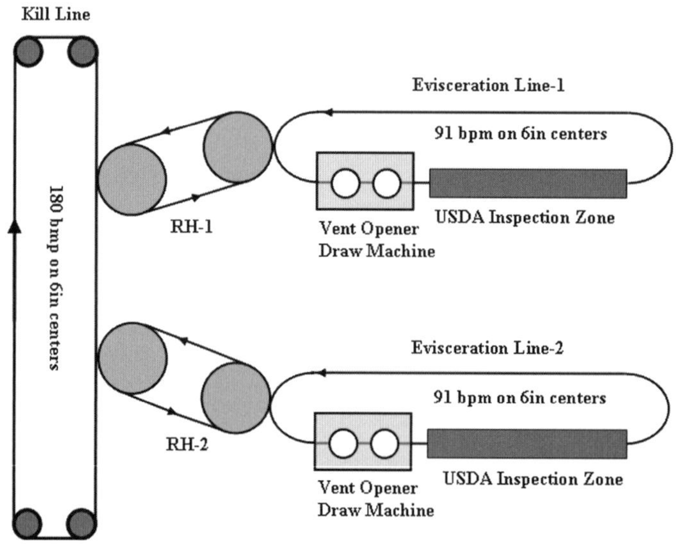

FIGURE 7.1 *Diagram of poultry processing lines, including a 180 bpm Kill Line, two Re-Hangers (RH), and two 91 bpm Evisceration Lines. Inspection stations are located in the USDA Inspection Zones on the Evisceration Line. (Full color version available on http://www.elsevierdirect.com/companions/9780123747532/)*

on-line at processing plants to accurately identify unwholesome carcasses, including those exhibiting conditions such as septicemia/toxemia (septox), airsacculitis, ascites, cadaver, and inflammatory process (IP), and defects such as bruises, tumors, sores, and scabs. This inspection process is subject to human variability, and also makes inspectors prone to developing fatigue and repetitive motion injuries.

With advances in science and improved food safety awareness, USDA food safety programs have seen further developments during the past two decades. With the 1996 final rule on Pathogen Reduction and Hazard Analysis and Critical Control Point (HACCP) systems (USDA, 1996), FSIS implemented the HACCP and Pathogen Reduction programs in meat and poultry processing plants throughout the country to prevent food safety hazards. HACCP systems for food safety focus on prevention and monitoring of potential hazards at critical control points throughout a food production process, instead of focusing only on product safety inspection. FSIS has also been testing the HACCP-Based Inspection Models Project (HIMP) in a few volunteer plants (USDA, 1997). In this project, food safety performance standards are set by FSIS and the processing plants bear primary responsibility for conducting inspections and processing so that their products satisfy FSIS standards, while FSIS inspectors perform carcass verification along and at the end of the processing line, before the birds enter the final chill step. FSIS inspectors no longer perform bird-by-bird inspection in these volunteer HIMP plants, which number 20 plants out of over 400 federally inspected plants nationwide.

7.3. DEVELOPMENT OF VIS/NIR SPECTROSCOPY-BASED POULTRY INSPECTION SYSTEMS

Since the early 1990s, significant advances have occurred in the development of automated poultry inspection systems. Leading research in this area has been done by USDA Agricultural Research Service (ARS) scientists, originally initiated by an FSIS request to ARS to develop methods for addressing issues in high speed poultry inspection. Methods based on visible/near-infrared (VIS/NIR) reflectance spectroscopy were first investigated as a tool for identifying carcass conditions based on spectral measurements. This initial work resulted in both spectroscopy-based on-line inspection systems and also selective filter-based multispectral imaging systems for identifying poultry conditions.

7.3.1. Spectral Analysis Using Laboratory-based Photodiode-array Detection Systems

Information about the color, surface texture, and chemical constituents of chicken skin and muscle tissue is carried in VIS/NIR light reflected from a carcass. Because unwholesome carcasses that are diseased or defective often have a variety of changes in skin and tissue, these carcasses can be detected with VIS/NIR reflectance techniques that require no physical contact during data acquisition. A photodiode-array (PDA) VIS/NIR spectrophotometer system was first developed (Chen & Massie, 1993) to measure chicken spectra in the laboratory. The system used a bifurcated fiber-optic assembly for sample illumination and spectral reflectance measurement (471 nm to 964 nm), in conjunction with quartz–tungsten–halogen lamps to provide the illumination. The end of the probe was positioned approximately 2 cm from the chicken surface, a distance that was determined during laboratory experiments to optimize the signal-to-noise ratio for the spectral measurements. Spectra were measured for wholesome carcasses and septicemic/cadaver carcasses, with an acquisition time of 2 second for each stationary measurement.

Analysis using principal component analysis (PCA) achieved classification accuracies of 93.3% and 96.2% for the chicken samples in the wholesome and unwholesome classes, respectively. Later experiments (Chen *et al.*, 1998) were conducted to measure the reflectance spectra of freshly slaughtered carcasses hung on track-mounted sliding shackles, to simulate processing line speeds of 60 and 90 bpm. The measurements were acquired for wholesome carcasses and for carcasses exhibiting symptoms of septicemia/toxemia, cadaver, airsacculitis, ascites, and tumors (these disease/defect conditions often cause birds to be removed from the processing line).

7.3.2. Pilot Scale On-line Photodiode-array Poultry Inspection System

Based on the laboratory PDA spectrophotometer system, a transportable pilot-scale VIS/NIR system (Chen *et al.*, 1995) was developed and taken to a chicken processing plant to conduct on-line VIS/NIR spectral measurements on a 70 bpm commercial evisceration line. Reflectance measurements were selectively triggered for individual wholesome and unwholesome carcasses specifically identified by a veterinary medical officer observing the birds on the processing line. Acquisition time for each spectral measurement was 0.32 s, targeting an area of approximately 10 cm^2 across the breast area of each bird as it passed in front of the fiber-optic probe.

Processing and analysis of the spectral data was performed off-line. Preprocessing of the 1024-point raw spectral data included smoothing by 9-point running mean and a second difference calculation. Reduction of the second difference spectra was performed by extracting every fifth point, producing 190-point spectra spanning 486.1 nm to 941.5 nm (2.4 nm spacing). PCA was performed on these reduced second difference spectra. The coefficients (PCA scores) of the first 50 principal components were used as inputs to a feed-forward back-propagation neural network for classification of the chicken carcasses. The neural network used 50 input nodes, seven nodes in a hidden layer, and two output nodes whose ideal output were either (0,1) or (1,0) to indicate wholesome or unwholesome bird condition. For the data set collected for 1750 chickens (1174 wholesome and 576 unwholesome) on the 70 bpm processing line, analysis resulted in an average classification accuracy of 95% (Chen *et al.*, 1998, 2000).

7.3.3. Charge-coupled Device Detector Systems for In-plant On-line Poultry Inspection

The VIS/NIR spectrophotometer system was updated to enable on-line chicken inspection at line speeds greater than 90 bpm. The PDA detection system was replaced with a charge-coupled device- (CCD-) based detection system, which allowed for much shorter data acquisition times for spectral measurement (Chao *et al.*, 2003). Additionally, moving away from the Microsoft DOS-based software drivers of the PDA detector allowed researchers to use up-to-date software based on the Microsoft Windows operating system that provided greater flexibility and modularity to implement real-time on-line acquisition, processing, and classification algorithms.

Testing of the updated CCD-based VIS/NIR chicken inspection system (Figure 7.2) for in-plant on-line poultry inspection was conducted in commercial poultry plants at line speeds of 140 and 180 bpm. In-plant testing of this system was guided by specific FSIS food safety performance standards defined under HIMP. One specific HIMP food safety performance standard requires removal of any bird exhibiting the systemic disease conditions of septicemia or toxemia, which are characterized by pathogenic microorganisms or their toxins in the bloodstream. As a result, system testing targeted classification of 450 wholesome and 426 unwholesome (specifically, systemically diseased) birds that were selected by an FSIS veterinary medical officer observing birds on the processing line. A 1024-point spectrum was acquired across the range of 431–943 nm for each individual bird, using a total data acquisition time of 60 ms per bird in order to process an accumulation of three consecutive spectral measurements of 20 ms each. At 140 bpm, classification accuracies of 95% for wholesome and 92% for unwholesome birds were achieved. At 180 bpm, classification accuracies of 94% and 92% for wholesome and unwholesome birds were achieved, respectively (Chao *et al.*, 2004).

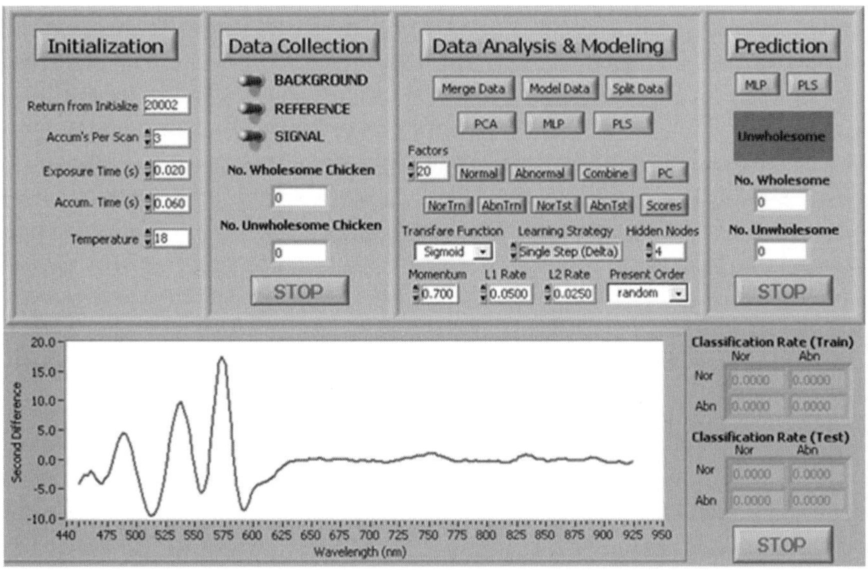

FIGURE 7.2

User interface of the CCD-based VIS/NIR chicken inspection system. (Full color version available on http://www.elsevierdirect.com/companions/9780123747532/)

7.3.4. Summary of Spectroscopy-based Poultry Inspection Systems

The spectroscopy-based inspection systems were designed to scan a limited area of each bird carcass, to measure the reflectance spectra in the VIS/NIR regions between 400 nm and 950 nm for detection of condemnable systemic disease conditions. In-plant testing demonstrated classification accuracies above 90% in differentiating wholesome and systemically diseased chickens. The upgrade from PDA- to CCD-based detection resulted in significantly improved data processing speeds due to the associated computer peripherals for data transfer. The use of CCD-based detection also provided significantly greater flexibility in the development of software controls for on-line inspection applications, especially for analysis algorithms such as neural network classification of spectra.

However, spectral classification utilizing the neural network approach requires a considerable volume of training data—e.g. in-plant on-line measurements of at least 500 wholesome and 500 unwholesome birds—in order for the classification model to reliably differentiate between wholesome and unwholesome birds. Because calibration is customized to a particular population of birds, re-calibration is necessary to accommodate different environmental and growth factors that affect bird condition (e.g. changes between seasons, diet) and processing variations between different processing plants (e.g. scalding parameters).

Consistent sample presentation is ideal for spectral classification—but difficult to achieve in the on-line processing environment. Ideally, spectral measurement would be performed with the probe positioned 2 cm from the surface of the bird breast, at the mid-breast area of the bird. In the processing plant, not only might there be variation in that probe-to-surface distance due to vibration and both forward/backward and side-to-side sway of the bird on the shackle, but also difficulty in using the fixed-position probe to accurately scan the mid-breast target area due to the external sensor system (used to trigger measurement for each bird) and variations in individual bird size and shape. Vibration and sway of the birds can be addressed with stability-enhancing equipment such as guide bars and synchronized belts on the processing line, but attempting to adjust probe position for bird-to-bird variations in size and shape would be an extremely difficult challenge.

Spectroscopy-based inspection using a fiber-optic probe is well suited for the commercial chicken processing environment since the probe assembly can be easily mounted on the line and can tolerate humidity and higher temperatures while the detector and computer system can be sheltered a short distance away within a climate-controlled enclosure. However, the spectroscopy-based inspection system is limited to small area measurements across each bird, and not necessarily with precision targeting of the measurement area. Systemic disease conditions can be detected in this way, but problems affecting only localized portions of a bird carcass, such as inflammatory process (IP) or randomly located defects or contamination (bruises, tumors, and fecal contamination), cannot be effectively identified without whole-surface carcass inspection.

7.4. DEVELOPMENT OF TARGET-TRIGGERED IMAGING FOR ON-LINE POULTRY INSPECTION

Following initial ARS development of spectroscopy-based on-line inspection systems, subsequent improvements in computer technology and optical sensing devices made possible the development of laboratory-based multispectral and hyperspectral imaging systems. Formerly used primarily for remote sensing applications, hyperspectral imaging technology was adapted for small-scale laboratory experiments by ARS researchers and others, once off-the-shelf computer systems became able to handle the huge hyperspectral data volumes. These laboratory systems were used to analyze hyperspectral image data for the development of multispectral methods suitable for addressing specific inspection applications. The resultant multispectral methods, using a limited number of wavelengths, were applied for target-triggered on-line implementation in separate multispectral imaging systems due to imaging and processing speed restrictions imposed by hardware limitations.

7.4.1. Dual-camera and Color Imaging

Based on the work of Chen & Massie (1993), who analyzed VIS/NIR spectral data by PCA, an intensified multispectral imaging system using six optical filters (at 542, 570, 641, 700, 720, and 847 nm) and neural network classifiers was developed in the laboratory for discrimination of wholesome poultry carcasses from unwholesome carcasses that included septicemia/toxemia and cadaver carcasses (Park & Chen, 1994). The accuracy for separation of wholesome carcasses from unwholesome carcasses was 89.3%. Following these results, the textural features based on the co-occurrence matrix of the multispectral images were analyzed (Park & Chen, 1996). Because the 542 nm and 700 nm wavelengths were found to be significant in separating the wholesome and unwholesome birds, a dual-camera imaging system using interference filters was then assembled for testing on a laboratory pilot-scale processing line (Park & Chen, 2000). This dual-wavelength system used two interference filters (20 nm bandpass), one centered at 540 nm and the other at 500 nm. Two black/white progressive scan cameras (TM-9701, PULNiX Inc., Sunnyvale, CA, USA) were positioned side-by-side, each fitted with one of the two filters. The dual-camera system acquired pairs of images for chickens on shackles moving at 60 bpm. Off-line image processing and input of the image intensity data to a feed-forward back-propagation neural network resulted in classification accuracies of 93.3% for the septicemia carcasses and 95.8% for cadaver carcasses (Chao *et al.*, 2000). At a commercial poultry processing plant, the dual-camera imaging system was used for on-line image acquisition on a 70 bpm evisceration line. The images of 13 132 wholesome and 1 459 unwholesome chicken carcasses were analyzed off-line and resulted in classification accuracies of 94% and 87% for wholesome and unwholesome carcasses, respectively (Chao, Chen *et al.*, 2002).

Symptoms of some unwholesome poultry conditions such as airsacculitis and ascites can be exhibited by the visceral organs of a bird. Since human inspectors on chicken processing lines often examine both the poultry viscera and the outer muscle and skin, chicken liver and heart samples were collected from wholesome carcasses and unwholesome septicemia, airsacculitis, and cadaver carcasses (40 birds for each of the four categories) to investigate the classification of bird condition based on color imaging of the viscera. With the available samples divided equally between training and validation data sets, combined color image features of the liver and heart for each individual bird were entered into a generalized neuro-fuzzy classification model that achieved 86.3% and 82.5% accuracies for the training and validation data sets, respectively (Chao *et al.*, 1999).

Although these results showed the potential of detecting individual diseases by color imaging of both carcass and visceral organs, this line of investigation was not developed further due to the relatively few plants using processing lines in which viscera and carcass can be suitably presented for imaging. Systems do exist in which the visceral organs are consistently presented on a tray alongside the carcass, but in most poultry plants, carcasses are hung on processing line shackles with the visceral organs automatically drawn and draped to the side in a randomly oriented manner.

7.4.2. Two-dimensional Spectral Correlation and Color Mixing

In general, the development of multispectral imaging techniques first requires analysis and selection of specific wavelengths to be implemented by said multispectral image techniques. Many studies have based wavelength selection on the use of chemometrics and multivariate analysis such as PCA. Alternative methods have also been investigated during the course of developing multispectral imaging methods for chicken inspection, including two-dimensional spectral correlation (2-D correlation) and color mixing.

Myoglobin is the major pigment in well-bled muscle-tissue. The color of muscle tissue is largely determined by the relative amounts of three forms of myoglobin at the surface: deoxymyoglobin, oxymyoglobin, and metmyoglobin. Generally, deoxymyoglobin appears purplish; oxymyoglobin—an oxygenated form—appears bright red; and metmyoglobin—the oxidized form of the previous two—appears brownish. Liu & Chen (2000, 2001) applied a 2-D correlation technique to spectral data collected for chicken breast meat samples to investigate spectral differences related to chicken meat conditions. These studies examined the changes in myoglobin proteins that occur during meat degradation and storage processes, and identified spectral absorptions associated with the molecular vibrations of specific myoglobin species (i.e. measurable changes due to inter-species reactions affecting relative amounts of the myoglobin forms found in different meat conditions). It was also realized that spectral absorptions could be affected by unique molecular vibrations resulting from the interactions of these myoglobin species with surrounding meat components such as water and lipids, not just from the molecular vibrations of the myoglobin species themselves. Visible wavebands identified in association with these myoglobin species included bands near 545 nm and 560 nm with oxymyoglobin, 445 nm with deoxymyoglobin, and 485 nm with metmyoglobin.

The development of imaging methods for chicken inspection has generally focused on methods to accentuate spectral and spatial features of carcasses, but not necessarily in connection to how these features are perceived through human vision. Color appearance models such as those ratified by the International Commission on Illumination (CIE) are often used in color imaging applications related to human color vision. Ding *et al.* (2005, 2006) investigated two-band color mixing for visual differentiation of the color appearance of several categories of chicken carcass condition, including wholesome, septicemia, and cadaver carcasses. Selection of waveband pairs to enhance differentiation between the categories was based on calculations of color difference and chromaticness difference indices from visible reflectance spectra of chicken samples. Simulation using the revised 1997 CIE color appearance model (CIECAM97) was performed to objectively evaluate the visual enhancement provided by an optical color-mixing device implementing the selected waveband pairs. It was found that single-category visual differentiation produced the best results when using pairs of filters, each 10 nm full width at half maximum (FWHM), as follows: 449 nm and 571 nm for wholesome carcasses, 454 nm and 590 nm for septicemia carcasses, and 458 nm and 576 nm for cadaver carcasses. Visually perceived differences between all the categories of chicken conditions (i.e. multi-category differentiation) could be enhanced by an optical color-mixing tool using filters centered at 454 nm and 578 nm.

7.4.3. Target-triggered Multispectral Imaging Systems

Initial development of a multispectral imaging chicken inspection system involved a three-channel common-aperture camera to simultaneously acquire spatially-matched three-waveband image data. The multispectral imaging system consisted of the common aperture camera (MS2100, DuncanTech, Auburn, CA, USA), a frame grabber (PCI-1428, National Instruments, Austin, TX, USA), an industrial computer, and eight 100W tungsten–halogen lights (Yang *et al.*, 2005). A color-separating prism split broadband light entering the camera lens into three optical channels, each of which passed through an interference filter placed before a CCD imaging array. Control of the camera settings, such as triggering mode, output bit depth, and the integration time of exposure and the analog gain at the CCD sensor before the image was digitized for each imaging channel, was accomplished using the CameraLink utility program (DuncanTech, Auburn, CA, USA). Signals from the CCD imaging arrays were digitized by the frame grabber. An 8-bit image was saved from each of the three channels; the selection of interference filters for the three channels was based on the results of previous studies (Liu & Chen, 2000, 2001; Ding *et al.*, 2005). Commercially available interference filters with center wavelengths at 461.75 nm (20.78 nm FWHM), 541.80 nm (18.31 nm FWHM), and 700.07 nm (17.40 nm FWHM) were used for the three-channel common aperture camera. Image data acquisition was performed for 174 wholesome, 75 inflammatory process, and 170 septicemia chicken carcasses on a pilot-scale processing line in the laboratory operating at a speed of 70 bpm. It was found that despite individually adjustable settings for gain and integration times for the three channels, simultaneously obtaining high-quality images across all three channels was difficult owing to a variety of factors, such as avoiding both image saturation at 700 nm and inadequate image intensity at 460 nm, due to the spectral characteristics of the tungsten–halogen illumination (increasing intensity with increasing wavelengths) and CCD detector sensitivity (low sensitivity under 500 nm). Integration times were finally settled at 5 ms, 10 ms, and 18 ms for the 700 nm, 540 nm, and 460 nm channels, respectively. Off-line image processing algorithms based on PCA and selection of region of interest (ROI) were developed as inputs to a decision tree classification model. This model was able to classify 89.6% of wholesome,

94.4% of septicemia, and 92.3% of inflammatory process chicken carcasses (Yang *et al.*, 2005).

Another common-aperture multispectral chicken inspection system was developed for detection of tumors on chicken carcasses (Chao, Mehl *et al.*, 2002). An enclosed illumination chamber was used to acquire multispectral images of individual chicken carcasses by the three-channel prism-based common-aperture camera (TVC3, Optec, Milano, Italy). The prism assembly of the camera system separated full spectrum visible light into three broadband channels (red, green, and blue). An 8-bit image (728 × 572 pixels each) was produced by one CCD for each channel and captured by a frame grabber (XPG-1000, Dipix, Ontario, Canada). The three CCDs were each preceded by a replaceable narrow band filter. The perfect image registration resulting from the use of the three CCDs allowed for true multispectral images of subjects in the illumination chamber. AC regulated 150W quartz–halogen source illumination was delivered to the chamber by a pair of fiber-optic line lights.

Selection of filter wavelengths to use the three-CCD system for tumor detection was based on hyperspectral analysis using a laboratory bench-top hyperspectral imaging system (Lu & Chen, 1998; Kim *et al.*, 2001) developed in-house. This hyperspectral imaging system used a CCD camera system (SpectraVideo, PixelVision, OR, USA) equipped with an imaging spectrograph (SPECIM Inspector, Spectral Imaging, Oulu, Finland), to capture a series of hyperspectral line-scan images from a linear field of view across the width of a conveyor belt. Each hyperspectral line-scan image consisted of 402 spatial pixels (spanning the width of sample presentation on the conveyor belt) and 120 spectral pixels spanning 420–850 nm. Samples on a conveyor belt were illuminated with light from a pair of 21V, 150W halogen lamps powered with a regulated DC voltage power supply (Fiber-Lite A-240P, Dolan-Jenner Industries, MA, USA). Eight chicken carcasses were placed on the conveyor belt, one at a time, and moved across the linear field of view while a series of line-scan images were acquired. These line-scan images were then compiled to form a complete hyperspectral image for each chicken carcass. Spatial ROIs from the hyperspectral images of eight chicken carcasses, each exhibiting tumors, were selected to include tumor areas and some surrounding normal skin tissue around the tumors. ENVI 3.2 software (Research Systems, Inc., CO, USA) was used to perform PCA. The principal component images for the tumor ROIs were visually examined to select the principal component showing the greatest contrast between tumor areas and normal skin. Identification of filters for multispectral detection of tumors was based on analysis for the major wavelengths contributing to the principal component producing the greatest visual contrast between tumors and normal skin on the chicken carcasses.

Two significant visible wavebands were noted from the weighted wavelength distribution corresponding to the eigenvector defined on chicken skin tumors that provided the best contrast between tumors and normal chicken skin: 475 nm and 575 nm. These wavebands correspond to metmyoglobin and oxymyoglobin bands (Liu & Chen, 2000). Because the far red region was previously found to be insensitive to surface defects on chickens (Park & Chen, 1994), bands in this region could therefore be utilized for masking and normalization of chicken carcass images. A filter centered at 705 ± 10 nm was chosen for this purpose, to be used with filters at 465 ± 10 nm and 575 ± 10 nm for the adaptable three-band CCD camera (the ± 10 nm designates the FWHM bandpass). Feature extraction from the variability of ratioed multispectral images, including mean, standard deviation, skewness, and kurtosis, provided the basis for fuzzy logic classifiers, which were able to separate normal from tumorous skin areas with increasing accuracies as more features were used. In particular, use of all three features gave successful detection rates of 91% and 86% for normal and tumorous tissue, respectively.

A high-resolution single-CCD imaging system utilized with an optical adaptor was also investigated for multispectral imaging inspection of wholesome vs. systemically diseased chicken carcasses (Yang *et al.*, 2006). This multispectral imaging system consisted of an image splitter (MultiSpec Imager, Optical Insights, LLC, Santa Fe, NM, USA), a back-illuminated CCD camera (SpectraVideo SV 512, PixelVision, Inc., Tigard, OR, USA), a PMB-004 shutter and cooler control board, a PMB-007 serial interface board, a PMJ-002 PCI bus data acquisition board, a LynxPCI frame grabber, a computer, and four 100W tungsten halogen lights. Four interference filters and an optical mirror assembly were used to create four waveband images of the target that were acquired simultaneously on a single CCD focal plane. The resulting 16-bit multispectral image contained four sub-images. The PixelView version 3.20 utility program (PixelVision, Inc., Tigard, OR, USA) was used to control camera settings, such as integration time and image acquisition.

Multispectral ROI features were developed to differentiate wholesome and systemically diseased chickens. Due to significant color differences between wholesome and systemically diseased chickens at 488 nm, 540 nm, and 580 nm, interference filters were selected at these wavebands for the multispectral imaging system; one additional filter was selected at 610 nm for image masking purposes. An algorithm was developed to find the ROI on the multispectral images. Classification thresholds for identifying wholesome and systemically diseased chickens were determined using a Classification and Regression Tree (CART) decision tree algorithm for 48 features per image that were defined by a combination of waveband, feature type, and classification area.

Multispectral images of a selected ROI for 332 wholesome and 328 systemic diseased chickens, using wavelengths at 488 nm, 540 nm, 580 nm, and 610 nm, were collected for image processing and analysis. The 610 nm image was used to create a mask to extract chicken images from background. Using a decision tree model, classification accuracies of 96.3% and 98.6% for wholesome and systemic diseased carcasses, respectively, were achieved.

7.5. DEVELOPMENT OF LINE-SCAN IMAGING FOR ON-LINE POULTRY INSPECTION

Although hyperspectral line-scan imaging was first used as a laboratory tool to develop target-triggered multispectral imaging systems, several key technological advances enabled the development of hyperspectral line-scan imaging for direct implementation in high-speed on-line inspection systems. In particular, the implementation of Electron-Multiplying Charge-Coupled-Device (EMCCD) detectors in camera systems and their use with imaging spectrographs made possible high-speed line-scan imaging systems capable of both hyperspectral and multispectral on-line imaging at the high speeds required by commercial poultry processing lines. As a result, both hyperspectral analysis for method development and multispectral implementation could be performed using the same on-line line-scan imaging system, greatly facilitating both method development and implementation.

7.5.1. Spectral Line-Scan Imaging System

Conventional development of multispectral inspection methods for on-line applications involves determination of specific spectral parameters using a hyperspectral imaging system or spectroscopy-based methods, followed by subsequent implementation of the parameters for use in a separate multispectral imaging system. The conversion and implementation of parameters from one system to another usually requires time-consuming cross-system calibration. The capability of a single system to operate in either hyperspectral or multispectral imaging mode can eliminate the need for cross-system calibration and ensure higher accuracy performance. This single-system approach was taken in the development of a line-scan imaging system capable of operating in either hyperspectral or multispectral imaging mode on a chicken processing line.

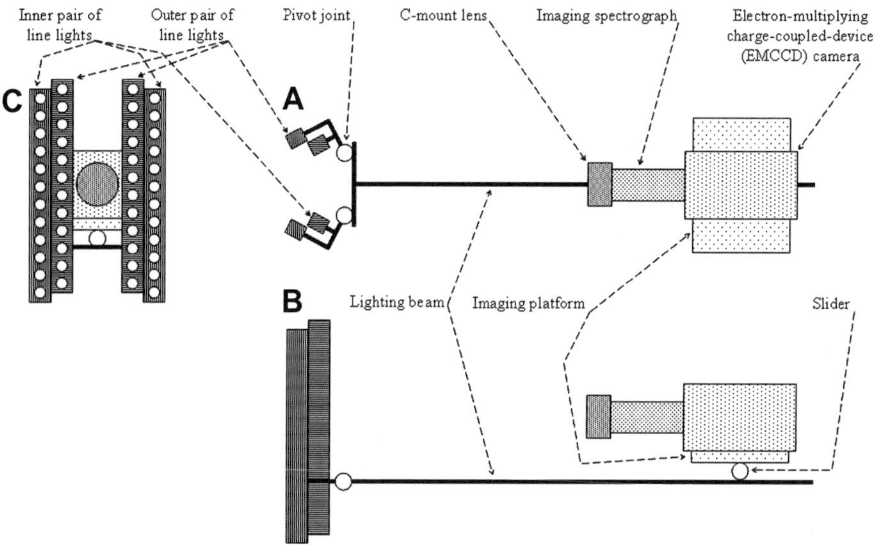

FIGURE 7.3 *Schematic of the hyperspectral/multispectral line-scan imaging system in (A) overhead view, (B) side view, and (C) front view*

Figure 7.3 shows the components of the hyperspectral/multispectral line-scan imaging system, including an EMCCD camera, an imaging spectrograph, a C-mount lens, and two pairs of high power, broad-spectrum white light emitting diode (LED) line lights. The EMCCD camera (PhotonMAX 512b, Roper Scientific, Inc., Trenton, NJ, USA) has 512 × 512 pixels and is thermoelectrically cooled to approximately −70 °C (via a three-stage Peltier device). An imaging spectrograph (ImSpector V10OEM, Specim/Spectral Imaging Ltd., Oulu, Finland), and a C-mount lens (Rainbow CCTV S6x11, International Space Optics, S.A., Irvine, CA, USA) are attached to the EMCCD imaging device. The 50 μm aperture slit of the spectrograph limits the instantaneous field of view (IFOV) of the imaging system to a thin line for line-scan imaging. Light from the IFOV is dispersed by a prism–grating–prism line-scan spectrograph and projected onto the EMCCD imaging device. The spectrograph creates a two-dimensional (spatial and spectral) image for each line-scan, with the spatial dimension along the horizontal axis and the spectral dimension along the vertical axis of the EMCCD imaging device. Thus, for hyperspectral imaging, a full spectrum is acquired for every pixel in each line scan (Figure 7.4). The spectral distribution of useful wavelengths and the size of the spatial image features to be processed determine the parameters for image binning, which reduces the number of image pixels and increases the signal-to-noise ratio by adding together photons from adjacent pixels in the detector array. More specific selection of wavelengths, spatial image size, and associated parameters such as binning can be optimized for

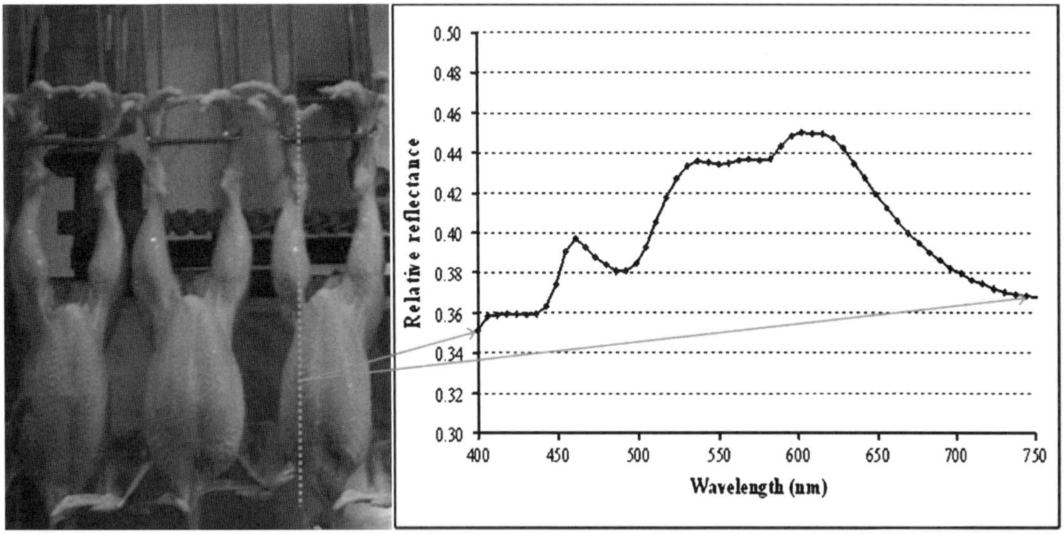

FIGURE 7.4 *Full-spectrum data is acquired for every pixel in each hyperspectral line-scan image. (Full color version available on http://www.elsevierdirect.com/companions/9780123747532/)*

either hyperspectral or multispectral imaging—for high-speed chicken processing, the capacity for short-exposure low-light imaging provided by the EMCCD detector is vital to successful on-line use in either mode. Pixels from the detector are binned by the high-speed shift register (which is built into the camera hardware) and transferred to the 16-bit digitizer, which has a rapid pixel-readout rate of approximately 10 MHz. The digitizer performs rapid analog-to-digital conversion of the image data for each line-scan image. The rapid image acquisition is followed by computer image analysis for real-time classification of wholesome and unwholesome pixels in the line-scan images of the chicken carcasses.

7.5.2. Hyperspectral Imaging Analysis

In hyperspectral imaging mode, a 55-band spectrum was acquired for each of the 512 spatial pixels in every hyperspectral line-scan image. The original hyperspectral line-scan image size (512 × 512 pixels) was reduced by 1 × 4 binning to produce line-scan images with a spectral resolution of 128 pixels (512 divided by 4) in the spectral dimension. Because the useful spectrum of light from the LED illumination did not span the entire width of the EMCCD detector, the first 20 and last 53 spectral bands were discarded, resulting in a final hyperspectral line-scan image size of 512 × 55 pixels. Hyperspectral images of wholesome and systemically diseased chickens, compiled from line-scans acquired on a 140 bpm commercial processing line, were analyzed off-line using MATLAB software (MathWorks, Natick, MA, USA) to determine ROI and spectral waveband parameters for use in multispectral wholesomeness inspection of the chickens.

For analysis, the 620 nm waveband was selected for masking purposes to remove the image background using a 0.1 relative reflectance threshold value. For any pixel in a hyperspectral line-scan, if its reflectance at 620 nm was below the threshold value, then that pixel was identified as background and its value at all wavebands was re-assigned to zero. The background-removed line-scan images were compiled to form images of chicken carcasses for a set of wholesome birds and a set of unwholesome birds. These images were analyzed to determine the parameters for an optimized ROI for use in differentiating wholesome and unwholesome birds. Within each bird image (Figure 7.5), the potential ROI area spanned an area from an upper border across the body of the bird to a lower border at the lowest non-background spatial pixel in each line scan, or to the last (512th) spatial pixel of the line-scan if there were no background pixels present at the lower edge. For each potential ROI, the average relative reflectance spectrum was calculated across all ROI pixels from all wholesome chicken images, and the average relative reflectance spectrum was also calculated across all ROI pixels from all unwholesome chicken images. The difference spectrum between the wholesome and unwholesome average spectra was calculated. This calculation was performed for all potential ROIs evaluated, which varied in size and

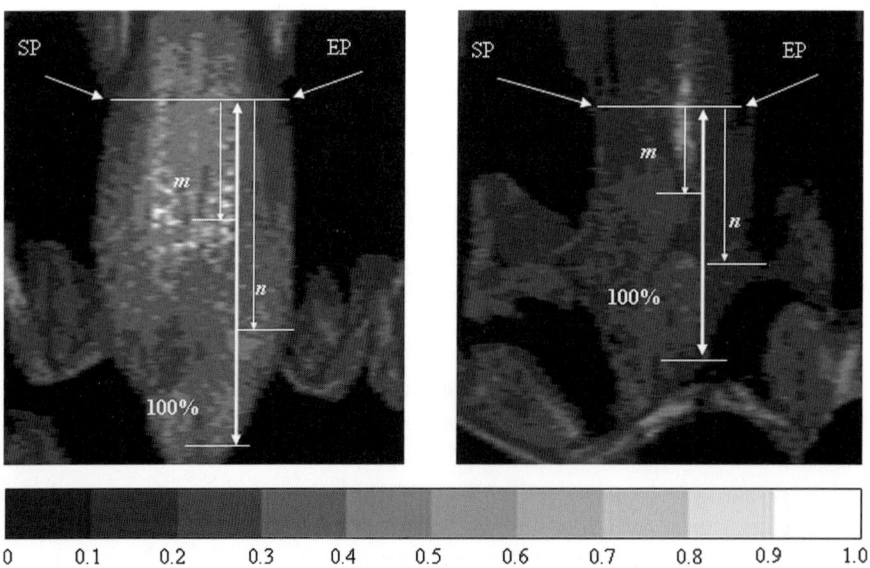

FIGURE 7.5 *Contour images of two chicken carcasses marked with example locations of the SP, EP,* m, *and* n *parameters used for locating the ROI*

were defined by the number of ROI pixels and their vertical coordinate locations within each line-scan. The optimized ROI was identified as the one that provided the greatest spectral difference between averaged wholesome pixels and averaged unwholesome pixels across all 55 wavebands.

A contour image of two example birds is shown in Figure 7.5, with the Starting Point and Ending Point (SP and EP, respectively) marked on each. Within each line-scan, possible ROI pixels begin at the SP–EP line and extend to the furthest non-background pixel below the SP–EP line, which in some cases coincides with the pixel at the far edge of the line-scan image. Parameters m and n indicate, as percentages of the pixel length between the SP–EP line and the furthest non-background pixel within each line-scan image, the location of the upper and lower ROI boundaries for ROIs under consideration. To optimize the ROI size and location, combinations of m and n were evaluated with values of m between 10% and 40% and values of n between 60% and 90%. For each possible ROI, the average spectrum was calculated across all ROI pixels from the 5 549 wholesome chicken carcasses, and the average spectrum was calculated across all ROI pixels from the 93 unwholesome chicken carcasses. The difference between the average wholesome and average unwholesome value at each of the 55 wavebands was calculated. Figure 7.6 shows the range of these 55 values for each possible ROI. Across all the possible ROIs, wavebands near 580 nm showed the

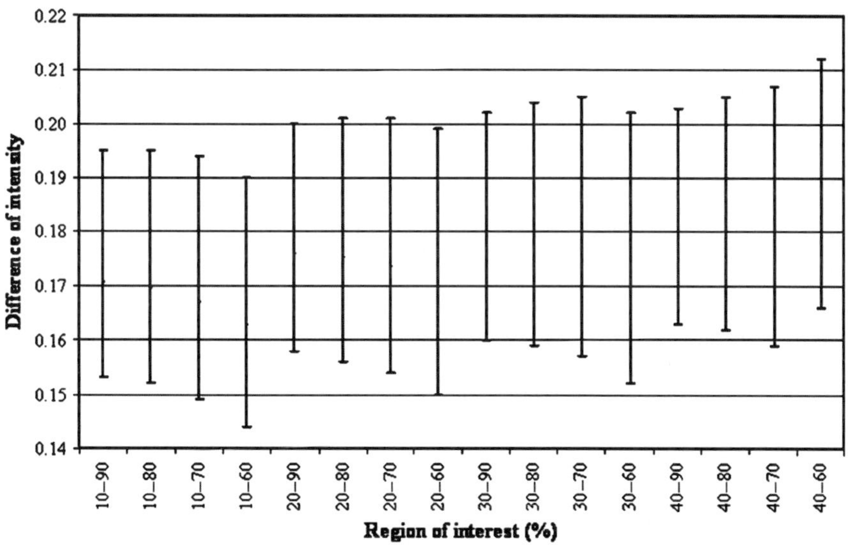

FIGURE 7.6 *Plot of the range of difference values between average wholesome and average unwholesome chicken spectra for ROIs evaluated during hyperspectral analysis to optimize the ROI selection for multispectral inspection of chickens*

highest difference between the average wholesome and average unwholesome spectra, and wavebands near 400 nm showed the lowest difference values. The 40–60% ROI showed the highest difference values overall, with the highest value of 0.212 occurring at 580 nm, and was thus the final ROI selection.

Using the optimized ROI, a single waveband was identified as being the waveband corresponding to the greatest spectral difference between averaged wholesome chicken pixels and averaged unwholesome chicken pixels, for differentiating wholesome and unwholesome chicken carcasses by relative reflectance intensity. The average wholesome and average unwholesome spectra from the optimized ROI were also examined for wavebands at which local maxima and minima occurred, to identify wavebands that might be used in two-waveband ratios for differentiating wholesome and unwholesome birds. The value of each potential band ratio was calculated for the average wholesome chicken pixels and for the average unwholesome chicken pixels. The two-waveband ratio showing the greatest difference in ratio value between average wholesome and average unwholesome chicken pixels was selected.

Figure 7.7 shows the average spectra for pixels within this optimized ROI from all line-scan images in the wholesome data set and in the unwholesome

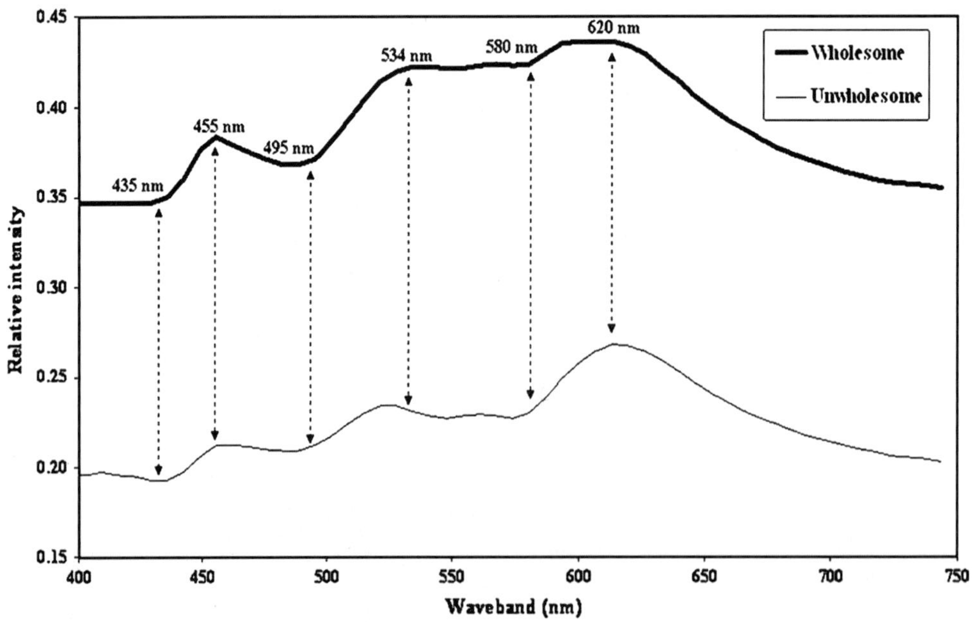

FIGURE 7.7 *The average ROI pixel spectrum for wholesome chickens and the average ROI pixel spectrum for unwholesome chickens, used to select wavebands for intensity- and ratio-based differentiation*

data set. Because the 580 nm band showed the greatest difference between the average wholesome and the average unwholesome spectra, this band was selected as the single waveband to be used for intensity-based differentiation of wholesome and unwholesome chicken carcasses. Six possible wavebands (also marked on Figure 7.7) were investigated for differentiation of wholesome and unwholesome chicken carcasses by a two-waveband ratio. Because visual examination showed noticeable differences between the average wholesome and average unwholesome spectral slopes in the three areas corresponding to 440–460 nm, 500–540 nm, and 580–620 nm, two-band ratios were investigated using these particular pairings. Two-band ratios for these pairings were calculated using the average wholesome reflectance, W, and average unwholesome reflectance, U, values. The differences in ratio value between wholesome and unwholesome were then calculated:

$$W_{440}/W_{460} - U_{440}/U_{460} = 0.003461$$

$$W_{500}/W_{540} - U_{500}/U_{540} = 0.038602$$

$$W_{580}/W_{620} - U_{580}/U_{620} = 0.115535$$

The last ratio, using the 580 nm and 620 nm wavebands, showed the greatest difference between the average wholesome and average unwholesome chicken spectra and was thus selected for use in differentiation by two-waveband ratio.

7.5.3. On-Line Multispectral Inspection

After hyperspectral analysis to select specific wavebands for multispectral inspection of chicken carcasses, the same line-scan imaging system was operated in multispectral imaging mode to use those selected wavebands for real-time inspection. Thus, there remained 512 pixels in the spatial dimension of the image but the pixels in the spectral dimension were further reduced from 55 to only two wavebands, with the elimination of unnecessary waveband data enabling even faster imaging speed. The ability of the spectral line-scan imaging system's EMCCD camera to use a very short integration time (0.1 ms) with a high gain setting, along with the selection of a limited number of pixels in the spectral dimension of the line-scan images, were vital to the system's successful on-line operation in multispectral imaging mode for differentiating wholesome and systemically diseased chickens at 140 bpm.

The capability to detect individual bird carcasses, classify the carcass condition, and generate a corresponding output useful for process control, all at speeds compatible with on-line operations, is required for effective multispectral imaging inspection for wholesomeness of chicken carcasses on a commercial processing line. LabVIEW 8.0 (National Instruments Corp., Austin, TX, USA) software was used to develop in-house inspection modules to control the spectral imaging system for performing these tasks in real-time. The following algorithm, based on the imaging system's line-by-line mode of operation, was developed to detect the entry of a bird carcass into the IFOV and classify the carcass as either wholesome or unwholesome using real-time multispectral inspection on a processing line.

Figure 7.8 shows a flowchart describing the line-by-line algorithm for multispectral inspection. First, a line-scan image was acquired that contained only raw reflectance values at the two key wavebands needed for intensity and ratio differentiation; these raw reflectance data were converted into relative reflectance data and background pixels were removed from the image (Figure 7.8, Box 8.1). The line-scan image was checked for the presence of the SP of a new bird (Figure 7.8, Box 8.2); if no SP was present, no further analysis was performed for this line-scan image and a new line-scan image was acquired. If the line-scan was found to contain an SP, then the ROI pixels were located (Figure 7.8, Box 8.3) and the decision output value of D_o was calculated for each ROI pixel in the line-scan image (Figure 7.8, Box 8.4), before a new line-scan image was acquired. With each new line-scan image acquired (Figure 7.8, Box 8.5), the ROI pixels were located, and the decision output value of D_o was calculated for each pixel, until the EP of that bird was detected (Figure 7.8, Box 8.6), indicating no additional line-scan images to be analyzed for that carcass. The average D_o value for the bird was calculated across all its ROI pixels (Figure 7.8, Box 8.9) and then compared to the threshold value (Figure 7.8, Box 8.10) for the final determination of wholesomeness or unwholesomeness for the bird carcass (Figure 7.8, Boxes 8.11 and 8.12). The decision output D_o calculation was based on fuzzy inference classifiers (Chao *et al.*, 2008) developed using mean and standard deviation values for ROI reflectance at the key wavebands during hyperspectral analysis of the wholesome and unwholesome sets of chicken images.

7.5.4. In-Plant Evaluation

Hyperspectral line-scan images of chickens were first acquired on a 140 bpm commercial chicken processing line for a total of 5 549 wholesome and 93 unwholesome chickens, with their conditions identified by an FSIS veterinary medical officer who observed the birds as they approached the illuminated IFOV of the imaging system. The 55-band hyperspectral data for the chicken carcasses were analyzed as described in Section 7.5.2 for ROI

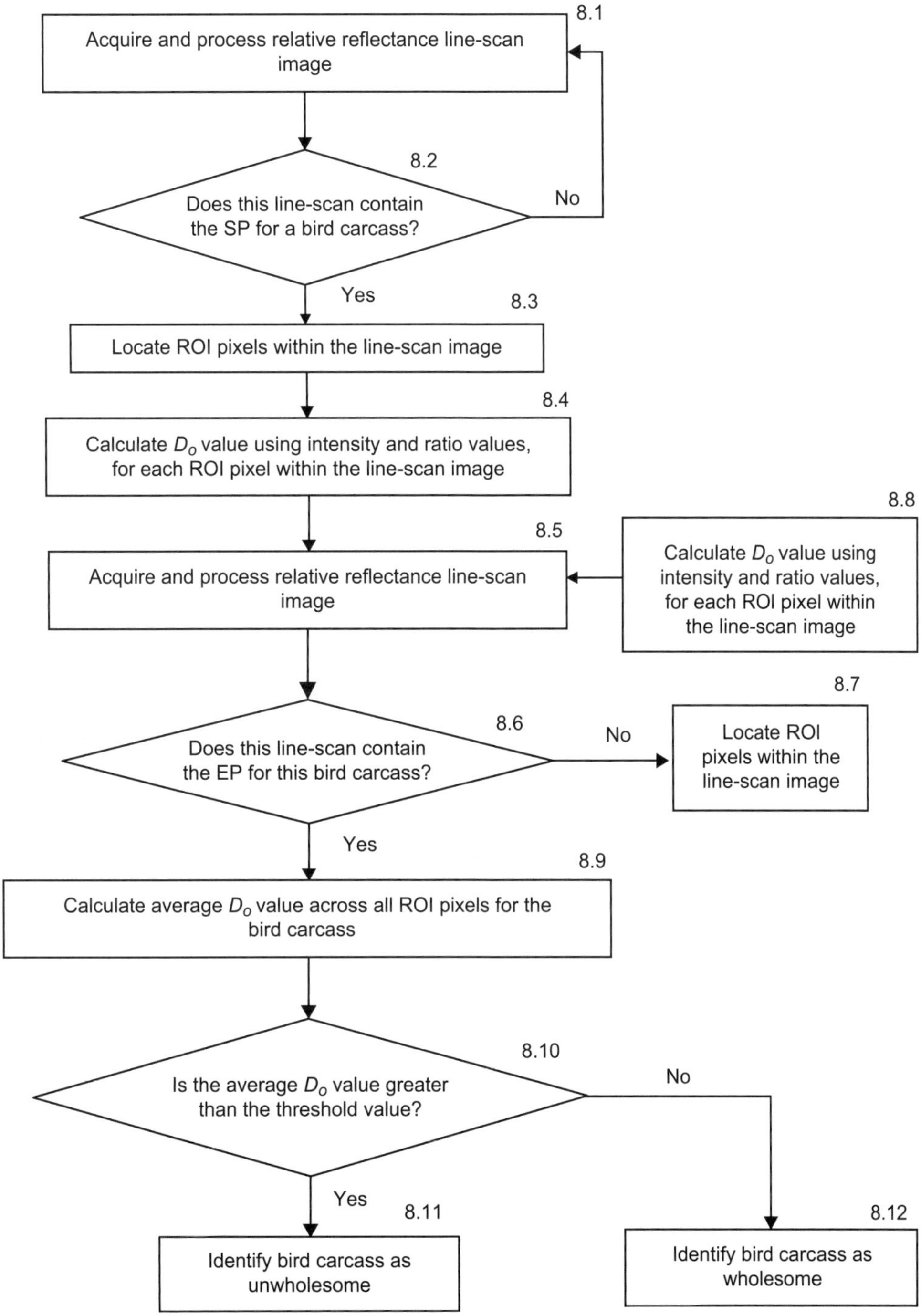

FIGURE 7.8 *A flowchart of the line-by-line algorithm for on-line multispectral wholesomeness inspection by the spectral line-scan imaging system*

optimization and for selection of one key wavelength and one two-waveband ratio, based on average spectral differences between wholesome and unwholesome birds. Multispectral imaging for on-line high-speed inspection in real time used only the two selected wavelengths for intensity- and ratio-based differentiation. LabView-based software modules were developed for detecting each bird and for implementing the on-line inspection algorithms. On-line multispectral inspection was tested on a commercial processing line over two 8-hour shifts during which over 100 000 birds were inspected by the imaging system. To verify system performance, an FSIS veterinary medical officer identified wholesome and unwholesome conditions of birds immediately before they entered the IFOV of the imaging system, during several 30–40 minute periods, for direct comparison with the classification results produced by the multispectral imaging system.

Figure 7.9 shows examples of chicken images acquired on-line, with the ROI pixels highlighted on each bird. During on-line operation, the inspection program automatically located the 40–60% ROI with the acquisition of each line-scan image. As shown, the ROI location was clearly affected by the size and position of the bird and thus varied between different birds. For a bird whose body extended past the lower edge of the image, such as the first bird in Figure 7.9, the ROI encompassed a rectangular area. In contrast, an irregularly shaped ROI resulted for birds positioned such that background pixels were present at the lower edge of the image.

The first image in Figure 7.10 (*top*) shows a masked image of nine chickens, highlighting all the ROI pixels for each bird. Using fuzzy inference classifiers (Chao *et al.*, 2007), two D_O values were calculated (ranging between 0 and 1) for each pixel in the ROI, one for the key waveband and one for the two-waveband ratio. On-line multispectral inspection averaged the D_O values for all ROI pixels for each bird, in order to classify the bird by comparison to the threshold value of 0.6. For illustration purposes, the second image in Figure 7.10 (*bottom*) highlights the results of classifying the individual pixels in the ROIs (instead of classifying whole birds), obtained by

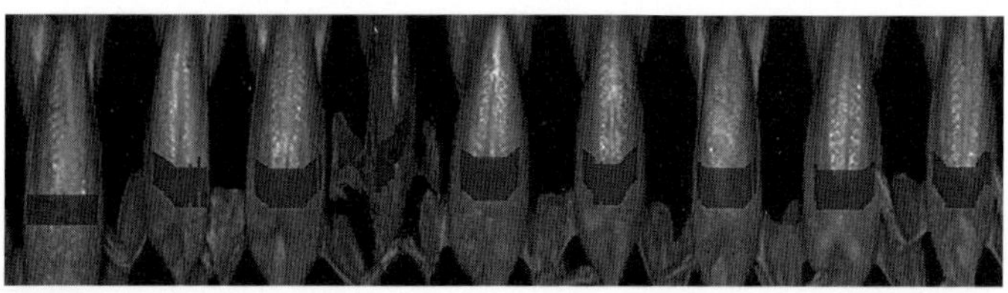

FIGURE 7.9 *Automated ROI identification highlighted on the images of nine chicken carcasses. (Full color version available on http://www.elsevierdirect.com/companions/9780123747532/)*

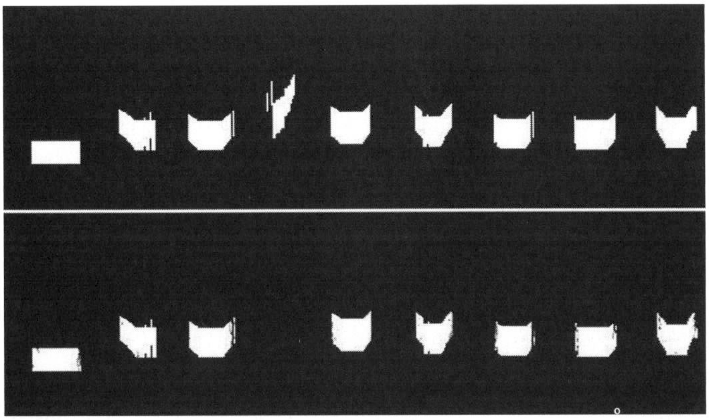

FIGURE 7.10 *A masked image (top) of nine chickens that highlights the ROI pixels to be analyzed for each chicken, and a second image (bottom) highlighting individual pixels within each ROI that were classified as wholesome*

averaging the two D_O values for each ROI pixel in the top image and comparing the average value with the 0.6 threshold value. In this illustrative example, the fourth chicken from the left is an unwholesome bird and all of its ROI pixels were individually identified as unwholesome, consequently not appearing at all in the second image.

Table 7.1 shows the total counts of chickens identified by the imaging system as being either wholesome or unwholesome during the two 8-hour shifts of on-line multispectral inspection. Numbers drawn from FSIS tally sheets, created by three inspection stations on the same processing line during the same inspection shifts, are shown for comparison. Although direct bird-to-bird comparison between the imaging inspection system and the inspectors was not feasible, the percentages indicate that the relative numbers of wholesome and unwholesome birds identified by the imaging

Table 7.1 Counts of wholesome and unwholesome birds identified on the processing line during inspection shifts by human inspectors and by the hyperspectral/multispectral line-scan imaging inspection system

	Line inspectors			Imaging inspection system		
Shift	**Wholesome**	**Unwholesome**	**Total**	**Wholesome**	**Unwholesome**	**Total**
1	53563 (99.84%)	84 (0.16%)	53647 (100%)	45305 (99.37%)	288 (0.63%)	45593 (100%)
2	64972 (99.89%)	71 (0.11%)	65043 (100%)	60922 (99.84%)	98 (0.16%)	61020 (100%)

inspection system and by the processing line inspectors were not significantly different.

System verification was also performed by an FSIS veterinary medical officer for several 30–40 minute periods within the inspection shifts. This consisted of bird-by-bird observation of chicken carcasses on the processing line immediately before they entered the IFOV of the imaging system; the imaging system output was compared with the veterinary medical officer's identifications. Over four verification periods during inspection shift 1, the imaging system correctly identified 99.3% of wholesome birds (16 056 of 16 174) and 95.4% of unwholesome birds (41 of 43). Over six verification periods during inspection shift 2, the imaging system correctly identified 99.8% of wholesome birds (27 580 of 27 626) and 97.1% of unwholesome birds (34 of 35). These verification period results, together with the whole-shift comparison results against tally sheets (Table 7.1), demonstrate that the hyperspectral/multispectral line-scan imaging inspection system can perform effectively on a 140 bpm high-speed commercial poultry processing line.

7.5.5. Commercial Applications

This work successfully demonstrates the potential of a hyperspectral/multispectral line-scan imaging system for effective on-line inspection of chickens. The spectral resolution of the imaging system was approximately 7 nm (FWHM). On the 140 bpm processing line, the imaging system was able to acquire approximately 50 hyperspectral (55-waveband) line-scan images per bird, for a spatial resolution of 0.35 mm in the hyperspectral images that were analyzed for waveband selection, with the ROI of any given bird spanning approximately 20–30 of those 50 line-scan images. During multispectral on-line inspection, the ROI per bird spanned approximately 40 multispectral (2-waveband) line-scan images, depending on the bird size. With about 4 000 pixels in the ROI to analyze for multispectral classification of a bird, the spatial resolution of the system is more than adequate for accurate and effective detection of unwholesome chickens at a speed of 140 bpm.

Automated on-line pre-sorting of broilers is an ideal application for this spectral line-scan imaging system. By detecting and diverting unwholesome birds exhibiting symptoms of systemic disease earlier on the processing line, production and efficiency can be improved—fewer unwholesome birds will be presented for inspection by human inspectors and fewer empty shackles (nearer 100% operating capacity) will occur during downstream processing. By diverting most unwholesome birds earlier, the reduced inspection workload for human inspectors can provide the opportunity for inspectors to address additional tasks beyond direct carcass inspection. The rejected birds are detected and diverted while still on the high-speed kill line, prior to automatic re-hanging on the evisceration line, which helps to reduce food safety risks from possible cross-contamination. For the small number of wholesome birds that might be misidentified as false positives by the automated inspection system, a processing plant can opt to re-inspect diverted birds and manually transfer any wholesome birds to the evisceration line.

For the purpose of pre-sorting young chickens on commercial processing lines, the spectral line-scan imaging technology has been recently reviewed and approved by the USDA FSIS Risk and Innovations Management Division. Commercialization of this system for industrial use will be the first application of spectral line-scan imaging technology for a food safety inspection task.

7.6. CONCLUSIONS

Due to increasing production needs and food safety concerns facing the poultry industry in the United States and worldwide, automated systems developed for safety inspection of poultry products on high-speed processing lines will be essential in the future. By enabling poultry producers and regulatory agencies to satisfy high-throughput production and inspection requirements more efficiently, science-based automated food inspection systems can help alleviate the pressures on human inspectors, improve production throughput, and grow public confidence in the safety and quality of the food production and distribution system. Development of automated non-destructive food safety inspection methods based on spectroscopy and spectral imaging have been one of the major ARS research priorities over the last decade.

VIS/NIR spectroscopy methods were first developed and demonstrated capable of over 90% accuracies on high-speed processing lines in differentiating wholesome chickens from unwholesome birds exhibiting systemic conditions; however, given the lack of spatial information, the field of application for VIS/NIR spectroscopy inspection systems was considered limited. In this light, expansion of the spectral techniques to multispectral imaging was sought, requiring investigation and development of wavelength selection methods such as 2-D spectral correlation, color mixing, and hyperspectral imaging analysis—with the addition of spatial information for whole bird carcasses, such wavelength selection was necessary to reduce image data volumes for practical application. Dual-camera and common-aperture systems for target-based multispectral imaging were developed, but encountered some problems with short-exposure image acquisition and processing speed during implementation on commercial processing lines.

Hyperspectral imaging was first used for spectral analysis to select wavelengths for implementation in automated multispectral imaging systems, and in itself was effective for laboratory-based research. The introduction of EMCCD cameras and their use with imaging spectrographs was a key development that enabled automated line-scan spectral imaging at the high speeds found on commercial processing lines, and was particularly important for allowing a single imaging system to perform both hyperspectral and multispectral imaging. With the transition from target-based imaging to line-scan imaging, algorithms such as line-scan target detection were a necessary development for effective on-line implementation. Not only could such algorithms streamline or simplify the processing-line imaging operations, for example by eliminating sensors formerly needed to trigger accurate imaging of individual birds on the line, but they also provided potential value-added applications that could be performed using the same image data—for example, quality inspection tasks such as assessing defects, size, shape, or weight attributes. The results of in-plant testing showed that the ARS line-scan spectral imaging system could successfully inspect chickens on high-speed processing lines operating at 140 bpm, accurately differentiating between wholesome and unwholesome birds. The system can be used for on-line pre-sorting of birds on commercial poultry processing lines, thereby increasing efficiency, reducing labor and costs, and producing significant benefits for poultry producers and processors.

NOMENCLATURE

2-D correlation	two-dimensional spectral correlation
ARS	Agricultural Research Service
bpm	birds per minute
CART	classification and regression tree
CCD	charge-coupled device
CIE	International Commission on Illumination
CIECAM	CIE color appearance model
EMCCD	electron-multiplying charge-coupled device
EP	ending point
FSIS	Food Safety and Inspection Service
FWHM	full width at half maximum

HACCP	Hazard Analysis and Critical Control Point
HIMP	HACCP-Based Inspection Models Project
IFOV	instantaneous field of view
IP	inflammatory process
LED	light emitting diode
NELS	New Line Speed
PCA	principal component analysis
PDA	photodiode array
PPIA	Poultry Product Inspection Act
ROI	region of interest
septox	septicemia/toxemia
SIS	Streamlined Inspection System
SP	starting point
USDA	United States Department of Agriculture
VIS/NIR	visible/near-infrared

REFERENCES

Chao, K., Chen, Y. R., Early, H., & Park, B. (1999). Color image classification system for poultry viscera inspection. *Applied Engineering in Agriculture, 15*(4), 363–369.

Chao, K., Park, B., Chen, Y. R., Hruschka, W. R., & Wheaton, F. W. (2000). Design of a dual-camera system for poultry carcasses inspection. *Applied Engineering in Agriculture, 16*(5), 581–587.

Chao, K., Mehl, P. M., & Chen, Y. R. (2002). Use of hyper- and multi-spectral imaging for detection of chicken skin tumors. *Applied Engineering in Agriculture, 18*(1), 78–84.

Chao, K., Chen, Y. R., Hruschka, W. R., & Gwozdz, F. B. (2002). On-line inspection of poultry carcasses by a dual-camera system. *Journal of Food Engineering, 51*(3), 185–192.

Chao, K., Chen, Y. R., & Chan, D. E. (2003). Analysis of VIS/NIR spectral variations of wholesome, septicemia, and cadaver chicken samples. *Applied Engineering in Agriculture, 19*(4), 453–458.

Chao, K., Chen, Y. R., & Chan, D. E. (2004). A spectroscopic system for high-speed inspection of poultry carcasses. *Applied Engineering in Agriculture, 20*(5), 683–690.

Chao, K., Yang, C. C., Chen, Y. R., Kim, M. S., & Chan, D. E. (2007). Hyper-spectral/multispectral line-scan imaging system for automated poultry carcass inspection applications for food safety. *Poultry Science, 86*, 2450–2460.

Chao, K., Yang, C. C., Kim, M. S., & Chan, D. E. (2008). High throughput spectral imaging system for wholesomeness inspection of chicken. *Applied Engineering in Agriculture, 24*(4), 475–485.

Chen, Y. R., & Massie, D. R. (1993). Visible/near-infrared reflectance and interactance spectroscopy for detection of abnormal poultry carcasses. *Transactions of the ASAE, 36*, 863–869.

Chen, Y. R., Huffman, R. W., Park, B., & Nguyen, M. (1995). A transportable spectrophotometer system for online classification of poultry carcasses. *Journal of Applied Spectroscopy, 50*(7), 910–916.

Chen, Y. R., Nguyen, M., & Park, B. (1998). Neural network with principal component analysis for poultry carcass classification. *Journal of Food Process Engineering, 21*, 351–367.

Chen, Y. R., Park, B., Huffman, R. W., & Nguyen, M. (1998). Classification of online poultry carcasses with back-propagation neural networks. *Journal of Food Process Engineering, 21*, 33–48.

Chen, Y. R., Hruschka, W. R., & Early, H. (2000). On-line trials of a chicken carcass inspection system using visible/near-infrared reflectance. *Journal of Food Process Engineering, 23*, 89–99.

Chen, Y. R., Chao, K., & Kim, M. S. (2002). Machine vision technology for agricultural applications. *Computers and Electronics in Agriculture, 36*, 173–191.

Ding, F., Chen, Y. R., & Chao, K. (2005). Two-waveband color-mixing binoculars for the detection of wholesome and unwholesome chicken carcasses: a simulation. *Applied Optics, 44*(26), 5454–5462.

Ding, F., Chen, Y. R., & Chao, K. (2006). Two-color mixing for classifying agricultural products for food safety and quality. *Applied Optics, 45*(4), 668–677.

Gowen, A. A., O'Donnell, C. P., Cullen, P. J., Downey, G., & Frias, J. M. (2007). Hyperspectral imaging—an emerging process analytical tool for food quality and safety control. *Trends in Food Science & Technology, 18*(12), 590–598.

Kim, M. S., Chen, Y. R., & Mehl, P. M. (2001). Hyperspectral reflectance and fluorescence imaging system for food quality and safety. *Transactions of the ASAE, 44*(3), 721–729, 2001.

Kim, M. S., Lefcourt, A. M., Chen, Y. R., Kim, I., Chan, D. E., & Chao, K. (2002). Multispectral detection of fecal contamination on apples based on hyperspectral imagery. Part II: Application of hyperspectral fluorescence imaging. *Transactions of the ASAE, 45*(6), 2039–2047.

Kim, M. S., Lefcourt, A. M., Chen, Y. R., & Kang, S. (2004). Hyperspectral and multispectral laser induced fluorescence imaging techniques for food safety inspection. *Key Engineering Materials, 270*, 1055–1063.

Lawrence, K. C., Park, B., Windham, W. R., & Mayo, C. (2003). Calibration of a pushbroom hyperspectral imaging system for agricultural inspection. *Transactions of the ASAE, 46*(2), 513–521.

Lawrence, K. C., Windham, W. R., Park, B., & Buhr, R. J. (2003). A Hyperspectral imaging system for identification of fecal and ingesta contamination on poultry carcasses. *Journal of Near-Infrared Spectroscopy, 11*(4), 269–281.

Liu, Y., & Chen, Y. R. (2000). Two-dimensional correlation spectroscopy study of visible and near-infrared spectral variations of chicken meats in cold storage. *Applied Spectroscopy, 54*(10), 1458–1470.

Liu, Y., & Chen, Y. R. (2001). Two-dimensional visible/near-infrared correlation spectroscopy study of thawing behavior of frozen chicken meats without exposure to air. *Meat Science, 57*(3), 299–310.

Lu, R., & Chen, Y. R. (1998). Hyperspectral imaging for safety inspection of foods and agricultural products. In Y. R. Chen (Ed.), *Pathogen detection and remediation for safe eating* (pp. 121–133). Bellingham, MA: SPIE—The International Society for Optical Engineering.

Lu, R., & Peng, Y. (2006). Hyperspectral scattering for assessing peach fruit firmness. *Biosystems Engineering, 93*(2), 161–171.

Mehl, P. M., Chao, K., Kim, M. S., & Chen, Y. R. (2002). Detection of defects on selected apple cultivars using hyperspectral and multispectral image analysis. *Applied Engineering in Agriculture, 18*(2), 219–226.

Noh, H., & Lu, R. (2007). Hyperspectral laser-induced fluorescence imaging for assessing apple quality. *Postharvest Biology and Technology, 43*(2), 193–201.

Park, B., & Chen, Y. R. (1994). Intensified multi-spectral imaging system for poultry carcass inspection. *Transactions of the ASAE, 37*, 1983–1988.

Park, B., & Chen, Y. R. (1996). Multispectral image co-occurrence matrix analysis for poultry carcass inspection. *Transactions of the ASAE, 39*(4), 1485–1491.

Park, B., & Chen, Y. R. (2000). Real-time dual-wavelength image processing for poultry safety inspection. *Journal of Food Process Engineering, 23*(5), 329–351.

Park, B., Lawrence, K. C., Windham, W. R., & Buhr, R. J. (2002). Hyperspectral imaging for detecting fecal and ingesta contaminants on poultry carcasses. *Transactions of the ASAE, 45*(6), 2017–2026.

Qin, J., & Lu, R. (2006). Hyperspectral diffuse reflectance for rapid, noncontact determination of the optical properties of turbid materials. *Applied Optics, 45*, 8366–8373.

USDA. (1996). Pathogen reduction: hazard analysis and critical control point (HACCP) systems: Final rule. *Federal Register, 61*, 38805–38989.

USDA. (1997). HACCP-based inspection models project (HIMP): Proposed rule. *Federal Register, 62*, 31553–31562.

USDA. (2005). Poultry products inspection regulations. 9 CFR 381.76. *Code of Federal Regulations, 9*(2), 456–465.

USDA. (2008). *Poultry Production and Value—2007 Summary.* Washington, DC: National Agricultural Statistics Service.

Yang, C. C., Chao, K., Chen, Y. R., & Early, H. L. (2005). Systemically diseased chicken identification using multispectral images and region of interest analysis. *Computers and Electronics in Agriculture, 49*(2), 255–271.

Yang, C. C., Chao, K., & Chen, Y. R. (2005). Development of multispectral imaging processing algorithms for food safety inspection on poultry carcasses. *Journal of Food Engineering, 69*(2), 225–234.

Yang, C. C., Chao, K., Chen, Y. R., Kim, M. S., & Early, H. L. (2006). Simple multispectral image analysis for systemically diseased. *Transactions of the ASAE, 49*(1), 245–257.

Yang, C. C., Chao, K., & Kim, M. S. (2009). Machine vision system for online inspection of freshly slaughtered chickens. *Sensing and Instrumentation for Food Quality and Safety, 3*(1), 70–80.

CHAPTER 8

Quality Evaluation of Fish by Hyperspectral Imaging

Paolo Menesatti [1], Corrado Costa [1], Jacopo Aguzzi [2]
[1] CRA-ING Agricultural Engineering Research Unit of the Agriculture Research Council, Monterotondo (Rome), Italy
[2] Institut de Ciències del Mar (ICM-CSIC), Barcelona, Spain

8.1. INTRODUCTION

CONTENTS

Quality is an important factor in enhancing competitiveness in agricultural or fish production. The concept of quality is related to safety, nutritional or nutraceutical value, and to organoleptic properties such as freshness. In order to ensure the appropriate food quality and safety for the health of consumers, legal requirements and new quality standards are constantly developed according to EU Directives (Knaflewska & Pospiech, 2007). Especially in Europe, there is an increasing interest in labeling the quality of agro-fish products for human consumption.

Quality evaluation, therefore, progressively began to be a central aspect in agro-food and fish production and industrial processing. In this context, it is important to consider that the term "quality" in commercial, scientific, and the related legislation fields may refer to different aspects for different operators. Moreover, the actual trend is to relate "quality" to each specific product type (species, origin, rearing technique) and each individual organism (Costa *et al.*, 2009c). For example, chemical composition differences in fish flesh between wild and farmed sea bass from Greece and Italy have been reported by Alasalvar *et al.* (2002) and Orban *et al.* (2002). According to this example, it is important to find efficient analytic methods to attribute a differential quality to captured or farmed stock fish with poorer meat quality.

In relation to fish products coming from aquaculture facilities or commercial fisheries, each category of product is characterized by size, shape, color, freshness, and finally by the absence of visual morphological defects

(Costa *et al.*, 2009c). In particular, appearance is an easily treatable criterion utilized to select the piece of product throughout the market chain from its production to its storage, marketing, and finally, to users (Kays, 1999). In that context, an important aspect of quality is related to the concept of freshness.

Freshness, in relation to fish quality, represents a pivotal aspect in its socioeconomic usage and economic value. Scientific methods for the evaluation of freshness may be conveniently divided into two categories: sensory and instrumental. Since the consumer is the ultimate judge of quality, most methods must be correlated with measures related to sight, touch or odor perception (Menesatti *et al.*, 2002, 2006; Menesatti, Urbani *et al.*, 2007). While sensory-based methods of measurements must be performed under carefully controlled scientific conditions in order to allow a trustworthy reproduction of results so that the effects of the testing environment, personal bias, etc., can be reduced (Huss, 1995), instrumental techniques are less subjective. The bias introduced by observer-based evaluation, as well as the surrounding environment, is comparatively higher.

Instrumental methods can be divided into biochemical–chemical methods, microbiological methods, and physical methods (Menesatti, Urbani *et al.*, 2007). The appeal of biochemical–chemical methods for fish quality evaluation is related to their ability to establish quantitative standards for freshness based on tolerance levels in chemical spoilage. The aim of microbiological examinations of fish products is to evaluate the possible presence of bacteria that are of public health concern and to maintain hygienic quality in terms of temperature and cleanliness during handling and processing. Methods of a microbiological nature are correlated with sensory quality evaluations of chemical compounds in relation to spoilage or to modifications associated with the industrial processing itself (e.g., the breakdown of amines or nucleotides in the canning process as a result of high temperatures). The microbiological aspects affecting fish quality are mostly related to public health and the obtained data on quality assessment do not provide information about freshness. Finally, there are physical methods that are based on the testing of softness/hardness of food texture. These methods are particularly appreciated for their rapid and non-destructive approach. One method in particular is based on the changes in the electrical properties of skin and tissue after fish death (Jason & Richards, 1975). Changes in conductance properties are associated with variations in meat quality post mortem in relation to bacterial spoilage. Also, the evaluation of firmness can be considered as an indicator of good quality, with a good correlation with sensory and chemical properties (Alasalvar *et al.*, 2001; Menesatti, Pallottino *et al.*, 2009). Texture in fish meat can be instrumentally measured by techniques based on puncturing, compression, cutting, or stretching (i.e., tension) (Menesatti & Urbani, 2004). Among all of these, the most widely used techniques are cutting force and compression (Sigurgisladottir *et al.*, 1999).

The analysis of the optical properties of food has recently been assuming a greater relevance in product and organoleptic assessments of quality within the physical methods. Spectrophotometric applications are particularly relevant in the outputs from meat under a light beam of a single spectral band and the components contribute to a more detailed and refined evaluation of quality characteristics, giving important indicators on the method of production of the meat and its origin (Menesatti, D'Andrea *et al.*, 2007). Spectrophotometric techniques are associated with a high analytic ability based on their non-destructiveness, relative simplicity, speed, and portability in operative environments during measurements. Also, the high level of automatization in information processing and hardware development (in terms of interfacing with other instrumental sensors) have progressively shifted scientific and applied interest in quality assessment procedures in meat production toward spectrophotometric applications.

In recent years near-infrared reflectance spectroscopy (NIRS) has been increasingly used as a non-destructive and rapid technique in the assessment of food quality (Chen & He, 2007; Xiccato *et al.*, 2004) and associated health issues (Kim *et al.*, 2002). NIRS has been also used in fish meat quality tests. The spectral variation among fish meat samples depends on the feeding regime of the fishes, as well as water quality, growth pattern, and muscular activity (Karoui *et al.*, 2007). NIRS has been successfully used in salmon, trout, cod, halibut, and sea bass in relation to chemical composition prediction, protein content, and levels of humidity (i.e., moisture) (Cozzolino *et al.*, 2002; Mathias *et al.*, 1987; Nortvedt *et al.*, 1998; Solberg & Fredriksen, 2001; Xiccato *et al.*, 2004).

Particular spectroscopic applications are conducted with the advanced technologies of VIS/NIR and NIR spectral imaging. These instruments are able to acquire spectral images at a high-density resolution (150–250 k-pixels) where each pixel possesses the entirety of the spectral information (VIS and NIR). Thus, this technique integrates conventional imaging and spectroscopy to obtain both spatial and spectral information from an object (Gowen *et al.*, 2007; Menesatti, Zanella *et al.*, 2009). Multi- or hyperspectral analysis ($\leq$10 and >10 spectral bands, respectively) is the new frontier of optical imaging. Hyperspectral imaging, within the VIS/NIRS techniques, is useful to analyse the spectra of inhomogeneous materials that contain a wide range of spectral (Mehl *et al.*, 2002) and spatial information (Park *et al.*, 2006). Hyperspectral images can be considered as hypercube matrices;

three-dimensional blocks of data made by two spatial plane coordinates and one wavelength dimension (Gowen *et al.*, 2007).

Multi- and hyperspectral optical imaging has been successfully used in vegetable and meat quality discrimination in recent years because of its high capability for the detailed analysis of food product structure (Menesatti, D'Andrea *et al.*, 2007). In fruit post-harvesting treatment, this technique has been successfully used for the detection of quality defects in cucumbers, tomatoes, pears, and apples (e.g., Li *et al.*, 2002; Liu *et al.*, 2006; Polder *et al.*, 2002). It has also been used in food applications in relation to the biochemical properties of sugar contents (Bellon *et al.*, 1993), moisture content (Katayama *et al.*, 1996), and acidity (Lammertyn *et al.*, 1998).

Referring to the use of hyperspectral imaging in fish production, few applications have been reported in the literature to date. Published data mostly refer to the detection of fat and water content in fish fillet products (ElMasry & Wold, 2008), for production line sampling (Wold *et al.*, 2006) or for fish freshness detection (Menesatti, Urbani *et al.*, 2007). In the near future, however, the technological evolution of photonics will reach a break-even point where spectroscopic technology could be broadly adopted given its low price and safety of use (Menesatti, D'Andrea *et al.*, 2007; Park *et al.*, 2004; Yang *et al.*, 2005). In this context, it is possible that hyperspectral imaging will provide a valid contribution in relation to the monitoring of the organoleptic and commercial properties of fish production during all steps along the production chain.

In this chapter we will discuss the use of hyperspectral imaging as a method to provide an objective and qualitative evaluation of fish freshness. We focus on establishing a correlation between the spectral reflectance of selected areas of the epidermis and the time of storage in standard refrigeration procedures. We will also discuss the possibility of finding objective parameters for the good prediction of fish freshness that consider products stored for more than three days as "non-fresh but still edible".

Case studies corresponding to two different analytic procedures will be described, including subjective ROI (region of interest) identification in hyperspectral images and morphometric superimposition for automated topographical hyperspectral image analysis. The first method is based on the subjective choice of the sampling areas within the hyperspectral images that bear the most interesting information according to a subjective criterion of observed evaluation. The operator can delimit the region to be analysed. For that region, an average value of spectral reflectance can be computed within the VIS/NIR or NIR ranges. The second method is based on the first method, and it represents an evolution of the technique, with the use of geometric morphometric tools for the superimposition of hyperspectral cubes from image pixels of different samples.

8.2. SUBJECTIVE ROI ON HYPERSPECTRAL IMAGES FOR FISH FRESHNESS IDENTIFICATION

Fifty wild chub mackerel (*Scomber japonicus*) and 80 hatchery-reared sea bass (*Dicentrarchus labrax*) were used in this case study. Chub mackerel were fished in the mid-low Adriatic Sea (Manfredonia). Sea bass were cultured intensively in concrete tanks (CT) or in sea cages (SC). All the fish were collected in May 2005 from three fish farms in the same southern Italian region (Puglia). Their rearing conditions were as follow:

Peschiere Tarantine farm: CT with a stocking condition at 19 °C with a density of 50 kg/m^3.

Panittica Pugliese farm: Both CT and SC; in CT with a stocking condition at 19.5 °C with a density of 35 kg/m^3.

Tortuga farm: SC (Ionian Sea).

Before harvesting, all fish were fasted for 24 hours. They were killed by immersion in chilled water and then covered with ice. All fish were transported to CRA-ING (Monterotondo, Rome) in a refrigerated unit maintaining a constant temperature of between 0 and 4 °C. Specimens were analysed at 1, 2, 4, and 6 days post mortem (d.p.m.).

Before the spectral scanner analysis the fish were taken from the refrigerator and left at room temperature for 30 min to eliminate the dry film on their skin created by the freezing process. A hyperspectral imaging system was used to integrate the spectroscopic and spatial imaging information of the fish. This system, in addition to the spatial information, can provide information at hyper/multiple wavelengths for each pixel of the sample.

The hyperspectral system used was composed of four parts (Figure 8.1): a sample transportation plate (spectral scanner DV, Padua, Italy); a collimated illumination device (Fiber-lite, Dolan-Jenner, MA, USA) composed of one 150W halogen lamp as the source light and one illumination opening in the optical fibre measuring 200 mm long and 2 mm width, positioned at 45° in relation to the transportation plate (i.e., bearing the sample) and presenting a minimum light divergence; and an imaging spectrograph (ImSpec V10, Specim Ltd, Oulu, Finland) coupled with

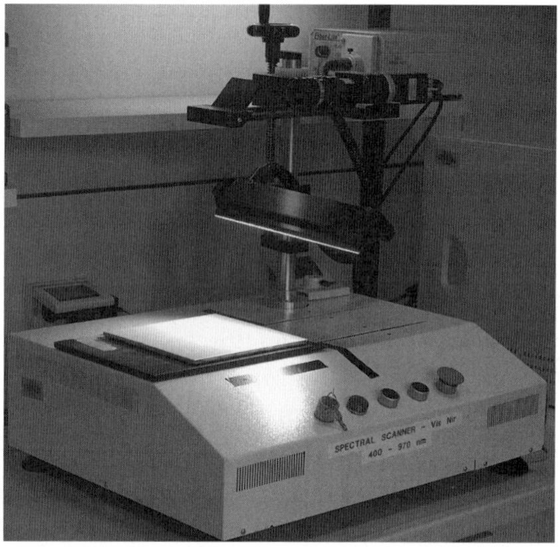

FIGURE 8.1 *The VIS/NIR hyperspectral system used. (Full color version available on http://www.elsevierdirect.com/companions/9780123747532/)*

a standard C-mount zoom lens and a Teli CCD monochrome camera (Toshiba-Teli CS8310BC, Japan).

The ImSpec is based on a patented prism–grating–prism (PGP) construction (a holographic transmission grating). The incoming line image (frame) was projected and dispersed onto the charge-coupled device (CCD). Each frame contained the line pixels in one dimension (spatial axis) and the spectral pixels in the other dimension (spectral axis), providing full spectral information for each line pixel. The reconstruction of the entire hyperspectral image of the sample was performed by scanning the sample line-by-line as the transportation plate moved it through the field of view. The resolution of the line image was 700 pixels by 10 bits. The number of frames (image resolution in Y-axes) was variable, from 10 to 500, depending on the speed and the accuracy of the transportation plate line scanning. The system was operated in a dark laboratory to minimize interference from ambient light. Other basic characteristics of the system were: spectral range, 400–970 nm; spectral resolution, 5 nm; dispersion, 90.9 nm/mm; sensor image size, 6.6 (spectral) $\times$ 8.8 (spatial) mm, corresponding to a standard 2/3 in. image sensor; spatial resolution, 15 line-pairs/mm; rms spot radius $<$60 μm within 2/3 image area; aberrations, insignificant astigmatism; slit width, 25; effective slit length, 9.8 mm; total efficiency (typical) $>$50%; and it was independent of polarization.

Spectral values were expressed in terms of relative reflectance (R), by applying the following equation:

$$R = \frac{r_s - r_b}{r_w - r_b} \tag{8.1}$$

where R is the relative reflectance of the sample at each wavelength; r_s is the absolute signal value (radiance) measured for the sample at each wavelength; r_b is the absolute signal value (radiance) measured at each wavelength for black (background noise); and r_w is the absolute signal value (radiance) measured at each wavelength for the standard white (100% of reflectance).

Hyperspectral images of the lateral side of the fish (from the area under the attachment of the first dorsal fin to the area above the end of the anal fin) were analysed with the software Spectral Scanner (ver. 1.4.1) (DV Optics, Padua, Italy). On each hyperspectral image a trained operator selected two ROIs (Figure 8.2) to measure the mean VIS/NIR spectral reflectance.

A supervised multivariate technique such as partial least squares-discriminant analysis (PLS-DA) was applied to observe freshness differences (<3d.p.m. vs. >3d.p.m.) in relation to mean spectral reflectance values. The PLS-DA (Sabatier *et al.*, 2003; Sjöström *et al.*, 1986) consists of a classic partial least squares (PLS) analysis regression where the response variable is a categorical one (Y-block; replaced by the set of dummy variables describing the categories) expressing the class membership of the statistical units (Aguzzi *et al.*, 2009; Costa *et al.*, 2008; 2009b). The PLS-DA does not allow for response variables other than those that define the groups of individuals, fresh (<3 d.p.m.) or non-fresh (>3 d.p.m.). The model includes a calibration phase and a cross-validation phase; during both phases the percentages of correct classification were calculated. The prediction ability in the test phase also depends on the number of latent variables (LV) used in the model. The

FIGURE 8.2 *Examples of ROIs applied by operators for the spectral image analysis*

optimal number of LV was chosen on the basis of the highest percentage of correct classification. The PLS-DA analysis provides the percentage of correct classification of the entire model as well as for the two classes considered. This analysis was performed using Matlab 7.1 (The Math Works, MA, Natick, USA) and PLS Toolbox 4.0 (Eigenvector Research Inc, Wenatchee, USA) for all the combinations of different preprocessing treatments (none, autoscale, mean center, Savitzky–Golay, ECC) and LV (2-20).

Figures 8.3 and 8.4 show the mean values in spectral reflectance at different d.p.m. (1, 2, 4, and 6) for chub mackerel and sea bass, respectively. In Figure 8.3 it is possible to observe that the mean reflectance

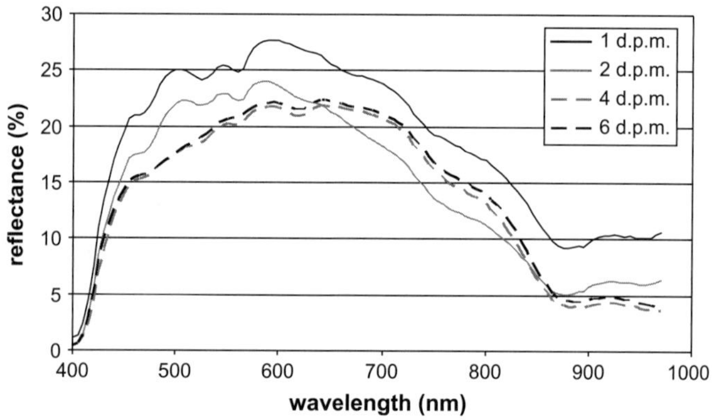

FIGURE 8.3 *Average spectral reflectance over consecutive days of freezing conservation as measured on chub mackerel side (d.p.m. = days post mortem)*

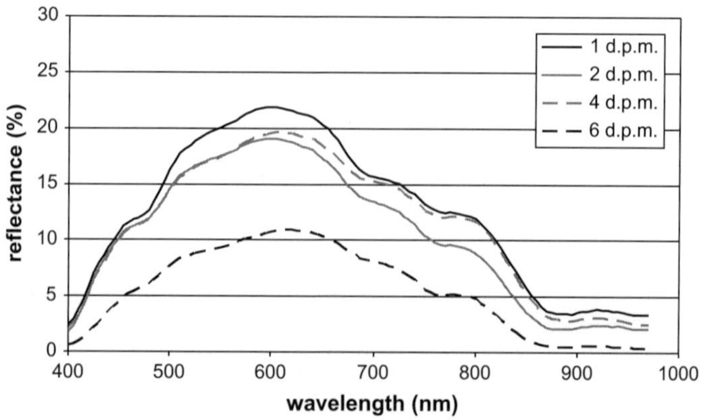

FIGURE 8.4 *Average spectral reflectance over consecutive days of freezing conservation as measured on sea bass side (d.p.m. = days post mortem)*

values for fresh individuals (<3 d.p.m) that are between 450 and 600 nm are well separated from values for the others. In Figure 8.4 (at 6 d.p.m.) the spectral reflectance values are always different from and lower than all the others. This first result shows how spectral information can be used to discriminate freshness status, although such information cannot be used for the topographic evaluation of which areas are more informative than others in the determination of freshness status. The selection of ROI is still subject to operator evaluation, its automation being at present difficult if the spectral–topographic contribution of the different areas is not established.

The present results agree well with observations on the progressive freshness deterioration of fish after death within 6 days. From the results, the skin brilliancy is a major element influencing the spectral analysis. This confirms the role of skin brilliancy in quality judgment on the integrity of the product. In fact, it is possible to notice a consistent reduction of spectral reflectance values within 450–650 nm in samples with more days of frozen conservation. Also, a modification in the spectral quality response occurs since the reflectance curve is more smoothed in fish that is less fresh. A similar trend, although with more variation, was also observed in the sea bass over a longer period of conservation time.

Results of the two PLS-DA models built for the two studied species are reported in Table 8.1. A high percentage of correct classification between

Table 8.1 Characteristics and principal results of the two PLS-DA models built from chub mackerel and sea bass reflectance data

	Chub mackerel	Sea bass
No. samples	50	80
No. units (X-block)	101	101
No. units (Y-block)	2	2
Preprocessing	None	None
Cross validation	Leave one out	Venetian blinds
No. LV	4	6
Cumulated variance X-block (%)	99.7	99.9
Cumulated variance Y-block (%)	25.3	22.7
Mean RMSEC	0.605	0.591
Mean RMSECV	0.628	0.626
Correct classification %	88.0	82.5

Note: No. units (Y-block) is the number of units (fresh ≤3 d.p.m.; non fresh >3 d.p.m.) to be discriminated by the PLS-DA, and No. LV is the number of latent vectors for each model.

fresh (≤3 d.p.m.) and non fresh (>3 d.p.m.) fish is reported at values of 88% for chub mackerel and 82.5% for sea bass. Using a multivariate approach on the hyperspectral results we efficiently show the ordering of samples according to the number of days after the death of the fish.

The present data indicate that better performing models possess lower numbers of LV (i.e., 4 and 6 for the chub mackerel and the sea bass, respectively). These samples were also those that did not undergo pre-processing. Our results can be explained by assuming that differences in the reported values of spectral reflectance intensity are sufficiently neat, especially in the range of visible wavelengths. Also, a relatively more simplified PSLDA model can still discriminate well among classes. The highest value of correct classification of the chub mackerel in comparison to the sea bass should be attributed to a clearer distinction in average spectral reflectance among samples of both considered classes. The variability in the spectral response is evidenced by the fact that the lowest value of cumulated variance in the Y-block depends not only upon samples' variability but also on the subjectivity in the ROI selection. Subjectivity problems in ROI selection were addressed by Peirs *et al.* (2002), who found different ROI values depending on observer attribution.

8.3. MORPHOMETRIC SUPERIMPOSITION FOR TOPOGRAPHICAL FISH FRESHNESS COMPARISON

In order to limit analytic errors in hyperspectral evaluation given the subjective choice of areas by the operator, an automatic topographic approach was developed. This represents a forward step in the analysis of quality, the importance of which has not been studied. In this case study, five specimens of rainbow trout (*Oncorhynchus mykiss*) were used that came from Azienda Agricola Sterpo (Rivignano, North-Eastern Italy). After collection the fish were killed by immersion in water and ice and then stored in refrigerated tanks for the duration of their transport to the laboratory facilities (CRA-ING, Rome). In the laboratory the fish were stored according to traditional market techniques, i.e., in an industrial refrigerator at 2 °C, and in polystyrene boxes with holes on all sides. The fish were also covered by ice both beneath and on top. Direct contact with ice, which causes potential damage to fish tissues, was prevented by using plastic parafilm. Each single trout was used three times at T0 = 1 d.p.m. (days post mortem), T1 = 3 d.p.m., T2 = 7 d.p.m., and finally T3 = 10 d.p.m. Before use in the spectral scanner, fish were taken from the refrigerator and left at room temperature for 30 min to eliminate the dry ice film on the fish surface.

The trout were scanned with the same spectral system used in the previous case study (see Figure 8.1). The hyperspectral VIS/NIR image acquisition time lasted about 8 s. For each acquired pixel in each image wavelength layer, the spectral reflectance value was measured and computed accordingly to Equation (8.1).

The image-warping protocol adapted for spectral matrixes was used to superimpose the RGB images of all sampled individuals taken on four different occasions (d.p.m.) (Costa *et al.*, 2009a). Images were warped to a standard view by fixing a set of reference points on the surfaces of the animal body. Using this method, the shape and color pattern of each individual was morphologically adjusted to the shape of the consensus configuration of the entire sample, as calculated via geometric morphometric tools. Geometric morphometric methods were developed to quantify and visualize deformations of morphometric points (landmarks) in a coordinate space of reference, as conceptualized by D'Arcy Thompson (1917). Landmarks are defined as homologous points that bear information on the geometry of biological forms (Bookstein, 1991). Using the consensus configuration of all specimens as the starting form, landmark configurations for each individual were aligned, translated, rotated, and scaled to a unit centroid size by the generalized Procrustes analysis (GPA) (Rohlf & Slice, 1990). Residuals from the fitting were modeled with the thin-plate spline interpolating function (Antonucci *et al.*, 2009; Bookstein, 1991; Costa *et al.*, 2006; Rohlf & Bookstein, 1990; Rohlf & Slice, 1990; Zelditch, *et al.*, 2004). This warping procedure involves standardizing the shape and size of each wavelength layer image with a generalized orthogonal least-squares Procrustes (GPA) superimposition (translation, scaling, and rotation) conducted on the set of 12 reference points (Figure 8. 5b) (Rohlf, 1999).

A supervised multivariate classification technique, such as PLS-DA, was used to observe freshness differences (<4 d.p.m. vs. >4 d.p.m.). Such an approach has never before been used in similar studies with hyperspectral methodology. Three different multivariate classification approaches (i.e., AP1, AP2, and AP3) were used for this occasion:

AP1: In order to evaluate the ROI topographic positioning based on the first 10 landmarks (ROI_L) and on the contribution of selected wavelengths, a data set was built for each tested individual at each time of sampling (d.p.m.) by considering each pixel for each wavelength layer at its topographic position as X-block variable. In order to reduce the

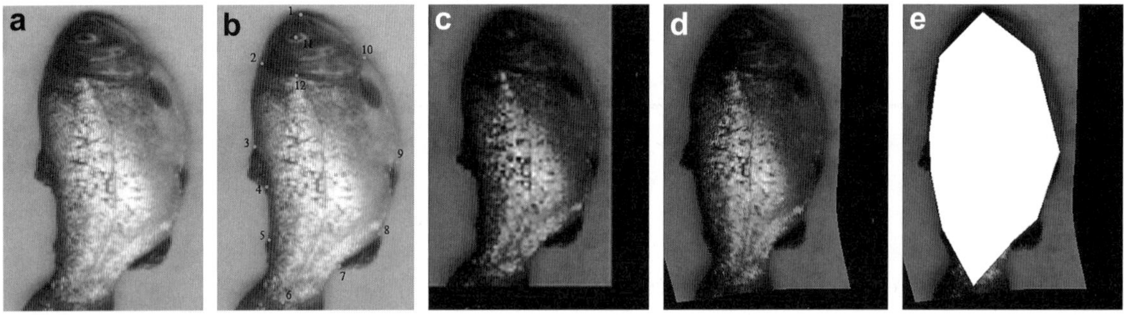

FIGURE 8.5 *Comparison of original fish image with various processed images: (a) original image; (b) image with 12 landmarks; (c) image of one hyperspectral layer (650 nm); (d) image after the warping procedure; (e) ROI based on the first 10 landmarks (ROI_L) positioned on the outline of the fish. (Full color version available on http://www.elsevierdirect.com/companions/9780123747532/)*

matrix dimension the images were resized 3 641 pixels (i.e., 1:0.3) and the number of wavelengths considered was 61 (500–800 nm; step-frequency: 5 nm).

AP2: In order to verify the classification capacity of this system each ROI_L pixel from each image of the fish at different d.p.m. was individually classified based on the dichotomic categorization "fresh"/"not-fresh" valid for the entire fish.

AP3: A reduced and most informative part of the hypercube either in terms of ROI (ROI_S) and in terms of wavelengths, both identified by the AP1 approach, was used, following the AP2 approach.

The results indicate that the consensus ROI_L encompassed 40 420 pixels when images were not rescaled. For a rescaling equal to 1:0.3, the ROI_L encompassed 3 641 pixels for each tested fish. All of these pixels present reflectance values for at least 61 wavelengths. Taken together, these represent a great quantity of data for each individual tested fish to be classified upon. This result is important since, for example, Farzam *et al.* (2008) showed how hyperspectral methodology requires huge calculation resources for data treatment.

Based on AP1, which considers both the spectral values of each pixel as well as its topographic position, more than 20 000 variables were obtained for each of the individuals, to be considered within the X-block (3 641*61 = 222 101). Based on AP2, by considering each pixel within the ROI_L, 72 820 samples were obtained. With AP3, by reducing the ROI_L at ROI_S and by considering only 17 variables, the samples number was reduced to 5 580

Table 8.2 Characteristics and principal results of the PLS-DA models

	AP1	AP2	AP3
No. samples	20	72 820	5 580
No. units (X-block)	222 101	61	33
No. units (Y-block)	2	2	2
No. LV	10	16	9
Preprocessing X-block	Baseline	Median center	Median center
Cumulated variance X-block (%)	95.28	99.95	99.91
Cumulated variance Y-block (%)	99.99	51.42	19.29
Mean sensitivity (%)	1	55.4	81.0
Mean specificity (%)	1	63.2	77.9
Mean classification error	0	0.39794	0.20645
Mean RMSEC	0.0061	0.49251	0.63524
Random probability (%)	50	50	50
Correct classification model (%)	100	66.77	79.4

Note: No. units (Y-block) is the number of units to be discriminated by the PLS-DA; No. LV is the number of latent vectors for each model; and random probability (%) is the probability of random assignment of an individual into a unit.

(Table 8.2). Data reduction is important for data treatment in quality analytic procedures (Farzam *et al.*, 2008). Accordingly, the present result set is based on considering this discrimination important for the industrial processing of cultured fish. These findings represent the first step in filling a gap in the existing technology in food industrial processing, as already identified by Menesatti, Zanella *et al.* (2009).

The PLS-DA results based on the three proposed approaches are reported in Table 8.2. As expected, AP1 reached a percentage of individuals correct classified of 100, since it is based on many variables and comparatively few samples. The AP2 and AP3 methods present a percentage of single pixel correct classification that is high (66.77% for AP2 and 79.40% for AP3). The selection of ROI_S and the concomitant reduction of the most significant variables, as in AP3, created an interesting increment in the percentage of correct pixel classification (from 66.77% in AP2, to 79.4% in AP3).

Considering AP1, the first LV explains the major variance for X-block and Y, achieving 83.62% and 48.54%, respectively. For this reason, we use the loadings for each pixel of LV1 to observe the main topographic contribution of the wavelengths. In Figure 8.6 the graphic output of AP1 is shown for the selected ROI_L, with the loading contribution for LV1 exceeding the 90th percentile for each wavelength layer. It can be noticed that the most

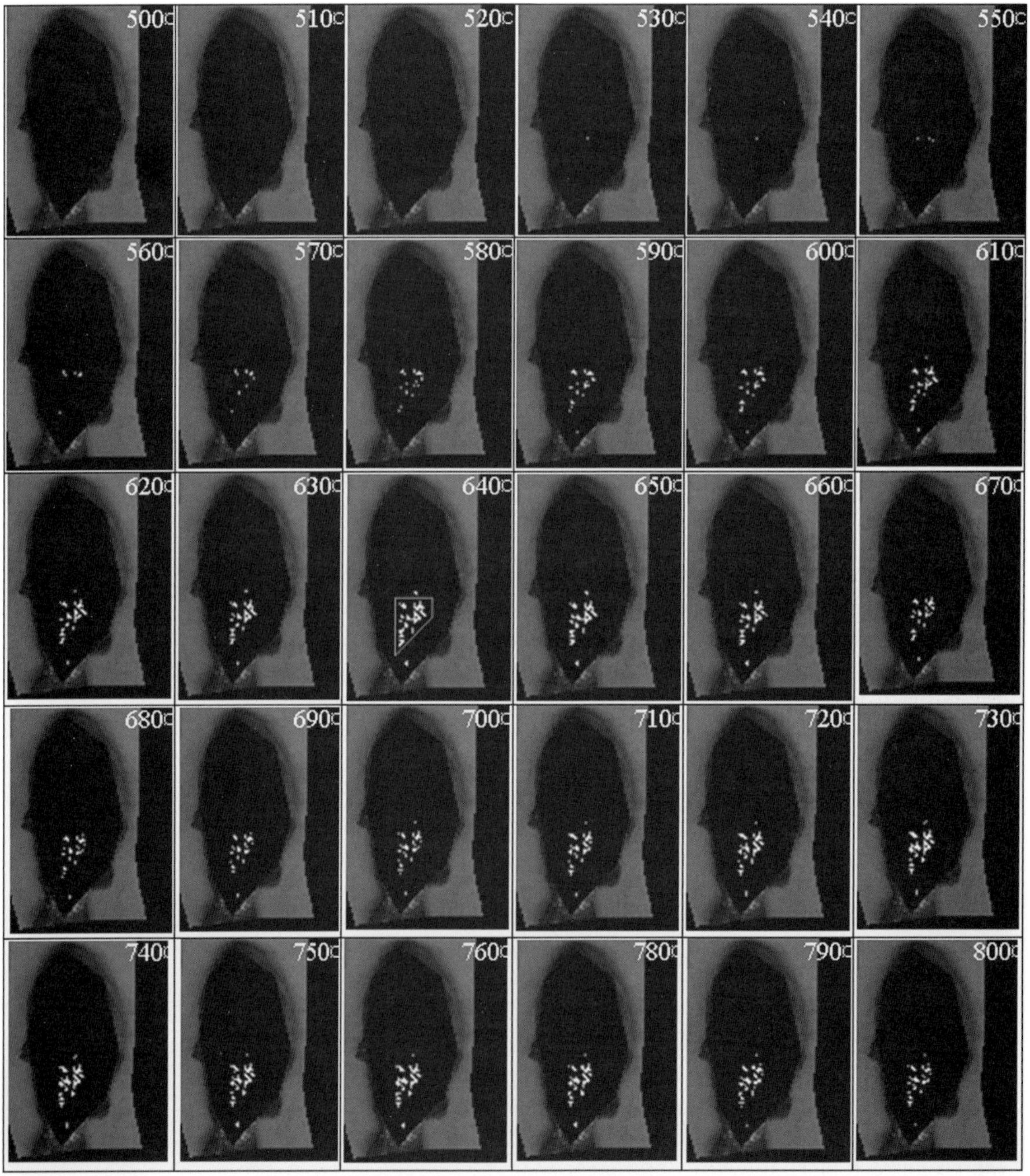

FIGURE 8.6 *Results of AP1 for the selected ROI_L for each wavelength layer (from 500 to 800 nm, on the top right of each image) with white pixels reported as the loadings LV1 contribution exceeding the 90th percentile. Area outlined in gray at 640 nm depicts the most informative region inside the ROI_L, named ROI_S.*

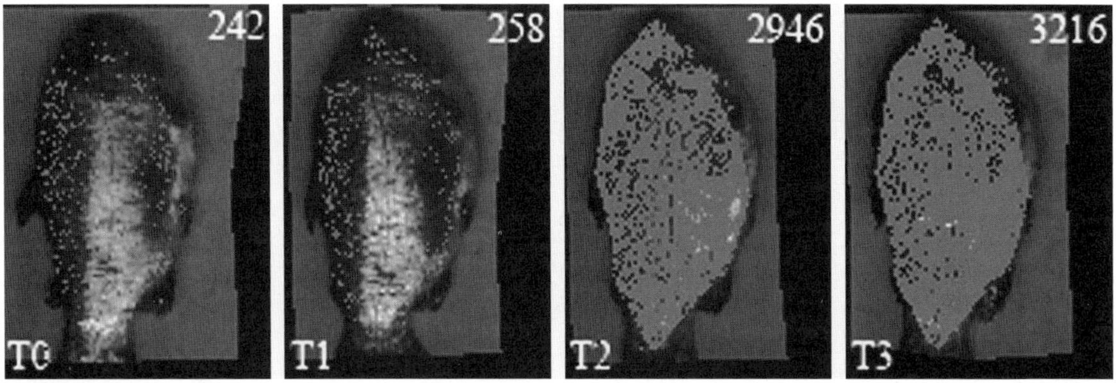

FIGURE 8.7 *Example of classification based on AP2 where one individual is pictured for one hyperspectral layer (i.e. at 650 nm) at different d.p.m. Pixels in red are those classified as "non-fresh". The number of red pixels is reported on the top right of each figure and the total number of pixels of the ROI is 3641. (Full color version available on http://www.elsevierdirect.com/companions/9780123747532/)*

important contribution is evident with wavelengths within the range 600–800 nm. From the topographic point of view, pixels that present higher levels of load contribution to the output of the PLS-DA are all located within the central area of the fish body, close to the anal fin, at the level of the lateral line (Figure 8.6). Pixels of this area, and associated with wavelengths within the 600–800 nm range, were chosen to implement a model of classification according to AP3.

According to AP2, each pixel per selected ROI_L in all fish was classified as fresh/non-fresh (Figure 8.7). With the same ROI_L used for all fish, because of previous superimposing with the geometric morphometry, all hyperspectral images are topographically comparable. Hence, each fish can be classified as fresh/non-fresh based on the number of pixels. A threshold of 50% in the ROI_L pixels was used to discriminate inclusion of the sample into one of the two chosen class-statuses. All of the fresh-category individuals were well classified and 9 out of 10 (90%) non-fresh individuals were correctly classified. Globally, with AP2 a percentage of correct classification of 95% was reached.

According to AP3, each pixel within the selected ROI_S was classified as fresh and non-fresh separately (Figure 8.8). Again, with the ROI_S being equal for each fish, all hyperspectral images can be topographically compared. Accordingly, each sample could be classified as fresh or non-fresh based on the area extension of pixels within that class. As stated before, in order to classify each individual into the two class-statuses a threshold of 50% in the ROI_S pixels was used. Globally, the AP3 percentage of correct classification was 95%.

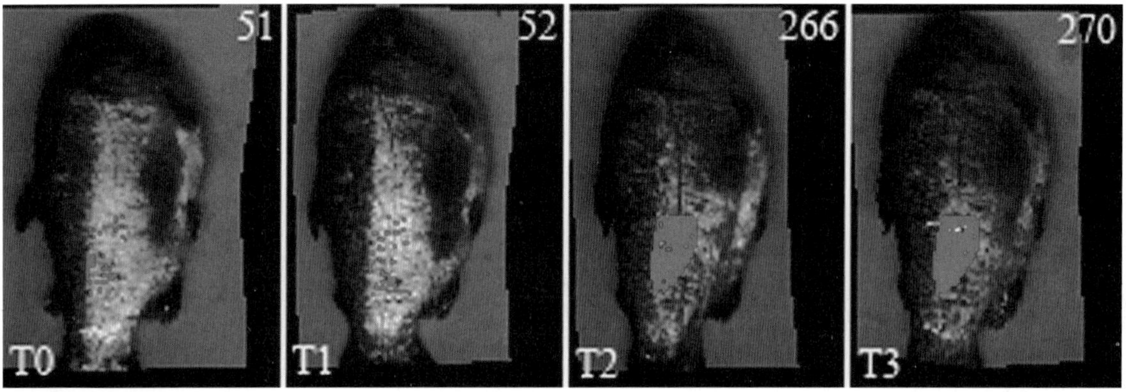

FIGURE 8.8 *Example of classification based on AP3 where one individual is pictured for one hyperspectral layer (i.e. at 650 nm) at different d.p.m. Pixels in red are those classified as "non-fresh". The number of red pixels is reported on the top right of each figure and the total number of pixels of the ROI is 279. (Full color version available on http://www.elsevierdirect.com/companions/9780123747532/)*

8.4. CONCLUSIONS

Hyperspectral imaging is a technique of high technological and methodological complexity, but with great application potential. In the market, fish freshness is defined and regulated by EU Directive No. 103/76, which classifies the product on the basis of quality parameters such as the consistency of the meat, the visual aspect (color of the eye and the gill, the brightness of the skin), and, finally, odor. It has been demonstrated that the quality of fish from both fishery and aquaculture can be evaluated using the hyperspectral video-image morphometric-based analysis.

In particular, two different methods were used on the acquired images that allow for both subjective and objective analysis. The first technique showed a greater efficiency in the assessment of fish freshness. The second technique represented an important methodological evolution of the first technique. Based on combined hyperspectral and geometric morphometric techniques, spectral information from pixels was associated with their topographic location for the first time. This novel approach is based on the *a priori* determination of which wavelength areas are more discriminating in relation to fish freshness, considering fish samples of non-homogeneous spectral quality.

In the second case study the proposed technique represents an important methodological development by combining hyperspectral imaging and geometric morphometric tools. This technique was applied in the hyperspectral field, resulting in an innovation allowing the association of topological spectral information. An automated method for the extraction of the fish outline should be implemented in the near future in association with hyperspectral processing in order to increase the efficiency of extraction of discriminant topologic information for quality assessment of non-homogeneous food samples.

NOMENCLATURE

Symbols

R	relative reflectance of the sample at each wavelength
r_s	absolute signal value (radiance) measured for the sample at each wavelength
r_b	absolute signal value (radiance) measured at each wavelength for the black (background noise)
r_w	absolute signal value (radiance) measured at each wavelength for the standard white (100% of reflectance)

Abbreviations

AP	approach
CT	concrete tanks
d.p.m.	days post mortem
GPA	generalized Procrustes analysis
LV	latent variables
NIR	near-infrared
NIRS	near-infrared reflectance spectroscopy
PGP	prism–grating–prism
PLS	partial least squares analysis
PLS-DA	partial least squares-discriminant analysis
RGB	red, green, blue
ROI	region of interest
ROI_L	ROI large (i.e. topographic positioning based on the first 10 landmarks)
ROI_S	ROI small (i.e. the proportion of ROI_L and the associated wavelengths that resulted as more informative)
SC	sea cages
VIS/NIR	visible/near-infrared

ACKNOWLEDGMENT

This work was funded by the project HighVision (DM 19177/7303/08) from the Italian Ministry of Agricultural, Food and Forestry Politics and by Programma Operativo Regionale – Puglia (Gesticom srl) and Friuli Venezia Giulia (Federcoopesca). Jacopo Aguzzi is a Postdoctoral Fellow of the "*JAE*" Program (Education and Science Ministry-MEC, Spain).

REFERENCES

Aguzzi, J., Costa, C., Antonucci, F., Company, J. B., Menesatti, P., & Sardá, F. (2009). Influence of diel behaviour in the morphology of decapod natantia. *Biological Journal of the Linnean Society, 96*, 517–532.

Alasalvar, C., Garthwaite, T., Alexis, M. N., & Grigorakis, K. (2001). Freshness assessment of cultured sea bream (*Sparus aurata*) by chemical, physical and sensory methods. *Food Chemistry, 72*, 33–40.

Alasalvar, C., Taylor, K. D. A., Oksuz, A., Shahidi, F., & Alexis, M. (2002). Comparison of freshness quality of cultured and wild sea bass (*Dicentrarchus labrax*). *Journal of Food Science, 67*, 3220–3226.

Antonucci, F., Costa, C., Aguzzi, J., & Cataudella, S. (2009). Ecomorphology of morpho-functional relationships in the family of Sparidae: a quantitative statistic approach. *Journal of Morphology, 270*, 843–855.

Bellon, V., Vigneau, J. L., & Leclercq, M. (1993). Feasibility and performance of a new, multiplexed, fast and low-cost fiber-optic NIR spectrometer for the on-line measurement of sugars in fruits. *Applied Spectroscopy, 47*(7), 1079–1083.

Bookstein, F. L. (1991). *Morphometric tools for landmark data*. Cambridge, UK: Cambridge University Press.

Chen, H., & He, Y. (2007). Theory and application of near infrared reflectance spectroscopy in determination of food quality. *Trends in Food Science & Technology, 18*, 72–83.

Costa, C., Aguzzi, J., Menesatti, P., Antonucci, F., Rimatori, V., & Mattoccia, M. (2008). Shape analysis of different populations of clams in relation to their geographical structure. *Journal of Zoology, 276*, 71–80.

Costa, C., Angelini, C., Scardi, M., Menesatti, P., & Utzeri, C. (2009a). Using image analysis on the ventral colour pattern in *Salamandrina perspicillata* (Savi, 1821) (Amphibia, Salamandridae) to discriminate among populations. *Biological Journal of the Linnean Society, 96*(1), 35–43.

Costa, C., Menesatti, P., Aguzzi, J., D'Andrea, S., Antonucci, F., Rimatori, V., Pallottino, P., & Mattoccia, M. (2009b). External shape differences between sympatric populations of commercial clams *Tapes decussatus* and. *T. philippinarum*. *Food and Bioprocess Technology*. doi: 10.1007/s11947–008–0068–8.

Costa, C., Menesatti, P., Paglia, G., Pallottino, F., Aguzzi, J., Rimatori, V., & Reforgiato Recupero, G. (2009c). Quantitative evaluation of Tarocco sweet orange fruit shape using opto-electronic elliptic Fourier based analysis. *Postharvest Biology and Technology, 54*, 38.

Costa, C., Tibaldi, E., Pasqualetto, L., & Loy, A. (2006). Morphometric comparison of the cephalic region of cultured *Acipenser baerii* (Brandt, 1869), *Acipenser naccarii* (Bonaparte, 1836) and their hybrid. *Journal of Ichthyology, 22*(1), 8–14.

Cozzolino, D., Murray, I., & Scaife, J. R. (2002). Near infrared reflectance spectroscopy in the prediction of chemical characteristics of minced raw fish. *Aquaculture Nutrition, 8*, 1–6.

ElMasry, G., & Wold, J. P. (2008). High-speed assessment of fat and water content distribution in fish fillets using online imaging spectroscopy. *Journal of Agricultural and Food Chemistry, 56*(17), 7672–7677.

FAO Fisheries Department, Fishery Information, Data and Statistics Unit (2007) FISHSTAT Plus: Universal software for fishery statistical time series. Version 2.32.

Farzam, M., Beheshti, S., & Raahemifar, K. (2008). Calculation of abundance factors in hyperspectral imaging using genetic algorithm. *Canadian Conference on Electrical and Computer Engineering*, 837–842.

Gowen, A. A., O'Donnell, C. P., Cullen, P. J., Downey, G., & Frias, J. M. (2007). Hyperspectral imaging: an emerging process analytical tool for food quality and safety control. *Trends in Food Science & Technology, 18*, 590–598.

Huss, H. H., (1995) Quality and quality changes in fresh fish. FAO Fisheries Technical Paper 348. http://www.fao.org/docrep/V7180E/V7180E00.HTM# Contents (accessed 26 January 2010).

Jason, A. C., & Richards, J. C. S. (1975). The development of an electronic fish freshness meter. *Journal of Physics E: Scientific Instruments, 8*, 826–830.

Karoui, R., Schoonheydt, R., Cecuypere, E., Nicolai, B., & De Baedemaeker, J. (2007). Front face fluorescence spectroscopy as a tool for the assessment of egg freshness during storage at a temperature of 12.2 degrees C and 87% relative humidity. *Analytica Chimica Acta, 582*, 83–91.

Katayama, K., Komaki, K., & Tamiya, S. (1996). Prediction of starch, moisture, and sugar in sweetpotato by near infrared transmittance. *HortScience, 31*(6), 1003–1006.

Kays, S. J. (1999). Preharvest factors affecting appearance. *Postharvest Biology and Technology, 15*, 233–247.

Kim, M. S., Lefcourt, A. M., Chao, K., Chen, Y. R., Kim, I., & Chan, D. E. (2002). Multispectral detection of fecal contamination on apples based on hyperspectral imagery. Part I: Application of visible and near-infrared reflectance imaging. *Transactions of the ASAE, 45*(6), 2027–2037.

Knaflewska, J., & Pospiech, E. (2007). Quality assurance systems in food industry and health security of food. Acta Scientiarum Polonorum. *Technologia Alimentaria, 6*(2), 75–85.

Lammertyn, J., Nicolaii, B., Ooms, K., De Smedt, V., & De Baerdemaeker, J. (1998). Non-destructive measurement of acidity, soluble solids, and firmness of Jonagold apples using NIR-spectroscopy. *Transactions of the ASAE, 41*(4), 1089–1094.

Li, Q., Wang, M., & Gu, W. (2002). Computer vision based system for apple surface defect detection. *Computers and Electronics in Agriculture, 36*(2), 215–223.

Liu, Y., Chen, Y. R., Wang, C. Y., Chan, D. E., & Kim, M. S. (2006). Development of hyperspectral imaging technique for the detection of chilling injury in cucumbers; spectral and image analysis. *Applied Engineering in Agriculture, 22*(1), 101–111.

Loy, A., Busilacchi, S., Costa, C., Ferlin, L., & Cataudella, S. (2000). Comparing geometric morphometrics and outlines fitting methods to monitor fish shape variability of *Diplodus puntazzo* (Teleostea, Sparidae). *Aquacultural Engineering, 21*(4), 271–283.

Lu, R., & Park, B. (2008). Hyperspectral and multispectral imaging for food quality and safety. *Sensing and Instrumentation for Food Quality and Safety, 2*(3), 131–132.

Mathias, J. A., Williams, P. C., & Sobering, D. C. (1987). The determination of lipid and protein in freshwater fish by using nearinfrared reflectance spectroscopy. *Aquaculture, 61*, 303–311.

Mehl, P. M., Chao, K., Kim, M. S., & Chen, Y. R. (2002). Detection of contamination on selected apple cultivars using hyperspectral and multispectral image analysis. *Applied Engineering in Agriculture, 18*(2), 219–226.

Menesatti, P., D'Andrea, S., & Costa, C. (2007). Spectral and thermal imaging for meat quality evaluation. In C. Lazzaroni, S. Gigli, & D. Gabina (Eds.), *New developments in evaluation of carcass and meat quality in cattle and sheep* (pp. 115–134). Wagening: Wageningen Academic Publishers, EAAP 123.

Menesatti, P., Paglia, G., Solaini, S., Zanella, A., Stainer, R., Costa, C., & Cecchetti, M. (2002). Non-linear multiple regression models to estimate the drop damage index of fruit. *Biosystems Engineering, 83*(3), 319–326.

Menesatti, P., Pallottino, F., Lanza, G., & Paglia, G. (2009). Prediction of blood orange MT firmness by multivariate modelling of low alterative penetrometric data set: a preliminary study. *Postharvest Biology and Technology, 51*, 434–436.

Menesatti, P., & Urbani, G. (2004). Prediction of the chilling storage time of fresh salt-water fishes by soft modelling (PLS) of low-alterative penetrometric test. Oral presentation International and European Agricultural Engineering Conference—AgEng2004—Workshop "Physical Methods in Agriculture", Leuven, Belgium, 12–16 September 2004. *Book of abstracts*, 940–941.

Menesatti, P., Urbani, G., Millozza, M., D'Andrea, S., Solaini, S., Paglia, G., & Niciarelli, I. (2006). *Prediction of the chilling storage time of fresh salt-water fishes by means of non-destructive techniques. Oral communication CIGR Section VI International Symposium on Future of Food Engineering*. Warsaw: Poland. 26–28 April 2006. Abstract 145.

Menesatti, P., Urbani, G., Pallottino, F., D'Andrea, S., & Costa, C. (2007). Non-destructive multi-parametric instruments for fish freshness estimation. *Instrumentation Viewpoint, 6*.

Menesatti, P., Zanella, A., D'Andrea, S., Costa, C., Paglia, G., & Pallottino, F. (2009). Supervised multivariate analysis of hyperspectral NIR images to evaluate the starch index of apples. *Food and Bioprocess Technology, 2*(3), 308–314.

Nicosevici, T., Garcia, R., & Gracias, N. (2007). Identification of geometrically consistent interest points for the 3D scene reconstruction. *Instrumentation Viewpoint, 6*.

Nortvedt, R., Torrissen, O. J., & Tuene, S. (1998). Application of near infrared transmittance spectroscopy in the determination of fat, protein, dry matter in Atlantic halibut fillets. *Chemometrics and Intelligent Laboratory Systems, 42*, 199–207.

Orban, E., Di Lena, G., Nevigato, T., Casini, I., Santaroni, G., Marxetti, A., & Caproni, R. (2002). Quality characteristics of sea bass intensively reared and from a lagoon as affected by growth conditions and the aquatic environment. *Journal of Food Science, 67*, 542–546.

Park, B., Lawrence, K. C., Windham, W. R., & Smith, D. (2006). Performance of hyperspectral imaging system for poultry surface fecal contaminant detection. *Journal of Food Engineering, 75*(3), 340–348.

Park B., Windham, W. R., Lawrence, K. C., Smith, D. P. (2004) Hyperspectral image classification for fecal and ingesta identification by spectral angle mapper. ASAE/CSAE Meeting, Ottawa, Ontario, Canada, ASAE Paper No. 043032.

Peirs, A., Scheerlinck, N., Perez, A. B., Jancsok, P., & Nicolai, B. M. (2002). Uncertainty analysis and modelling of the starch index during apple fruit maturation. *Postharvest Biology and Technology, 26*(2), 199–207.

Polder, G., Heijden, G., & Young, I. (2002). Spectral image analysis for measuring ripeness of tomatoes. *Transactions of the ASAE, 45*, 1155–1161.

Rohlf, F. J. (1999). Shape statistics: Procrustes superimpositions and tangent spaces. *Journal of Classification, 16*, 197–223.

Rohlf, F. J., & Bookstein, F. L. (1990). *Proceedings of the Michigan morphometric workshop. Special Publication No. 2*. Ann Arbor, MI: University of Michigan Museum of Zoology.

Rohlf, F. J., & Slice, D. (1990). Extensions of the Procrustes method for the optimal superimposition of landmarks. *Systematic Zoology, 39*, 40–59.

Sabatier, R., Vivein, M., & Amenta, P. (2003). Two approaches for discriminant partial least square. In M. Schader, *et al.* (Eds.), *Between data science and applied data analysis*. Berlin: Springer.

Sigurgisladottir, S., Hafsteinsson, H., Jonsson, A., Lie, O., Nortvedt, R., Thomassen, M., & Torrissen, O. (1999). Textural properties of raw salmon fillets as related to sampling method. *Journal of Food Science, 64*, 99–104.

Sjöström, M., Wold, S., & Söderström, B. (1986). PLS discrimination plots. In E. S. Gelsema, & L. N. Kanals (Eds.), *Pattern recognition in practice II*. Amsterdam: Elsevier.

Solberg, C., & Fredriksen, G. (2001). Analysis of fat and dry matter in capelin by near infrared transmission spectroscopy. *Journal of Near Infrared Spectroscopy, 9*, 221–228.

Thompson D'AW. (1917). *On growth and form*. London: Cambridge University Press.

Wold, J. P., Johansen, I. R., Haugholt, K. H., Tschudi, J., Thielemann, J., Segtnan, V. H., Narum, B., & Wold, E. (2006). Non-contact transflectance near infrared imaging for representative on-line sampling of dried salted coalfish (bacalao). *Journal of Near Infrared Spectroscopy, 14*(1), 59–66.

Xiccato, G. A., Trocino, F., & Tulli and Tibaldi, E. (2004). Prediction of chemical composition and origin identification of European sea bass (*Dicentrarchus labrax* L.) by near infrared reflectance spectroscopy (NIRS). *Food Chemistry, 86*, 275–281.

Yang, C. C., Chen, Y. R., & Chao, K. (2005). Development of multispectral image processing algorithms for identification of wholesome, septicemic, and inflammatory process chickens. *Journal of Food Engineering, 69*(2), 225–234.

Zelditch, M. L., Swiderski, D. L., Sheets, H. D., & Fink, W. L. (2004). *Geometric morphometrics for biologists: a primer.* San Diego, CA: Elsevier Academic Press.

CHAPTER 9

Bruise Detection of Apples Using Hyperspectral Imaging

Ning Wang[1], Gamal ElMasry[2]

[1] *Department of Biosystems and Agricultural Engineering, Oklahoma State University, Stilwater, Oklahoma, USA*

[2] *Agricultural Engineering Department, Suez Canal University, Ismailia, Egypt*

CONTENTS

9.1. INTRODUCTION

Apple is one of the most widely cultivated tree fruits today. In the United States, apple fruits are the third most valuable fruits following grapes and oranges. In 2007, the USA produced 4.2 tons of apples with a value of about $2.5 billion (source: National Agricultural Statistics Service, USDA). Hence, apple has been recognized as an important economic crop.

Apple fruit has a beautiful appearance, special fragrance, rich taste, crunchy texture, and, most importantly, many healthy constituents, such as vitamins, pectin, and fiber. It is rated as the second most consumed fruit, both fresh and processed, after orange. High quality and safety of the fruit are always the consumers' top preference and are the goals that apple producers and the processing industry continually pursue. However, due to the complexity of apple handling, including harvest, packaging, storage, transportation, and distribution, a large percentage of apples are wasted cach year due to damage of various kinds. Bruise damage is a primary cause of quality loss and degradation for apples destined for the fresh market. Apples with bruise damage are not accepted by consumers. Bruising also affects the quality of processed apple products.

From an orchard to a supermarket, apples are subjected to various static and dynamic loads that may result in bruise damage. Brown *et al.* (1993) reported that apple bruises are largely caused by picking, bin hauling, packing, and distribution operations. Fresh market apples usually require

Hyperspectral Imaging for Food Quality Analysis and Control

harvest and packing by hand. Apples for processing are commonly handled mechanically which may lead to extensive bruising. Improper packaging methods can result in severe bruises, especially for apples that need to travel a long distance. The collisions among fruits and between fruits and their packaging can be intensified during transportation. Thus, it is very important to avoid bruise damage by improving apple handling processes and identifying bruises at an early stage before the apples are sent to the fresh market or to processing lines.

Apples are inspected at many handling stages by inspectors. Based on the quality, apples are graded into different classes. For example, USDA defines as the highest grade, "USA Extra Fancy", apples that are mature but not overripe, clean, fairly well formed, free from decay, diseases, and internal/external damage including bruises. The lowest grade is defined as "USA Utility", which is apples that are mature but not overripe, not seriously deformed and free from decay, diseases, serious damage caused by dirt or other foreign matter, broken skins, bruises, brown surface discoloration, russeting, sunburn or sprayburn, limb rubs, hail damage, drought spots, scars, stem or calyx cracks, visible water core, bitter pit, disease, and insects. Bruise damage is commonly evaluated based on the size and depth of bruise. Fresh apples are graded according to the size of the bruised area and the number of bruised areas, while apples for processing are mainly chosen based on the percentage of bruised area on the whole surface.

Apple bruise damage is due to impact, compression, vibration, or abrasion during handling. The level of bruise damage depends on the hardness/firmness of an apple. When a force is over the tolerance limit of an apple, bruise damage is formed. An impact bruise results from dropping the fruit onto a hard surface, such as conveyors and packing boxes. It can also happen during transportation when a vehicle runs on a rough road. An impact bruise may not be visible immediately when the impact applies; the symptom appears after a certain period of time. A compression bruise can be generated due to over-packing fruits in a package or a weak-loading capability of the package. Many methods and procedures have been developed and adopted during apple handling to reduce bruise damage.

Bruise damage can be observed as the discoloration of flesh, usually with no breach of the skin. The applied force causes physical changes of texture and/or chemical changes of color, smell, and taste. Two basic effects of apple bruise can be distinguished, namely browning and softening of fruit tissue. Although bruise damage is not visible initially, it may develop very fast, especially when inappropriate environmental conditions are applied during storage, transportation, and distribution.

9.2. GENERAL METHODS TO DETECT BRUISE DAMAGE

Effectively identifying and classifying apples with bruise damage is important to ensure the fruit quality. However, due to the invisibility of the symptom at the early stage when bruising occurs, it is very difficult to identify fruits with bruise damage. In addition, bruises usually have no breach on the surface. For apples with dark and brownish color, e.g. the Red Delicious variety, the bruise area is not obvious even after a long time (Figure 9.1).

Bruise detection has been predominantly performed manually in the past, and in some current sorting applications is carried out by people trained in the standards of the quality characteristics of the fruit. In most apple packing stations workers are standing along the apple conveyors visually inspecting passing apples and removing rotten, injured, diseased, bruised, and other defective fruits. After a few hours of continuous inspection, their efficiency reduces rapidly which lead to incorrect and inconsistent grading. New automated bruised detection technology is in demand.

It has always been a challenging task to detect bruise damage, which usually takes place under the fruit skin. Detection accuracy is greatly affected by many factors such as time, bruise type, bruise severity, fruit variety, and fruit pre- and post-harvest conditions (Lu, 2003). Much research has been conducted to overcome these difficulties. Wen & Tao (1999) developed a near-infrared (NIR) vision system for automating apple defect inspection using a monochrome CCD camera attached with a 700 nm long-pass filter. A chlorophyll absorption wavelength at 685 nm and two wavelengths in the NIR band were found to provide the best visual separation of the defective area from the sound area of Red Delicious, Golden Delicious, Gala, and Fuji apples (Mehl *et al.*, 2004). Shahin *et al.* (2002) examined new (1 day) and old (30 days) bruises in Golden and Red Delicious apples using line-scan x-ray imaging and artificial neural network (ANN) classification. They found that new bruises were not adequately separated using this methodology. The

FIGURE 9.1 *Apple fruits after bruising on: (left) red, (center) green, and (right) reddish background colors. (Full color version available on http://www.elsevierdirect.com/companions/9780123747532/)*

preliminary tests of Leemans *et al.* (1999) proposed Bayesian classification to avoid misclassification among different defects and sound surface of apples. Kleynen *et al.* (2005) stated that russet defects and recent bruises were badly segmented because they presented a color similar to the healthy tissue. Thus, 3-CCD color cameras are not fully adapted to defect detection in fruits since they are designed to reproduce human vision. They found the three most efficient wavelength bands centered at 450, 750 and 800 nm. The 450 nm spectral band brought significant information to identify slight surface defects like russet, while the 750 and 800 nm bands offered a good contrast between the defect and the sound tissue. These wavebands were well suited to be used for detecting internal tissue damage like hail damage and bruises. Bennedsen & Peterson (2005) and Throop *et al.* (2005) developed an automatic inspection system and succeeded in identifying the bruise area on apples using three wavebands at 540, 740 and 950 nm.

Unfortunately, all of the above-mentioned attempts were conducted to detect bruises 24 hours after occurrence and on varieties with one uniform background color. Problems arose if the bruises appeared on a variety with a homogeneous, multicolored background and in the early stages when the edges between a bruise and its surrounding area are often poorly defined (Zwiggelaar *et al.*, 1996). Since bruising take place beneath the peel, it is difficult to detect visually or with any regular color imaging methods, especially those bruises on a dark-colored background. Dark-colored apple skin can easily obscure human vision or mislead automatic color sorting systems (Gao *et al.*, 2003). Since bruises are most likely to appear at any stage of handling, the challenge is to detect these early occurring bruises as soon as possible to avoid any possibility of invasion. Furthermore, bruises are affected by apple variety and bruise severity, and they change with time and at different rates, even for the same apple fruit. Therefore, an effective detection system must have the capability to detect bruises, both new and old, for different background colors (Lu, 2003). All these factors make bruise detection very difficult when needed at an early stage as well as on multicolored backgrounds. To overcome these difficulties, the image contrast needs to be enhanced by selecting the most suitable spectral images accompanied by arithmetic manipulations to isolate bruises from normal surfaces.

Recently, thermal imaging technology has become technologically and economically feasible for food quality applications. It has shown great potential for the detection of bruise and other disease damage. Baranowski & Mazurek (2008) based their research on a hypothesis that internal defects and physiological disorders of fruit lead to changes of tissue thermal properties. They used a pulsed-phase thermography (PPT) system to collect thermal images after apple fruits are subject to the pulsed heat sources. The results show that the PPT method can not only locate bruise damage, but also evaluate the intensity of the bruise damage. However, the complexity of developing thermal imaging systems for processing line conditions and avoiding noise and interference from the surrounding environment limits their practical deployment.

9.3. HYPERSPECTRAL IMAGING TECHNOLOGY

Spectral reflectance imaging originated from the fields of chemistry and remote sensing and has been widely used for assessing quality aspects of agricultural produce (Kavdir & Guyer, 2002). Hyperspectral imaging can be utilized as the basis for developing such systems due to its high spectral and spatial resolution, non-invasive nature, and capability for large spatial sampling areas. With the development of optical sensors, hyperspectral imaging integrates spectroscopy and imaging techniques to provide spectral information as well as spatial information for the measured samples. The hyperspectral imaging technique has been implemented in several applications, such as the inspection of poultry carcasses (Chao *et al.*, 2001; Park *et al.*, 2004), defect detection or quality determination on apples, eggplants, pears, cucumbers, and tomatoes (Cheng *et al.*, 2004; Kim *et al.*, 2004; Li *et al.*, 2002; Liu *et al.*, 2006; Polder *et al.*, 2002) as well as estimation of physical, chemical, and mechanical properties in various commodities (Lu, 2004; Nagata *et al.*, 2005; Park *et al.*, 2003; Peng & Lu, 2005). Research has also been reported on applying hyperspectral imaging technology to apple bruise detection. The main procedures in these applications are presented in the following sections.

9.3.1. Establishing Hyperspectral Imaging Systems for Apple Bruise Detection

The hyperspectral imaging systems used for apple bruise detection are very similar in general. They are composed of five components: an imaging spectrograph coupled with a standard zoom lens, an illumination unit, a camera, a movable/stationary fruit holder, and a personal computer. The major difference is whether the tested sample is still or moving. Figures 9.2 and 9.3 show examples of hyperspectral imaging systems for still and moving samples, respectively.

9.3.1.1. Imaging spectrograph

The imaging spectrograph is a line-scan device which is capable of producing full contiguous spectral information with high-quality spectral and spatial resolution. It is combined with any area camera to produce hyperspectral

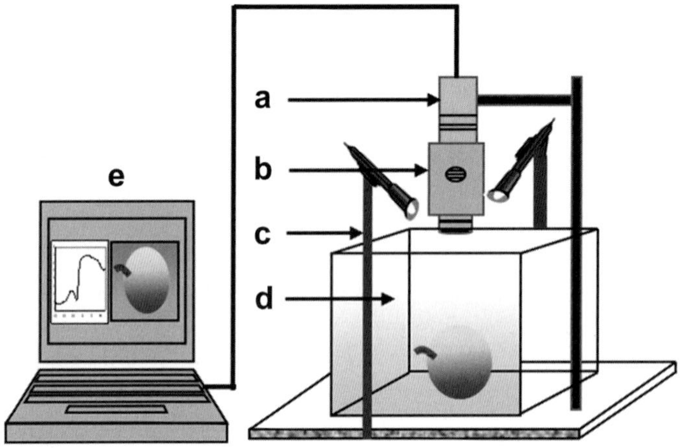

FIGURE 9.2 *Hyperspectral imaging system for still samples: (a) a camera; (b) an imaging spectrograph with a standard zoom lens; (c) an illumination unit; (d) a test chamber; and (e) a computer with image acquisition software (after ElMasry* et al., *2007. © Elsevier 2007). (Full color version available on http://www.elsevierdirect.com/companions/9780123747532/)*

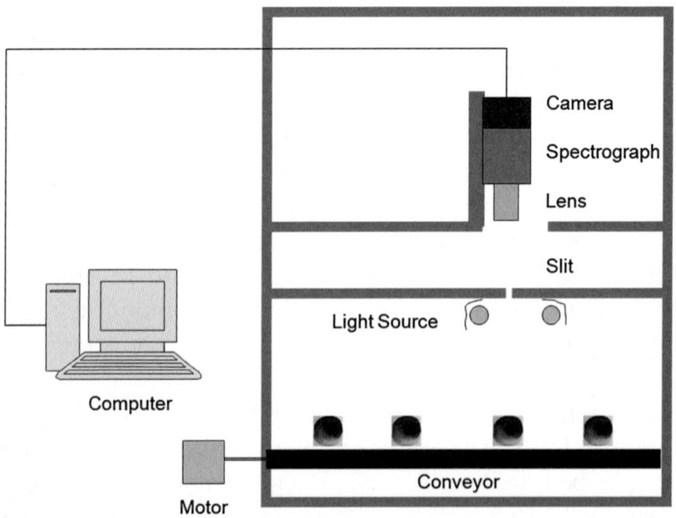

FIGURE 9.3 *Hyperspectral imaging system for moving samples (after Xing & De Baerdemaeker, 2005. © Elsevier 1995). (Full color version available on http://www.elsevierdirect.com/companions/9780123747532/)*

images. Typical commercially available spectrograph is the ImSpector® series manufactured by Specim Imaging Ltd, Finland. The spectrographs in the series have different spectral ranges from 200 nm to 12 000 nm. For example, ImSpector VNIR V10 works for a spectral range of 400–1000 nm, while ImSpector NIR V17E has a spectral range of 900–1700 nm. Users can select the model of ImSpector spectrographs based on the required wavelength range and the characteristics of target objects.

The selection of spectral resolution is also very important. The selection criterion is to include the minimum amount of data in the later processes while maintaining useful information. The benefits are the reduction of the amount of the data to be processed and improvement of signal-to-noise ratio due to noise and interference. Once the resolution is selected, a binning process can be implemented by grouping or averaging adjacent pixels in the spectral images. Many commercial systems allow users to select different binning ratios.

9.3.1.2. Camera detectors

The ImSpectors are mainly designed to work with area scan cameras. When a light beam reflected from the target objects hits the imaging spectrograph, it is dispersed according to wavelengths while preserving its spatial information. The dispersed light beams are then mapped to the camera detector array. For each scan, the spectrograph–camera assembly results in a two-dimensional image (a spectral axis and a spatial) of the scanned line. In order to obtain an area image, an additional spatial dimension can be created by moving the target object with a precisely controlled conveyor system (Figure 9.3). Lu (2003) used a controllable roller to rotate the tested sample with a speed synchronized with the imaging system. The additional spatial dimension can also be formed by moving the spectrograph and camera assembly by a stepper motor within the field of view, while keeping the tested sample still (Figure 9.2). After finishing the scans on the entire fruit, the spatial-hyperspectral matrices were combined to construct a three-dimensional spatial and spectral data space (x, y, z), where x and y are the spatial dimensions and z is the spectral dimension.

When selecting the camera attached to the spectrograph, besides the factors considered for regular imaging systems, the spectral sensitivity of the camera needs to be carefully considered. For example, the spectral range of the ImSpector VNIR V10 is 400–1000 nm. The sensitivity of silicon-based CCD (charge-coupled device) camera detectors is typically excellent within the visible (VIS) range, but may tail off at the NIR range (800–1000 nm). Hence, the collected image data are often found noisy at the two far ends of the spectral range. Special considerations are needed based on the requirements of the applications.

Recently, CMOS (complementary metal-oxide semiconductor) camera detectors have been adopted by a hyperspectral imaging system with the advantages of lower cost, lower power consumption, and capability of random access to the individual pixels. However, same as CCD, CMOS camera detectors are also silicon-based. Their sensitivity also drops in the infrared (IR) range. CMOS detectors are also subject to higher noise which may affect their sensitivity, especially in the IR range. When only IR is the spectral range of interest, the ImSpector N17E with a spectral range of 900–1700 nm can be paired with an InGaAs (indium gallium arsenide) camera which has a high sensitivity and dynamic range in IR range (Lu, 2003).

9.3.1.3. Illumination unit

In order to acquire high-quality spectral images, the illumination unit needs to be designed carefully so that its spectral emission, intensity, and scattering/reflection pattern of the light source will match the requirements of the imager and spectrograph. In many applications, DC quartz–halogen lamps with an adjustable power controller are used. A light diffused tent or frame can be used to ensure uniform lighting within the field of view (FOV) of the hyperspectral imaging system.

9.3.1.4. Movable/stationary fruit holder

Based on the types of spectrograph and camera assembly, the fruit holder can be selected to be a conveyor driven by a precisely controlled stepper motor or a simple stationary holder. If the conveyor is used, its speed has to be synchronized with the imaging system.

9.3.1.5. Personal computer

A computer is an imperative component in the hyperspectral imaging system. It controls the spectral image acquisition, binning process, and stepper motor. Due to the huge amount of image data generated by hyperspectral imaging acquisition, the computer needs to have a large RAM (e.g. >2 GB), a large hard drive, and a fast processing speed.

9.3.2. Preprocessing of Hyperspectral Images

The raw spectral–spatial images acquired from the hyperspectral imaging system need to be preprocessed before proceeding to bruise detection algorithms. To reduce the size of the data set, the background of the image is first removed using simple thresholding methods. During the spectral image acquisitions, it is very common that the spectral data at the two ends of the spectral range are very noisy, and thus are often chopped off and excluded from the following processes. Only the stable data set is used for further analysis. To improve the image quality, a low-pass filter is used to smooth both spatial and spectral data.

The acquired hyperspectral images need to be corrected with a white and a dark reference. The dark reference is used to remove the effect of dark current of the CCD detectors, which are thermally sensitive. The corrected image (R) is then defined using Equation (9.1):

$$R = \frac{R_0 - D}{W - D} \times 100 \tag{9.1}$$

where R_0 is the recorded hyperspectral image, D the dark image (with 0% reflectance) recorded by turning off the lighting source with the lens of the camera completely closed, and W is the white reference image (Teflon white board with 99% reflectance). These corrected images are used to extract information about the spectral properties of normal and bruised surfaces for optimizing defect identification, selection of effective wavelengths, and segmentation purposes.

9.3.3. Wavelength Selection Strategy

A hyperspectral imaging system produces a huge amount of spectral-image data. It demands significant computer resource and computation power to process the data. The time required to process the data is usually too long for any real-time applications. In addition, a lot of redundant data often exist in the data set which may reduce the power of bruise detection. Hence, instead of using the whole data set, a few effective wavelengths are identified so that the image data at the selected wavelengths are the most influential on apple bruise detection. The other wavelengths, which have no discrimination power, should be eliminated from analysis.

There is no standard method to select the significant wavelengths from the whole spectrum. A variety of strategies have been used to select effective wavelengths for bruise detection, such as general visual inspection of the spectral curves and correlation coefficients (Keskin *et al.*, 2004), analysis of spectral differences from the average spectrum (Liu *et al.*, 2003), correlelogram analysis (Xing *et al.*, 2006), stepwise regression (Chong & Jun, 2005), principal component analysis (Xing & De Baerdemaeker, 2005), principal component transform and minimum noise fraction transform (Lu, 2003), and partial least squares (PLS) and stepwise discrimination analyses (ElMasry *et al.*, 2007). The outcome of these strategies is a set of multiple feature waveband images reduced from the high-dimensional raw spectral images which can be used in image classification algorithms to identify bruised apple fruits.

9.3.4. Bruise Detection Algorithms

As mentioned previously, bruise damage is usually hard to detect based on color features, even after a certain period of time following its occurrence. Xing & De Baerdemaeker (2005) used shape deformation found in spectral images to identify bruised apples. Apples with no damage (sound apples) are spherical and smooth on the surface. When an apple is bruised, after a period of time, the damaged areas may grow larger and flatter, affecting the smooth curvature of the surface. This phenomenon was used in a principal component analysis (PCA) to identify feature multiple waveband images. An image processing and classification algorithm was developed based on PCA scores to classify sound or bruised apples with an accuracy of about 77.5% for impact-bruised apples.

It has also been mentioned that after bruising, the tissue of the damaged area will change physically and chemically. The spectral information acquired by the hyperspectral imaging system is well suited to this task. Lu (2003) applied principal component (PC) transform and minimum noise fraction transform (MNF) methods to detect the bruised areas. For each raw image, multiplication of the first and third PC images was performed. In the resultant image, the bruises, both old and new, would always appear to be darker than normal tissue. Bruises were normally present in the third MNF image, either dark or bright. By comparing the mean pixel values for the two groups of areas corresponding to those identified in the MNF images, true bruises were identified (Figure 9.4). Lu (2003) also found that the difference in reflectance between normal and bruised apples was greatest between 900 nm and 1400 nm. With the developed algorithms, Lu (2003) concluded that the detection accuracy was low when bruises were less than four hours old and became higher (88.1%) one day after bruises were induced.

Artificial neural networks (ANN) have proven to be very effective in the identification and classification of agricultural produce (Bochereau *et al.*, 1992; Jayas *et al.*, 2000), where non-coherence or non-linearity often exists. Kavdir & Guyer (2002, 2004) developed a back-propagation neural network (BPNN) with the textural features extracted from the spatial distribution of color/gray levels to detect defects (leaf roller, bitter pit, russet, puncture, and bruises) in Empire and Golden Delicious apples. ElMasry *et al.* (2008) developed feed-forward back-propagation ANN models for a hyperspectral imaging system to select the optimal wavelength(s), classify the apples, and detect firmness changes due to chilling injury. The model could be modified to apply to bruise detection.

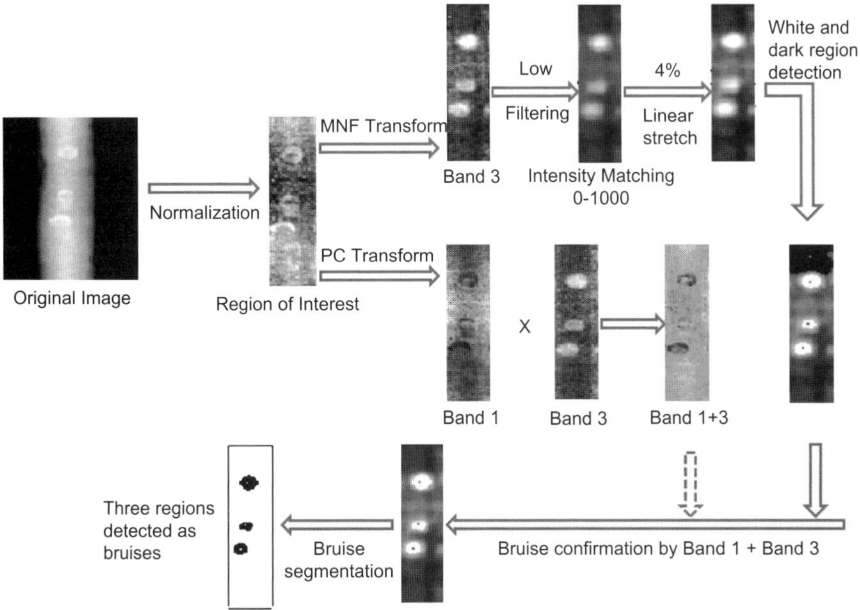

FIGURE 9.4 *Flowchart of the procedures for bruise detection using principal component transform and minimum noise fraction transform (MNF) methods (after Lu, 2003. © American Society of Agricultural and Biological Engineers 2003). (Full color version available on http://www.elsevierdirect.com/companions/9780123747532/)*

9.4. AN EXAMPLE OF A HYPERSPECTRAL SYSTEM DEVELOPED FOR EARLY DETECTION OF APPLE BRUISE DAMAGE

In this section, research work on early bruise detection will be presented in detail. The goal is to show a systematic program of work on developing a hyperspectral imaging system for early apple bruise detection. This work will provide a reference for further study by other researchers.

The main objective of this research was to investigate the potential of a hyperspectral imaging system that could be used for the early detection (<12 h) of bruises on different background colors of McIntosh apples. The research was conducted through (1) establishment of a hyperspectral imaging system with a spectral region from 400 nm to 1000 nm to detect bruises on different background colors (green, red, and green-reddish) of McIntosh apples; (2) the determination of the effective wavelengths for bruise detection by developing a statistical wavelength selection technique to identify and segregate both new and old bruises from the normal surface; and (3) the development of the algorithms to distinguish and isolate a bruised area from the sound surface.

9.4.1. Apple Sample Preparation and Hyperspectral System Setup

Apples were provided by the Horticulture Research and Development Centre of Agricultural and Agri-Food Canada, Saint-Jean-sur-Richelieu, Quebec, in the autumn of 2005. During the experiment, the apples were stored at 3 °C. Thirty fruits free from disease, defects, and blemishes were carefully selected to be used as a training group. Fruits were removed from the storage and left at room temperature (20 ± 1 °C) for 24 hours, after which bruises were created. McIntosh apples, as shown in Figure 9.1, were characterized by a green ground color, a darker red blush color, as well as transition colors between the blush and the ground color. The blush (red), intermediate (reddish) and ground (green) distribution on the apple surface varied with apple maturity.

A uniform bruise was produced in the middle area between the stem and calyx on each fruit by dropping a 250 g flat steel plate from 10 cm height on the fruit. This created a bruise of approximately 14–18 mm in diameter. Bruises were tested at different times (1 h, 12 h, 24 h, 3 days) from bruising to evaluate the ability of the hyperspectral imaging system to differentiate the bruised from normal skin and to define a time threshold at which bruises could be recognized.

A laboratory hyperspectral imaging system was established, as shown in Figure 9.5. It was composed of the following four components: an illumination unit with two 50W halogen lamps mounted at an angle of 45° to illuminate the camera's field of view, a fruit holder surrounded by a cubic tent made from white nylon fabric to diffuse the light and provide a uniform lighting condition, an ImSpector V10E spectrograph coupled with a standard C-mount zoom lens, and a CCD camera (PCO-1600, PCO Imaging, Germany). The assembly dispersed the incoming line of light into the spectral and spatial matrices and then projected them onto the CCD. The optics, including the spectrograph and the camera, had high sensitivity in the spectral range of 400 to 1000 nm. The exposure time was adjusted to 200 ms throughout the whole test. The distance from lens to the fruit surface was fixed at 40 cm. The camera–spectrograph assembly was provided with a stepper motor to move this unit through the camera's field of view to scan the fruit line by line. After finishing the scans on the entire fruit, the spatial-by-spectral matrices were combined to construct a 3-D spatial and spectral data space (x, y, λ), where x and y are the spatial

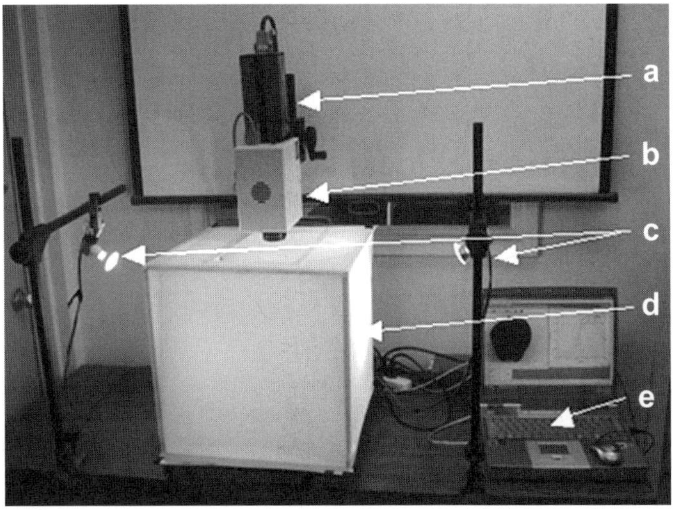

FIGURE 9.5 *The hyperspectral imaging system: (a) a CCD camera; (b) a spectrograph with a standard C-mount zoom lens; (c) an illumination unit; (d) a light tent; and (e) a PC supported with the image acquisition software (Full color version available on http://www.elsevierdirect.com/companions/9780123747532/)*

dimensions and λ is the spectral dimension. Images were binned during acquisition in the spatial direction to provide images with a spatial dimension of 400×400 pixels with 826 spectral bands from 400 to 1000 nm. The hyperspectral imaging system was controlled by a PC supported with a Hypervisual Imaging Analyzer® (ProVision Technologies, Stennis Space Center, MS, USA) for spectral image acquisition, binning, and camera and motor control.

9.4.2. Hyperspectral Image Processing

9.4.2.1. Preprocessing of hyperspectral images

All the acquired hyperspectral images were processed and analyzed using Environment for Visualizing Images (ENVI 4.2) software (Research Systems Inc., Boulder, CO, USA). The acquired images were corrected with a white and a dark reference. These corrected images were used to extract information about the spectral properties of normal and bruised surfaces for optimizing defect identification, selection of effective wavelengths and segmentation purposes. About 2000 pixels were manually selected from each corrected image as a region of interest (ROI). The average reflectance spectrum from the ROI of the normal surface of each background color (red, green, and reddish)

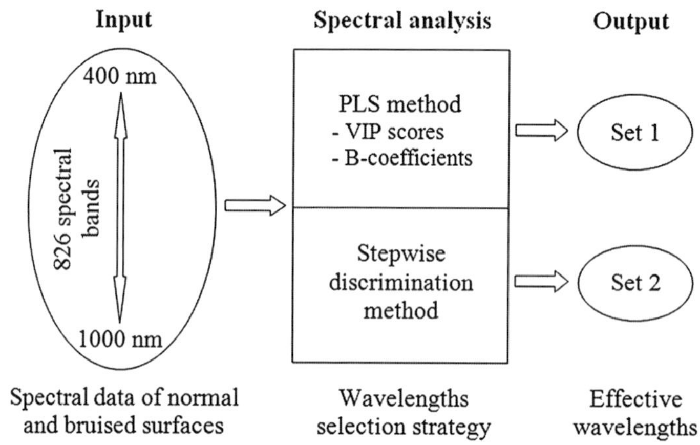

FIGURE 9.6 *Layout of dimensionality reduction for effective wavelengths selection*

was calculated by averaging the spectral value of all pixels in the ROI. In addition, the average spectra of the bruised region at different age of bruising (1 h, 12 h, 24 h, 3days) were calculated by averaging the spectral values of all pixels in the ROI of the bruised region.

9.4.2.2. Wavelength selection strategy

Partial least squares (PLS) and stepwise discrimination analyses were the two selection strategies used in this study to reduce high dimensionality of the spectral data and provided only a few essential wavelengths representing the whole spectrum. As shown in Figure 9.6, the input of the two methods was the raw spectral data extracted from both normal and bruised surfaces. Set 1 was the effective wavelengths selected using PLS with the variable importance in projection (VIP) scores (see Equation 9.5), while Set 2 was the effective wavelengths resulted from stepwise discrimination analysis described below.

In the first method of wavelength selection, PLS analysis was conducted between normal and bruised spectra using SAS® statistical software (SAS Institute Inc., NC, USA). PLS was implemented to transfer a large set of highly correlated and often collinear experimental data into independent latent variables or factors. When applied to spectra, the aim of PLS analysis was to find a mathematical relationship between a set of independent variables, the X matrix ($N_{samples} \times K_{wavelengths}$), and the dependent variable, the Y matrix ($N_{samples} \times 1$). The surface type (normal and/or bruised) represented the dependent variable (Y); meanwhile, the 826 wavelengths represented the independent variables or the predictors (X). Typically, most of the variance could be captured with the first few latent variables while the remaining latent variables described random noise or linear dependencies between the wavelengths/predictors.

The PLS algorithm (Osborne *et al.*, 1997) determined a set of orthogonal projection axes W, called PLS-weights, and wavelength scores T. For direct projection using the matrix of wavelength loadings (P'), $W^* = W\,(P'^*W)^{-1}$ was used:

$$T = XW^* \tag{9.2}$$

Then, regression coefficients β were obtained by regressing Y onto the wavelength scores T as follows:

$$Y = T\beta \tag{9.3}$$

If the number of PLS factors was a, the PLS model would be:

$$\widehat{Y} = XW_a^*\beta = T_a\beta \tag{9.4}$$

where $\widehat{Y}$ is the predicted surface type (normal or bruised) depending on the PLS-weights (W_a) and regression coefficient (β).

The relative importance of wavelengths in the model with respect to surface type (Y) could be reflected by new scores called variable importance in projection (VIP) scores according to the following formula:

$$VIP_k = \sum_{j=1}^{a}(w_{jk}^2.SSR_j)\,\frac{L}{SST} \tag{9.5}$$

where SSR is the residual sum-of-squares, SST is the total sum-of-squares of Y variable, and L is the total number of the examined wavelengths (826 spectral bands). VIP scores of each wavelength could be considered as selection criteria. Wavelengths with higher VIP scores were considered more relevant in classification (Bjarnestad & Dahlman, 2002). Based on the studies conducted by Olah *et al.*, (2004), predictors/wavelengths could be classified according to their relevance in explaining Y as: VIP > 1.0 (highly influential), $0.8 <$ VIP < 1.0 (moderately influential) and VIP < 0.8 (less influential). In this study, all wavelengths at which the VIP scores were above a threshold of 1.0 (highly influential wavelengths) were considered important and were compared with those extracted from stepwise discrimination methods to be used for classification processes.

The second method for wavelengths selection was implemented using stepwise discrimination. Although the stepwise discrimination method had some constraints, especially in case of the multicollinearity, it was used to confirm the selected wavelength from the VIP method. Stepwise discrimination is a standard procedure for variable selection, which is based on the procedure of sequentially introducing the predictors (wavelengths) into the model one at a time. In this method, the number of predictors retained in the final model is determined by the levels of significance assumed for inclusion and exclusion of predictors from the model. This test was conducted by SAS® statistical software using a level of significance value of 0.15 for entering and excluding predictors from the model.

Finally, to determine the potential of the selected wavelengths for bruise discrimination, PCA was conducted on the reflectance spectral data using only these optimal wavelengths instead of the full wavelength range. PCA is a projection method for extracting the systematic variations to generate a new set of orthogonal variables.

9.4.2.3. Image processing algorithms

The first step of the bruise detection algorithm is to create a binary mask to produce an image containing only the fruit, avoiding any interference from the background that could reduce discrimination efficiency. Imaging at 500 nm was used for this task because the fruit appeared opaque compared with the background and can be segmented easily by global thresholding. Secondly, images at the effective wavelengths identified from VIP and stepwise discrimination selection methods were averaged using ENVI, and this averaged image would be the basis for bruise area identification. In the ordinary RGB images, recent bruises are badly segmented because color is presented similar to the healthy tissue (Gao *et al.*, 2003; Kleynen *et al.*, 2005; Shahin *et al.*, 2002). On the contrary, with the averaged image in the NIR region the bruise area is well contrasted. In these images, a bruise's pixels were generally darker than the sound tissue's pixels.

In most cases the simple thresholding was not able to identify all of the defective area, due to variations in the graylevel within the defective area and the surrounding surface (Bennedsen & Peterson, 2005). The solution to this problem is to use an adaptive thresholding. Whereas the conventional thresholding uses a global threshold for all pixels, the adaptive thresholding changes the threshold dynamically over the image. In addition, multilevel adaptive thresholding selects individual thresholds for each pixel based on the range of intensity values in its local neighborhood. This allows for thresholding of an image whose global intensity histogram does not contain distinctive peaks. This more sophisticated version of thresholding can deal with a strong intensity gradient or shadows. This technique is successful in tackling the problems of noise and large difference in intensity in averaged images. So, the principle segmentation was carried out using a multilevel adaptive threshold method, which would select levels based on a histogram of the graylevels in the average image. The threshold was found by statistically examining the intensity values of the local neighborhood of each pixel. The statistic that is most appropriate includes the mean of the local intensity distribution. The size of the neighborhood has to be large enough to cover sufficient variations among pixels, otherwise a poor threshold is chosen. Hence, the average between the minimal and the maximal graylevel in the neighborhood was considered. If there were no defects in the image the resulting segmented image would be blank. Finally, the noise was removed by median filtering, in addition to erosion and dilation operations as shown in Figure 9.6. All image processing operations were performed using MATLAB 7.0 (Release 14, The MathWorks Inc., Natick, MA, USA) with the image processing toolbox.

9.4.3. Spectral Characteristics of Normal and Bruised Surfaces and Wavelength Selection

Figure 9.7(a)–(d) shows the reflectance spectra in the VIS (400–700 nm) and NIR (700–1000 nm) ranges for a typical McIntosh apple collected from ROIs of different background colors. Also, the average spectra of ROIs representing bruises at different ages (1 h, 12 h, 24 h, and 3 days) were illustrated. The presence of water in the fruit caused a rise at the characteristic absorption bands that appear as localized minima. The samples containing higher moisture contents had lower reflectivity across their spectra. In spite of background color, the absorption curves of McIntosh apples were rather smooth across the entire spectral region and had three broadband valleys around 500, 680, and 960 nm in addition to small valley at 840 nm. The absorption valleys around 500 and 680 nm represent carotenoids and chlorophyll pigments which represent the color characteristics in the fruit (Abbott *et al.*, 1997). The absorption valleys in the NIR range at 840 and 960 nm represent sugar and water absorption bands, respectively.

On the other hand, the reflectance from a bruised surface, even from recently bruised ones, was consistently lower than that from the normal tissue over the entire spectral region. These results are in agreement with the findings of several authors (Geola & Pieper, 1994; Zwiggelaar *et al.*, 1996). The difference in reflectance between the bruised and unbruised tissue on red and reddish apples was the greatest in the NIR region, while it decreased dominantly in the visible region, and the spectral images had higher levels of noise with low reflectance especially in the case of red and reddish background colors. Furthermore, the reflectance changed over time and the same pattern was observed for bruises after 12 h, 24 h and 3 days, which had much

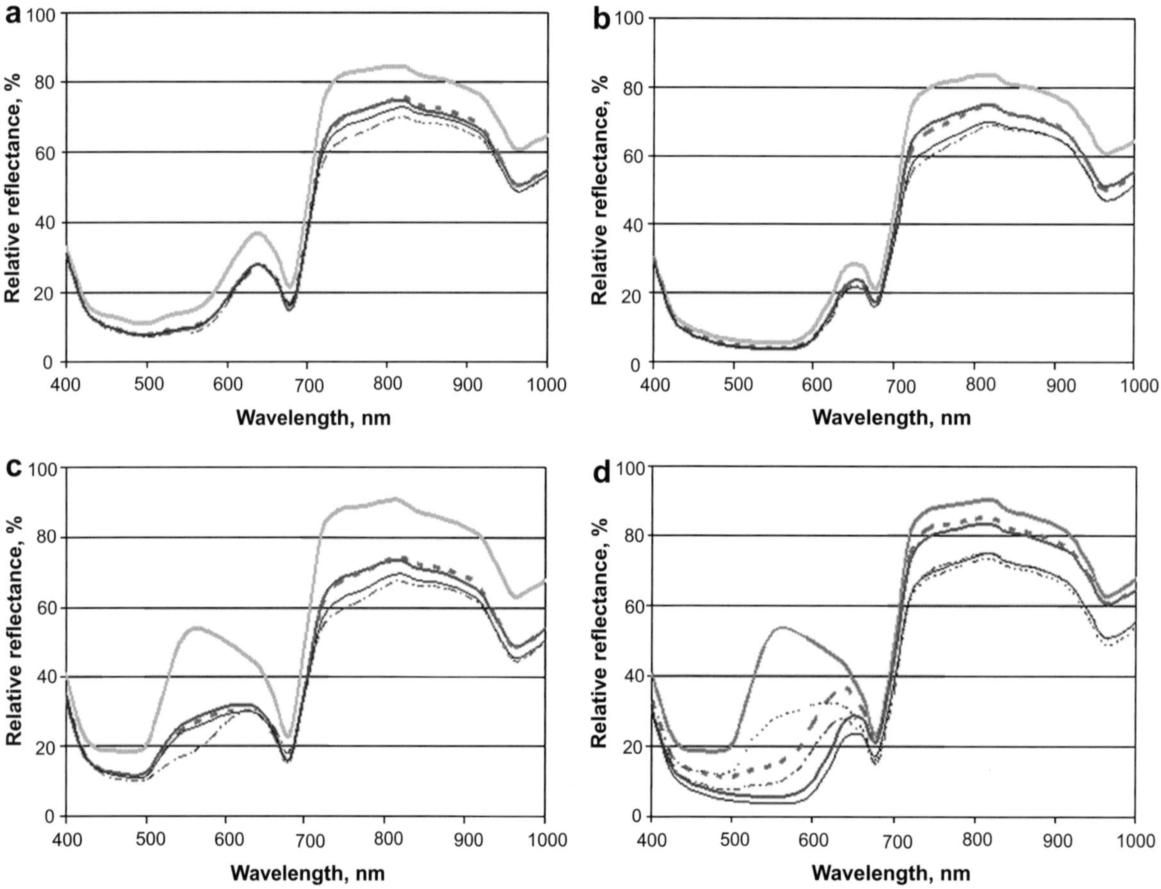

FIGURE 9.7 *Visible and NIR spectral characteristic curves extracted from the ROI pixels of the hyperspectral image representing normal and bruised tissue from McIntosh apple with (a) reddish background color, (b) Red background color, (c) green background color: (▬▬) normal green, (▬▬) bruise after 1 h, (■ ■ ■) bruise after 12 h, (–·–·–·) bruise after 24 h, and (——) bruise after 3 days; and (d) bruises after 1 h at different background colors: (▬▬) normal green, (▬▬) normal red, (■ ■ ■) normal reddish, (–·–·–·) 1 h after bruise on normal red, (——) 1 h bruise on normal green, (·········) 1hour bruise on reddish. (Full color version available on http://www.elsevierdirect.com/companions/9780123747532/)*

lower reflectance than normal tissue in the NIR region. Generally, at all wavelengths, most of the decreases in bruise reflectance occurred within the few hours after bruising. In order to detect this, the effect of background should be removed. Thus, Figure 9.7(d) represents all reflectance curves of the bruised surface on different normal surfaces. Because the main concern was early detection, the bruises at 1 h are illustrated. If the system is able to detect the bruises at this stage, then they could be detected later as well. It is obvious that the spectral signature of the bruise after one hour is almost the same as in the NIR region in all background colors; meanwhile a big variation is observed in the visible region. Generally, the visual inspection of the reflectance characteristic curves indicates that the NIR region would be more appropriate for detecting both recent and old bruises than the VIS region where there is no discrimination between normal and bruised surfaces.

The effective wavelengths identified from the stepwise discrimination method lie in the important region selected by the VIP method. It is obvious that there is a coincidence between the two wavelength selection strategies. Based on the previous spectral data analysis and the coincidence between the two methods of wavelength selection (Set 1 and Set 2), three wavelengths, 750, 820, and 960 nm, were chosen for bruise detection purposes. An obvious advantage of working in the NIR range is that the problem caused by color variations on normal surfaces can be circumvented. PCA was conducted on the reflectance spectral data using only these optimal wavelengths instead of the full wavelength range. The PC scores are illustrated based on variance explained by each PC. The first two components explained 93.95 % (PC1: 70.01 % and PC2: 23.94%) of the variance between normal and bruised spectral data. It is clear that the selected wavelength has a great discrimination power for bruise detection in different background colors.

9.4.4. Bruise Detection Algorithm and Validation

Due to their high performance in the classification of the spectral data to the two groups (normal and bruise) despite the color of the apples, the selected wavelengths were used to form multispectral images for bruise recognition. The images at the effective wavelengths (750, 820, 960 nm) were averaged using ENVI with the help of the binary mask to exclude the background that could interfere with the results. Figure 9.7 presents a complete picture of the whole process from acquiring the hyperspectral image through the wavelength selection until identification of the bruised area in the fruit surface.

As shown in Figure 9.8, the color image shows little difference between bruise and normal surrounding skin as this bruise has the same appearance in the visible spectrum. Whereas in the images at the effective wavelengths, the color difference between bruise and normal surface does exist clearly, owing to the fact that both the normal surface and the bruise have different spectral signatures in the near infrared zone. In addition, the NIR responses have the advantage of free-color influences. Previous studies have reported that, though a lot of biological materials show similar color appearance in the VIS spectrum, the same pigmentation could have a different appearance in the NIR spectrum (Kondo *et al.*, 2005). Moreover, in the NIR region, organic substances (like glucose, fructose, and sucrose) absorb the electromagnetic

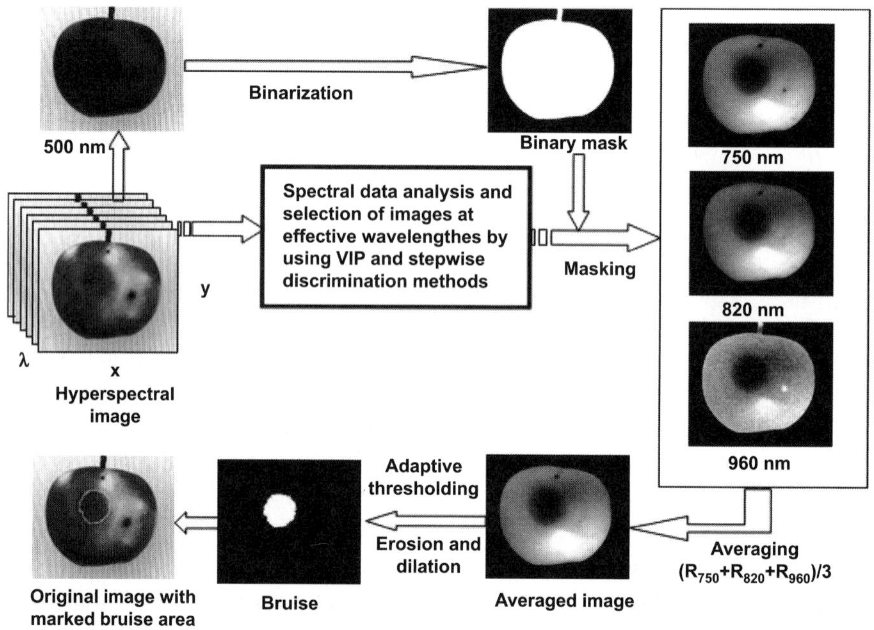

FIGURE 9.8 *Flow chart of the key steps involved in bruise detection algorithm. (Full color version available on http://www.elsevierdirect.com/companions/9780123747532/)*

radiation and the bonds of these organic molecules change their vibrational energy when irradiated by NIR frequencies and exhibit absorption peaks through the spectrum (Carlomagno *et al.*, 2004).

In some cases, the original images might contain natural scars. These scars may not appear clearly in both multispectral and averaged images. Median filtering, dilation, and erosion processes were used to remove the noise resulting from separate pixels and small spots that may carry the same spectral signature as bruise. Finally, the bruised region was marked on the original image for visualization as shown at the left bottom image in Figure 9.8.

It was also noticed that due to the natural wax of apples and their circular shape, regular reflectance produces a glared or specular area. These specular regions were generally quite small compared to the surface of the apple in the images. They appeared in the spectral images in the NIR region with high reflectance values caused by specular reflection of the illumination source at the apple surface. These specular regions predominantly show the spectral power distribution of the light source (Polder *et al.*, 2000). When multilevel adaptive thresholding was implemented, these areas were discarded from the final segmented images.

Apple bruise is normally caused by impact. Under impact conditions, the stresses overcome the cell wall strength, and when this break occurs, enzymes are released to cause the browning which characterizes the bruise. When a bruise occurs, cell wall destruction and chemical changes in the fruit tissue may change the light scatter in the bruised area, leading to a difference in reflectance when compared to non-bruised fruit (Kondo *et al.*, 2005). Furthermore, the bruised region increases with time, especially from its edges, so that the algorithm has to be sensitive for this increase.

To validate the results of the above-mentioned algorithm, bruise area was estimated as number of pixels of the bruised region. Bruises were created by the same manner mentioned above in a new group consisting of 20 apple fruits collected in a different batch from the training group. Hyperspectral images were acquired and calibrated as described earlier and only the images at the effective wavelengths (750, 820, and 960 nm) were used for bruise area estimation. The validation results showed that when time elapsed, the estimated area of the bruised region increased, thus reflecting the validity of this algorithm for bruise detection even in its early stage. The error noticed in some measurements in terms of estimated bruise area could be attributed to the relative difference in fruit position during image acquisition.

In comparison with other similar research, the results of this investigation indicate that this technique can be used to effectively detect bruises on apple surfaces in the early stage of bruising. High performance was reached for apples presenting recent (1 h) and old (> 3 days) bruises. The information in the spectral range of 400–1000 nm can be used for early bruise detection as those in higher spectral range (>1000 nm) (Lu, 2003). Since the efficiency of the method was demonstrated on a multicolor apple variety presenting high color variability, this procedure has the potential for being extended to other varieties.

9.5. CONCLUSIONS

Hyperspectral imaging techniques can provide not only spatial information, as regular imaging systems, but also spectral information for each pixel in an image. This information will form a 3-D "hypercube" which can be analyzed to ascertain minor and/or subtle physical and chemical features in fruits. Thus, a hyperspectral image can be used to detect physical and geometric characteristics such as color, size, shape, and texture. It can also be used to extract some intrinsic chemical and molecular information (such as water, fat, and protein) from a product.

The sign of apple bruise damage is physical and chemical change in comparison with sound fruits. Hyperspectral imaging technology has been showing its potential for detecting apple bruises effectively. However, the speed, cost, and processing power required make the technique more suited for research than practical applications. In some applications the outcomes of a hyperspectral imaging system have been used as a reference to develop multispectral imaging systems for specific applications. New spectral imaging systems with lower costs, wider spectral range, and better dynamic range are becoming commercially available. These factors, in combination with the increasing power of computer technology, will propel the hyperspectral imaging technology into a new and broader arena of practical applications.

NOMENCLATURE

Symbols

a	number of PLS factors
β	regression coefficients
D	dark image (with 0% reflectance)
L	total number of the examined wavelengths
P'	wavelength loadings
R	corrected image
R_0	recorded hyperspectral image
SSR	residual sum-of-squares
SST	total sum-of-squares
T	wavelength scores
W	white reference image
W_a	PLS weights
$\hat{Y}$	predicted surface type

Abbreviations

ANN	artificial neural network
BPNN	back-propagation neural network
CCD	charge-coupled device
CMOS	complementary metal-oxide-semiconductor
DC	direct current
FOV	field of view
IR	infrared
MNF	minimum noise fraction transform

NIR	near infrared
PC	principal component
PCA	principal component analysis
PLS	partial least squares
PPT	pulsed phase thermography
RGB	red, green, blue
ROI	region of interest
USDA	Department of Agriculture of the United States
VIP	variable importance in projection

REFERENCES

Abbott, J. A., Lu, R., Upchurch, B. L., & Stroshine, R. L. (1997). Technologies for non-destructive quality evaluation of fruits and vegetables. *Horticultural Review, 20*, 1–120.

Baranowski, P., & Mazurek, W. (2008). Chosen aspects of thermographic studies on detection of physiological disorders and mechanical defects in apples. *The Proceedings of the 9th International Conference on Quantitative InfraRed Thermography (QIRT 2008)*. July 2–5, 2008, Cracow, Poland.

Bennedsen, B. S., & Peterson, D. L. (2005). Performance of a system for apple surface defect identification in near-infrared images. *Biosystems Engineering, 90*(4), 419–431.

Bjarnestad, S., & Dahlman, O. (2002). Chemical compositions of hardwood and softwood pulps employing photoacoustic fourier transform infrared spectroscopy in combination with partial least-squares analysis. *The Analyst (Chemistry), 74*, 5851–5858.

Bochereau, L., Bourgine, P., & Palagos, B. (1992). A method for prediction by combining data analysis and neural networks: application to prediction of apple quality using near infra-red spectra. *Journal of Agricultural Engineering Research, 51*(2), 207–216.

Brown, G. K., Schulte, N. L., Timm, E. J., Armstrong, P. R., & Marshall, D. E. (1993). Reduce apple bruise damage. *Tree Fruit Postharvest Journal, 4*(3), 6–10.

Carlomagno, G., Capozzo, L., Attolico, G., & Distante, A. (2004). Non-destructive grading of peaches by near-infrared spectrometry. *Infrared Physics & Technology, 46*(1), 23–29.

Chao, K., Chen, Y. R., Hruschka, W. R., & Park, B. (2001). Chicken heart disease characterization by multi-spectral imaging. *Transactions of the ASAE, 17*(1), 99–106.

Cheng, X., Chen, Y. R., Tao, Y., Wang, C. Y., Kim, M. S., & Lefcourt, A. M. (2004). A novel integrated PCA and FLD method on hyperspectral image feature extraction for cucumber chilling damage inspection. *Transactions of the ASAE, 47*(4), 1313–1320.

Chong, L. G., & Jun, C. H. (2005). Performance of some variable selection methods when multicollinearity is present. *Chemometrics and Intelligent Laboratory Systems, 78*(1), 103–112.

ElMasry, G., Wang, N., Vigneault, C., Qiao, J., & ElSayed, A. (2007). Early detection of apple bruises on different background colors using hyperspectral imaging. *LWT – Food Science and Technology, 41*(2), 337–345.

ElMasry, G., Wang, N., & Vigneault, C. (2008). Detecting chilling injury in Red Delicious apple using hyperspectral imaging and neural networks. *Postharvest Biology and Technology. Postharvest Biology and Technology, 52*(1), 1–8.

Gao, X., Heinemann, P.H., Irudayaraj, J. (2003). Non-destructive apple bruise on-line test and classification with Raman spectroscopy. *ASAE Paper No. 033025*. The 2003 Annual Meeting of the American Society of Agricultural and Biological Engineers (ASABE), Las Vegas, Nevada, USA, July 27–30, 2003.

Geola, F., & Pieper, U. M. (1994). A spectrophotometer method for detecting surface bruises on "Golden Delicious" apples. *Journal of Agricultural Engineering Research, 58*(1), 47–51.

Jayas, D. S., Paliwal, J., & Visen, N. S. (2000). Multi-layer neural networks for image analysis of agricultural products. *Journal of Agricultural Engineering Research, 77*(2), 119–128.

Kavdir, I., & Guyer, D. E. (2002). Apple sorting using artificial neural networks and spectral imaging. *Transaction of the ASAE, 45*(6), 1995–2005.

Kavdir, I., & Guyer, D. E. (2004). Comparison of artificial neural networks and statistical classifiers in apple sorting using textural features. *Biosystems Engineering, 89*(3), 331–344.

Keskin, M., Dodd, R. B., Han, Y. J., & Khalilian, A. (2004). Assessing nitrogen content of golf course turfgrass clippings using spectral reflectance. *Applied Engineering in Agriculture, 20*(6), 851–860.

Kim, I., Kim, M. S., Chen, Y. R., & Kong, S. G. (2004). Detection of skin tumors on chicken carcasses using hyperspectral fluorescence imaging. *Transactions of the ASAE, 47*(5), 1785–1792.

Kleynen, O., Leemans, V., & Destain, M. F. (2005). Development of a multi-spectral vision system for the detection of defects on apples. *Journal of Food Engineering, 69*(1), 41–49.

Kondo, N., Chong, V. K., Ninomiya, K., Nishi, T., & Monta, M. (2005). Application of NIR-color CCD camera to eggplant grading machine. *ASABE Paper* No. 056073. The 2005 Annual Meeting of ASABE, Tampa, Florida, USA, July 17–20, 2005.

Leemans, V., Magein, H., & Destain, M. F. (1999). Defect segmentation on "Jonagold" apples using colour vision and a Bayesian classification method. *Computers and Electronics in Agriculture, 23*(1), 43–53.

Li, Q., Wang, M., & Gu, W. (2002). Computer vision based system for apple surface defect detection. *Computers and Electronics in Agriculture, 36*(2), 215–223.

Liu, Y., Chen, Y. R., Wang, C. Y., Chan, D. E., & Kim, K. S. (2006). Development of hyperspectral imaging technique for the detection of chilling injury in cucumbers; spectral and image analysis. *Applied Engineering in Agriculture, 22*(1), 101–111.

Liu, Y., Windham, W. R., Lawrence, K. C., & Park, B. (2003). Simple algorithms for the classification of visible/near-infrared and hyperspectral imaging spectra of chicken skins, feces, and fecal contaminated skins. *Applied Spectroscopy, 57*(12), 1609–1612.

Lu, R. (2003). Detection of bruises on apples using near infrared hyperspectral imaging. *Transactions of the ASABE, 46*(2), 523–530.

Lu, R. (2004). Multispectral imaging for predicting firmness and soluble solids content of apple fruit. *Postharvest Biology and Technology, 31*(1), 147–157.

Mehl, P. M., Chen, Y. R., Kim, M. S., & Chan, D. E. (2004). Development of hyperspectral imaging technique for the detection of apple surface defects and contaminations. *Journal of Food Engineering, 61*(1), 67–81.

Nagata, M., Tallada, J. G., Kobayashi, T., & Toyoda, H. (2005). NIR hyperspectral imaging for measurement of internal quality in strawberries. *ASABE Paper* No. 053131. The 2005 Annual Meeting of ASABE, Tampa, Florida, USA, July 17-20, 2005.

Olah, M., Bologa, C., & Oprea, T. I. (2004). An automated PLS search for biologically relevant QSAR descriptors. *Journal of Computer-Aided Molecular Design, 18*, 437–449.

Osborne, S. D., Jordan, R. B., & Künnemeyera, R. (1997). Method of wavelength selection for partial least squares. *The Analyst (Chemistry), 122*, 1531–1537.

Park, B., Abbott, J. A., Lee, K. J., Choi, C. H., & Choi, K. H. (2003). Near-infrared diffuse reflectance for quantitative and qualitative measurement of soluble solids and firmness of Delicious and Gala apples. *Transactions of the ASAE, 46*(6), 1721–1731.

Park, B., Windham, W. R., Lawrence, K. C., & Smith, D. P. (2004). Hyperspectral image classification for fecal and ingesta identification by spectral angle mapper. *ASAE Paper* No. 043032. The 2004 Annual Meeting of ASAE/CSAE, Ottawa, Ontario, Canada, August 1–4, 2004.

Peng, Y., & Lu, R. (2005). Modeling multispectral scattering profiles for prediction of apple fruit firmness. *Transactions of the ASABE, 48*(1), 235–242.

Polder, G., Van der Heijden, G. W., & Young, I. T. (2000). Hyperspectral image analysis for measuring ripeness of tomatoes. *ASAE Paper* No. 003089. The 2000 Annual Meeting of ASABE, Milwaukee, Wisconsin, USA, July 9–12, 2000.

Polder, G., Van der Heijden, G. W., & Young, I. T. (2002). Spectral image analysis for measuring ripeness of tomatoes. *Transactions of the ASAE, 45*(4), 1155–1161.

Shahin, M. A., Tollner, E. W., McClendon, R. W., & Arabnia, H. R. (2002). Apple classification based on surface bruises using image processing and neural networks. *Transactions of the ASAE, 45*(5), 1619–1627.

Throop, J. A., Aneshansley, D. J., Anger, W. C., & Peterson, D. L. (2005). Quality evaluation of apples based on surface defects: development of an automated inspection system. *Postharvest Biology and Technology, 36*(1), 281–290.

Wen, Z., & Tao, Y. (1999). Building a rule-based machine-vision system for defect inspection on apple sorting and packing lines. *Expert Systems with Applications, 16*, 307–313.

Xing, J., & De Baerdemaeker, J. (2005). Bruise detection on "Jonagold" apples using hyperspectral imaging. *Postharvest Biology and Technology, 37*(1), 152–162.

Xing, J., Bravo, C., Moshou, D., Ramon, H., & De Baerdemaeker, J. (2006). Bruise detection on "Golden Delicious" apples by VIS/NIR spectroscopy 2006. *Computers and Electronics in Agriculture, 52*, 11–20, 2006.

Zwiggelaar, R., Yang, Q., Garcia-Pardo, E., & Bull, C. R. (1996). Use of spectral information and machine vision for bruise detection on peaches and apricots. *Applied Engineering in Agriculture, 63*(4), 323–332.

CHAPTER 10

Analysis of Hyperspectral Images of Citrus Fruits

Enrique Moltó [1], José Blasco [1], Juan Gómez-Sanchís [2]

[1] *Instituto Valenciano de Investigaciones Agrarias (IVIA), Centro de Agroingeniería, Moncada (Valencia), Spain*

[2] *Intelligent Data Analysis Laboratory (IDAL), Electronic Engineering Department. Universidad de Valencia, Burjassot (Valencia), Spain*

CONTENTS

10.1. INTRODUCTION

Citrus are the most cultivated fruit in the world. An annual production of more than 89 million tonnes testifies to the importance of this fruit within the world economy. Production is principally aimed at two differentiated markets, that of the citrus juice industry and processed fruit, and that of citrus fruits for consumption as fresh produce, with the latter accounting for some 65% of total production. The sector makes enormous efforts to guarantee high product quality, especially when the citrus fruits are consumed as fresh fruit. For such purposes, computer vision can be used to automatically assess the quality of each individual fruit (Brosnan & Sun, 2004; Chen *et al.*, 2002; Sun, 2007) and has been incorporated on a widespread scale in commercial automatic inspection systems.

The automatic inspection systems that are currently available in the market are capable of performing an efficient analysis of the size and color of each fruit. The most advanced systems can even detect skin surface damage. However, one of the main problems facing these automatic systems is the identification of damage types, depending on which, the economic consequences can be markedly different. For this purpose, defects found on citrus peel can be classified into two categories: severe and slight.

Severe defects, for instance, are those that evolve over time, such as those caused by different types of fungi: the rotten fruit can be neither packaged nor stored in a cold chamber since the damage will gradually

Hyperspectral Imaging for Food Quality Analysis and Control

CONTENTS

increase depending on the temperature and humidity conditions. Slight defects reduce the commercial value of the fruit by causing aesthetic damage, but do not stop it being used in the internal market or the processing industry.

Another severe defect is citrus canker disease, caused by bacteria that affect leaves, stems, and fruit of citrus trees, including lime, orange, and grapefruit (Schubert *et al.*, 2001). This disease is extremely persistent when it becomes established in an area. Citrus orchards must be destroyed in an attempt to eradicate the disease. Since it does not affect all citrus-growing regions, the detection of this damage is very important in order to avoid the spread of the infection to canker-free areas.

Green rot, caused by *Penicillium digitatum*, leads to most damage to citrus fruits during the postharvest and marketing processes (Eckert & Eaks, 1989). Economic losses generated by this fungus are enormous, amounting in overall terms to between 10% and 15% of total product value. As mentioned before, a small number of infected fruits can spread the infection to a whole consignment. This problem is made worse if the fruit is stored for a long period of time or during long-term transportation when exported. For this reason, the detection of fungi will be discussed in one section of this chapter.

Artificial vision systems try to imitate human perception of color. Given that biological products present a wide variety of textures and colors, it occasionally happens that the color of a damaged area of the peel in one fruit might be the same as the color of a healthy peel of a different fruit. This problem is even further complicated when the surface of the fruit is not uniformly lit, as occurs when lighting quasi-spherical objects like citrus fruits.

Defects have different reflectance spectra in certain areas of the electromagnetic spectrum. Gaffney (1973) studied different types of external citrus fruit damage and characterized their reflectance spectrum in the visible region, demonstrating how different types of defect can be distinguished by using spectrometric methods.

As the cost of electronic equipment continuously decreases, it is now possible to tackle the problem of fruit inspection with ever more efficient technology. In general, these approaches use different areas of the electromagnetic spectrum to highlight the differences between the stains that appear on the image and the normal color of the peel. The next technological advance involves the use of hyperspectral image processing, which allows reflectance of defects and other regions of interest in particular wavelengths to be studied.

10.2. A FIRST APPROACH TO AUTOMATIC INSPECTION OF CITRUS: MULTISPECTRAL IDENTIFICATION OF BLEMISHES ON CITRUS PEEL

Most commercial machines only discriminate between blemished and unblemished fruit. Advances in electronics have led to improvements in the capabilities of the machines currently available. Nowadays, near-infrared (NIR) information can be combined with visible (VIS) imaging in electronic fruit sorters to discriminate between fruits and background, since the reflectance of the skin at this spectral area is higher than the background, thus generating a high contrast between them, and allowing measurement of the size of individual fruit more accurately than using color images (Aleixos *et al.*, 2002).

Before hyperspectral systems were easily available in terms of cost, several authors attempted to broaden the scope of the visible information in order to build automatic citrus sorters. For instance, Blasco *et al.* (2007a) developed a multispectral system to identify skin defects on citrus skin. Experiments were carried out using images of commercial fruit (Navelina and Valencia orange varieties and Marisol, Clemenules, and Fortune mandarins) provided by a local manufacturer. Blemishes were identified by an expert, and then labeled. Images of each fruit were acquired with four different systems: a conventional color camera under white illumination, a NIR camera, a near-ultraviolet (UV) camera, and a conventional color camera under ultraviolet illumination to induce fluorescence (UVFL). This fluorescence method is currently used to manually detect decay in citrus packing houses, as normally the essential oils of the citrus peel are reduced as a result of a decay process.

Defects were classified as severe (anthracnose, stem-end injury, green mold, and medfly egg deposition) or slight (rind-oil spots, presence of scales, scarring, thrips, chilling injury, sooty mold, and phytotoxicity). Figure 10.1 shows different images of a fruit affected by green mold acquired using the different cameras. The images are different because the acquisition systems were placed in different inspection chambers.

The experiments showed that only two types of defects, anthracnose and sooty mold, could be detected in NIR (Figure 10.2) and only stem-end injuries were detected in UV (Figure 10.3). Induced UV fluorescence images were only useful for detection of fruit affected by thrips, scarring, and decay caused by green mold. However, an important finding was that no false detections were generated when processing these images. Figure 10.4 shows

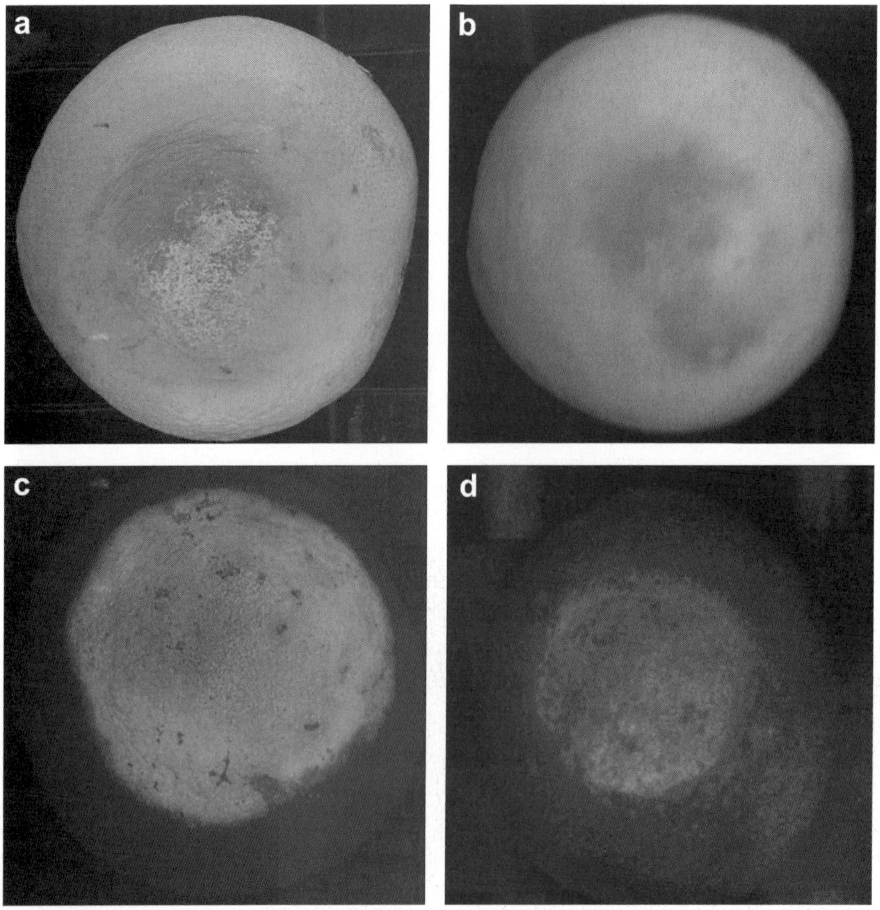

FIGURE 10.1
*Different images of the same fruit as affected by green mold in (*from left*) visible, near-infrared, fluorescence and ultraviolet illumination. (Full color version available on http://www.elsevierdirect.com/companions/9780123747532/)*

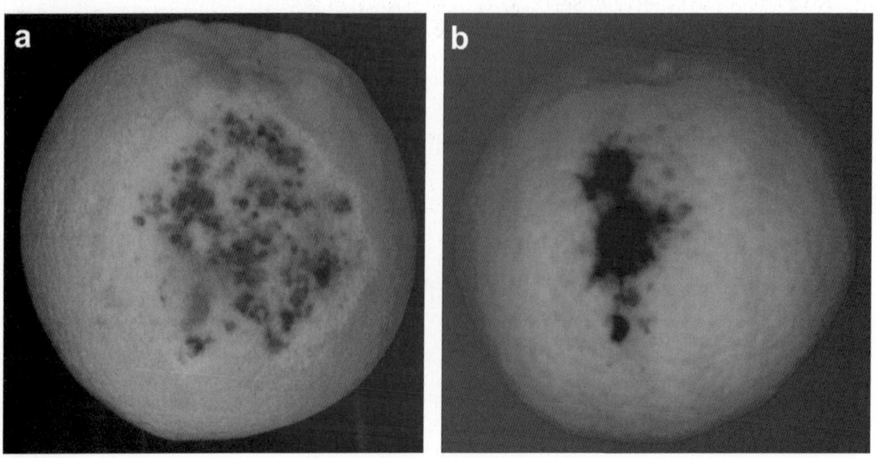

FIGURE 10.2
NIR images of fruits as affected by anthracnose (a) and sooty mold (b)

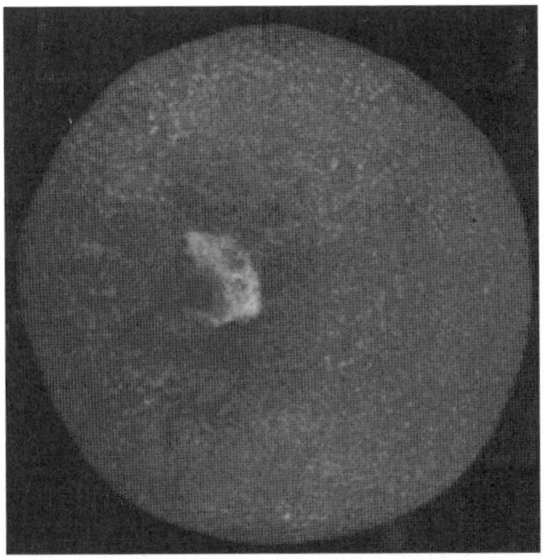

FIGURE 10.3 *UV image of a fruit as affected by stem-end injury*

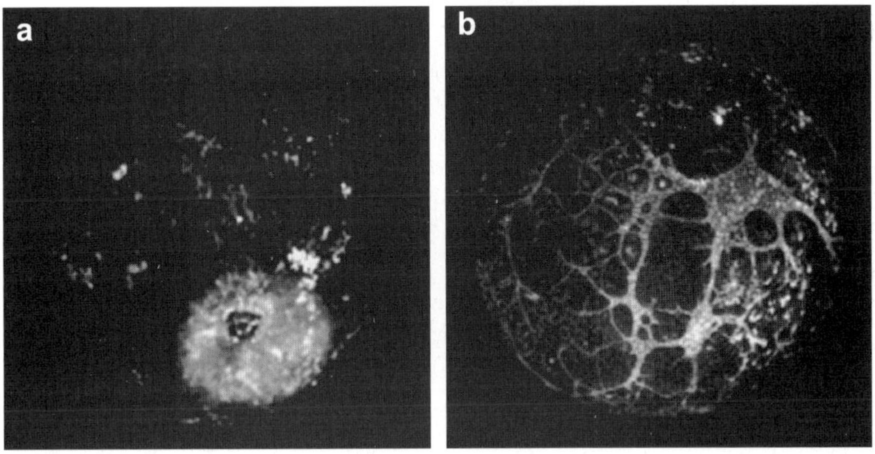

FIGURE 10.4 *FL image of a fruit as affected by green mold (a) and thrips (b)*

two graylevel UVFL images, one of a fruit affected by green mold and the other of a fruit affected by thrips.

By introducing NIR, UV, and UVFL images into the analysis, the success rate increased from 65% to 86% owing to an improvement in the identification of anthracnose and decay caused by green mold. However, decay detection which averts the need for UV radiation to induce fluorescence is still a challenge in which hyperspectral imaging can play an important role. This work was enhanced by including several morphological parameters of the defects, reaching a success of 86% in classifying the defects (Blasco *et al.*, 2009).

10.3. CONSIDERATIONS ON HYPERSPECTRAL IMAGE ACQUISITION FOR CITRUS

Hyperspectral image analysis involves processing a large number of monochromatic images of the same scene at different wavelengths, enabling simultaneous analysis of the spatial and spectral information (Figure 10.5). The set of monochromatic images that are captured constitute a hyperspectral image. Hyperspectral image acquisition systems have two main parts: a light-sensitive system (the camera) and a system that enables wavelength selection (often a tunable filter).

As a hyperspectral image is made up of a large collection of monochromatic images at different wavelengths, the hyperspectral image contains much more extensive information than that provided by a single monochromatic image or a conventional color image (which is the combination of three broad-band monochromatic images). The number of monochromatic images depends on the resolution of the system used and they are combined by forming a cube in which two dimensions are spatial (pixels) and the third one is the spectrum of each pixel. Without adequate processing, such vast amounts of data, despite being one of the main advantages of hyperspectral systems, can complicate the extraction of useful information since much of the information obtained is redundant, or by its nature cannot be used to distinguish between regions with similar characteristics.

It should also be borne in mind that raw hyperspectral images provide information about the radiance of the object. However, conventional machine inspection/assessment is generally based on the observed reflectance of the object. For this reason, image compensation methods should be used to determine the reflectance of the object from the observed radiance. The image compensation method used depends on the way in which the image is captured. If the hyperspectral image is captured from a satellite, for example for crop yield prediction, then the effects of atmospheric scattering need to be taken into consideration (Shaw & Burke, 2003). On the other hand, if the scene is lit in a controlled manner, for example, in the case of a lighting chamber for an automatic inspection machine, and the approximate shape of the object is known beforehand, then compensation can be performed using a white reference and a digital elevation model that takes into account the effect of the geometry of the object on the reflection of the radiation.

Many statistical techniques can be used to condense the information provided by hyperspectral images. These techniques include principal

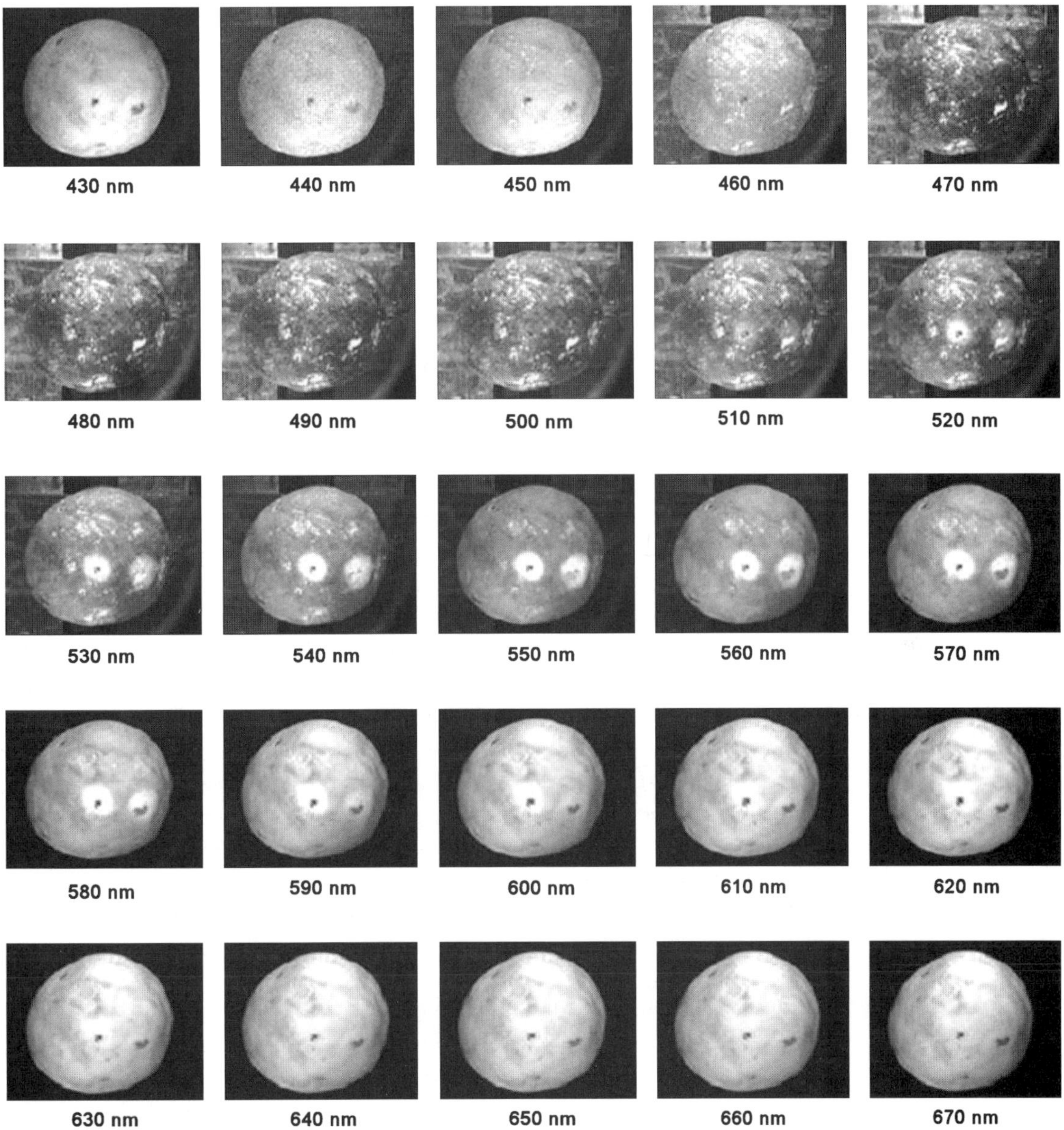

FIGURE 10.5 *A series of monochromatic, narrow band images of an orange with a defect caused by medfly egg deposition, which form a hyperspectral image*

component analysis (Jolliffe, 1986) and linear discriminant analysis (Cheng *et al.*, 2004).

10.3.1. Illumination

Fruit often has very varied colors and textures. For this reason, lighting used in an inspection system based on artificial vision has a major impact on the final results of the image analysis (Du & Sun, 2004). An inefficient lighting system can prevent the detection of defects, with defective areas being confused with healthy ones and vice-versa. The appearance of bright spots due to specular reflection, or the existence of poorly lit areas (shadows), are common sources of noise which conceal the damage or give false-positive results. On the other hand, the choice of a source with an unsuitable radiation spectrum can alter the perception of the colors or hide any damage (Bennedsen *et al.*, 2005). When correct lighting is used the quality of the end result of the image analysis is maximized, with the analysis being more cost-efficient as the time required in the preprocessing stages for noise elimination or image correction is reduced (Chen *et al.*, 2002).

Attempts have been made to avoid specular reflection in some studies by locating the camera to receive the light from the source at an angle of 45° (Papadakis *et al.*, 2000), but this technique does not work well with spherical objects. The other possibility is to create spatially diffuse and spectrally uniform lighting. One possible solution to the problems that arise as a result of the reflection of light on quasi-spherical objects is based on applying reflectance models with the assumption of constant curvature (Tao & Wen, 1999). However, these are very rigid models for the inspection of citrus fruit with their noticeably different curvature radii. Another solution that has traditionally been suggested for this problem involves eliminating analyses of those areas that appear less well lit (Blasco *et al.*, 2003), but this means that a significant area of the fruit will not undergo analysis. Some works assume that the pixels that belong to the peripheral areas of the object and the pixels that appear in the center of it can be segmented into different classes and later grouped together (Blasco *et al.*, 2007a). The drawback to this solution is that with the increase in the number of classes during segmentation, there is a fall in the hit rate (Duda *et al.*, 2000).

In order to correct the effects of the lack of spatial uniformity of illumination, many authors use a white reference (Kleynen *et al.*, 2005). However, this approach does not take into account the particular geometry of the citrus fruits. One way of diffusely lighting objects consists of introducing them below hemispherical lighting domes. This lighting method is particularly useful for objects that are almost spherical. The light source determines the

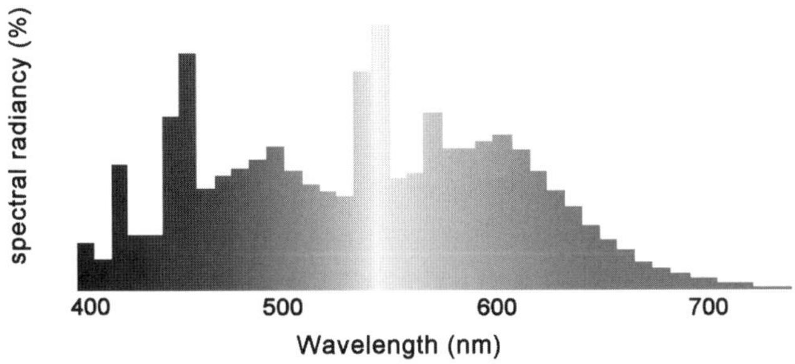

FIGURE 10.6 *Emission spectrum of daylight-type fluorescent tubes. (Full color version available on http://www.elsevierdirect.com/companions/9780123747532/)*

spectral range that can be studied in each particular case. For example, if work to be performed is in the infrared region, daylight-type fluorescent tubes are inappropriate as they exhibit low efficiency at these wavelengths (Figure 10.6). However, tungsten filament and halogen lamps present high luminous efficiency in the NIR region (Figure 10.7). It is also very important to maintain a constant radiation flow, avoiding any flickering or temporary drops in radiation. The lamps should therefore be operated by high frequency electronic ballasts (in the case of fluorescent lamps), or stabilized power sources (in the case of halogen lamps).

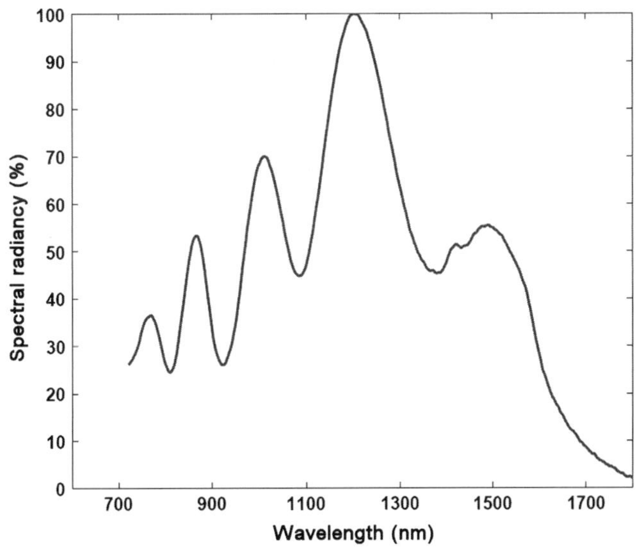

FIGURE 10.7 *Emission spectrum of halogen lamps.*

The emission spectrum of all sources varies with temperature, so it is important to take into consideration the time required for temperature stabilization (known as the pre-heating time). This is defined as the time required for the spectral response to stabilize. The heating effect of the lamps is of particular significance in the acquisition of hyperspectral images, as the relative amount of emission between wavelengths is variable until the temperature of the lighting source reaches a steady state.

10.3.2. Hardware for Hyperspectral Image Acquisition

Electronic systems for hyperspectral image acquisition need a filter system for selection of the incident radiation wavelength. Various types of filter can be used, with the most interesting being tunable, of which the AOTF (Acoustic–Optic Tunable Filters) and LCTF (Liquid Crystal Tunable Filter) are the most common (Poger & Angelopoulou, 2001). Both are used to capture hyperspectral images. Operation of the AOTF is based on the piezoelectric properties of the materials (Bei *et al.*, 2004), while operation of the LCTF is based on a combination of Lyot filters, capable of electronically controlling the interference between the ordinary and extraordinary beams of the incident electromagnetic radiation (Hetchs, 1998).

The filters are constructed to cover a specific wavelength range. When a wider wavelength range (for example visible and NIR) needs to be covered, several filters have to be combined. In these cases, a filter exchange system is required that does not alter the perspective of the scene. Imprecise camera handling or incorrect filter positioning can prevent correct image overlapping. In order to achieve this objective, Gómez-Sanchís et al. (2008a) developed a filter exchange system comprising a container and guide track system. The container can be moved over the guide tracks between two end points, enabling the filters to be easily moved between two positions.

The most common light-sensitive elements are based on the use of CCDs (charge-coupled devices). Conventional silicon CCDs are NIR-sensitive up to approximately 1000 nm. As the focus of the image varies considerably between well separated wavelengths, optics becomes an important part of hyperspectral image acquisition systems. This is particularly important when working in a VIS/NIR system. For example, a focused image close to 400 nm wavelength will appear out of focus at wavelengths close to 800 nm due to the high chromatic scattering that conventional optics produce. To avoid this problem low-scatter lenses are required to work simultaneously in both the VIS and NIR spectrum. Such optics also must exhibit a practically uniform transmittance throughout the targeted spectral range.

10.4. DESCRIPTION AND TUNING OF A HYPERSPECTRAL SYSTEM FOR CITRUS FRUIT INSPECTION

Gómez-Sanchís *et al.* (2008a) used a dome in which the light sources were located at its base and the light was directed upwards so that the radiation was reflected and reached the fruit from all directions (Figure 10.8). The internal part of the aluminium dome was coated in white paint which maximized the reflectivity of the surface and had a rough surface that created a more diffuse illumination. Additionally, they used LCTF filters to generate monochromatic images and low-scatter lenses to reduce focal problems. The light-sensitive element was a conventional silicon CCD camera.

10.4.1. Correcting Integration Time at Each Wavelength

Efficiency of liquid crystal tunable filters depends on the band to be tuned. For this reason, it is very important to quantify this effect and propose corrections. For this purpose an optical test bench, with a calibrated light source, a spectrometer, and the necessary optical elements are required.

The procedure is as follows. Filters are tuned to each of the frequencies for which they are to be characterized, thereby obtaining the transmission spectrum of each filter. These data are then compared with the light source spectrum to determine the absolute transmittance of each filter.

Figure 10.9 shows the results of applying these methods to two tunable filters (CRI, Varispec VIS-07 and Varispec NIR-07). VIS-07 (Figure 10.9a)

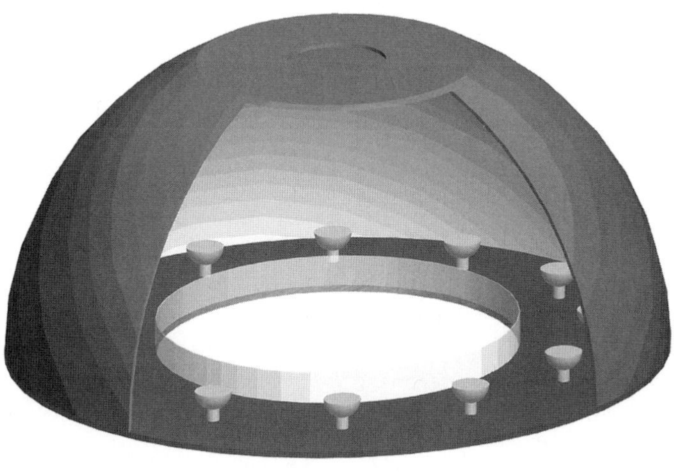

FIGURE 10.8 *Hemispherical lighting dome. (Full color version available on http://www.elsevierdirect.com/companions/9780123747532/)*

exhibits a very low transmittance (less than 5%) in the bands below 460 nm. Additionally, for wavelengths lower than 460 nm, the filter exhibits low frequency selectivity, allowing the passage of a considerable amount of radiation of other neighboring wavelengths. The NIR filter presents a continuously increasing transmittance as is shown in Figure 10.9b.

Each part of the hyperspectral vision system (lighting system, camera, optics, and filter) exhibits a different spectral efficiency. For these differences to be homogenized and for the complete system to have a uniform spectral efficiency, integration times can be assigned inversely proportional to the efficiency of the system at each wavelength. In this way, a higher integration time can be employed in those bands that exhibit low efficiency. If this correction is not performed, the intensity differences that appear in the images may not always be due to radiance coming from the object, but to the effect of the different efficiencies of the system for each wavelength.

One method that can be used to implement this correction comprises the acquisition of images of a white reference for each wavelength, increasing

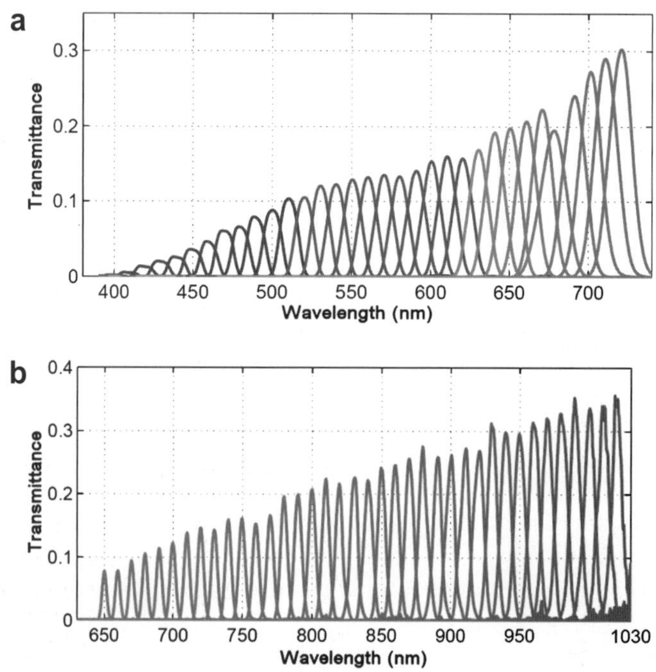

FIGURE 10.9 *Comparison of the real transmittances of (a) VIS-07 filter and (b) NIR-07 filter. (Full color version available on http://www.elsevierdirect.com/companions/9780123747532/)*

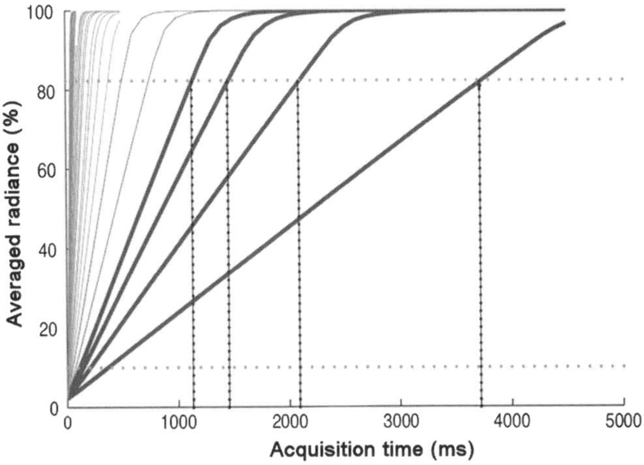

FIGURE 10.10 *Graphic display of the average radiance of a white reference versus image integration time. (Full color version available on http://www.elsevierdirect.com/companions/9780123747532/)*

the integration time of the image between 0 ms up to the saturation of the camera sensor. The average level of all the image pixels (average radiance) is estimated for each image obtained. In this way, curves are obtained that depend on the average radiance and integration time.

In order to determine integration times per band, a least squares linear fit can be performed for each of the curves (associated with each band) in their linear area. One possible criterion is based on selecting the integration time for each band that provides 85% radiance of the dynamic range in the fitted curve. Figure 10.10 shows a graph with the integration times and averaged radiance for each band. It can be seen that in the bands in which the filter exhibits lower efficiency, a higher integration time is needed to reach the 85% level.

10.4.2. Spatial Correction of the Intensity of the Light Source

When illuminating a scene, spatial variations of the radiation intensity may appear in the shot of the scene. One way of compensating for these variations is based on calculation of the ratio between the radiance of the fruit surface $R(\lambda)$ and that of the light source $I_T(\lambda)$.

$$\rho_{xy}(\lambda) = \frac{R(\lambda)}{I_T(\lambda)} \quad (10.1)$$

where $\rho_{xy}(\lambda)$ is the corrected monochromatic image.

These values are not directly measurable by a hyperspectral vision system, but they can be deduced from the use of a white reference (Bajcsy & Kooper, 2005). The equation used to correct the spatial variations of the light source is expressed as follows:

$$\rho_{xy}(\lambda) = \rho_{ref}(\lambda)\frac{R_{xy}(\lambda) - R_{dark}(\lambda)}{R_{white}(\lambda) - R_{dark}(\lambda)} \tag{10.2}$$

where $\rho_{ref}(\lambda)$ is the certified reflectance of the reference white, $R_{xy}(\lambda)$ the uncorrected image, $R_{dark}(\lambda)$ the image obtained by the system with no illumination, and $R_{white}(\lambda)$ the monochromatic image obtained by the hyperspectral vision system corresponding to the reference white.

In this way, in addition to correcting the spatial variations in light source intensity, local correction (for each of the pixels of the scene) of the effect of different efficiencies caused by different parts of the hyperspectral vision system is performed. Figure 10.11 shows the effect of simultaneously correcting three RGB bands (B = 480 nm, G = 550 nm, and R = 640 nm) in a hyperspectral image of a mandarin. It can be seen that, after correction, the fruit appears more uniformly lit.

Despite this correction, a gradual darkening can be observed from the center outwards towards the peripheral areas of the fruit. The spherical geometry of citrus fruits introduces a significant limitation to the correct determination of the reflectance of a particular point, owing to the fact that the radiation reflected by the citrus fruit towards the camera depends on the curvature at that point. Thus, the correction described corrects the spatial variations caused by the light source, but does not take into account those variations due to the geometry of the fruit, since the white reference used is flat but the fruit quasi-spherical.

10.4.3. Correction of Effects Due to the Spherical Shape of the Citrus Fruit

The effect of the reflection of the light on the spherical geometry of citrus fruits has also to be corrected in order to ensure that the radiance observed at any point is independent of its position.

Assuming that the fruit has a Lambertian surface (which reflects the light in an identical manner in all directions, regardless of the position), the light received by the observer depends on the angle of incidence ϕ between the beam of direct light and the direction of the normal vector to the surface (Foley, 1996). The illumination used in a citrus fruit inspection system, $I_T(\lambda)$, can be modeled as the overlaying of two components; the diffuse component, $I_F(\lambda)$, which lights the object indirectly through multiple reflections, and the

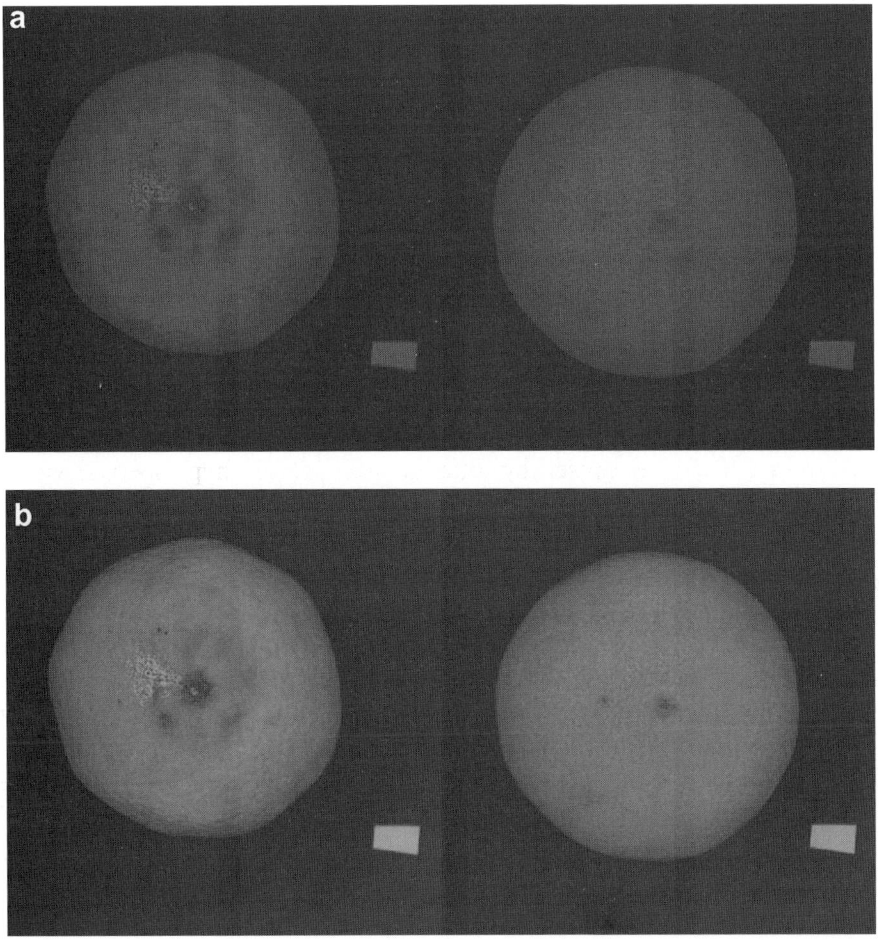

FIGURE 10.11 *RGB images (640 nm, 550 nm, and 480 nm) of two mandarins before white reference correction (a) and after white reference correction (b). (Full color version available on http://www.elsevierdirect.com/companions/9780123747532/)*

direct component, $I_D(\lambda)$, which comes directly from the light source and is modulated by the angle, ϕ. Then, the illumination model can be described by the following equation:

$$I_T(\lambda) = I_D(\lambda)\cos(\phi) + I_F(\lambda) \tag{10.3}$$

A parameter α_D is then defined which relates the proportion of direct light and diffuse light with the total average, I, given in the equations below. This parameter has its values between 0 and 1, depending on the characteristics of the lighting system. It can be determined by obtaining the ratio between the average light detected by the camera sensor at the points on the perimeter of the fruit in the image and the average total light received by this sensor from the whole fruit. The light reflected by the points situated on the perimeter of the citrus fruit is only diffuse light, since at these points ϕ is close to 90°.

$$I_D = \alpha_D I \tag{10.4}$$

$$I_F = (1 - \alpha_D)I \tag{10.5}$$

Combining Equations (10.4) and (10.5) with Equation (10.3), it can be derived that the behavior of the illumination can be modeled by using the equation below:

$$I_T(\lambda) = I(\lambda)[\alpha_D \cos(\phi) + (1 - \alpha_D)] \tag{10.6}$$

which gives the following geometric correction factor:

$$\varepsilon_g = [\alpha_D \cos(\phi) + (1 - \alpha_D)] \tag{10.7}$$

Integrating Equation (10.1) with the illumination model of Equation (10.6), the following equation is obtained, which expresses the result of correcting image $\rho_{xy}(\lambda)$ as affected by the geometry of the citrus fruit, $\rho(\lambda)$.

$$\rho(\lambda) = \frac{\rho_{xy}(\lambda)}{[\alpha_D \cos(\phi) + (1 - \alpha_D)]} \tag{10.8}$$

In order to apply this correction and to estimate the real reflectance of a particular point on the fruit, the angle ϕ should be calculated for each of the pixels in the image. For this purpose a digital elevation model (DEM) is developed, which consists of performing a 3-D modelling of the fruit from a 2-D image. Once the model is constructed and all the elevations of each pixel are estimated, the geometric correction factor, ϵ_g, can be calculated. An example of DEM for citrus fruit can comprise the following steps:

1. Determination of the pixels belonging to the fruit. This can be solved by defining a threshold in one of the monochromatic images which exhibits a high contrast between the fruit and the background.
2. Determination of the center of the fruit and of the start points of the meridians of an interpolation grid. The center of the fruit (P_G) is calculated from the coordinates of the pixels belonging to the fruit. Equidistant points from the perimeter (P_i) can be selected to mark the start of the meridians.
3. Obtaining the elevations of the interpolation grid, and calculating the maximum height of the fruit (h_c). The maximum height can be the

average distance between the fruit center and the N_{Pi} points to be obtained by the equation below:

$$h_C = \frac{1}{N_{Pi}} \sum_{i=1}^{N_{Pi}} \|P_i P_G\| \tag{10.9}$$

4. The interpolation grid nodes are obtained by subdividing each of the radii $\|P_iP_G\|$ into 16 sub-radii r_{ij} (j = 1...16, i = 1.... N_{Pi}) and determining the coordinates of each sub-radius. Once the interpolation grid nodes are determined, the height is estimated by modelling ellipses in the N_{Pi} transversal planes, with semi-axes $\|P_iP_G\|$ and h_c as follows

$$\frac{{r_{ij}}^2}{\|P_iP_G\|^2} + \frac{{h_{ij}}^2}{h_c^2} = 1 \tag{10.10}$$

By repeating this process for the N_{Pi} transversal planes the interpolation grid of the fruit is determined.

5. Obtaining the elevation of each pixel by interpolation. The elevation of each of the pixels of the citrus fruit can then be obtained from the nodes by bilinear interpolation. Figure 10.12 shows the result of modelling the elevation of a fruit.

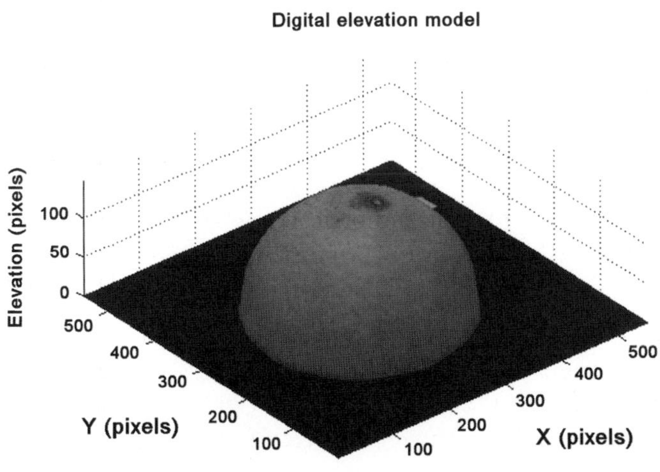

FIGURE 10.12 *Result of applying the digital elevation model to an RGB image (R = 640 nm, G = 550 nm, B = 480 nm) of a Clemenules mandarin. (Full color version available on http://www.elsevierdirect.com/companions/9780123747532/)*

Now ϕ can be calculated to obtain the actual reflection of the fruit surface independently from the sphericity. From the geometric parameters of the fruit, the factor ε_g for each pixel of the fruit can then be obtained. The angle ϕ is obtained from the spatial coordinates of each pixel (extracted from the digital elevation model) by using the following equation, which constitutes one of the transformation equations of spherical coordinates:

$$\tan(\phi) = \frac{\sqrt{(x^2 + y^2)}}{h_{xy}} \qquad (10.11)$$

where x, y, h_{xy} are the three Cartesian coordinates of each pixel of the fruit.

This method has been employed by Gómez-Sanchís *et al.* (2008b) to correct the images of 40 mandarins, of which 20 belong to the Clemenvilla variety (which generally have a uniform spherical shape) and 20 to the Clemenules variety (whose shape is slightly less uniform). Figure 10.13 shows two images of the same fruit obtained at a wavelength of 640 nm. The image on the left shows the citrus fruit before correction, while the image on the right shows the same fruit after correction. On the left, peripheral areas of the citrus fruit appear darkened in comparison with the center of the fruit, though the peel of this fruit is in fact uniform. On the right, much more uniform intensity levels can be seen throughout the surface.

Figure 10.14 shows the reflectance profile of the section of the image in Figure 10.13. As shown in Figure 10.14, on the original image, the reflectance of each pixel of the section (blue) exhibits a bell shape, because the shape of the fruit modulates the amount of radiation that the camera receives. After correction, the profile of the reflectance values (in red) is considerably flattened.

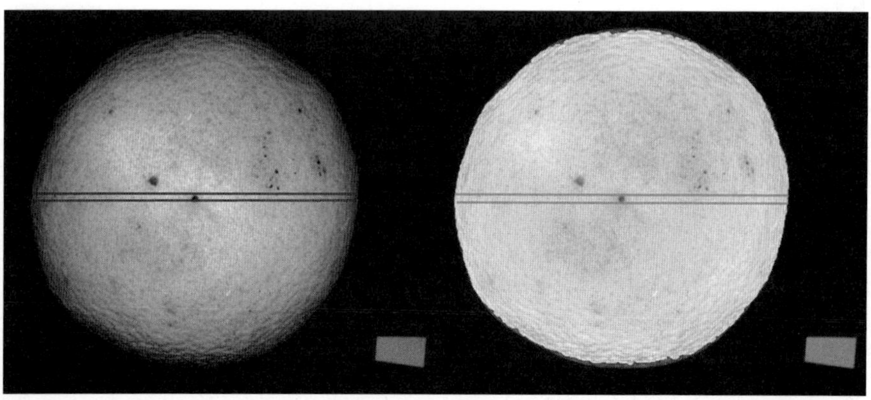

FIGURE 10.13
Uncorrected image (left) and corrected image (right) of a mandarin (Clemenvilla) at 640 nm. (Full color version available on http://www.elsevierdirect.com/companions/9780123747532/)

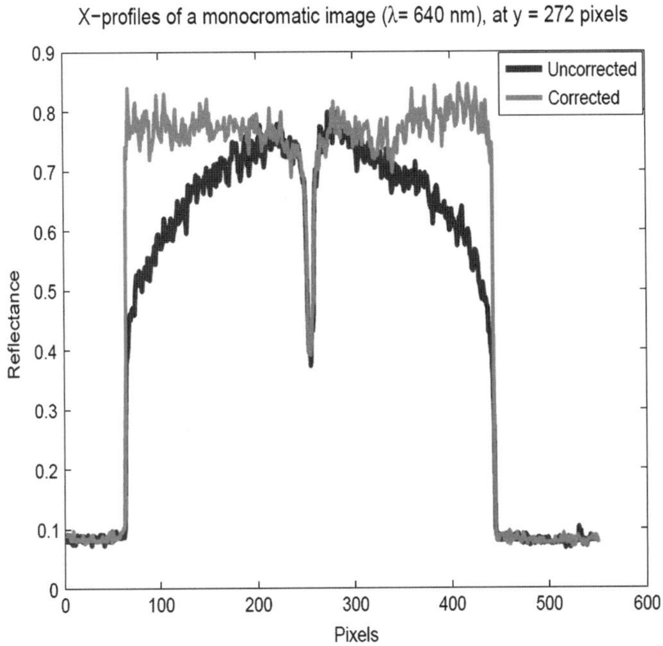

FIGURE 10.14 *Reflectance profile of the section of the image shown in Figure 10.13. Blue = reflectance before correction; red = corrected reflectance. (Full color version available on http://www.elsevierdirect.com/companions/9780123747532/)*

Figure 10.15 shows the average spectrum of different 5 × 5 pixel windows, belonging to four areas of a Clemenules mandarin before and after correction. A high degree of variability of the four spectra can be observed in the Figure 10.15(a), though they in fact belong to similar areas of the skin but situated in different regions (more and less peripheral). Figure 10.15(b) shows the spectra corresponding to the same areas, but calculated from the corrected images. A notable reduction in the variability of the spectra can be observed.

10.5. AUTOMATIC EARLY DETECTION OF ROTTEN FRUIT USING HYPERSPECTRAL IMAGE ANALYSIS

Early detection of severe diseases in citrus fruits is important for the citrus industry because a small number of infected fruits can spread the infection to other fruits. Though early detection facilitates the execution of a series of effective actions against fungal infestation, it is very difficult for the human eye to detect the initial stages of decay.

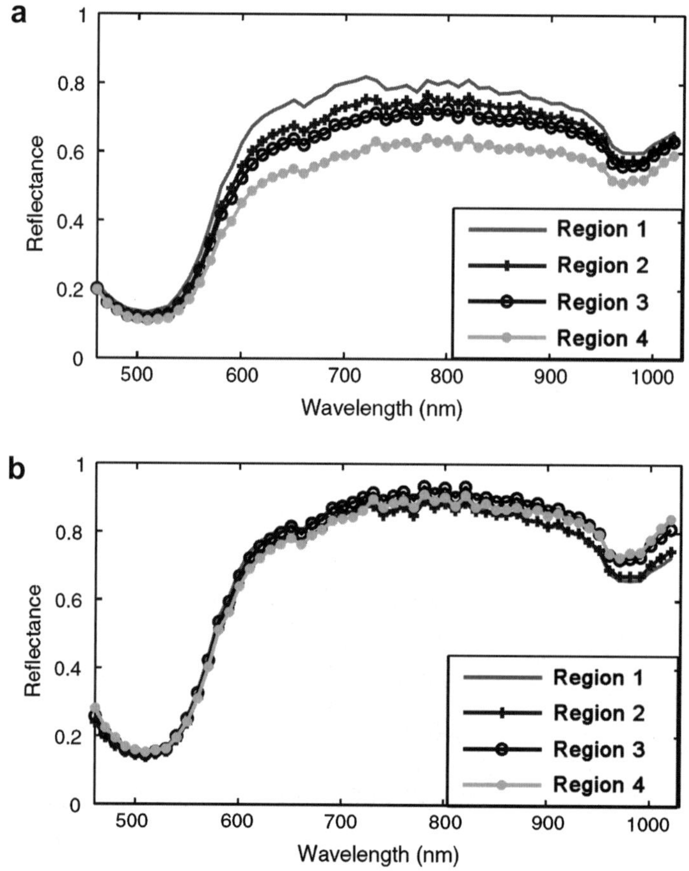

FIGURE 10.15 *Averaged uncorrected (a) and corrected (b) spectra of a 5 × 5 pixel window of four different regions of the image. (Full color version available on http://www.elsevierdirect.com/companions/9780123747532/)*

In current packing houses, for detecting decay caused by fungi, trained operators visually examine the fruit as it passes under ultraviolet light in order to detect those fruits that exhibit phenomena of fluorescence caused by the essential oils (Latz & Ernes, 1978) released after the fungal attack. However, this method is potentially harmful for the operator and is very labor-intensive.

One possible solution to this problem is through the development of automatic computer vision systems able to detect this damage. Gómez-Sanchís *et al.* (2008a) described one approach by using hyperspectral imaging. They used mandarins that were artificially infected with *P. digitatum* spores. A sequence of 57 monochromatic images was obtained from

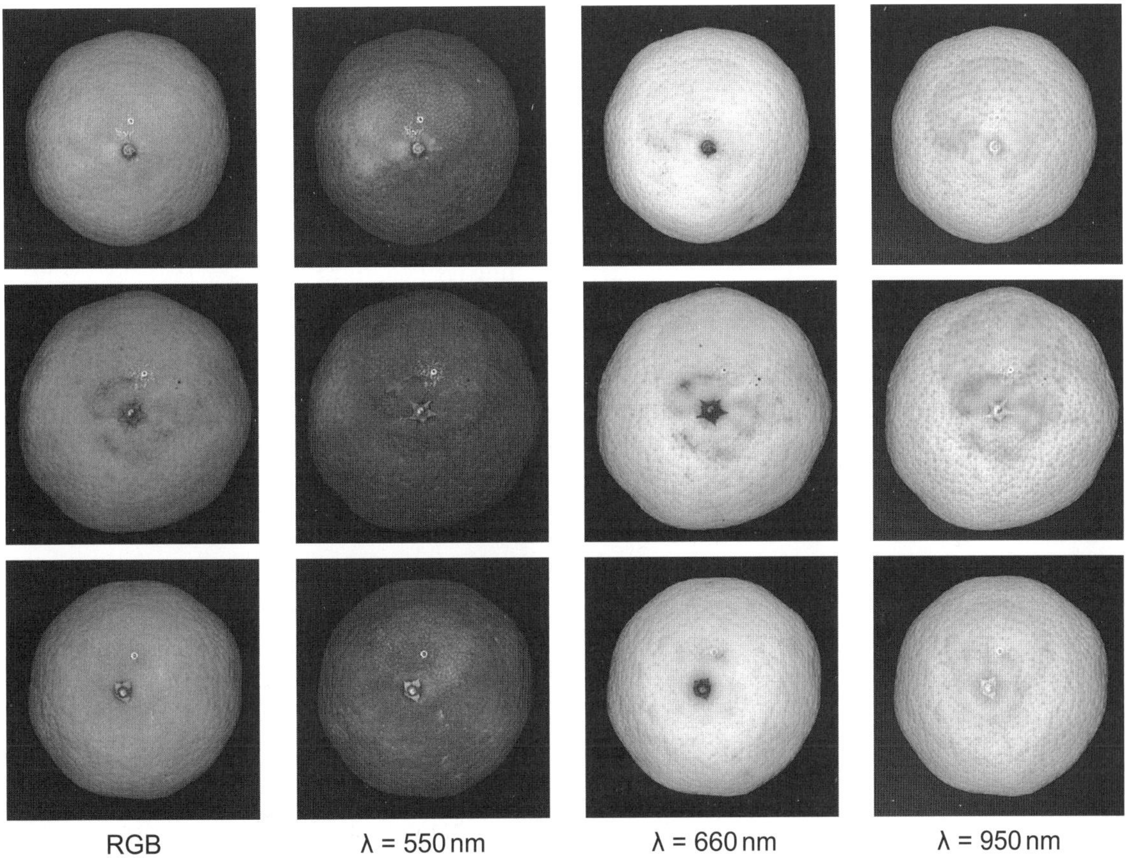

FIGURE 10.16 *RGB and monochromatic images (550 nm, 660 nm and 950 nm) of various mandarins* (cv. Clemenules). *(Full color version available on http://www.elseiverdirect.com/companions/9780123747532/)*

460 nm to 1020 nm with a spectral resolution of 10 nm as soon as the rot began to appear. Figure 10.16 shows examples of such monochromatic images (550 nm, 660 nm, and 950 nm), and RGB images of these mandarins. It can be observed how the damage is barely visible in the RGB images.

Given the large amount of information that hyperspectral images provide, it is often important to discard redundancies. A fast way to tackle this problem consists of eliminating bands that contain redundant information. Several methods are available in literature:

- Correlation analysis (CA), which consists of calculating the coefficients of correlation of each band with the variable class, and selecting those bands which have a higher correlation (Lee *et al.*, 2007).

- The mutual information function (MI) between each band and the variable class, which measures the interdependence between characteristics instead of evaluating the existence of linear relations between the variables, as in the case of linear correlation (Martínez-Sotoca and Pla, 2006).
- Stepwise multivariate regression (SW), which is based on the fact that if a variable is not important for the model then, when including it in the model, its corresponding coefficient of regression should not be significantly different from zero. SW offers two variants, depending on whether one begins including all the variables in the model, and excluding them at each step (backward stepwise) or, if one begins without variables, and including new ones at each step (forward stepwise). The search finishes when there are no improvements from one inclusion/exclusion step to the next (Yang *et al.*, 2004).
- Genetic algorithms (GA), which use a cost function in order to assess the importance of the groups of spectral bands that exist in each generation (iteration). The individuals (groups of spectral bands) with a higher cost function value are those that have a higher probability of being propagated to the next generation. When the overall hit rate provided by a linear discriminant analysis (LDA) algorithm is used as the cost function, this selection method is given the name GALDA (GA+LDA).

The variation ranges of all the variables should be made uniform to enable a comparison of the methods. The four selection methods can be programmed to iteratively increase the number of selected bands to determine an optimal set. These bands can be used to classify a labeled set of pixels using a classification method such as LDA or CART (classification and regression trees) (Breiman *et al.*, 1998). In this way, the success rate can be obtained according to the number of bands selected.

An example of using the above methods is given in the work carried out by Gómez-Sanchis *et al.* (2008a), in which each pixel containing 57 reflectance values (one for each band) was assigned to a class by an expert. These classes were named "sound peel", "damaged peel", "peel with spores" (peel with the characteristic green spores of *P. digitatum*), and "stem end". The labeled set of pixels was divided into two subsets: one training subset comprising 120 000 samples (40% of the total) and a validation subset comprising 180 000 samples (60% of the total). The first subset was used to construct the selection models of characteristics and classification, while the second one was used to assess the performance of these models.

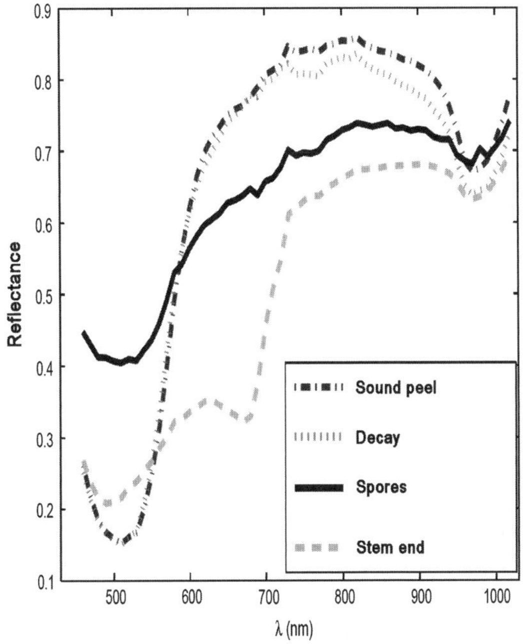

FIGURE 10.17 *Average spectra by class in the training subset for the Clemenules variety. (Full color version available on http://www.elseiverdirect.com/companions/9780123747532/)*

Figure 10.17 shows the average spectra of each of the aforementioned classes in the training subset. It can be seen that the main difficulty in both varieties for a classifier lies in distinguishing between the classes "sound peel" and "damaged peel" as a consequence of the high degree of overlapping that the average spectra of these classes exhibited.

The images made up from the bands selected using the methods described above were segmented using LDA and CART classifiers. Each pixel was classified into one of the four classes previously described in order to determine which fruit showed signs of rot. A citrus fruit with more than 5% of pixels classified as belonging to one of two classes of rotten peel was assigned to the class "decayed fruit" and the rest to "sound fruit". Success rate was defined as the percentage of fruit correctly classified. Figure 10.18 shows the evolution of the average success rate with respect to the number of selected bands. Figure 10.18(a) shows the results obtained using the LDA classifier and Figure 18(b) the results using CART (in both cases GALDA was the reduction method with the highest percentage of correct pixel classification). The maximum success rate in fruit sorting was approximately 92% for LDA, using the 57 bands of data sets, while the success rate rose to 95%

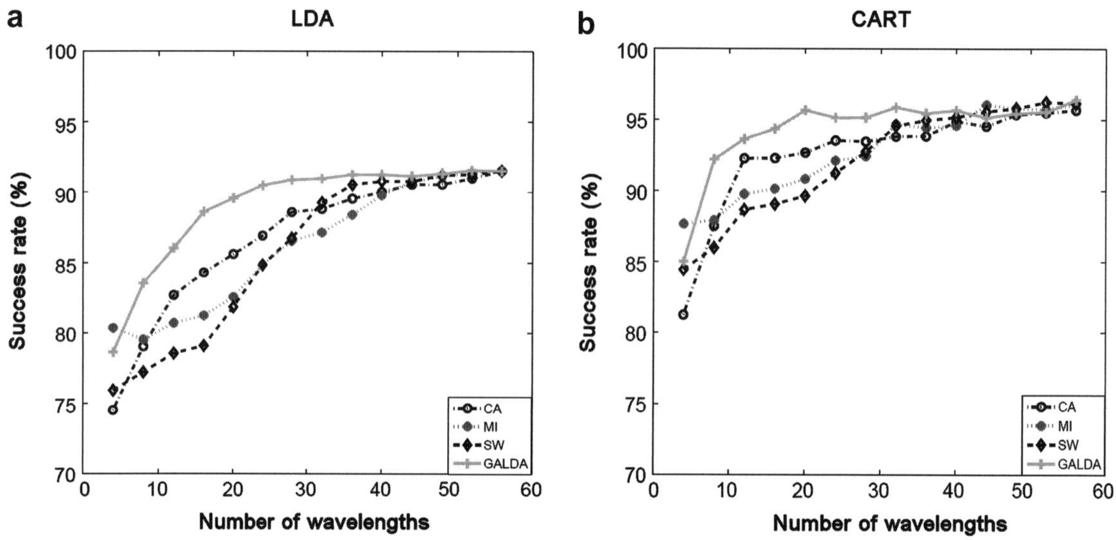

FIGURE 10.18 *Evolution of the average success rate using the classifiers based on LDA (a) and CART (b), with the bands obtained with the four selection methods employed. (Full color version available on http://www.elseiverdirect.com/companions/9780123747532/)*

using only 20 bands with CART. In the latter case, the addition of more bands to the classification model did not increase the success rate. This work demonstrated that a hyperspectral sorting machine can be envisaged to substitute current manual removal of rotten fruit. However, real-time requirements have not yet been achieved.

However, fungal diseases are not the only ones targeted by hyperspectral image systems. As mentioned before, detection of citrus canker, an important bacterial disease, has been addressed by researchers. Recently Qin *et al.* (2009) measured the reflectance of grapefruits with cankerous and five other common diseases in the spectral region between 450 nm and 930 nm. They developed an algorithm to detect and classify canker lesions from sound peel and other diseases with an overall correct classification of canker of 92%. The results obtained show that canker lesions on the peel of the grapefruit were observed at all wavelengths covered by the hyperspectral imaging system, being more distinctive from the fruit surface in the spectral region between 600 nm and 700 nm. Similar conclusions were reported by Balasundaram *et al.* (2009), who determined that the highest discriminating wavelengths were comprised between 500 and 800 nm.

10.6. CONCLUSIONS

Some of the most important aspects that need to be taken into consideration when developing a hyperspectral inspection system for citrus include the geometry of the fruit, the emission spectrum of the lighting source and their interaction. Because many citrus fruits are almost spherical, each point of their surface reflects the electromagnetic radiation differently towards the camera. This causes a gradual darkening of the image especially the further pixels from the light source, which is a phenomenon that must be artificially corrected. In addition, the variation of the efficiency of the filters with the wavelength should be also taken into consideration in order to enable the appropriate corrections to obtain true reflectance images.

Hyperspectral systems are an important tool for the quality inspection of citrus fruits, offering the possibility of designing machines for the automatic identification of blemishes. This is particularly important for early rot detection, one of the major problems faced by this sector. However, a realistic implementation of such systems probably still requires an important effort in adequately reducing the number of input bands.

NOMENCLATURE

Symbols

α_D	constant that relates the proportion of direct and diffuse light with the total light
ϵ_g	geometric correction factor of the image
ϕ	angle of incidence between the beam of direct light and the direction of the normal on the surface
λ	working wavelength
$\rho_{xy}(\lambda)$	corrected monochromatic image corrected using the reference
$\rho_{ref}(\lambda)$	certified reflectance of the reference
$\rho(\lambda)$	the image corrected geometrically
h_c	height of the fruit
$I(\lambda)$	total light in the system
$I_F(\lambda)$	diffuse component of the light
$I_D(\lambda)$	direct component of the light
$I_T(\lambda$	radiance of the light source
ms	milliseconds
nm	nanometers

N_{Pi}	number of P_i points
P_G	centroid of the fruit
P_i	equidistant points from the perimeter used to perform the interpolation grid ($I = 0..Np_i$)
r_{ij}	subradii calculated between consecutive $\|P_iP_G\|$ radii ($j = 0..16$; $i = 0..Np_i$)
$R(\lambda)$	radiance of the fruit
$R_{xy}(\lambda)$	uncorrected image
$R_{dark}(\lambda)$	image obtained with no illumination
$R_{white}(\lambda)$	monochromatic image of the reference

Abbreviation

AOTF	acoustic–optic tunable filter
CA	correlation analysis
CART	classification and regression trees
CCD	charge-coupled device
DEM	digital elevation model
GA	genetic algorithms
GALDA	genetic algorithms based on LDA
LDA	linear discriminant analysis
LCTF	liquid crystal tunable filter
MI	mutual information
NIR	near-infrared
RGB	red, green, blue
SW	stepwise regression
UV	ultraviolet
UVFL	ultraviolet-induced fluorescence

REFERENCES

Aleixos, N., Blasco, J., Navarrón, F., & Moltó, E. (2002). Multispectral inspection of citrus in real-time using machine vision and digital signal processors. *Computers and Electronics in Agriculture, 33*, 121–137.

Bajcsy, P., & Kooper, R. (2005). Prediction accuracy of color imagery from hyperspectral imagery. In Sylvia S. Shen, & Paul E. Lewis (Eds), Algorithms and Technologies for Multispectral, Hyperspectral, and Ultraspectral Imagery XI. *Proceedings of SPIE* 5806–34.

Balasundarama, D., Burks, T. F., Bulanona, D. M., Schubert, T., & Lee, W. S. (2009). Spectral reflectance characteristics of citrus canker and other peel conditions of grapefruit. *Postharvest Biology and Technology, 51*, 220–226.

Bei, L., Dennis, G., Miller, H., Spaine, T., & Carnahan, J. (2004). Acousto–optic tunable filters: fundamentals and applications as applied to chemical analysis techniques. *Progress in Quantum Electronics, 28*, 67–87.

Bennedsen, B. S., Peterson, D. L., & Tabb, A. (2005). Identifying defects in images of rotating apples. *Computers and Electronics in Agriculture, 48*(2), 92–102.

Blasco, J., Aleixos, N., Gómez-Sanchís, J., & Moltó, E. (2007a). Citrus sorting by identification of the most common defects using multispectral computer vision. *Journal of Food Engineering, 83*(3), 384–393.

Blasco, J., Aleixos, N., Gómez-Sanchís, J., & Moltó, E. (2009). Recognition and classification of external skin damages in citrus fruits using multispectral data and morphological features. *Biosystems Engineering, 103*(2), 137–145.

Blasco, J., Aleixos, N., & Moltó, E. (2003). Machine vision system for automatic quality grading of fruit. *Biosystems Engineering, 85*(4), 415–423.

Blasco, J., Aleixos, N., & Moltó, E. (2007b). Computer vision detection of peel defects in citrus by means of a region oriented segmentation algorithm. *Journal of Food Engineering, 81*(3), 535–543.

Breiman, L., Friedman, J., Olshen, R., & Stone, J. (1998). *Classification and regression trees*. Boca Raton, FL: CRC Press.

Brosnan, T., & Sun, D. W. (2004). Improving quality inspection of food products by computer vision—a review. *Journal of Food Engineering, 61*, 3–16.

Chen, Y. R., Chao, K., & Kim, M. S. (2002). Machine vision technology for agricultural applications. *Computers and Electronics in Agriculture, 36*(2), 173–191.

Cheng, X., Chen, Y., Tao, Y., Wang, C., Kim, M. S., & Lefcourt, A. (2004). A novel integrated PCA and FLD method on hyperspectral image feature extraction for cucumber chilling damage inspection. *Transactions of ASAE, 47*(4), 1313–1320.

Du, C. J., & Sun, D. W. (2004). Recent development in the applications of image processing techniques for food quality evaluation. *Trends in Food Science & Technology, 15*, 230–249.

Duda, R. O., Hart, P. E., & Stork, D. G. (2000). *Pattern classification* (2nd ed.). Hoboken, NJ: Wiley–Interscience.

Eckert, J. W., & Eaks, I. L. (1989). *The citrus industry, Vol. 5: Postharvest disorders and diseases of citrus*. Berkeley, CA: University of California Press.

Foley, J., Van Dam, A., Feiner, S., & Hughes, J. (1996). *Computer graphics: principles and practice*. Reading, MA: Addison–Wesley.

Gaffney, J. (1973). Reflectance properties of citrus fruit. *Transactions of the ASAE, 16*(2), 310–314.

Gómez-Sanchís, J., Gómez-Chova, L., Aleixos, N., Camps-Valls, G., Montesinos-Herrero, C., Moltó, E., & Blasco, J. (2008a). Hyperspectral system for early detection of rottenness caused by *Penicillium digitatum* in mandarins. *Journal of Food Engineering, 89*(1), 80–86.

Gómez-Sanchis, J., Moltó, E., Camps-Valls, G., Gómez-Chova, L., Aleixos, N., & Blasco, J. (2008b). Automatic correction of the effects of the light source on spherical objects: an application to the analysis of hyperspectral images of citrus fruits. *Journal of Food Engineering, 85*(2), 191–200.

Hetchts, E. (1998). *Optics* (3rd ed.). Reading, MA: Addison Wesley Longman.

Jolliffe, I. T. (1986). *Principal component analysis*. New York, NY: John Wiley & Sons.

Kleynen, O., Leemans, V., & Destain, M. F. (2005). Development of a multi-spectral vision system for the detection of defects on apples. *Journal of Food Engineering, 69*(1), 41–49.

Latz, H. W., & Ernes, D. A. (1978). Selective fluorescence detection of citrus oil. components separated by high-pressure liquid chromatography. *Journal of Cromatography, 166*, 189–199.

Lee, D. J., Archibald, J. K., Xu, X. Q., & Zhan, P. C. (2007). Using distance transform to solve real-time machine vision inspection problems. *Machine Vision and Applications, 18*(2), 85–93.

Martínez-Sotoca, J., & Plá, F. (2006). Hyperspectral data selection from mutual information between image bands. *Lecture Notes in Computer Science, 4109*, 853–861.

Papadakis, S. E., Abdul-Malek, S., Kandem, R. E., & Yam, K. L. (2000). A versatile and inexpensive technique for measuring color of foods. *Food Technology, 54*(12), 48–51.

Poger, S., & Angelopoulou, E. (2001). *Multispectral Sensors in Computer Vision. Technical Report*. Hoboken, NJ: Stevens Institute of Technology.

Qin, J., Burks, T. F., Ritenour, M. A., & Bonn, W. G. (2009). Detection of citrus canker using hyperspectral reflectance imaging with spectral information divergence. *Journal of Food Engineering, 93*, 183–191.

Schubert, T. S., Rizvi, S. A., Sun, X. A., Gottwald, T. R., Graham, J. H., & Dixon, W. N. (2001). Meeting the challenge of eradicating citrus canker in Florida—again. *Plant Disease, 85*(4), 340–356.

Shaw, G., & Burke, H. (2003). Spectral imaging for remote sensing. *Lincoln Laboratory Journal, 14*(1), 3–28.

Sun, D. W. (2007). *Computer vision technology for food quality evaluation*. San Diego, CA: Elsevier Academic Press.

Tao, Y., & Wen, Z. (1999). An adaptive spherical image transform for high-speed fruit defect detection. *Transactions of the ASAE, 42*(1), 241–246.

Yang, C., Everitt, J. H., & Bradford, J. M. (2004). Airborne hyperspectral imagery and yield monitor data for estimating grain sorghum yield variability. *Transactions of the ASAE, 47*(3), 915–924.

CHAPTER 11

Visualization of Sugar Distribution of Melons by Hyperspectral Technique

Junichi Sugiyama, Mizuki Tsuta
National Food Research Institute, Tsukuba, Ibaraki Japan

CONTENTS

11.1. INTRODUCTION

In Japan, automated sweetness sorting machines for peaches, apples, and melons based on near-infrared (NIR) spectroscopy techniques have been developed and are now in use in more than 172 packing houses (Hasegawa, 2000). However, parts of a fruit sorted by the machine as sweet may sometimes taste insipid because of an uneven distribution of the sugar content. Visualization of the sugar content of a melon is expected to be useful not only for evaluation of its quality but also for physiological analysis of the ripeness of a melon. There have been several attempts to obtain a distribution map of the constituents of agricultural produce (Bertrand *et al.*, 1996; Ishigami & Matsuura, 1993; Robert *et al.*, 1991, 1992; Taylor & McClure, 1989). However, a quantitatively labeled distribution map has not yet been obtained.

On the other hand, recently, device and personal computer (PC) technology have advanced greatly. Cooled charge-coupled device (CCD) imaging cameras with a wide dynamic range have been introduced, which makes quantitative measurements possible. Modern PCs can easily accept and/or process a large volume of data. Taking advantage of these, the conventional NIR spectroscopy technique, which is a technique for point measurement, could be extended to two-dimensional measurements. This chapter thus discusses the development of a technique for visualization of the sugar content of a melon by applying NIR spectroscopy to each pixel in an image.

11.2. VISUALIZATION BY VISIBLE WAVELENGTH REGION

11.2.1. Melons

Maturity levels of melons at which they are harvested significantly affect their sugar content distribution. Sugiyama (1999) used a hyperspectral technique to compare three ripeness stages (unripe, mature, and fully mature) of Andes melons harvested in Tsuruoka, Yamagata Prefecture, Japan, in 1998. Unripe melons were harvested 6 days earlier, and fully mature ones 5 days later than the mature melons. The mature melons were harvested 55 days after pollination. Two melons at each stage, that is, a total of six melons, were investigated. There were cracks on the bottom of the fully mature melons because of overripening. Each sample was sent to the laboratory the day after the harvest using a special delivery service. Experiments were carried out in a dark room at 25 °C.

11.2.2. NIR Spectroscopy

11.2.2.1. Measurement of spectra and sugar content

In order to determine the wavelength that has a high correlation with sugar content, spectra between 400 and 1100 nm in the flesh of a melon were analyzed using a NIR spectrometer (NIRS 6500, FOSS NIR Systems, Silver Spring, MD, USA). This wavelength range covers the spectrum of the CCD camera used for the imaging application. A cylindrical sample with a diameter of 20 mm was extracted from the equator of the melon by a stainless steel cylinder with a knife edge at one end (Figure 11.1). A spectrum of the sample's inner surface was obtained using a fiber-type detector with an interactance mode (Kawano *et al.*, 1992). The wavelength interval was 2 nm and the number of scans was 50. Then, the measured portion was cut into a 2 mm-thick slice with a kitchen knife and squeezed with a portable garlic crusher to measure the Brix sugar content using a digital refractometer (PR-100, ATAGO, Yorii, Saitama, Japan). The measurements of the spectrum and the sugar content were repeated similarly at various depths within the melon.

11.2.2.2. Wavelength selection by NIR spectroscopy

Figure 11.2 shows the simple correlation coefficient calculated by a standard regression model with no data pretreatment between the absorbance at each wavelength and the sugar content. Each line was calculated from 22 slices made from two cylindrical samples which were extracted from a melon. It is clear that the wavelength of 676 nm exhibits the maximum absolute

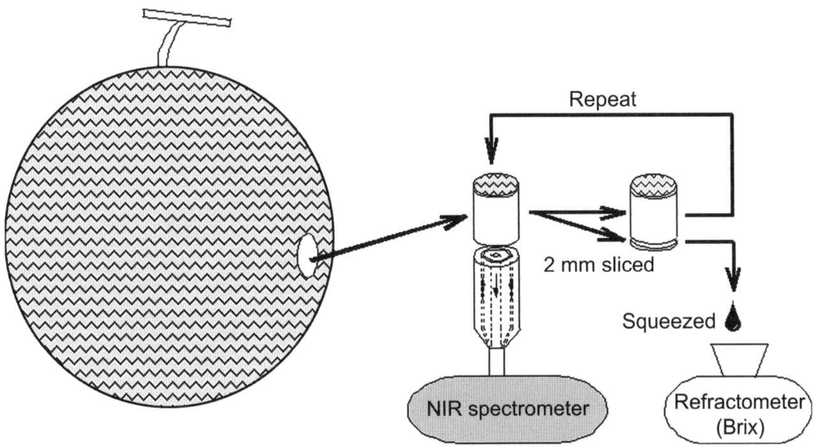

FIGURE 11.1 *Sample preparation for measurements of NIR spectra and the sugar content*

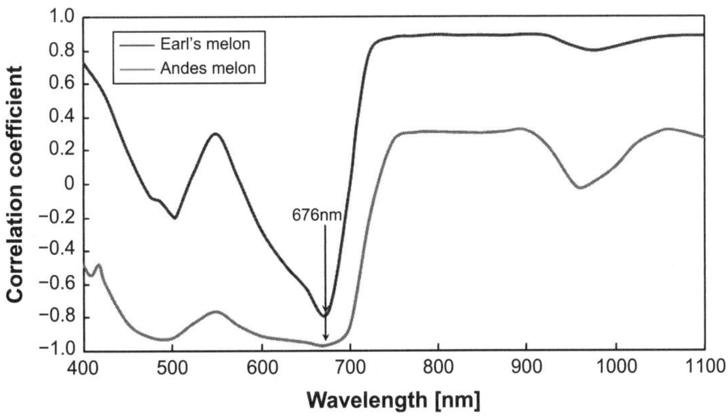

FIGURE 11.2 *Correlation coefficients at each wavelength between the absorbance and the Brix sugar content*

correlation, although it is inversely correlated with the sugar content for both the Andes and Earl's varieties of melon. Because of the inverse correlation, it seems that 676 nm is not a direct absorption band of sugar but a wavelength of secondary correlation (Osborne & Feam, 1986) with a component inversely proportional to the sugar content. It is actually close to the absorption band of chlorophyll (Nussberger *et al.*, 1994; Watada *et al.*, 1976) and there are some implications for other produce (Izumi *et al.*, 1990; Morita *et al.*, 1992). With the fact that absorbance at 676 nm is inversely correlated with the sugar content for its visualization, the interpretation of 676 nm is also important in terms of the physiological aspects and must be further studied.

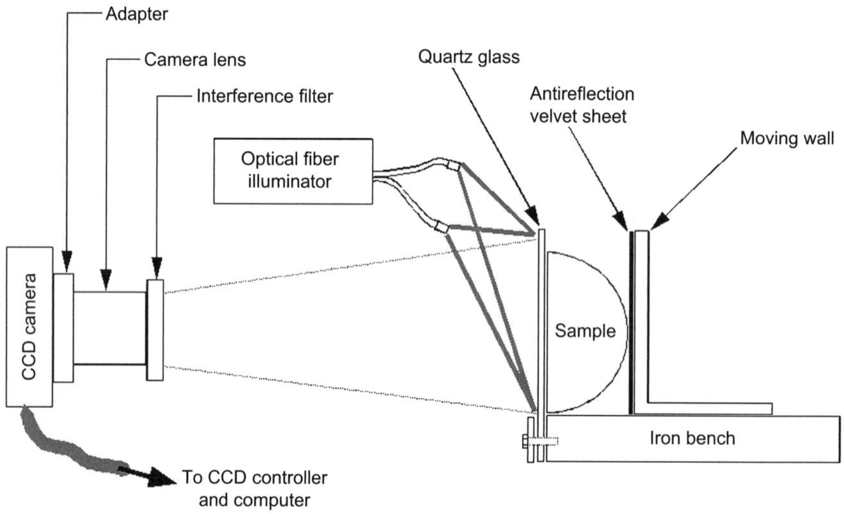

FIGURE 11.3 *Configuration of an apparatus for spectroscopic image acquisition*

11.2.3. Imaging Spectroscopy

11.2.3.1. Instrumentation

Figure 11.3 shows the configuration of the imaging apparatus used to obtain the spectroscopic images. Although a monochrome CCD camera normally has an 8-bit (256 steps) analog-to-digital (A/D) resolution, a cooled CCD camera (CV-04II, Mutoh Industries Ltd., Tokyo, Japan) with a 16-bit (65 536 steps) A/D resolution was adopted. The advantage of the high A/D resolution is that each pixel can function as a detector of the NIR spectrometer for quantitative analysis. The CCD camera has a linear intensity characteristic ($\gamma = 1$) and no antiblooming gate for quantitative analysis. To decrease the electrical dark current noise of the CCD camera, both double-stage thermoelectric cooling and water cooling were utilized. A camera lens (FD28 mm F3.5 S.C., Canon, Tokyo, Japan) with an interference filter (J43192, Edmond Scientific, Tokyo, Japan) was installed through the camera adapter (Koueisya, Kawagoe, Saitama, Japan). The interference filter had band-pass characteristics of 676 nm at the central wavelength, which was determined in the NIR spectroscopic experiment (see 11.2.2.2), and 10 nm at half-bandwidth. The illuminator (LA-150S, Hayashi Watch-Works, Tokyo, Japan) had a tungsten–halogen bulb driven by direct current to reduce optical noise. The source light was introduced into two fiber-optic probes, illuminating a sample from two different positions so as not to create any shadows or direct reflection. A sample was placed perpendicularly on the quartz glass maintaining a constant focal distance between the CCD camera and the sample (Figure 11.3). The sample was supported by the moving wall covered with a black antireflection velvet sheet.

11.2.3.2. Image of half-cut melon for sugar distribution map

Each melon (six in total) was cut vertically in half with a kitchen knife. Spectroscopic images of the surface of a half-cut melon at 676 nm were taken with an aperture of 16 (F16) and an exposure period of 0.5 seconds. The cooling temperature of the CCD camera was $-15\,^{\circ}\mathrm{C}$. The size of the image was 768×512 pixels. After obtaining a vertical image of the half-cut melon, the melon was cut in a horizontal plane, and a horizontal image of the quarter-cut melon was captured under the same conditions as described above.

11.2.3.3. Partial image for sugar content calibration

After obtaining the aforementioned images, two cylindrical samples with a diameter of 20 mm were extracted from the equator of the same melon. In the same manner as in the NIR spectroscopic experiment (Figure 11.1), an image of the surface was taken at 676 nm using the CCD camera under the same conditions as for the half-cut melon described previously. Then a 2 mm-thick slice was cut off and squeezed for the measurement of sugar content. These procedures were repeated until the rind appeared.

11.2.3.4. Noise corrections

Images acquired using a CCD camera include (i) thermal noise due to dark current thermal electrons, (ii) bias signals to offset the CCD slightly above zero A/D counts, (iii) sensitivity variations from pixel to pixel on the CCD, and (iv) lighting variations on the sample's surface. In order to compensate for the above, the following noise corrections (Fukushima, 1996; Morita *et al.*, 1992; SBIG, 1998) were carried out:

$$\text{Processed image} = \frac{\text{raw image} - \text{dark frame}}{\text{flat field} - \text{dark frame of flat field}} \times M \qquad (11.1)$$

In Equation (11.1), the dark frame is the image acquired under the same conditions as the raw image except for the absence of lighting. Subtracting it from the raw image allows corrections for (i) thermal noise and (ii) bias signals. On the other hand, the flat field is obtained by taking an exposure of a uniformly lit "flat field" such as a Teflon board. After subtracting the dark frame of the flat field, in the same way as the numerator, the ratio between the two images is compensated for the effect of (iii) sensitivity variations and

(iv) lighting variations. M is the intensity value averaged over all pixels of the flat field after dark frame subtraction (= denominator in Equation 11.1). The multiplier M restores the ratio of the images to the image intensity level.

All of these image processes were carried out using software (CCD Master, Mutoh Industries Ltd., Tokyo, Japan) compatible with the CCD camera.

11.2.3.5. Conversion from intensity into sugar content

Each pixel of the image processed using Equation (11.1) has 16 bits, that is, 65 536 levels of intensity. The method for converting the intensity into sugar content on an image data was developed in accordance with NIR spectroscopy. Based on the fact that the functional group of chemical compounds responds to near-infrared radiation, NIR spectroscopy can measure the amount of a specific constituent from its absorbance at several wavelengths (Osborne & Feam, 1986). Absorbance A can be defined as follows:

$$A = \log(I_S/I) \tag{11.2}$$

where I_s is the intensity of reflection of a white standard board; I is the intensity of reflection of the sample.

Because I_S and I correspond to the denominator and the numerator in the first term of Equation (11.1), respectively, Equations (11.3) and (11.4) are introduced.

$$I_S = \text{flat field} - \text{dark frame of flat field} \tag{11.3}$$

$$I = \text{raw image} - \text{dark frame} \tag{11.4}$$

Considering Equations (11.1 to 11.4), absorbance A in the spectroscopic image can be expressed as follows:

$$\begin{aligned} A &= \log(I_S/I) \\ &= \log(M/\text{processed image}) \\ &= \log(M/R) \end{aligned} \tag{11.5}$$

where R is the intensity of the reflection of each pixel in a processed image, and M is the average intensity of reflection of the flat field.

On the other hand, the NIR spectroscopic experiment indicated that the absorbance at 676 nm was correlated with the sugar content. The same relationship in the image system of this experiment could be confirmed by using the following procedure: (i) calculation of the average intensity of a partial image of 20 mm diameter, (ii) conversion of the average intensity into absorbance using Equation (5), and (iii) plotting the relationship between

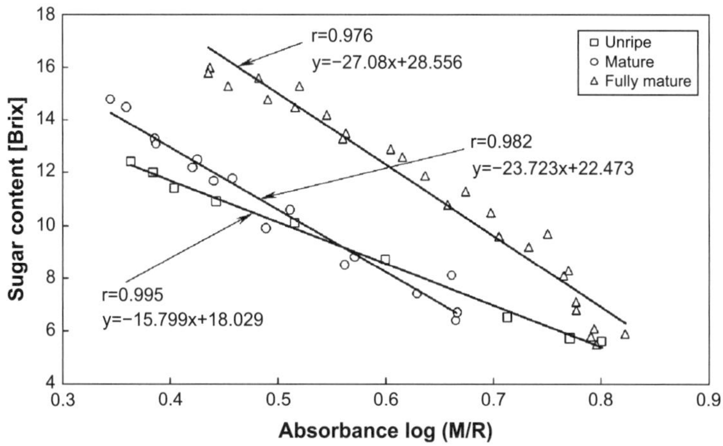

FIGURE 11.4 *Calibration curves between the sugar content and the absorbance at 676 nm by the imaging system*

the absorbance and sugar content for each partial image. A total of six melons, two for each stage of ripeness, were analyzed, and the representative results for unripe, mature, and fully mature melons are shown in Figure 11.4. The number of symbols in Figure 11.4 indicates the number of the sliced samples subjected to measurements of the absorbance and sugar content. Each calibration curve is slightly different from the others because the light condition had been adjusted for each sample to avoid direct reflection (glittering) on rugged portions. This adjustment changed the lighting intensity level, which is not corrected by Equation (11.1), and subsequently affected the calibration curves. However, it was confirmed that the image system can reveal the sugar content using the calibration curves for each sample.

11.2.4. Visualization of Sugar Distribution

The image of a half-cut melon for drawing a sugar distribution map was corrected for noise using Equation (11.1). The processed image was converted into an absorbance image using Equation (11.5). Then, an image of sugar content was calculated by applying the calibration curve to each pixel of the absorbance image. These actual procedures, from retrieval of the processed image to saving the sugar content image, were carried out using an original program written in Visual Basic (Microsoft, Redmond, WA, USA). Finally, the sugar content image was visualized with a linear color scale by the visualization software (AVS/Express Viz, Advanced Visual Systems, Waltham, MA, USA). Figure 11.5 shows the results of visualization of the sugar content corresponding to unripe, mature, and fully mature melons,

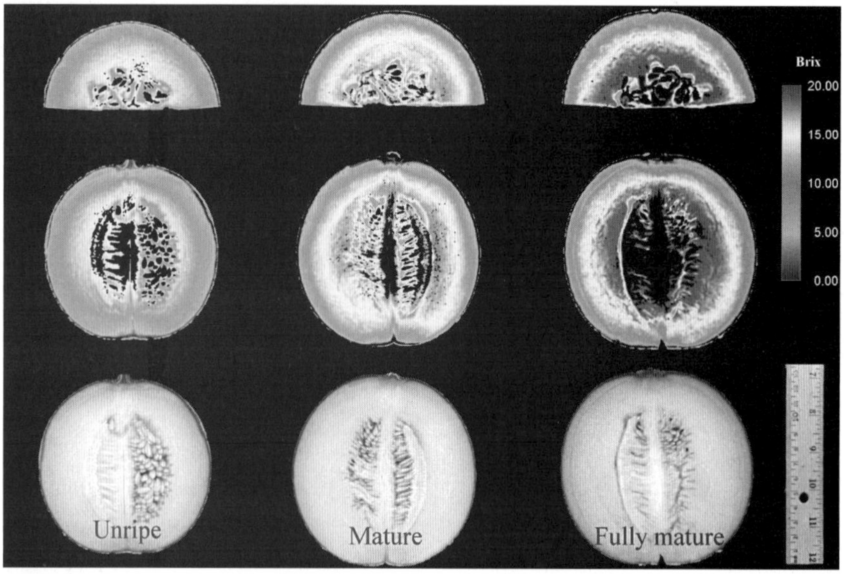

FIGURE 11.5 *Sugar distribution map for unripe, mature, and fully mature melons. (Full color version available on http://www.elsevierdirect.com/companions/9780123747532/)*

respectively. Since the measurements were carried out just after the harvest, the flesh of each melon was sufficiently hard that it was actually difficult to tell the differences in sugar content by the naked eye. However, as a result of the visualization, the sugar content at each stage of ripeness was clarified. In particular, in the mature and fully mature melons the distribution of the sugar content varies among different parts of the fruit, indicating the importance of the part of the fruit sampled in the conventional measurement of the sugar content with a refractometer. In addition, as shown in the mature melon, the upper part had higher sugar content than the bottom part. These results suggest that the visualization technique by NIR imaging could become a useful new method for quality evaluation of melons. Moreover, there is a good possibility that applying several wavelengths for the calibration curve could allow visualization of many more constituents of other agricultural products.

11.3. VISUALIZATION BY SUGAR ABSORPTION WAVELENGTH

The former method cannot be applied to a red-flesh melon because it depends not on the absorption band of the sugar, but on the color information at 676 nm. Therefore, Tsuta *et al.* (2002) developed a universal method for visualization of sugar content based on the absorption band of the sugar in the NIR wavelength region.

11.3.1. Melons

Two green-flesh melons (Delicy) and three red-flesh melons (Quincy) were prepared for NIR spectroscopy and another red-flesh melon (Quincy) for imaging. They were obtained from a store and left overnight in a dark room at 25 °C before the experiment. The experiments were carried out in the same room.

11.3.2. NIR Spectroscopy for Sugar Absorption Band

11.3.2.1. Measurement of spectra and sugar content

To specify the absorption band of sugar, a NIR spectrometer (NIRS 6500, FOSS NIRSystems, Silver Spring, MD, USA) and a digital refractometer (PR-100, ATAGO, Yorii, Saitama, Japan) were utilized (Figure 11.6). Pretreatment of the acquired spectra and a multi-linear regression (MLR) analysis were carried out using spectral analysis software (VISION, FOSS NIRSystems, Silver Spring, MD, USA).

A 25 mm-diameter cylindrical sample (Figure 11.6a) was extracted from the equator of a melon using a stainless steel cylinder with a knife edge at one end. A spectrum of the sample's inner surface was obtained using a fiber-optic probe (Figure 11.6b) of the NIR spectrometer in the interactance mode

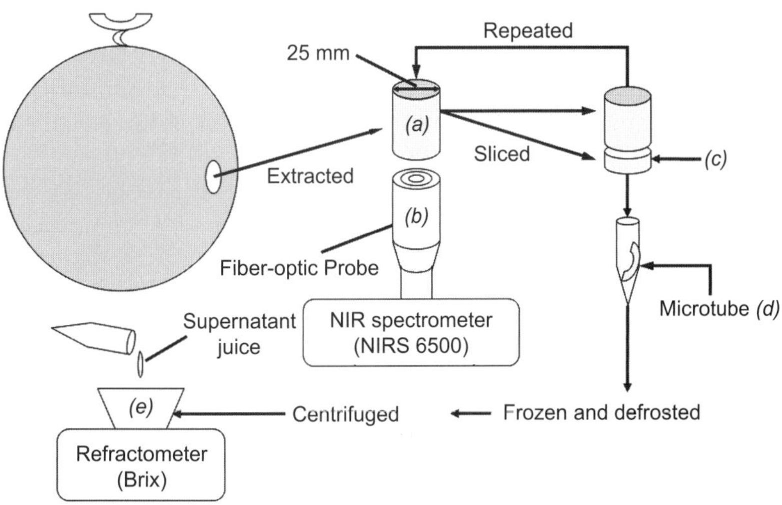

FIGURE 11.6 *NIR spectroscopy for evaluation of sugar content of melons*

(Kawano *et al.*, 1992). The wavelength interval was 2 nm, and the number of scans was 50. The measured portion was then cut into a 1 mm-thick slice (Figure 11.6c) using a handicraft cutter and put into a 1.5 ml microtube (Figure 11.6d) to be frozen and defrosted. This process was intended to break the cell walls of the portion in order to extract a sufficient amount of juice for measuring the sugar content (Martinsen & Schaare, 1998). The portion was then centrifuged for 10 min at 10 000 rpm to extract juice. The °Brix sugar content of the juice was measured using the digital refractometer (Figure 11.6e). A set of the spectrum and the sugar content measurements for every 1 mm-thick slice was repeated from the inner surface toward the rind. Each raw spectrum was converted into a second-derivative spectrum to decrease the effect of spectral baseline shifts (Iwamoto *et al.*, 1994; Katsumoto *et al.*, 2001). An MLR analysis was carried out for all of the data sets to acquire the calibration curve for the sugar content and the second-derivative spectra.

11.3.2.2. Calculation of second-derivative spectrum

Derivative methods are important pretreatment methods in NIR spectroscopy. The second-derivative method is most often used because of its following merits (Iwamoto *et al.*, 1994; Katsumoto *et al.*, 2001):

1. Positive peaks in a raw spectrum are converted into negative peaks in a second-derivative spectrum.
2. The resolution is enhanced for the separation of overlapping peaks and the emphasis of small peaks.
3. The additive and multiplicative baseline shifts in a raw spectrum are removed.

By applying the truncated Taylor series expansion, a second-derivative spectrum can be calculated as follows (Morimoto *et al.*, 2001):

$$f^2(x) = \frac{f(x + \Delta x) - 2 \times f(x) + f(x - \Delta x)}{\Delta x^2} \tag{11.6}$$

where $f(x)$ is the spectral function at x and $f^2(x)$ is the second-derivative function at x. Actual spectral data, however, take discrete values because of the limited wavelength resolution of NIR spectrometers. Therefore, a second-derivative spectrum is calculated as follows in NIR spectroscopy (Katsumoto *et al.*, 2001):

$$d^2A_i = A_{i+k} - 2 \times A_i + A_{i-k} \tag{11.7}$$

where A_i is an absorbance at i nm, d^2A_i is a second-derivative absorbance at i nm, and k is a distance between the neighboring wavelengths, which is called a derivative gap. Equation (11.7) shows that absorbances at three wavelengths of i, $i+k$ and $i-k$ are enough for calculating the second-derivative absorbance at i nm. It also indicates that the imaging system can acquire a second-derivative spectroscopic image using three band-pass filters.

11.3.2.3. Absorption band of sugar by NIR spectroscopy

One hundred and fifty-seven spectra were obtained as a result of NIR spectroscopy, and the MLR analysis of the spectra revealed that the second-derivative absorbances at 874 and 902 nm were highly correlated with the sugar content as shown in Table 11.1. The correlation was maintained at a high level of more than 0.99, whereas the derivative gap changed from 20 to 36 nm. The derivative gap was selected to decrease the number of band-pass filters for the imaging system. Conventionally, six band-pass filters are necessary to acquire two second-derivative spectroscopic images. However, when the derivative gap of 28 nm was adopted, only four band-pass filters, that is, 846, 874, 902, and 930 nm, were sufficient for the analysis. This is because 874 and 902 nm overlapped between two second derivatives (indicated by bold italic digits in Table 11.1). When 28 nm was selected as the derivative gap, the calibration curve was as follows:

$$°\text{Brix} = 21.93 - 410.76\, d^2A_{902} + 1534.76\, d^2A_{874} \quad (11.8)$$

Table 11.1 Relationship among the derivative gap, correlation, and necessary band-pass filters

		Necessary band-pass filters (nm)					
Gap (nm)	***R***	**For d^2A_{874}**			**For d^2A_{902}**		
4	0.976	870	874	878	898	902	906
8	0.975	866	874	882	894	902	910
12	0.983	862	874	886	890	902	914
16	0.988	858	874	890	886	902	918
20	0.990	854	874	894	882	902	922
24	0.991	850	874	898	878	902	926
28	0.991	846	***874***	***902***	***874***	***902***	930
32	0.991	842	874	906	870	902	934
36	0.990	838	874	910	866	902	938
40	0.988	834	874	914	862	902	942

Note: Bold type denotes overlapping wavelengths.

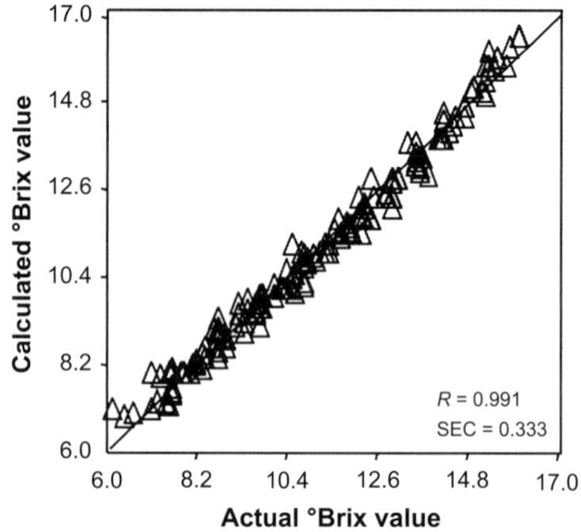

FIGURE 11.7 *Calibration by NIR spectroscopy*

The curve had a high correlation with the sugar content ($R = 0.991$), and the standard error of calibration was 0.333 (Figure 11.7). The second-derivative absorbance at 902 nm had an inverse correlation with the sugar content, which indicated that the raw absorbance had a positive correlation with it. In addition, several publications (Ito *et al.*, 1996; Kawano & Abe, 1995; Kawano *et al.*, 1992, 1993; Temma *et al.*, 1999) indicated that 902 nm is one of the typical absorption bands of sugar components. On the other hand, 874 nm can be considered as the reference wavelength to compensate for different surface conditions or some other influences. As a result, four band-pass filters of 846, 874, 902, and 930 nm were adopted for the imaging system.

11.3.3. Imaging Spectroscopy for Sugar Absorption Band

11.3.3.1. Instrumentation

Figure 11.8 shows the configuration of the apparatus for obtaining spectroscopic images. The cooled CCD camera (CV-04 II, Mutoh Industries Ltd., Tokyo, Japan) had a 16-bit (65 536 steps) A/D resolution, a linear intensity characteristic ($\gamma = 1$), and no antiblooming gate, so that each pixel could function as a detector of an NIR spectrometer for quantitative analysis. To decrease the electrical dark current noise of the CCD camera, both double-stage thermoelectric cooling and water cooling were utilized. A filter adapter with a filter holder (Koueisha, Kawagoe, Saitama, Japan) and a camera lens (FD28 mm F3.5 S. C., Canon, Tokyo, Japan) were installed in the CCD camera. The filter holder had four holes to which four filters could be fitted. The four filters in this experiment (BWE*x*; $x = 846$,

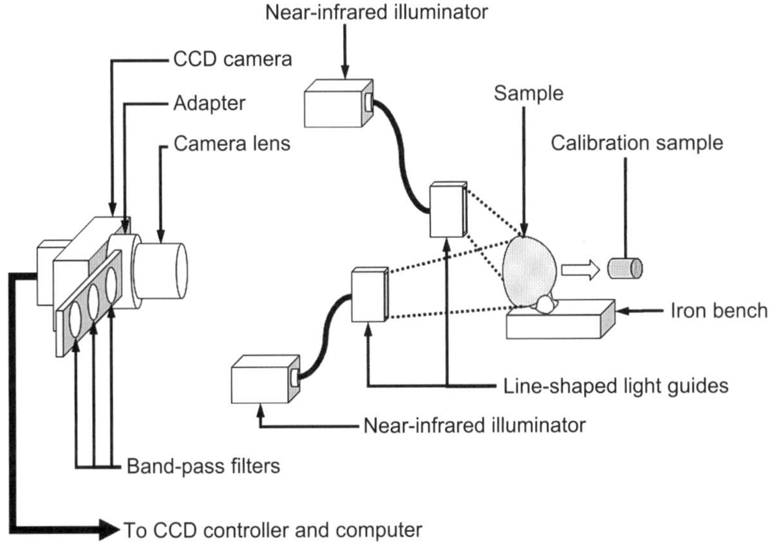

FIGURE 11.8 *Imaging system for spectroscopic image acquisition at different wavelengths*

874, 902, 930, Koshin Kogaku Filters, Hatano, Kanagawa, Japan) were designed to have band-pass characteristics of *x* nm at the central wavelength, and their details are shown in Table 11.2. The wavelengths of 902 and 874 nm were determined in the NIR spectroscopic experiment (see 11.3.2.3), and the others were their neighboring wavelengths selected to calculate the second-derivative absorbances. The near-infrared illuminator (LA-100IR, Hayashi Watch-Works, Tokyo, Japan) irradiated only NIR light because a NIR reflecting mirror was installed around a tungsten–halogen bulb, and a high-pass filter, which transmits only light above 800 nm, was attached to the irradiation hole. The source light was introduced into line-shaped light guides through a fiber-optic probe, illuminating a sample from two different positions in order not to create any shadows or direct reflection. Previously, a quartz glass had been placed on the surface of the sample to maintain a constant focal distance between the CCD camera and the sample (Sugiyama, 1999). In this experiment, however, a direct reflection image of the CCD camera on the glass was observed because the intensities of the sample and unnecessary images were both low; these were enhanced by a long exposure period. Therefore, the quartz glass was not adopted in this experiment; instead, the sample was placed on an iron bench facing the camera.

Table 11.2 Characteristics of the band-pass filters

Model	Central wavelength (nm)		Bandwidth (nm)
	Specified value	Measured value	
BWE846φ30	846.0 ± 2.0	847.5	13.3
BWE874φ30	874.0 ± 2.0	875.8	13.6
BWE902φ30	902.0 ± 2.0	900.0	16.0
BWE930φ30	930.0 ± 2.0	928.5	16.0

11.3.3.2. Acquisition of the spectroscopic images

Using the imaging system, whole images of the surface of a half-cut sample were taken at 846, 874, 902, and 930 nm at an exposure period of 3.7 s. The binning mode of 2×2, that is, four pixels of the CCD camera were combined to function as one pixel, was applied to acquire a higher sensitivity (Fukushima, 1996). The temperature of the CCD camera was maintained at −20 °C. The size of the image was 384×256 pixels after binning. After a half-cut image had been captured, two 25 mm-diameter cylindrical samples were extracted from the equator of the same melon. These cylindrical samples were used to acquire a sugar content calibration curve based on the imaging system. In the same manner as in the case of a half-cut sample, images of the surface of the cylindrical samples were taken, after which a 1 mm-thick slice was obtained and the °Brix sugar content was measured as described for the NIR spectroscopic experiment (see Figure 11.6). Image capture and measurement of the sugar content were repeated until the rind appeared.

11.3.3.3. Image processing for calibration

The obtained raw images of the cylindrical samples include (1) thermal noise, (2) bias signals to offset the CCD slightly above zero A/D counts, (3) sensitivity variations from pixel to pixel on the CCD, and (4) lighting variations on the sample's surface (Fukushima, 1996). To compensate for the above effects, noise and shading corrections were carried out for all images. The average intensity of the images of the cylindrical sample was converted into the average absorbance based on the spectroscopy theory (Figure 11.9). These processes were described in Section 11.2.3.5. Once the average absorbance at each wavelength was obtained, the second-derivative absorbances at 902 and 874 nm were calculated as follows (Katsumoto *et al.*, 2001; Morimoto *et al.*, 2001) (see 11.3.2.2):

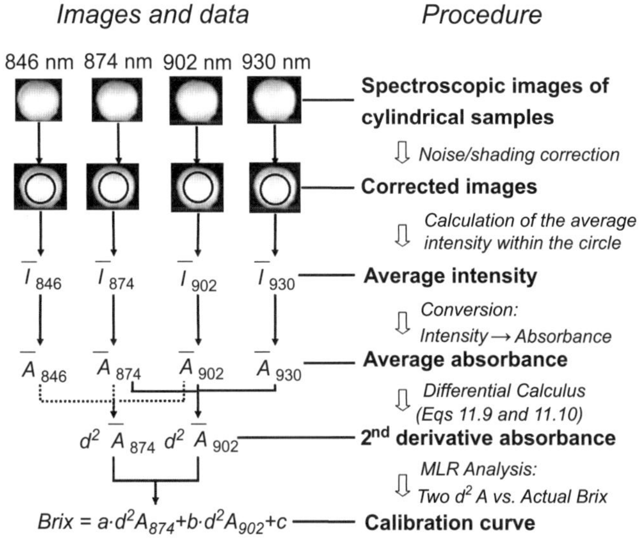

FIGURE 11.9 *Image processing procedure for calibration*

$$d^2A_{902} = A_{930} - 2 \times A_{902} - A_{874} \quad (11.9)$$

$$d^2A_{874} = A_{902} - 2 \times A_{874} - A_{846} \quad (11.10)$$

where A_i is the absorbance at i nm and d^2A_i is the second-derivative absorbance at i nm. Then, MLR analysis using these second-derivative absorbances was carried out to acquire the calibration curve for the sugar content of the imaging system.

11.3.3.4. Calibration by the imaging system

A calibration curve for the second-derivative absorbance and the sugar content of the imaging system was obtained by MLR analysis of 33 slices from the cylindrical samples (Figure 11.10), which is given below:

$$^\circ\text{Brix} = 19.01 - 438.84\, d^2A_{902} + 70.32\, d^2A_{874} \quad (11.11)$$

The second-derivative absorbance at 902 nm had an inverse correlation with the sugar content in Equation (11.11), which is the same as in Equation (11.8). Equation (11.11) had a high correlation of $R = 0.891$, and the standard error of calibration was 1.090. Therefore it can be considered that the imaging system adopted has sufficient capability to visualize the sugar content.

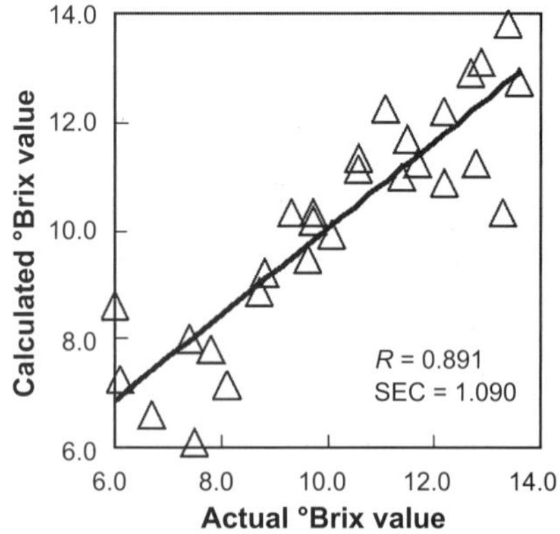

FIGURE 11.10 *Calibration for absorbance and the sugar content by imaging*

11.3.4. Visualization of Sugar Distribution by Its Absorption Band

The intensity of each pixel on the half-cut sample image was converted into the second-derivative absorbances at 902 and 874 nm in the same manner as in the case of cylindrical samples (Figure 11.11). The acquired calibration curve was applied to these two second-derivative absorbances at each pixel in order to calculate the sugar content. The sugar content was then visualized by mapping the value with a linear color scale. Original image

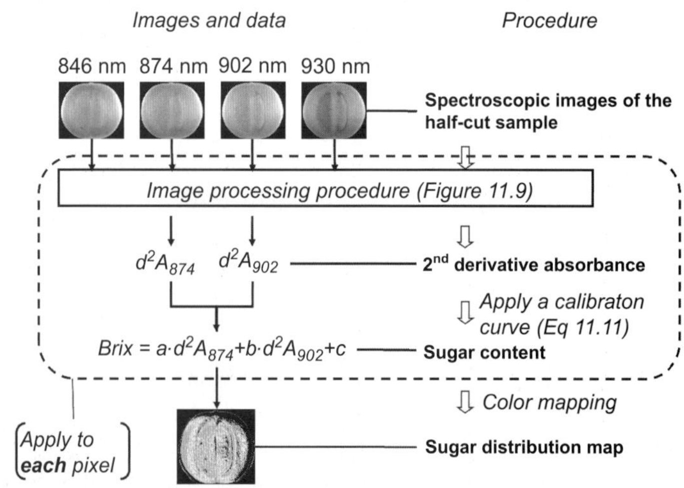

FIGURE 11.11 *Visualization procedure. (Full color version available on http://www.elsevierdirect.com/companions/9780123747532/)*

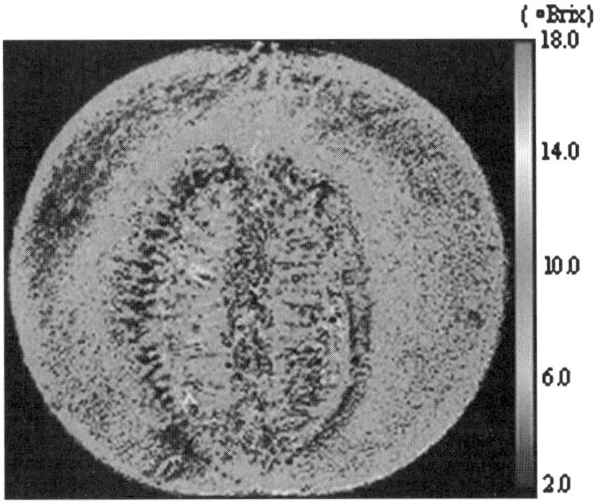

FIGURE 11.12 *Sugar distribution map of a half-cut red-flesh melon. (Full color version available on http://www.elsevierdirect.com/companions/9780123747532/)*

processing software was utilized to process images and to construct a sugar distribution map.

A sugar distribution map of the half-cut red-flesh melon was constructed by applying Equation (11.11) to each pixel of the processed images (Figure 11.11). In Figure 11.12, sugar contents ranging from 2 to 18 °Brix were assigned a linear color scale. The color changes gradually from blue to red as the sugar content increases. Although it was difficult to differentiate the sugar distribution by the naked eye, Figure 11.12 shows that the sugar content increases from the rind to near the seeds. It also indicates that the central upper part of the sample was sweeter than the bottom part, which is the reverse of the general notion in Japan. These results suggest that NIR imaging could become a useful method for evaluating the distribution of sugar in melons. In addition, further studies may lead to the application of this method not only to various varieties of melons but also to other constituents of other agricultural products because it does not depend on color information.

11.4. CONCLUSIONS

The relationship between the sugar content and absorption spectra can be investigated by using a near infrared (NIR) spectrometer to visualize the sugar content of a melon. The absorbance at 676 nm, which is close to the absorption band of chlorophyll, exhibited a strong inverse correlation with the sugar content. A high-resolution cooled charge-coupled device (CCD) imaging camera fitted with a band-pass filter of 676 nm was used to capture the spectral absorption image. The calibration method was used for converting the absorbance values on the image into the °Brix sugar content in accordance with NIR spectroscopy techniques. When this method was applied to each pixel of the absorption image, a color distribution map of the sugar content could be constructed.

In addition, a method for visualizing the sugar content based on the sugar absorption band was also developed. This method can avoid bias caused by the color information of a sample. NIR spectroscopic analysis revealed that each of the two second-derivative absorbances at 874 and 902 nm had a high correlation with the sugar content of melons. A high-resolution cooled CCD camera with band-pass filters, which included the above two wavelengths, was used to capture the spectral absorption image of a half-cut melon. A color distribution map of the sugar content on the surface of the melon was constructed by applying the NIR spectroscopy theory to each pixel of the acquired images. As a result, NIR spectroscopy theory can be extended to imaging applications with a high resolution CCD camera. Constructing the calibration method by the imaging system is the key point of this method because it is impossible to measure the actual sugar content of each pixel. Because an indium gallium arsenide (InGaAs) camera that can detect longer wavelengths (900–1600 nm) is available nowadays, wider applications can be expected using hyperspectral imaging techniques.

NOMENCLATURE

Symbols

A	absorbance
A_i	absorbance at i nm
d^2A_i	second-derivative absorbance at i nm
$f(x)$	spectral function at wavelength of x
$f^2(x)$	second-derivative spectral function at wavelength of x
I	intensity of reflection of a sample
Is	intensity of reflection of a white standard board
k	derivative gap
M	average intensity value of pixels of flat field after dark frame subtraction
R	intensity of the reflection of each pixel in a processed image

Abbreviation

A/D	analog-to-digital
CCD	charge-coupled device
InGaAs	indium gallium arsenide
NIR	near-infrared
PC	personal computer

REFERENCES

Bertrand, D., Robert, P., Novales, B., & Devaux, M. (1996). Chemometrics of multichannel imaging. In A. M. C. Davies, & P. Williams (Eds.), *Near infrared spectroscopy: the future waves* (pp. 174–178). Chichester, UK: NIR Publications.

Fukushima, H. (1996). How to use the CCD camera. In *Reikyaku CCD nyuumon* (pp. 73–132). Tokyo: Seibundou Sinkousha.

Hasegawa, Y. (2000). Merits and demerits of the automated sweetness sorting techniques. *Fresh Food System, 30*, 74–77.

Ishigami, K., & Matsuura, H. (1993). Studies on the sugar distribution and composition of muskmelon fruit. *Bulletin of Shizuoka Agricultural Experiment Station, 37*, 33–40.

Ito., M., Iida, J., Terashima, A., & Kishimoto, T. (1996). Non-destructive sugar content measuring method (Japanese Patent). JP 08–327536 A.

Iwamoto, M., Kawano, S., & Uozumi, J. (1994). Data processing method. In *Kin-Sekigai Bunkouhou Nyuumon* (pp. 62–95). Tokyo, Japan: Saiwai Shobou.

Izumi, H., Ito, T., & Yoshida, Y. (1990). Seasonal changes in ascorbic acid, sugar and chlorophyll contents in sun and shade leaves of satsuma mandarin and their interrelationships. *Journal of the Japanese Society for Horticultural Science, 59*, 389–397.

Katsumoto, Y., Jiang, J., Berry, R. J., & Ozaki, Y. (2001). Modern pretreatment methods in NIR spectroscopy. *Near Infrared Analysis, 2*, 29–36.

Kawano, S., & Abe, H. (1995). Development of a calibration equation with temperature compensation for determining the Brix value in intact peaches. *Journal of Near Infrared Spectroscopy, 3*, 211–218.

Kawano, S., Fujiwara, T., & Iwamoto, M. (1993). Nondestructive determination of sugar content in satsuma mandarin using near infrared (NIR) transmittance. *Journal of the Japanese Society for Horticultural Science, 62*, 465–470.

Kawano, S., Watanabe, H., & Iwamoto, M. (1992). Determination of sugar content in intact peaches by near infrared spectroscopy with fiber optics in interactance mode. *Journal of the Japanese Society for Horticultural Science, 61*, 445–451.

Martinsen, P., & Schaare, P. (1998). Measuring soluble solid distribution in kiwifruit using near-infrared imaging spectroscopy. *Postharvest Biology and Technology, 14*, 271–281.

Morimoto, S., McClure, W. F., & Stanfield, D. L. (2001). Hand-held NIR spectrometry. Part I: An instrument based upon gap-second derivative theory. *Applied Spectroscopy, 55*, 182–189.

Morita, K., Shiga, T., & Taharazako, S. (1992). Evaluation of change in quality of ripening bananas using light reflectance technique. *Memoirs of Faculty of Agriculture, Kagoshima University, 28*, 125–134.

Nussberger, S., Dekker, J. P., Kuhlbrandt, W., van Bolhuis, B. M., van Grondelle, R., & van Amerongen, H. (1994). Spectroscopic characterization of three different monomeric forms of the main chlorophyll a/b binding protein from chloroplast membranes. *Biochemistry, 33*, 14775–14783.

Osborne, B. G., & Feam, T. (1986). Near infrared data handling and calibration by multiple linear regression. In *Near infrared spectroscopy in food analysis*. Harlow, UK: Longman Scientific & Technical.

Robert, P., Bertrand, D., Devaux, M. F., & Sire, A. (1992). Identification of chemical constituents by multivariate near infrared spectral imaging. *Analytical Chemistry, 24*, 664–667.

Robert, P., Devaux, M. F., & Bertrand, D. (1991). Near infrared video image analysis. *Sciences des Aliments, 11*, 565–574.

SBIG. (1998). *Product Catalog*. Santa Babara, CA: SBIG Astronomical Instruments.

Sugiyama, J. (1999). Visualization of sugar content in the flesh of a melon by near-infrared imaging. *Journal of Agricultural and Food Chemistry, 47*, 2715–2718.

Taylor, S. K., & McClure, W. F. (1989). NIR imaging spectroscopy: measuring the distribution of chemical components. In M. Iwamoto, & S. Kawano (Eds.), *Proceedings of the 2nd International NIRS Conference* (pp. 393–404). Tokyo Japan: Korin.

Temma, M., Hanamatsu, K., & Shinoki, F. (1999). Development of a compact near-infrared apple-sugar-measuring instrument and applications. In H. Tsuyuki (Ed.), *Proceedings of the 15th Symposium on Non-destructive Measurements* (pp. 113–117). Ibaraki, Japan: Japanese Society for Food Science and Technology.

Tsuta, M., Sugiyama, J., & Sagara, Y. (2002). Near-infrared imaging spectroscopy based on sugar absorption band for melons. *Journal of Agricultural and Food Chemistry, 50*, 48–52.

Watada, A. E., Norris, K. H., Worthington, J. T., & Massie, D. R. (1976). Estimation of chlorophyll and carotenoid contents of whole tomato by light absorbance technique. *Journal of Food Science, 41*, 329–332.

CHAPTER 12

Measuring Ripening of Tomatoes Using Imaging Spectrometry

Gerrit Polder, Gerie van der Heijden
Wageningen UR, Biometris, Wageningen, The Netherlands

CONTENTS

12.1. INTRODUCTION

12.1.1. Tomato Ripening

Tomatoes, with an annual production of 60 million tons, are one of the main horticultural crops in the world, with 3 million hectares planted every year. Tomatoes (*Lycopersicum esculentum*) are widely consumed either raw or after processing.

Tomatoes are known as health-stimulating fruits because of the antioxidant properties of their main compounds (Velioglu *et al.*, 1998). Antioxidants are important in disease prevention in plants as well as in animals and humans. Their activity is based on inhibiting or delaying the oxidation of biomolecules by preventing the initiation or propagation of oxidizing chain reactions (Velioglu *et al.*, 1998). The most important antioxidants in tomato are carotenes (Clinton, 1998) and phenolic compounds (Hertog *et al.*, 1992). Amongst the carotenes, lycopene dominates. The lycopene content varies significantly with ripening and with the variety of the tomato and is mainly responsible for the red color of the fruit and its derived products (Tonucci *et al.*, 1995). Lycopene appears to be relatively stable during food processing and cooking (Khachik *et al.*, 1995; Nguyen & Schwartz, 1999). Epidemiological studies have suggested a possible role for lycopene in protection against some types of cancer (Clinton, 1998) and in the prevention of cardiovascular disease (Rao & Agarwal, 2000). Blum *et al.* (2005) suggest that a hypocholesterolemic effect can be inhibited by lycopene. The second important carotenoid is β-carotene, which is about 7% of the total carotenoid content (Gould, 1974). The amount of carotenes as well as their antioxidant activity is significantly influenced by the tomato variety (Martinez-Valverde *et al.*, 2002) and maturity (Arias *et al.*, 2000; Lana & Tijskens, 2006).

Ripening of tomatoes is a combination of processes including the breakdown of chlorophyll and build-up of carotenes. Chlorophyll and carotenes have specific, well-known reflection spectra. Using knowledge of the known spectral properties of the main constituent compounds, it may be possible to calculate their concentrations using spectral measurements. Arias *et al.* (2000) found a good correlation between color measurements using a chromameter and the lycopene content measured by high-performance liquid chromatography (HPLC). In order to be able to sort tomatoes according to the distribution of their lycopene and chlorophyll content, a fast on-line imaging system is needed that can be placed on a conveyor-belt sorting machine.

12.1.2. Optical Properties of Tomatoes

Optical properties of objects in general are based on reflectance, transmittance, absorbance, and scatter of light by the object. The ratio of light reflected from a surface patch to the light falling onto that patch is often referred to as the bi-directional reflectance distribution function (BRDF) (Horn, 1986) and is a function of the incoming and outgoing light direction. The BRDF depends on the material properties of the object. Material properties vary from perfect diffuse reflection in all directions (Lambertian surface), to specular reflection mirrored along the surface normal, and are wavelength-dependent.

The physical structure of plant tissues is by nature very complex. In Figure 12.1 a broad outline of possible interactions of light with plant tissue is given. Incident light which is not directly reflected interacts with the structure of the different cells and the biochemicals within the cells. The biochemical chlorophyll, the major component in the plant's photosynthesis system, is especially important for the color of a plant. Chlorophyll strongly absorbs the red and blue part of the spectrum and it reflects the green part, hence causing the observed green color. The absorbed light energy is used for carbon fixation, but a portion of the absorbed light can be emitted again as light at a lower energy level, i.e. of higher wavelength. This process is called fluorescence. Fluorescence is much lower in intensity than reflection and is difficult to distinguish from regular reflection under white light conditions. So in general diffuse reflectance is responsible for the observed color of the product. The more cells are involved in reflectance, the more useful is the chemometric information that can be extracted from the reflectance spectra.

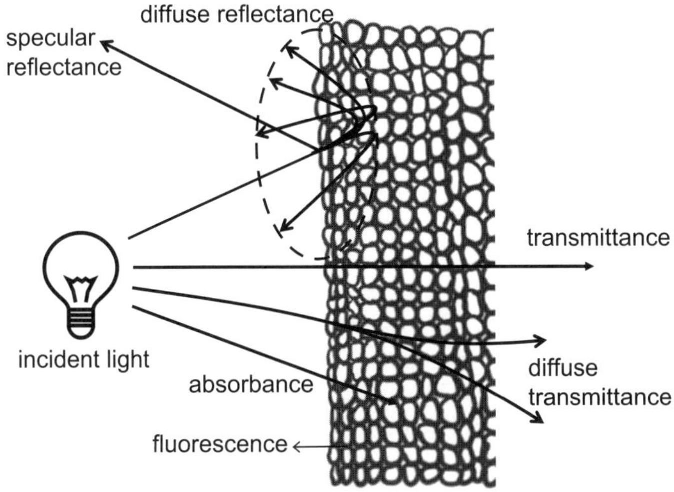

FIGURE 12.1 *Incident light on the tissue cells of tomatoes results in specular reflectance, diffuse reflectance, (diffuse) transmittance, and absorbance. These strongly depend on properties such as tomato variety and maturity and the wavelength of the light*

Instead of measuring diffuse reflectance, it is also possible to measure transmittance. In that case chemometric information of the whole interior of a tomato can be determined, but high incident light intensities are needed. Also, spatial information is disturbed by the scattering of light in the object.

Abbott (1999) gives a nice overview of quality measurement methods for fruits and vegetables, including optical and spectroscopic techniques. According to Birth (1976), when harvested food, such as fruits, are exposed to light, depending on the kind of product and the wavelength of the light, about 4% of the incident light is reflected at the outer surface, causing specular reflection. The remaining 96% of incident light is transmitted through the surface into the cellular structure of the product where it is scattered by the small interfaces within the tissue or absorbed by cellular constituents.

12.2. HYPERSPECTRAL IMAGING COMPARED TO COLOR VISION

12.2.1. Measuring Tomato Maturity Using Color Imaging

Traditionally, the surface color of tomatoes is a major factor in determining the ripeness of tomato fruits (Arias *et al.*, 2000). A color-chart standard has been specifically developed for the purpose of classifying tomatoes in 12 ripeness classes (The Greenery, Breda, The Netherlands). For automatic sorting of tomatoes, RGB color cameras are used instead of the color chart (Choi *et al.*, 1995). RGB-based classification, however, strongly depends on recording conditions. Next to surface and reflection/absorption characteristics of the tomato itself, the light source (illumination intensity, direction, and spectral power distribution), the characteristics of the filters, the settings of the camera (e.g. aperture), and the viewing position, all influence the final RGB image. Baltazar *et al.* (2008) added the concept of data fusion of acoustic impact measurements to colorimeter tests. A Bayesian classifier considering a multivariate, three-class problem reduces the classification error of single colorimeter measurements considerably. Schouten *et al.* (2007) also added firmness measurements to the tomato ripening model. They state that, in practice, knowledge of the synchronization between color and firmness might help growers to adapt their growing conditions to their greenhouse design so as to produce tomatoes with a predefined color–firmness relationship. Also, color measurements of tomatoes should suffice to assess the quality once the synchronization is known according to Schouten *et al.* (2007). Lana *et al.* (2006) used RGB measurements to build a model in order to describe and simulate the behavior of the color aspects of tomato slices as a function of the ripening stage and the applied storage temperature.

12.2.2. Measuring Tomato Maturity Using Hyperspectral Imaging

Van der Heijden *et al.* (2000) has shown that color information in hyperspectral images can be made invariant to recording conditions as described above, thus providing a powerful alternative to RGB color cameras. In this way, a hyperspectral imaging system and spectral analysis would permit the sorting of tomatoes under different lighting conditions. Polder *et al.* (2002) compared ripeness classification of hyperspectral images with standard RGB images. Hyperspectral images had been captured under different lighting conditions. By including a gray reference in each image, automatic compensation for different light sources had been obtained. Five tomatoes (Capita F1 from De Ruiter Seeds, Bergschenhoek, The Netherlands) in ripeness stage 7 (orange) were harvested. The ripeness stage was defined using a tomato color chart standard (The Greenery, Breda, The Netherlands), which is commonly used by growers. Each day over a time period of 5 days, color RGB images and hyperspectral images were taken of the five fruits on a black velvet background. The imaging spectrograph used in the experiment was the ImSpector (Spectral Imaging Ltd., Oulu, Finland) type V7 with a spectral range of 396 to 736 nm and a slit size of 13 μm resulting in a spectral resolution of 1.3 nm. The hyperspectral images were recorded using halogen lamps with a relatively smooth emission between 380 and 2000 nm.

Full-size hyperspectral images are large. If the full spatial resolution of the camera (1320 × 1035 pixels) for the *x*-axis and spectral axis was used, and with 1320 pixels in the *y*-direction, a single hyperspectral image would be 3.6 GB (using 16 bits/pixel). Due to limitations in lens and ImSpector optics, such a hyperspectral image is oversampled and binning can be used to reduce the size of the image without losing information (Polder *et al.*, 2003a).

After image preprocessing in which different tomatoes are labeled separately and specular parts in the image are excluded, 200 individual pixels were randomly taken from each tomato. In the case of the RGB image each pixel consists of a vector of red, green, and blue reflection values, whereas each pixel in the hyperspectral images consists of a 200-dimensional vector of the reflection spectrum between 487 and 736 nm.

Each consecutive day is treated as a different ripeness stage. Using linear discriminant analysis (LDA) (Fukunaga, 1990; Ripley, 1996) pixels were classified into the different ripeness stage (days) using cross-validation.

Scatter plots of the LDA mapping to two canonical variables for the RGB (Figure 12.2) and hyperspectral images (Figure 12.3) show considerable overlap at the different time stages for RGB; for the hyperspectral images this overlap is considerably reduced. The error rates for five individual tomatoes are tabulated in Table 12.1. From this table, it can be seen that the error rate varies from 0.48 to 0.56 with a standard deviation of 0.03 for RGB. For hyperspectral images the error rate varies from 0.16 to 0.20 with a standard deviation of 0.02. It should be noted that Table 12.1 shows the results for individual tomato pixels. When moving from pixel classification to object classification, only one tomato RGB image was misclassified, whereas each hyperspectral image was properly classified. Object classification was performed by a simple majority vote (i.e. each object was assigned to the class with the highest frequency of individually assigned pixels). These results show that for classifying ripeness of tomato, hyperspectral images have a higher discriminating power compared to regular color images.

In hyperspectral images there is variation that is not caused by object properties such as the concentration of biochemicals, but by external aspects, such as aging of the illuminant, the angle between the camera and the object surface, and light and shading. Using the Shafer reflection model (Shafer, 1985), hyperspectral images can be corrected for variation in illumination and sensor sensitivity by dividing for each band the reflectance at

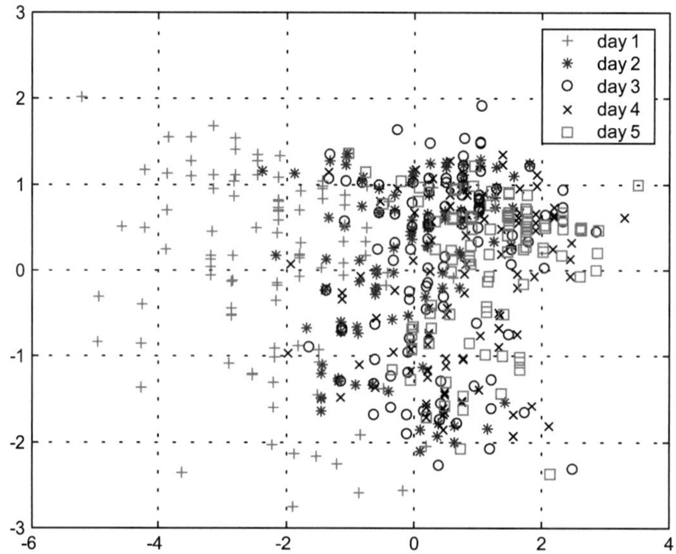

FIGURE 12.2 *Scatter plot of the first and second canonical variables (CV) of the LDA analysis of the RGB images. Classes 1 to 5 represent the ripeness stages of one tomato during the five days after harvest, respectively. (Full color version available on http://www.elsevierdirect.com/companions/9780123747532/)*

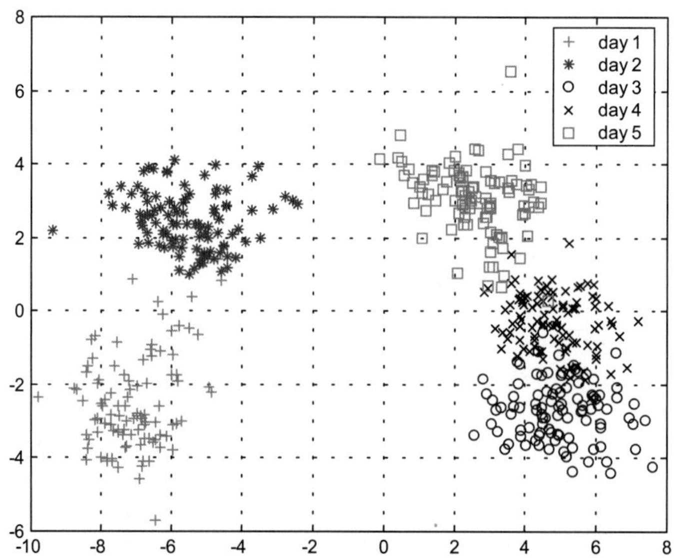

FIGURE 12.3 *Scatter plot of the first and second canonical variables (CV) of the LDA analysis of the hyperspectral images. Classes 1 to 5 represent the ripeness stages of one tomato during the five days after harvest, respectively. (Full color version available on http://www.elsevierdirect.com/companions/9780123747532/)*

Table 12.1 Error rates for RGB and hyperspectral pixel classification of five individual tomatoes

Tomato	Error rate for RGB	Error rate for hyperspectral
A	0.50	0.18
B	0.56	0.20
C	0.48	0.18
D	0.54	0.16
E	0.48	0.20
Mean	0.51	0.19
Standard deviation	0.03	0.02

every pixel by the corresponding reflectance of a white or grey reference object. The images are now color-constant. When the spectra are also normalized (e.g. by dividing for every pixel the reflectance at each band by the sum over all bands), the images become independent for object geometry and shading. In order to test the classification performance under different recording conditions, Polder *et al.* (2002) used four different light sources, namely:

- tungsten–halogen light source;
- halogen combined with a Schott KG3 filter in front of the camera lens;
- halogen with an additional TLD58W (Philips, The Netherlands) fluorescence tube; and
- halogen with an additional blue fluorescence tube (Marine Blue Actinic, Arcadia, UK).

As the aim was to classify the tomatoes correctly, irrespective of the light source used, classification was carried out on color-constant and normalized color-constant images which were calculated using the spectral information of a white reference tile. Table 12.2 shows the error rates. These results indicate that hyperspectral images are reasonably independent of the light source.

Variations in lighting conditions such as intensity, direction and spectral power distribution, are the main disturbing factors in fruit sorting applications. Traditionally, these factors are kept constant as much as possible. This is very difficult, since illumination is sensitive to external factors such as temperature and aging. In addition, this procedure does not guarantee identical results using various machines, each equipped with different

Table 12.2 Error rates for individual pixels of hyperspectral images captured with different illumination sources, using raw, color-constant, and color-constant normalized spectra. The training pixels were captured with halogen illumination

Illumination	Raw	Color-constant	Normalized color constant
Halogen	0.19	0.19	0.19
Kg3 filter	0.80	0.35	0.36
Halogen/TLD	0.41	0.35	0.34
Halogen/blue	0.42	0.36	0.33

cameras and light sources. Calibration of machines is tedious and error-prone. By using color-constant hyperspectral images the classification becomes independent of recording conditions such as the camera and light source, as long as the light source is regularly measured (e.g., by recording a small piece of white or gray reference material in every image). It should be noted that comparing tomatoes with very limited maturity differences was a rather demanding problem. From Table 12.2 it can be seen that, although the error rate increases from 0.19 to 0.36 when using different light sources, it is still considerably below the 0.51 for RGB under the same light source. Nevertheless, an error rate of 0.36 is still very high. The main reasons for this high error rate are the rather small differences in maturity (one-day difference) and non-uniform ripening of the tomato. If tomatoes are classified as whole objects, using majority voting of the pixels, all tomatoes are correctly classified based on the hyperspectral images, and only one tomato is wrongly classified using the RGB images. Another aspect is that the assumption of uniform ripening of a single tomato is not fully valid and that different parts of the same tomato may have a slightly different maturity stage.

Tomatoes are spherical objects with a shiny, waxy skin. Since high intensity illumination is required for hyperspectral imaging, it is almost impossible to avoid specular patches on the tomato surface. Pixels from these specular patches do not merely show the reflection values of the tomato, but also exhibit the spectral power distribution of the illumination source. To avoid disturbance from this effect, preprocessing the images is needed to discard these patches. In the normalized hyperspectral image, the color difference due to object geometry has also been eliminated. When using normalized images, the color is independent of the surface normal, the angle of incident light, the viewing angle, and shading effects, as long as sufficient light is still present and under the assumption of non-specularity. The results indicate that the normalized hyperspectral images yield at least the same results as, if not better than, the color-constant hyperspectral images.

Since a tomato fruit is a spherical object, the above-mentioned effects play a role in the images. Because the training pixels were randomly taken from the whole fruit surface, the positive effect of normalization could possibly be achieved in the color-constant images using linear discriminant analysis. In situations where the training pixels are taken from positions on the tomato surface that are geometrically different from the validation pixels, it is expected that normalized hyperspectral images would give a better result than color-constant spectra.

Since the normalized images do not perform worse than the color-constant images, in general normalization is preferred, which corrects for differences in object geometry. However care should be taken not to include specular patches. The accuracy of hyperspectral imaging appeared to suffer slightly if different light sources were used. Under all circumstances, however, the results were better than those for RGB color imaging under a constant light source. This opens possibilities to develop a sorting machine with high accuracy that can be calibrated to work under different conditions of light source and camera.

12.2.3. Classification of Spectral Data

In Section 12.2.2 Fisher linear discriminant analysis (LDA) was used for classification of the RGB and spectral data. This classification method is straightforward and fast, and suitable for comparing classification of RGB images with hyperspectral images. However, other classifiers might perform better.

An experiment was conducted (Polder, 2004) to compare the Fisher LDA (fisherc) with the nearest mean classifier (nmc) (Fukunaga, 1990; Ripley, 1996) and the Parzen classifier (parzenc) (Parzen, 1962). The optimum smoothing parameter h for the Parzen classifier was calculated using the leave-one-out Lissack & Fu estimate (Lissack & Fu, 1972). Depending upon the size of the training set and the tomato analyzed, the value of h was between 0.08 and 0.19.

The data used in the above experiment (Polder, 2004) are a random selection of 1000 pixels from hyperspectral images of five tomatoes in five ripeness classes (total 25 images) as described in Section 12.2.2. For each classifier the classification error (error on the validation data) and the apparent error (error on the training data) as a function of the size of the training data were examined. The 1000 original pixels per tomato were split

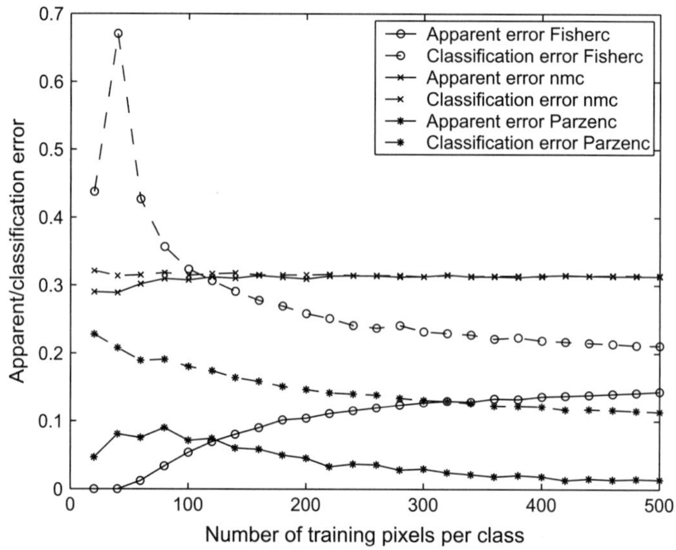

FIGURE 12.4 *Classification error and apparent error for Fisher LDA*

up in two parts of 500 pixels each for training and validation. The number of training pixels was varied between 20 and 500 pixels per class in steps of 20 pixels. The total experiment was repeated three times with each time a new random selection of 1000 pixels from each tomato. The average errors from these experiments are plotted in Figure 12.4.

From Figure 12.4, it can be seen that the nearest mean classifier (nmc) is less suitable for these data. The Parzen classifier performs much better than Fisher LDA. A drawback of the Parzen is that it is very expensive in terms of computing power and memory usage when this classifier is trained. For real-time sorting applications, however, classification speed is more important than training speed. For these three classifiers, classification speed depends mainly on the dimensionality of the data and hardly on the kind of classifier. In practice, calibration of the sorting system is regularly needed. Training the classifier is part of the calibration; therefore a classifier that can be quickly trained is preferable to slower ones.

Processing time for training the Fisher classifier with 500 pixels per class (2 500 total) was 12 seconds, for the nearest mean classifier this was less than 100 ms. Training the Parzen classifier took more than 400 seconds.

Another important conclusion that can be drawn from Figure 12.4 is that the number of training objects needs to be sufficiently high. When for instance 40 pixels are used for training the Fisher LDA classifier, the apparent error is zero, while the classification error is almost 0.7. This is due to the fact that when fewer training samples are used, the classifier is completely trained to the noise in the data. And when this trained classifier is applied to new data with other noise terms, the new noise causes the classifier to fail. For the Parzen classifier this effect is less distinct but it is clear that the classification error is smaller when a large number of training pixels is used.

12.3. MEASURING COMPOUND DISTRIBUTION IN RIPENING TOMATOES

As mentioned earlier, ripening of tomatoes is a combination of processes, including the breakdown of chlorophyll and build-up of carotenes. Polder *et al.* (2004) developed methods for measuring the spatial distribution of the concentration of these compounds in tomatoes using hyperspectral imaging. The spectral data were correlated with compound concentrations, measured by HPLC.

Tomatoes were grown in a greenhouse and harvested at different ripening stages, varying from mature green to intense red color, and scored by visual evaluation performed by a five-member sensory panel. The ripeness stage was determined using a tomato color chart standard (The Greenery, Breda, The Netherlands). The number of tomatoes used in the experiment was 37. After washing and drying the tomatoes thoroughly, hyperspectral images were recorded. Immediately after the recording of each tomato four circular samples of 16 mm diameter and 2 mm thickness were extracted from the outer pericarp, and after determination of the sample fresh weight, the samples were frozen in liquid nitrogen and stored for later HPLC processing to measure the lycopene, lutein, β-carotene, chlorophyll-*a* and chlorophyll-*b* concentrations. The hyperspectral images were made color-constant and normalized as described in Section 12.2.2. Savitsky-Golay smoothing (Savitsky & Golay, 1964) was used to smooth the spectra. The procedure was combined with first-order derivatives to remove the baseline of the spectra. Partial least square regression (PLS) (Geladi & Kowalski, 1986; Helland 1990) was used to relate the spectral information to the concentration information of the different compounds in the tomatoes. A bottom view hyperspectral image of each tomato was captured. In this image the center part is ignored because of possible specular reflection. In order to compare the variation in spectra-predicted concentrations with the variation in measured HPLC concentration, eight circular patches were defined on the tomato. The size of these patches was about the same as the size of the sample patches used in the HPLC analysis. From each of the eight patches, 25 spectra were extracted for the PLS regression. The total number of spectra extracted this way per tomato was 200. These spectra form the X-block in the PLS regression and cross-validation. The size of the contiguous blocks was also chosen to be 200. In this way the cross-validation acts as leave-one-out cross-validation on the whole tomatoes. In Figure 12.5 the hyperspectral predicted lycopene concentration is plotted against the observed concentration measured by HPLC. The root mean square error of prediction (RMSEP) for lycopene was 0.17. The RMSEP for the other compounds were 0.25, 0.24, 0.31, and 0.29 for lutein, β-carotene, chlorophyll-*a* and chlorophyll-*b*, respectively. This indicates that hyperspectral imaging allows us to estimate the compound concentration in a spatial preserving way. The PLS model is trained on a random selection of pixels. After the model has been trained it can be applied to the spectra of all pixels. The result is an image with gray values that stand for a certain concentration. The variation in gray values gives an idea about the spatial distribution of the compounds. Figure 12.6 shows the spatial distribution of the compounds on tomatoes with a manually scored maturity class of 2, 8, and 6, respectively.

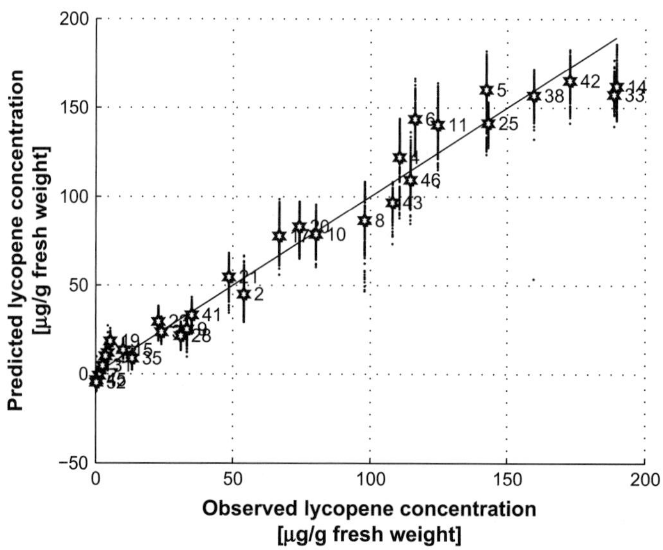

FIGURE 12.5 *Spectral predicted against real (HPLC) lycopene concentration of the tomato pixels. The mean of the pixels denoting the average concentration per tomato is indicated with a star*

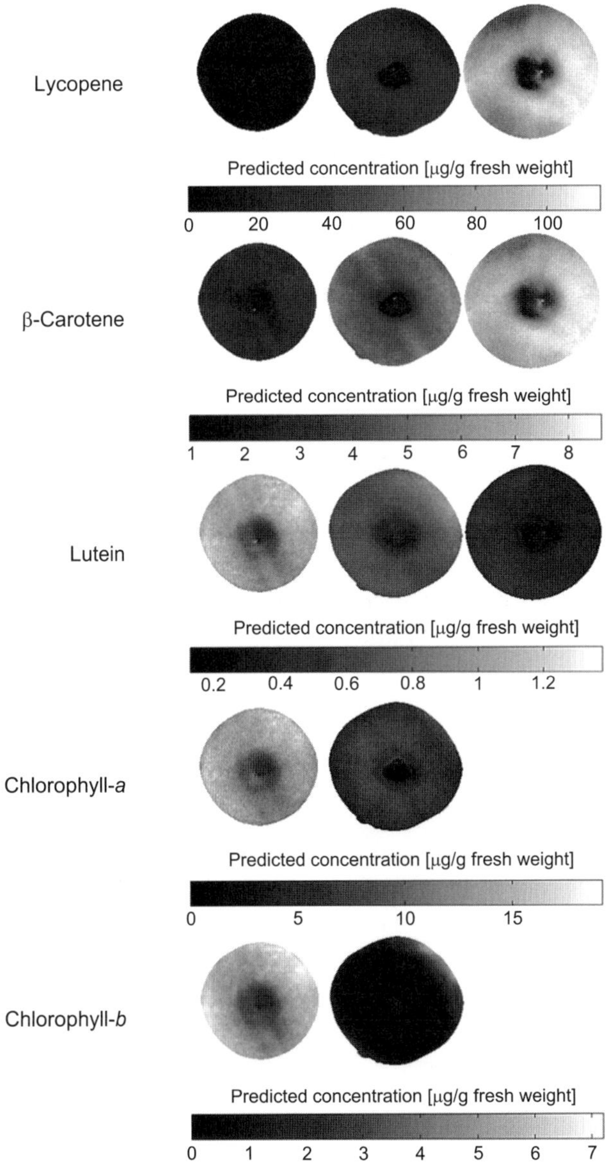

FIGURE 12.6 *Concentration images of the spatial distribution of compounds in three tomatoes. The corresponding maturity classes are 2, 6, and 8. The second and third tomato show non-uniform ripening on the edge of the images*

12.4. ON-LINE UNSUPERVISED MEASUREMENT OF TOMATO MATURITY

Much research found in the literature, including that described earlier in this chapter, is based on supervised techniques, where a regression or classification model is trained on hyperspectral images of tomatoes with known compound concentrations, expert score or other reference data. When this system is implemented in a real-time sorting machine two major steps can be distinguished in the total process: the calibration step and the sorting step.

- The first step is *calibrating* the system. Calibration refers to assessing the relationship between the hyperspectral data and the concentration of the compound of interest, for example lycopene. In our case the calibration objects are tomatoes of different maturity over the whole range of ripeness classes. Calibration of the system needs to be done each time something changes in the total system. This can be a change in sensors or light sources due to aging, or a new batch of tomatoes of different origin or variety. A standard procedure for calibration is to compare hyperspectral data with reference measurements such as those obtained with HPLC, expert score or color chart. Using the hyperspectral images and the result of the reference measurements a mathematical model is built, for instance regression (e.g. PLS) or classification (e.g. LDA).
- The second step in the total process is the real-time *sorting* step. This step needs to be very fast to produce sorting machines that are able to sort enough objects (tomatoes) per second in order to be economically feasible. Currently color-sorting machines are on the market which can sort up to 12 tomatoes per second in eight parallel lanes. For a hyperspectral sorting system the speed requirements are similar. In the sorting step, hyperspectral images of the tomatoes are first captured. These images are then mapped to an output result using the model that was calculated in the first step. Standard real-time imaging techniques can be applied on these images in order to calculate sorting criteria.

Calibration of hyperspectral images using chemical reference measurements is time-consuming and expensive and hampers practical applications. Thus the question arises whether a reference method is really needed in the calibration step, in order to train a regression model. In other words can unsupervised classification or regression be performed? For an initial calibration the answer is no, because a relationship is needed between the measured spectra and compound concentrations. However, for on-line calibration which corrects for changes in sensors or light sources, or a new batch of tomatoes of different origin or variety, this method might be suitable. If signals are to be separated (in our case the reflectance spectra of different compounds) from a set of mixed signals, without the aid of information, blind source separation (BSS) is the procedure commonly used. One of the most widely used methods for blind source separation is Independent Component Analysis (ICA) (Hyvärinen & Oja, 2000). Polder *et al.* (2003b) examined the applicability of ICA for on-line calibration purposes. An experimental laboratory setup was used to unravel the spectrum of the tomatoes in order to separately measure specific compounds using ICA. The results of this analysis are compared to compound concentrations measured by HPLC. The analysis was performed on the same dataset as detailed in Section 12.2.2. The ICA algorithm results in a number of independent component spectra and a mixing matrix which denotes the concentration of each component in the source spectrum, comparable to the scores and loadings in principal components analysis (PCA). It appeared that 99% of the variation was retained within the first two independent components. This indicates that probably only two major independent components can be found. When attempts were made to estimate more independent components the ICA algorithm did not converge.

HPLC analysis showed that lycopene and chlorophyll are the compounds with the highest concentration in the process of tomato ripening. The signals of the independent components (IC) that were found resemble more or less the actual absorption spectra of lycopene and chlorophyll, but there is some discrepancy (Figure 12.7 and 12.8). The transition between high and low lycopene absorption is round 550 nm in the real measured data, where as in IC-1 this transition is shifted to 600 nm. In IC-2 the chlorophyll absorption peek at 670 nm is clearly visible, but the high absorption around 430 nm in the reference spectra is shifted to 510 nm in IC-2. These shifts are possibly caused by other unknown compounds, or the effect of the solvent on the reference spectra. Besides ICA, a regular PCA was also performed. The relationship between the actual spectra and the principal components (PC) is slightly less clear: PC-1 has an extra peak at 670 nm compared to IC-1 and the actual lycopene spectrum. This gives the impression that ICA is more suitable for finding compound concentrations than PCA.

Since the ICA algorithm starts with a random weight vector, the optimization can stick in a local maximum. It appeared that in 80% of the cases the result was similar to that in Figure 12.7, in 20% of the cases the

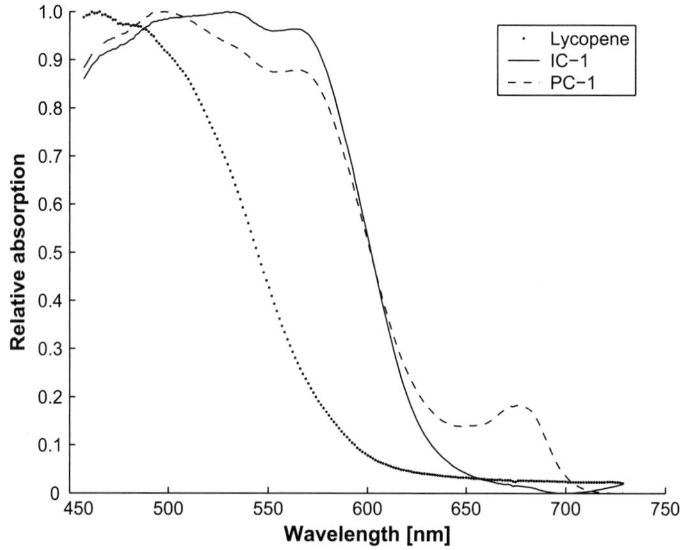

FIGURE 12.7 *Relative absorption spectrum of lycopene in acetone, IC-1, and PC-1. The spectra are scaled between 0 and 1*

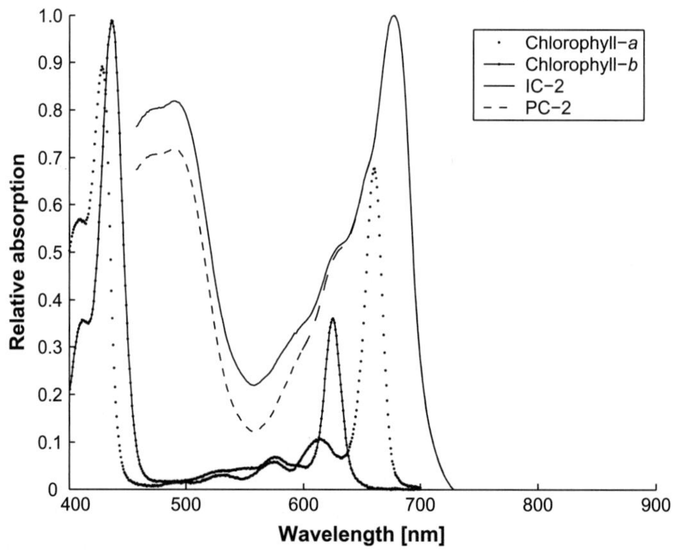

FIGURE 12.8 *Relative absorption spectrum of chlorophyll-a and chlorophyll-b in diethyl ether, IC-2, and PC-2. The spectra are scaled between 0 and 1*

independent components more or less resembled the principal components. The variation within these two solutions was almost zero. Therefore two clusters of solutions were found with small intra-cluster variation. The decision on which of the two solutions is the proper one can be ascertained by repeating the ICA algorithm several times and choosing the solution with the highest frequency, or by comparing the solution with the principal components, or the real compound spectra.

In Figure 12.9 independent component (IC) concentrations from the mixing matrix and the PCA scores are plotted as a function of the actual concentration of lycopene and chlorophyll measured with HPLC. In Figure 12.9, each point is one of the randomly selected pixels, and the numbers are the labels of the individual tomatoes. Tomatoes with zero concentration of one of the compounds were excluded from the figure. The chlorophyll concentration was obtained by summing the chlorophyll-*a* and chlorophyll-*b* concentrations. It can be seen that there is not much difference between the graphs, which is expected because there is also not so much difference between the ICs and PCs. The variation within IC-1 is slightly less then the variation in PC-1, indicating that ICA gives a better solution than PCA.

It can also be observed that the IC-1 is indeed related to lycopene and IC-2 to chlorophyll. However, the found concentration values of the independent components are not the real concentration values of the compounds. To relate the values found with real compound concentrations, a first-order linear fit of the mixing matrix on the logarithm of the HPLC concentrations was performed as an initial calibration. The performance of the on-line ICA calibration was tested using a leave-one-out cross-validation. For the lycopene concentration, the predicted percentage variation Q^2 was 0.78 for IC-1, while for the chlorophyll concentration Q^2 was 0.80 for IC-2. For the supervised method (Section 12.3) these values were 0.95 and 0.73, respectively.

By multiplying the independent components with all the pixels of the hyperspectral images, after restoring the spatial relationship between pixels, images of the distribution of concentration of the independent components can be obtained. Figure 12.10 shows concentration images of six tomatoes ranging from raw to overripe. Increase of the independent component IC-1 and decrease of the independent component IC-2, can clearly be seen in this figure. Spatial variation in the distribution of independent components is caused by non-uniform ripening. Real-time image analysis techniques on these two-dimensional concentration images can be applied in order to distinguish between uniform and non-uniform ripened tomatoes.

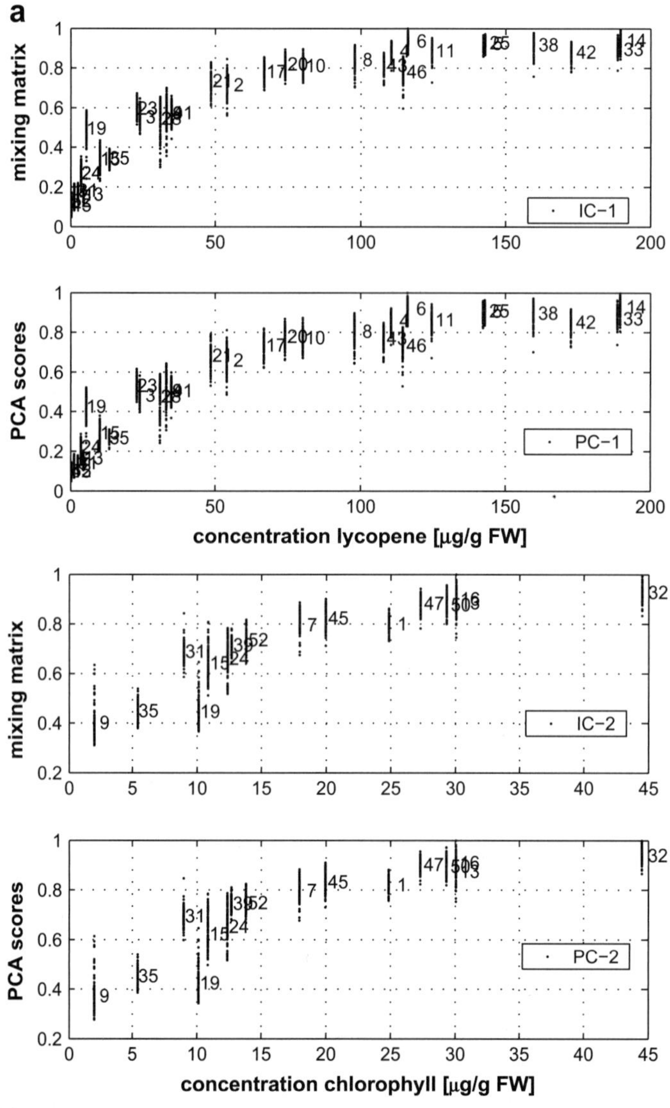

FIGURE 12.9 *Concentration of IC-1 and IC-2 from the mixing matrix and PCA scores as a function of concentrations of (a) lycopene and (b) chlorophyll determined by HPLC*

The described system can be implemented in a practical quality sorting system. A big advantage of this system compared to supervised systems is that fewer reference data for the calibration are needed. This makes this system easier, faster, and cheaper to use. However, for estimating concentrations of compounds, some sort of supervised calibration is still required.

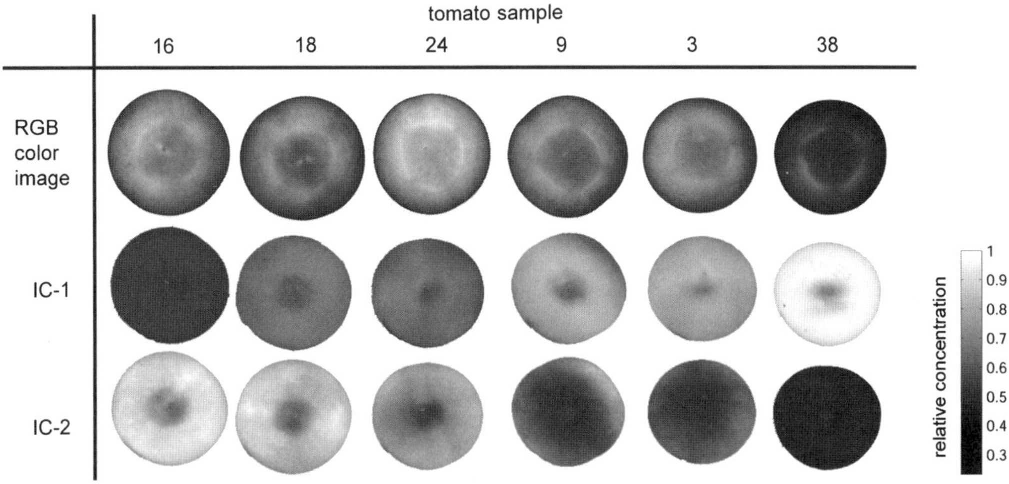

FIGURE 12.10 *Concentration images of IC-1 and IC-2 of six tomatoes ranging from raw to overripe. The labels correspond to the manual scored ripeness. (Full color version available on http://www.elsevierdirect.com/companions/9780123747532/)*

12.5. HYPERSPECTRAL IMAGE ANALYSIS FOR MODELING TOMATO MATURITY

12.5.1. Spectral Data Reduction

As discussed in Section 12.2, for sorting tomatoes, hyperspectral imaging is superior to RGB color imaging with three "spectral" bands. However, hyperspectral images with 200–300 bands are huge. Capturing and analyzing such data sets currently costs more computing power than that available in real-time sorting applications. Therefore an experiment was conducted to study the effect of reducing the number of bands, and ways to select bands that give the greatest discrimination between classes.

The data used in this experiment are the same as in Section 12.2. The Parzen classifier was used for classification. Table 12.3 shows the error rates

Table 12.3 Error rates for tomatoes 1 to 5 for a varying number of wavelength bands (features), using Parzen classification

	Error rate for tomato					
Spectra	**1**	**2**	**3**	**4**	**5**	**Processing time [s]**
186 bands (color constant normalized)	0.11	0.10	0.11	0.12	0.11	430
Smoothed (Gaus $\sigma = 2$)	0.09	0.10	0.12	0.09	0.08	418
Subsampled to 19 bands	0.08	0.10	0.09	0.07	0.08	120

for all five tomatoes. The original spectra, smoothed spectra, and spectra subsampled with a factor of 10 were analyzed. The processing time is the mean of the elapsed time needed for training the Parzen classifier per tomato.

It can be seen from Table 12.3 that the error slightly decreases when the spectra are smoothed, and decreases even more when the spectra are subsampled. From this it can be concluded that the spectra of the tomatoes are so smooth that the number of bands can very well be reduced by a factor of 10. Due to correlation between neighboring bands, reflection values are more or less the same. Hence taking means averages out the noise and increases performance. Besides, a lower dimensionality makes the classifier more robust. Since most biological materials have smooth reflection spectra in the visible region, it is expected that spectral subsampling or binning can be used in many real-time sorting applications. When subsampling or binning is carried out during image recording, both the acquisition and processing speed can be significantly improved. Further subsampling without selecting specific wavelengths does not improve the classification. An experiment was conducted with the number of bands being gradually reduced. Figure 12.11 shows the classification error as a function of the number of bands used. For this experiment the optimum number of bands is about 20.

When the number of bands can be reduced further, to three, four or five bands, other types of multispectral cameras can be used. Examples of these cameras are the four- or nine-band MultiSpec Agro-Imager (Optical Insights,

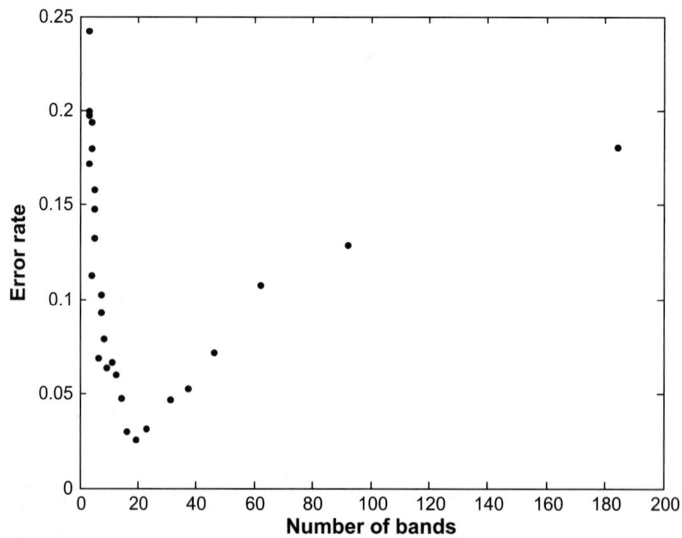

FIGURE 12.11 *Classification error as function of the number of bands used in the spectra*

Santa Fe, NM, USA) (Nelson, 1997) which can be equipped with user-selectable narrow-band filters. Hahn (2002) successfully applied the multi-spectral imager for predicting unripe tomatoes with an accuracy of over 85%. The Quest-Innovations Condor-1000 MS5 parallel imager is a high-quality smart CCD/CMOS (complementary metal-oxide semiconductor) multi-spectral camera with five spectral bands (www.quest-innovations.com). However, blind selection of broad-band filters does not give the optimal result. In order to successfully apply those cameras with a limited number of filters, it would be nice to have a method to select the optimal band-pass filters from the hyperspectral images. Optimal can be defined as selecting those bands which give a maximum separation between classes.

The technique of selecting the bands (features) is known as feature selection, and has been studied for several decades (Cover & Campenhout, 1977; Fu, 1968; Mucciardi & Gose, 1971). Feature selection consists of a search algorithm for finding the space of feature subsets, and an evaluation function which inputs a feature subset and outputs a numeric evaluation. The goal of the search algorithm is to minimize or maximize the evaluation function.

For selecting the best discriminating subset of k bands from a total of K bands, the number of possible combinations (n) is given by:

$$n = \binom{K}{k} = \frac{K!}{(K-k)!k!}$$

An exhaustive search is often computationally not practical since n can be large. In our case, with $K = 19$ and $k = 4$, n is 3 876 which is not very large, but when K increases, n will rapidly become too large. A feature selection method that avoids the exhaustive search and guarantees to find the global optimum is based on the branch and bound technology (Narendra & Fukunaga, 1977). This method can avoid an exhaustive search by using intermediate results for obtaining bounds on the final evaluation value. It only works, however, with monotonic evaluation functions.

An experiment was performed to test the branch and bound method, and the simple individual, forward and backward feature selection methods. As a criterion function, the sum of the estimated Mahalanobis distances was used (Ripley, 1996). The Mahalanobis distance is a monotonic criterion and therefore also suitable for the branch and bound algorithm. Again the same data as in Section 12.2 were used. Although for each tomato the five ripeness classes are different, the actual ripeness in each class is undefined. Also the initial ripeness for each tomato can be different. Therefore the tomatoes cannot be combined in the feature selection procedure.

Table 12.4 Sum of estimated Mahalanobis distances for different feature selection algorithms

Feature selection algorithm	Sum of estimated Mahalanobis distances	Computing time per tomato [s]
Individual	0.19	5
Branch and bound	0.13	1 200
Forward	0.14	20
Backward	0.15	55

The goal was to select four bands, for instance for the AgroImager (Nelson, 1997), with filters having a bandwidth of 10 nm. Such a setup can easily be implemented in a practical sorting application.

In Table 12.4 the results of the tested feature selection procedures are listed. The computing time per tomato was 5 s for the individual feature selection method, 20 s for forward feature selection, 55 s for backward feature selection and 1 200 s for the branch and bound algorithm. It appeared that, depending on the feature selection procedure and the optimization criterion, different bands are selected. The branch and bound algorithm gives the lowest error for all tomatoes, but the bands selected per individual tomato differ more from each other than with the other methods. This indicates that the found selection is rather specific for the tomato used in the selection procedure. This might also indicate that it will perform worse when this selection is applied to other tomatoes. Also the criterion function used influences the selected bands. Further optimization might be possible by using smaller or broader bands.

When this method is used for selecting filters for implementation in a three- or four-band multispectral camera with fixed filters, it is important to carry out the feature selection on the full range of possible objects that must be sorted in the future. This might not always be possible because the spectrum of the fruit is influenced by the variety and environmental conditions, which are subject to change over the years. Whether this is a problem can only be established on a large dataset covering all relevant variations. The gain in speed when switching from a 200-band hyperspectral system to a 4-band multispectral system comes at the expense of loss of flexibility.

12.5.2. Combining Spectral and Spatial Data Analysis

Hyperspectral imaging is also known by the term imaging spectroscopy. It has the advantage compared with point spectroscopy, that spatial information is available in addition to spectral information. From an image analysis point of view the information content per pixel increases from grayscale, to color, to multispectral, to hyperspectral images. In addition to spectral analysis of the pixels, image analysis can be applied to extract more information using the spatial relationship between the pixels. There are several approaches to combine spectral and spatial information. Without giving a complete taxonomy of all available methods, these approaches can be subdivided into sequential, parallel, and integrated methods.

12.5.2.1. Sequential spectral and spatial classifiers

Spatial information can be used for preprocessing the hyperspectral images in order to select those pixels that are required for further (spectral) analysis. Image processing on the sum of the spectral band images or on a single selected band image with high signal-to-noise ratio can already distinguish, for instance, between object, background, and specular parts. The result of subsequent spectral classification or regression can be a labeled image with the different (maturity) classes, or a gray value image with perhaps concentration values.

A simple form of spatial postprocessing is to use a "pepper and salt" filter (Ritter & Wilson, 2000) on a spectrally classified image to remove isolated (probably wrongly classified) pixels. When spectral regression is used to obtain a gray value image or "chemical" images, where the spatial distribution of the concentration of a certain compound in the object is displayed, spatial postprocessing on these images can be used to extract object features such as uniformity of concentration. In Figure 12.12 these steps are depicted in a flowchart.

12.5.2.2. Parallel spectral and spatial classifiers

Instead of performing the image and spectral processing sequentially they can be performed in parallel. In this way the same input data are used for parallel operating classifiers. After spectral and spatial classification, the results of both classifiers will be combined. The whole process can be carried out in an iterative way until the combined classifier gives a stable result. An example of this approach is depicted in Figure 12.13. This approach, described by Paclik *et al.* (2003), was used to classify material in eight-band multispectral images of detergent laundry powders acquired by scanning electron microscopy.

To investigate the feasibility of this approach for our application, an experiment was conducted using the method described by Paclik *et al.* (2003). The data in this experiment were from the hyperspectral imaging of four tomatoes of different maturity (Figure 12.14). The visually scored maturity using a tomato color chart standard (The Greenery, Breda, The Netherlands)

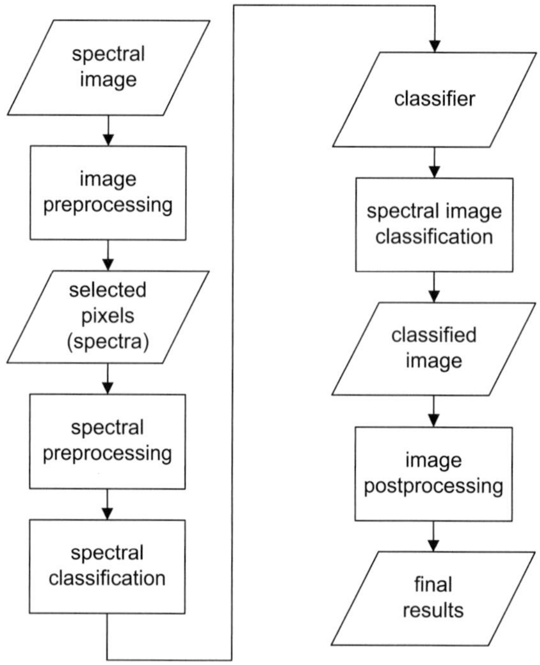

FIGURE 12.12 *Flowchart of hyperspectral image classification steps, where image processing and spectral processing are performed sequentially*

was 1 (green), 4 (green–orange), 8 (orange–red), and 12 (red), respectively. The size of the hyperspectral image was 128 × 128 pixels, with each pixel consisting of 80 wavelength bands, between 430 and 900 nm. The idea was to test whether the classification of ripeness using this combined classifier could be improved. The processing started with an initial segmentation to separate the background and the specular parts into different classes (total six) for each tomato. Improvements could be seen; for instance in tomato 2 (Figure 12.14, *upper right*), which is a combination of green and orange pixels. Also the classification of the specular reflection, which was initially based on a simple threshold of the sum of all bands, might be improved when using a combined classifier on the whole hyperspectral image.

Fisher classification is used as a spectral classifier, with the wavelength bands as features. In order to lessen computing time, the number of bands was reduced by a factor of four by convolving the spectrum with a Gaussian window ($\sigma = 1.5$) and subsequent subsampling. The first test was performed using only the spectral classifier without a spatial classifier.

Figure 12.15 shows the initial labeling and the result after 50 and 500 iterations. Figure 12.16 shows the label changes as a function of the iteration

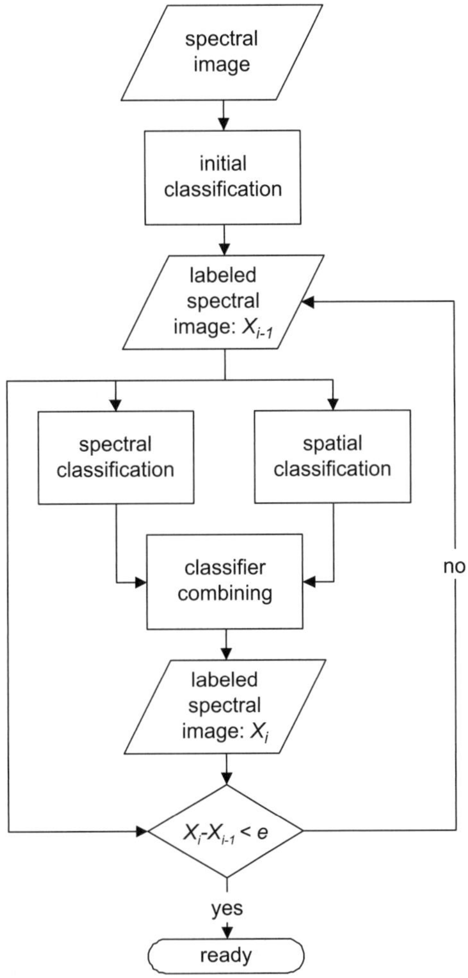

FIGURE 12.13 *Flowchart of hyperspectral image classification steps, where image processing and spectral processing are combined*

number. The results indicate that a repeated spectral classifier does not converge to a stable solution. After 500 iterations the specular class is grown into tomato 2, and the tomato 2 class is grown into the background.

The question now is whether a stable solution can be reached when the spectral classifier is combined with a spatial classifier. This was tested by adding a Parzen spatial classifier using the *x*, *y* coordinates as features. Since the features of the spatial classifier are independent of the features of the spectral classifier, the probabilities can simply be multiplied. The resulting labeling after 10, 25, and 500 iterations is shown in Figure 12.17.

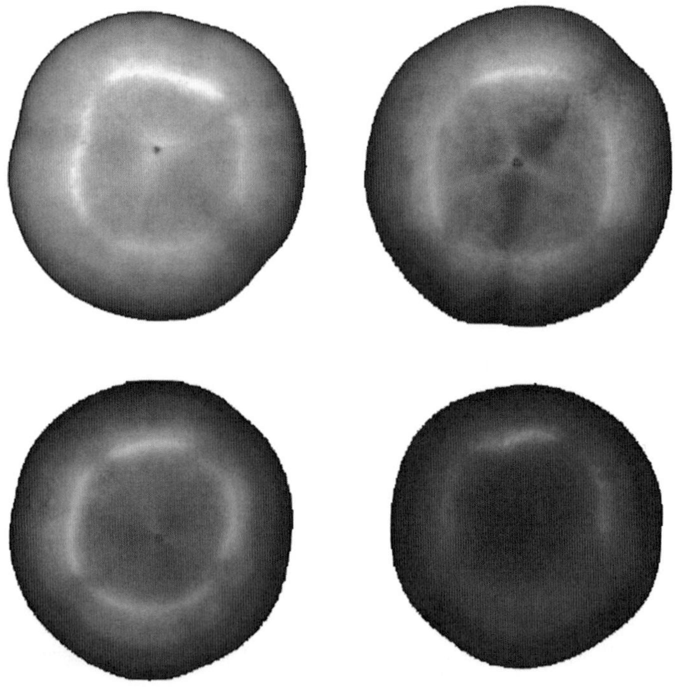

FIGURE 12.14 *RGB image of four tomatoes of different maturity. (Full color version available on http://www.elsevierdirect.com/companions/9780123747532/)*

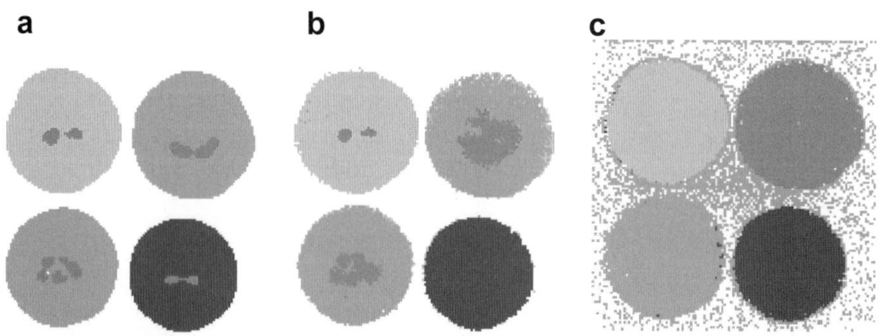

FIGURE 12.15 *Comparison of a spectral classifier (Fisher): (a) initial labeling; (b) labeling after 50 iterations, (c) labeling after 500 iterations. (Full color version available on http://www.elsevierdirect.com/companions/9780123747532/)*

Figure 12.18 shows the label changes as a function of the iteration number. Compared with Figure 12.16 the number of label changes converges to ≈1000, but there is still a considerable amount of noise. By examining the classification results in Figure 12.17, it can be noted that after 500 iterations

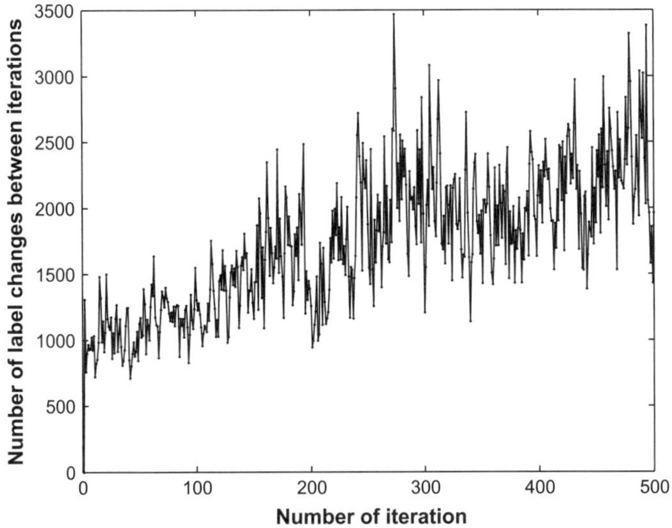

FIGURE 12.16 *The number of label changes as a function of the iteration number, for a repeated spectral classifier*

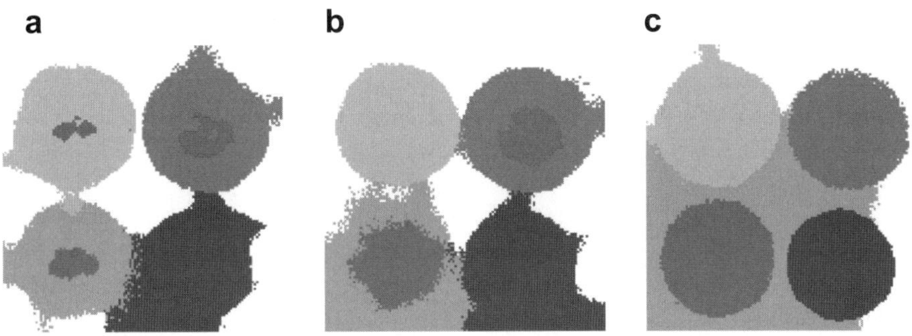

FIGURE 12.17 *Combined spectral/spatial classifier, after (a) 10, (b) 25, and (c) 500 iterations. (Full color version available on http://www.elsevierdirect.com/companions/9780123747532/)*

the specular class is grown into tomato 3 and the tomato 3 class is grown into the background. The results make it clear that adding a spatial classifier does not necessarily improve classification results in this case. Additional experiments, with other spatial classifiers and features, such as the spatial distance transform, and a combination of the x, y coordinates with the distance transform, did not improve the results.

From this experiment it may be concluded that for this kind of data with a large number of bands, and a very high signal-to-noise ratio, this method does not improve classification results, in contrast to cases with a low number of wavelength bands or a lot of noise in the images, as in the experiment described by, for example, Paclik *et al.* (2003).

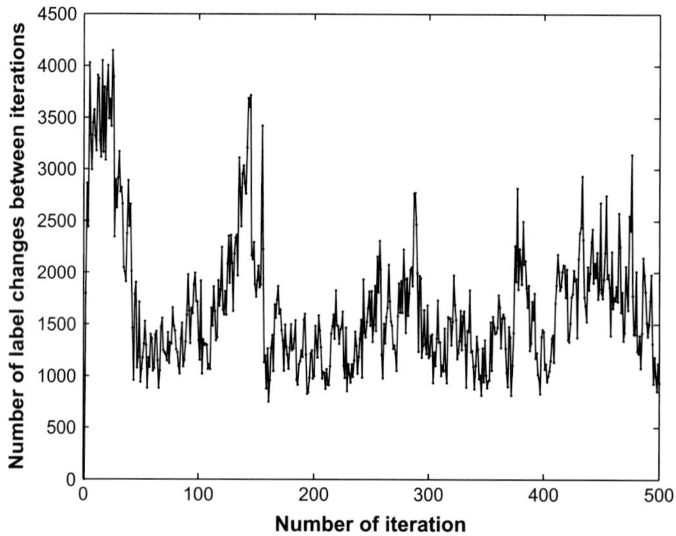

FIGURE 12.18 *The number of label changes as a function of the iteration number, for a repeated spectral/spatial classifier*

12.5.2.3. Integrated spectral and spatial classifiers

Instead of performing the image and spectral processing separately, either sequentially or in parallel, they can be integrated in one classifier. In this way the spatial information is used to influence the results of the spectral classifier or vice versa.

Combined multispectral–spatial classifiers were studied in the early and mid-1980s, in most cases for the analysis of earth observational data. Examples are the ECHO (Extraction and Classification of Homogeneous Objects) classifier from Kettig & Landgrebe (1976), and Landgrebe (1980), contextual classification from Swain *et al.* (1981) and from Kittler & Föglein (1984).

A fully Bayesian approach of image restoration where the contextual information is modeled by means of Markov Random Fields was introduced by Geman & Geman (1984). This is, however, a very time-consuming approach. The Iterated Conditional Modes (ICM) from Besag (1986), can be regarded as a special case of Geman & Geman (1984), and has been used successfully for multispectral images (see e.g. Frery *et al.*, 2009). Another example is the spatially guided fuzzy C-means (SG-FCM) method by Noordam *et al.* (2002, 2003). This method uses unsupervised clustering of spectral data which is guided by *a priori* shape information.

In order to check whether the integrated approach has added value for the tomato application, an experiment was performed in which hyperspectral images of six close-ripeness classes of one tomato were classified with the ECHO classifier. The results were compared with a standard per pixel maximum likelihood classifier on the spectra.

The ECHO classifier is an early example of a combined classifier. This algorithm is a maximum likelihood classifier that first segments the scene into spectrally homogeneous objects. It then classifies the objects utilizing both first- and second-order statistics, thus taking advantage of spatial characteristics of the scene, and doing so in a multivariate sense. Full details can be found in Landgrebe (1980). The ECHO classifier assumes that there are homogeneous regions in the image. This algorithm was tested on hyperspectral images with 80 bands of one tomato in six maturity stages (6 days). It is assumed that the ripening is uniform, so that each image is a different class. In Figure 12.19 the results of the ECHO classifier are given, and Figure 12.20 shows the result of a maximum likelihood classifier. As can be seen from Figure 12.19, the differences are marginal and a simple morphological filter, such as a "pepper and salt removal" (Ritter & Wilson, 2000) applied after the maximum likelihood classifier will remove the noise pixels and give a result similar to the ECHO classifier.

The analysis in this section was performed on a Pentium 4 PC running at 2 GHz with 512 Mb memory, using Matlab (The Mathworks Inc., Natick, MA, USA) and the Matlab PRTools toolbox (Faculty of Electrical Engineering, Mathematics and Computer Science, Delft University of Technology, The Netherlands) (Van der Heijden *et al.*, 2004). The ECHO and Maximum

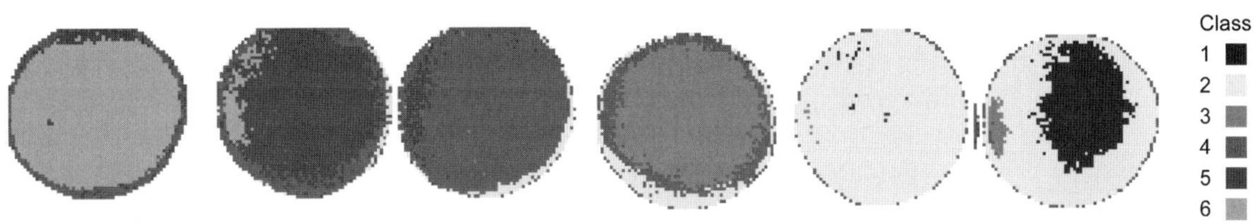

FIGURE 12.19 *Six ripeness stages of tomatoes classified with the ECHO classifier. (Full color version available on http://www.elsevierdirect.com/companions/9780123747532/)*

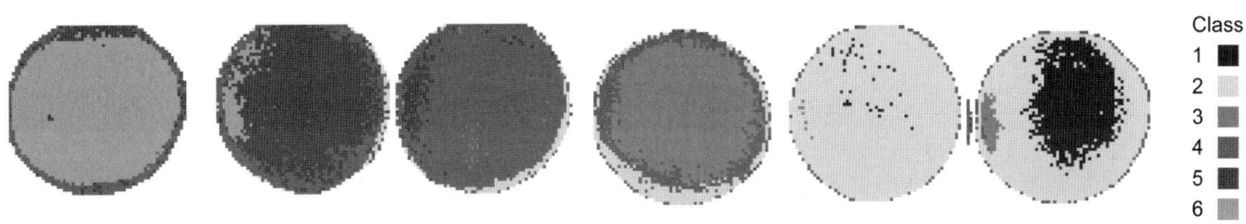

FIGURE 12.20 *Six ripeness stages of tomatoes classified with the maximum likelihood classifier. (Full color version available on http://www.elsevierdirect.com/companions/9780123747532/)*

Likelihood classifications were carried out using MultiSpec (Purdue Research Foundation, West Lafayette, IN, USA).

12.6. CONCLUSIONS

Currently image analysis and spectroscopy are used in real-time food-sorting machines. For image analysis, mostly gray value or RGB color cameras are used. Spectroscopy is most often implemented using a point sensor, which accumulates the reflection, transmission or absorption of light on the whole object.

The combination of both techniques in the form of hyperspectral imaging makes it possible to measure the spatial relationship of quality-related biochemicals, which can improve the sorting process. Currently, however, the large amount of data that needs to be acquired and processed hampers practical implementation. Characterizing the system and its optical components gives information about the actual resolution of the image, which often is much lower than the resolution of the camera sensor. This makes it possible to reduce the data in the camera, using binning, which improves both acquisition and processing speed. Although the amount of data is significantly reduced this way, it still remains too large for real-time implementation.

Spectral data reduction as described in this chapter makes it possible to select wavelength bands with maximum discriminating power. These wavelength bands can be implemented in a multi-band camera with custom filters. These cameras do not significantly differ from RGB cameras in speed, and practical implementation in real-time sorting machines is currently feasible. However, the optimal set of wavelength bands can change in time due to changes in fruit variety, environmental conditions, or simply aging of the illumination. When that occurs, adaption of the camera filters will be difficult and expensive.

Another approach is to use an imaging spectrograph in combination with a camera with pixel addressing. Instead of acquiring complete spectra for each pixel, only wavelength bands of interest are grabbed from the sensor. On-chip binning can be used to determine the bandwidth of these bands. In this way a kind of on-line configurable filter is available, with the advantages of the multi-band camera systems, and the system is now more flexible. It can easily be adapted to changing external conditions. And when allowed by ever-increasing computing power, more bands can be used if needed. Standard CCD cameras are not suitable for pixel addressing, but CMOS image sensors are. Pixels in these sensors can be addressed, which allows fast acquisition of regions or wavelength bands of interest, as described above. Some years ago these sensors were rather noisy, but their quality is rapidly increasing. Another advantage of CMOS sensors compared to CCD sensors is their high dynamic range. For hyperspectral imaging, with large intensity differences over the spectral range, this is a major advantage.

Taking all these developments into account, real-time food sorting machines based on these techniques can be expected in the near future. These machines could measure the spatial distribution of biochemicals which are related to food quality. Besides the applications described in this chapter, many other applications can be considered: for example, the detection of small rotten spots or other defects in apples, which are difficult to assess in traditional color images, or the measurement of taste of fruit, based on its compounds.

NOMENCLATURE

BRDF	bi-directional reflectance distribution function
BSS	blind source separation
CCD	charge-coupled device
CMOS	complementary metal-oxide semiconductor
CV	canonical variable
ECHO	extraction and classification of homogeneous objects
HPLC	high-performance liquid chromatography.
IC	independent component
ICA	independent component analysis
ICM	iterated conditional modes
LDA	linear discriminant analysis
NMC	nearest mean classifier
PC	principal component
PCA	principal components analysis
PLS	partial least square regression
Q^2	predicted percentage variation
RGB	red, green, blue
RMSEP	root mean square error of prediction
SG-FCM	spatially guided fuzzy C-means

REFERENCES

Abbott, J. A. (1999). Quality measurement of fruits and vegetables. *Postharvest Biology and Technology, 15*(3), 207–225.

Arias, R., Tung Ching, L., Logendra, L., & Janes, H. (2000). Correlation of lycopene measured by HPLC with the L*, a*, b* color readings of a hydroponic tomato and the relationship of maturity with color and lycopene content. *Journal of Agricultural and Food Chemistry, 48*(5), 1697–1702.

Baltazar, A., Aranda, J. I., & González-Aguilar, G. (2008). Bayesian classification of ripening stages of tomato fruit using acoustic impact and colorimeter sensor data. *Computers and Electronics in Agriculture, 60*(2), 113–121.

Besag, J. E. (1986). On the statistical analysis of dirty pictures. *Journal of the Royal Statistical Society B, 48*(3), 259–302.

Birth, G. S. (1976). How light interacts with foods. In *Quality detection in foods* (pp. 6–11). St Joseph, MI: American Society for Agricultural Engineering.

Blum, A., Monir, M., Wirsansky, I., & Ben-Arzi, S. (2005). The beneficial effects of tomatoes. *European Journal of Internal Medicine, 16*(6), 402–404.

Choi, K. H., Lee, G. H., Han, Y. J., & Bunn, J. M. (1995). Tomato maturity evaluation using color image analysis. *Transactions of the ASAE, 38*(1), 171–176.

Clinton, S. K. (1998). Lycopene: chemistry, biology, and implications for human health and disease. *Nutrition Reviews, 56*(2), 35–51.

Cover, T. M., & Campenhout, J. V. (1977). On the possible orderings in the measurement selection problem. *IEEE Transactions on Systems, Man, and Cybernetics, 7*, 657–661.

Frery, A. C., Ferrero, S., & Bustos, O. H. (2009). The influence of training errors, context and number of bands in the accuracy of image classification. *International Journal of Remote Sensing, 30*(6), 1425–1440.

Fu, K. S. (1968). *Sequential methods in pattern recognition and machine learning*. New York, NY: Academic Press.

Fukunaga, K. (1990). *Introduction to statistical pattern recognition* (2nd ed.). San Diego, CA: Academic Press.

Geladi, P., & Kowalski, B. R. (1986). Partial least squares regression: a tutorial. *Analytica Chimica Acta, 185*, 1–17.

Geman, S., & Geman, D. (1984). Stochastic relaxation, Gibbs distributions, and the Bayesian restoration of images. *IEEE Transactions on Pattern Analysis and Machine Intelligence (PAMI), 6*(6), 721–741.

Gould, W. (1974). Color and color measurement. In *Tomato production processing and quality evaluation* (pp. 228–244). Westport, CT: Avi Publishing.

Hahn, F. (2002). Multi-spectral prediction of unripe tomatoes. *Biosystems Engineering, 81*(2), 147–155.

Helland, I. S. (1990). Partial least-squares regression and statistical-models. *Scandinavian Journal of Statistics, 17*(2), 97–114.

Hertog, M. G. L., Hollman, P. C. H., & Katan, M. B. (1992). Content of potentially anticarcinogenic flavonoids of 28 vegetables and 9 fruits commonly consumed in the Netherlands. *Journal of Agricultural and Food Chemistry, 40*(12), 2379–2383.

Horn, B. K. P. (1986). *Robot vision*. Cambridge, MA: MIT Press.

Hyvärinen, A., & Oja, E. (2000). Independent component analysis: algorithms and applications. *Neural Networks, 13*(4–5), 411–430.

Kettig, R. L., & Landgrebe, D. A. (1976). Computer classification of remotely sensed multispectral image data by extraction and classification of homogeneous objects. *IEEE Transactions on Geoscience Electronics, GE-14*(1), 19–26.

Khachik, F., Beecher, G. R., & Smith, J. C. (1995). Lutein, lycopene, and their oxidative metabolites in chemoprevention of cancer. *Journal of Cellular Biochemistry, 22*, 236–246.

Kittler, J., & Föglein, J. (1984). Contextual classification of multispectral pixel data. *Image and Vision Computing, 2*(1), 13–29.

Lana, M. M., & Tijskens, L. M. M. (2006). Effects of cutting and maturity on antioxidant activity of fresh-cut tomatoes. *Food Chemistry, 97*(2), 203–211.

Lana, M. M., Tijskens, L. M. M., & van Kooten, O. (2006). Modelling RGB color aspects and translucency of fresh-cut tomatoes. *Postharvest Biology and Technology, 40*(1), 15–25.

Landgrebe, D. A. (1980). The development of a spectral–spatial classifier for earth observational data. *Pattern Recognition, 12*(3), 165–175.

Lissack, T., & Fu, K. S. (1972). *A separability measure for feature selection and error estimation in pattern recognition*. School of Electrical Engineering, Pardue University.

Martinez-Valverde, I., Periago, M. J., Provan, G., & Chesson, A. (2002). Phenolic compounds, lycopene and antioxidant activity in commercial varieties of tomato (*Lycopersicum esculentum*). *Journal of the Science of Food and Agriculture, 82*(3), 323–330.

Mucciardi, A. N., & Gose, E. E. (1971). A comparison of seven techniques for choosing subsets of pattern recognition properties. *IEEE Transactions on Computers, C-20*, 1023–1031.

Narendra, P., & Fukunaga, K. (1977). A branch and bound algorithm for feature subset selection. *IEEE Transactions on Computers, 26*(9), 917–922.

Nelson, L. J. (1997). Simple, low-noise multispectral imaging for agricultural vision and medicine. *Advanced Imaging, 12*(11), 65–67.

Nguyen, M. L., & Schwartz, S. J. (1999). Lycopene: chemical and biological properties. *Food Technology, 53*(2), 38–45.

Noordam, J. C., van der Broek, W. H. A. M., & Buydens, L. M. C. (2002). Multivariate image segmentation with cluster size insensitive Fuzzy C-means. *Chemometrics and Intelligent Laboratory Systems, 64*(1), 65–78.

Noordam, J. C., van der Broek, W. H. A. M., & Buydens, L. M. C. (2003). Unsupervised segmentation of predefined shapes in multivariate images. *Journal of Chemometrics, 17*, 216–224.

Paclik, P., Duin, R. P. W., van Kempen, G. M. P., & Kohlus, R. (2003). Segmentation of multi-spectral images using the combined classifier approach. *Image and Vision Computing, 21*, 473–482.

Parzen, E. (1962). On the estimation of a probability density function and the mode. *Annals of Mathematical Statistics, 33*, 1065–1076.

Polder, G. (2004). *Spectral imaging for measuring biochemicals in plant material*. PhD Thesis, Delft University of Technology.

Polder, G., Van der Heijden, G. W. A. M., Keizer, L. C. P., & Young, I. T. (2003a). Calibration and characterization of imaging spectrographs. *Journal of Near Infrared Spectroscopy, 11*(3), 193–210.

Polder, G., Van der Heijden, G. W. A. M., & Young, I. T. (2002). Spectral image analysis for measuring ripeness of tomatoes. *Transactions of the ASAE, 45*(4), 1155–1161.

Polder, G., Van der Heijden, G. W. A. M., & Young, I. T. (2003b). Tomato sorting using independent component analysis on spectral images. *Real-Time Imaging, 9*(4), 253–259.

Polder, G., Van der Heijden, G. W. A. M., Van der Voet, H., & Young, I. T. (2004). Measuring surface distribution of carotenes and chlorophyll in ripening tomatoes using imaging spectrometry. *Postharvest Biology and Technology, 34*(2), 117–129.

Rao, A. V. R., & Agarwal, S. (2000). Role of antioxidant lycopene in cancer and heart disease. *Journal of the American College of Nutrition, 19*(5), 563–569.

Ripley, B. D. (1996). *Pattern recognition and neural networks*. Cambridge, UK: Cambridge University Press.

Ritter, G. X., & Wilson, J. N. (2000). *Handbook of computer vision algorithms in image algebra* (2nd ed.). Boca Raton, FL: CRC Press.

Savitsky, A., & Golay, M. J. E. (1964). Smoothing and differentiation of data by simplified least squares procedures. *Analytical Chemistry, 36*, 1627.

Schouten, R. E., Huijben, T. P. M., Tijskens, L. M. M., & van Kooten, O. (2007). Modelling quality attributes of truss tomatoes: linking color and firmness maturity. *Postharvest Biology and Technology, 45*(3), 298–306.

Shafer, S. A. (1985). Using color to separate reflection components. *Color Research Applications, 10*(4), 210–218.

Swain, P. H., Vardeman, S. B., & Tilton, J. C. (1981). Contextual classification of multispectral image data. *Pattern Recognition, 13*(6), 429–441.

Tonucci, L. H., Holden, J. M., Beecher, G. R., Khachik, F., Davis, C. S., & Mulokozi, G. (1995). Carotenoid content of thermally processed tomato-based food-products. *Journal of Agricultural and Food Chemistry, 43*(3), 579–586.

Van der Heijden, F., Duin, R. P. W., de Ridder, D., & Tax, D. M. J. (2004). *Classification, parameter estimation and state estimation: an engineering approach using Matlab*. Chichester, UK: John Wiley & Sons.

Van der Heijden, G. W. A. M., Polder, G., & Gevers, T. (2000). Comparison of multispectral images across the Internet. *Internet Imaging, 3964*, 196–206.

Velioglu, Y. S., Mazza, G., Gao, L., & Oomah, B. D. (1998). Antioxidant activity and total phenolics in selected fruits, vegetables, and grain products. *Journal of Agricultural and Food Chemistry, 46*(10), 4113–4117.

CHAPTER 13

Using Hyperspectral Imaging for Quality Evaluation of Mushrooms

Aoife A. Gowen, Masoud Taghizadeh, Colm P. O'Donnell

Biosystems Engineering, School of Agriculture, Food Science and Veterinary Medicine, University College Dublin, Belfield, Dublin, Ireland

CONTENTS

13.1. INTRODUCTION

White mushrooms (*Agaricus bisporus*) are one of Ireland's most important agricultural crops, with an export value exceeding €100 million in 2008 (Bord Bia, 2009). *Agaricus bisporus* is valued for its white appearance, and browning of the mushroom cap is an indicator of poor quality (Green *et al.*, 2008). Mushrooms commonly exhibit surface browning due to physical impact during picking, packaging, and distribution (Figure 13.1). Browning and bruising of the mushroom surface lead to reduced shelf-life and lower financial returns to producers, therefore there is a need for objective evaluation of mushroom quality to ensure that only high-quality produce reaches the market (Gonzalez *et al.*, 2006). Conventional mushroom quality grading methods are based on their luminosity or L-value. Gormley & O'Sullivan (1975) correlated L-values with sensory analysis in order to develop an objective mushroom grading scale (see Table 13.1). However, due to the contact nature of this approach it is not feasible for on-line use for routine quality measurement. Consequently, the mushroom industry generally relies on subjective and labour-intensive human inspection.

Spectroscopy examines the scattering and absorption of light energy from various regions of the electromagnetic spectrum, including the ultraviolet (UV), visible (VIS) and near-infrared (NIR) wavelength regions. Low cost sensors have been developed to detect UV–VIS–NIR light reflected from, transmitted through, and emitted from various materials. NIR sensing technology is well established as a non-destructive tool in food analysis for raw

Hyperspectral Imaging for Food Quality Analysis and Control

FIGURE 13.1 *Stages in mushroom harvesting and transportation* (left to right)*: growing, harvesting, transportation. (Full color version available on http://www.elsevierdirect.com/companions/9780123747532/)*

Table 13.1 Mushroom quality based on L-value

L-value	Quality
>93	Excellent
90–93	Very good
86–89	Good
80–85	Reasonable
69–79	Poor—not acceptable for wholesale
<69	Very poor—not acceptable for retail

Source: *Gormley & O'Sullivan, 1975*

material testing, quality control, and process monitoring, mainly due to the advantages that it allows over traditional methods, e.g. speed, little/no sample preparation, capacity for remote measurements (using fiber-optic probes) and prediction of chemical and physical properties from a single spectrum. VIS–NIR spectroscopy has been used for identification of bruise damage (Esquerre *et al.*, 2009) and prediction of moisture content of fresh mushrooms (Roy *et al.*, 1993). In the case of bruise damage identification, the most important spectral changes were found to occur in the visible part of the spectrum, indicating that this region would be useful for quality evaluation of mushrooms.

Spectrometers integrate spatial information to give an average spectrum for each sample studied; their inability to capture internal component distribution within food products may lead to discrepancies between predicted and measured compositions. Furthermore, spectroscopic assessments with relatively small point-source measurements do not contain spatial information, which is important to many food inspection applications. On the other hand, red, green, blue (RGB) color vision systems, which capture spatial information, find widespread use in food quality control for the detection of surface defects and grading operations. Applications of such machine vision systems have been investigated for monitoring quality in mushrooms. Heinemann *et al.* (1994) investigated the utility of a monochrome camera for mushroom grading (in terms of size, shape, color, veil opening, and stem cut), reporting an average misclassification rate of 20% which compared favourably with the ability of human inspectors. Van de Vooren *et al.* (1992) applied various image analysis techniques to obtain morphological parameters from grayscale images of different mushrooms cultivars, using just four parameters which enabled classification of 80% of the cultivars studied. More recently, Vizhanyo & Felfoldi (2000) reported a technique to distinguish between diseased mushrooms and those that had experienced natural browning by transforming a color image into CIELAB a^* and b^* color axes, with 81% of the diseased region on a test material being correctly classified. The imaging and spectroscopic methods outlined above have shown to perform well for mushroom quality prediction. In addition, Aguirre *et al.* (2009) used grayscale images to examine browning and brown spotting in mushrooms.

Conventional RGB vision systems may be useful for many food sorting operations, but they tend to be poor identifiers of surface features sensitive to wavebands other than RGB, such as low but potentially harmful concentrations of contamination on foods. To overcome this, multispectral imaging systems have been developed to combine images acquired at a number (usually <10) of narrow wavebands, sensitive to features of interest on the object. Hyperspectral imaging (HSI) expands the potential of multispectral imaging, enabling images at a larger number of wavebands (typically >100) with greater resolution to be examined. In this way, HSI combines the advantages of imaging and spectroscopy. Wavelength ranges typically employed in hyperspectral imaging for food control range from the visible through to near-infrared regions (~400–2500 nm). HSI offers many advantages over traditional analytical methods: it is a non-contact, non-destructive method, which enables multi-component information to be obtained from a sample. Moreover, the ability to identify the spatial distribution of multiple chemical and physical components in a sample makes HSI stand out over traditional analytical methods. As a result of these unique advantages, there is considerable interest in developing on-line monitoring tools for mushrooms based on HSI (Gowen *et al.*, 2007). This work is part of a study that aims to use hyperspectral imaging for the rapid assessment of white mushroom quality.

13.2. HYPERSPECTRAL IMAGING OF MUSHROOMS

13.2.1. Hyperspectral Imaging Equipment

The hyperspectral imaging data described in the following sections were obtained using a pushbroom line-scanning HSI instrument (DV Optics Ltd.,

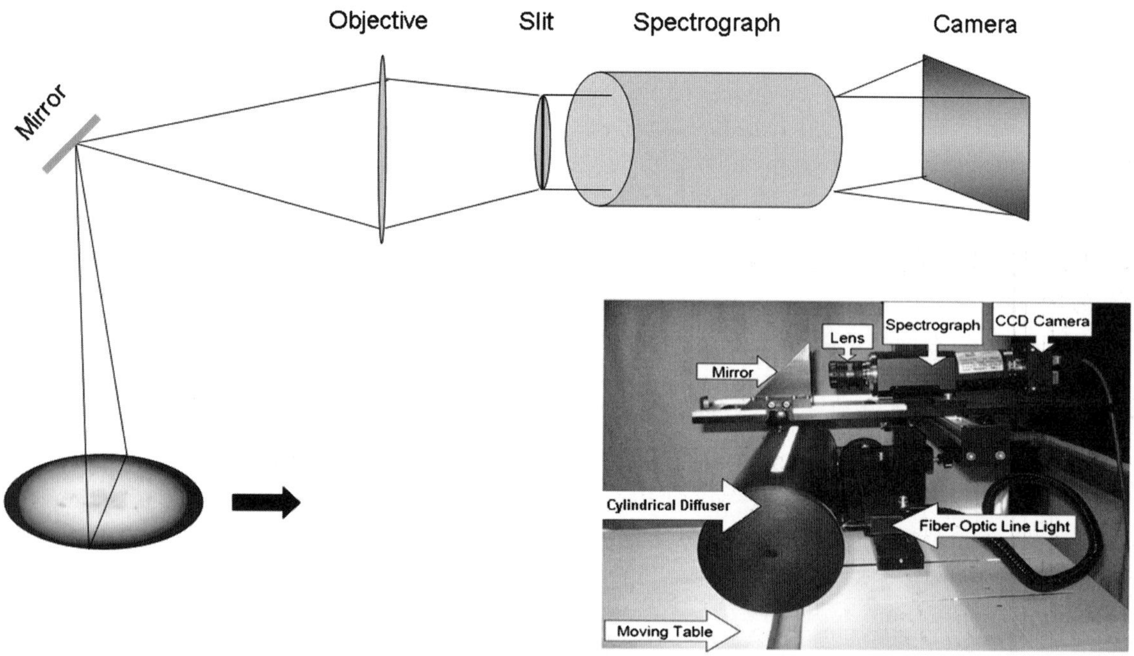

FIGURE 13.2 *Pushbroom hyperspectral imaging system employed in the research. (Full color version available on http://www.elsevierdirect.com/companions/9780123747532/)*

Padua, Italy), operating in the VIS–NIR (400–1000 nm) wavelength range. As shown in Figure 13.2, the main components of this instruments are a translation stage, illumination source (150W halogen lamp) attached to a fiber-optic line light positioned parallel to the translation stage and covered with a cylindrical diffuser, mirror, objective lens (16 mm focal length), spectrograph (Specim V10E, Spectral Imaging Ltd, Oulu, Finland), detector (CCD camera, Basler A312f, effective resolution of 580 × 580 pixels by 12 bits), acquisition software (SpectralScanner, DV Optics, Padua, Italy), and computer. The noise characteristics of the sensor were investigated by acquiring 50 scans of the calibration tile over a time period of one hour. Signal-to-noise ratio was the lowest at the upper (950–1000 nm) and lower (400–445 nm) wavelength limits; in these regions the noise level exceeded 1% of the signal. This is due to decreased CCD detector sensitivity in these regions. Because of this noise, subsequent analysis of spectra was performed only on data in the 445–945 nm wavelength range.

A two-point reflectance calibration was carried out as follows: the bright response (*W*) was obtained by collecting a hyperspectral image or hypercube from a uniform white ceramic tile, the reflectance of which was calibrated against a tile of certified reflectance (Ceram Research Ltd, UK); while the dark response ("dark") was acquired by turning off the light source, completely covering the lens with its cap and recording the camera response. The corrected reflectance value (R) was calculated from the measured signal (I) on a pixel-by-pixel basis as shown in Equation 13.1:

$$Ri = (Ii - \text{dark})/(Wi - \text{dark}) \tag{13.1}$$

where i is the pixel index, i.e. $i = 1, 2, 3, \ldots, n$ and n is the total number of pixels. Therefore reflectance units have a range of 0 to 1.

Mushrooms were imaged individually, mounted on a specially designed mushroom holder incorporating a black paper background.

13.2.2. Spectral Variation Arising from Mushroom Shape

Curvature inherent in their morphology introduces spectral variability in hyperspectral images of many agricultural products, e.g. apples, wheat kernels, and mushrooms. This can be seen in a typical hyperspectral image of the surface of a mushroom, as shown in Figure 13.3. In order to assess the effect of curvature, the hyperspectral image of this mushroom (Fig. 13.3a) was grouped into regions of spectral similarity using k-means clustering (Gowen & O'Donnell, 2009), and the resultant regions, as shown in Figure 13.3(b), form concentric ovals, decreasing in reflectance intensity from the centre of the mushroom to its edge. Mean and standard deviation spectra from each region are shown in Figure 13.3(c). It is clear that the amplitude of the spectra decreases as the mushroom edge is approached, with

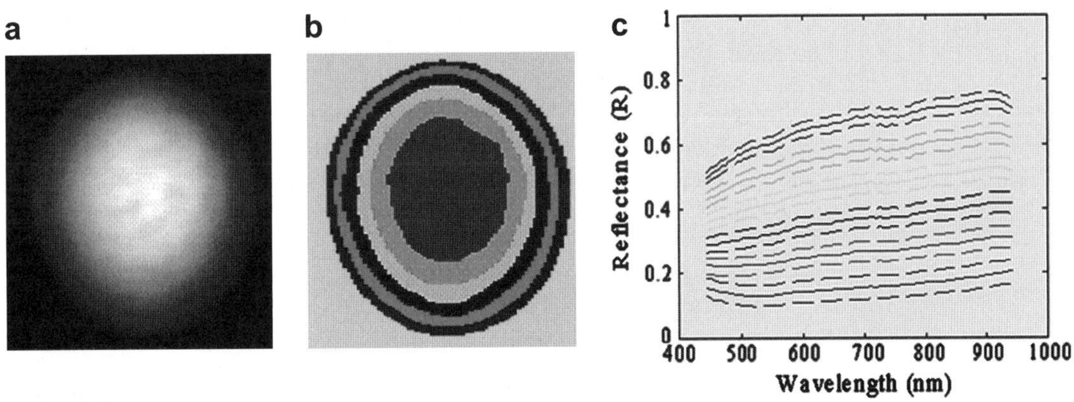

FIGURE 13.3 *Typical hyperspectral image of the surface of a mushroom: (a) mean intensity image; (b) segmentation of mean intensity image into regions of similar light intensity using k-means clustering; (c) corresponding mean and standard deviation reflectance spectra for each region in (b) showing the effect of curvature on spectral response. (Full color version available on http://www.elsevierdirect.com/companions/9780123747532/)*

the overall spectral profile for each region having a similar shape. The extreme edge region has a very low signal and may include some background pixels. This effect on the spectra is caused in part by the relative difference in path length from different points of the curved mushroom surface to the detector: points on the mushroom surface that are nearer to the detector result in higher intensity reflectance counts than points that are further away, such as those on the edge. Non-uniform lighting over the curved surface adds to the spectral variation in regions of similar composition. The inherent curvature of the mushroom surface is problematic for classification of damage on the mushroom surface by direct analysis of reflectance intensity images; for example, regions of similar composition at the edge and center of the mushroom could potentially be classified as different due to the differences in their spectral amplitude.

With the aim of decreasing spectral variability introduced by sample morphology (as is the case for mushrooms), it is desirable to apply spectral or spatial preprocessing to the hyperspectral image data. Pixel spectra obtained from each region shown in Figure 13.3(c) were subjected to two commonly used chemometric pretreatments: multiplicative scatter correction and standard normal variate (SNV) preprocessing (Burger & Geladi, 2007). Multiplicative scatter correction (MSC) corrects the observed spectrum (S) with reference to an ideal or "reference" spectrum (S_{ref}), assuming that (in the linear case) the observed spectrum is a combination of the reference spectrum with some additive and multiplicative noise:

$$S = a + b^* S_{\text{ref}} + \text{error} \tag{13.2}$$

The constants a and b may be estimated by least squares regression and the corrected spectrum ($S_{\text{corrected}}$) can be calculated as follows:

$$S_{\text{corrected}} = (S - a)/b \tag{13.3}$$

In the case of hyperspectral images of individual mushrooms, the mean spectrum of the mushroom may be used as a reference spectrum in the MSC correction. Unlike MSC, SNV does not require a reference spectrum; instead each spectrum in the hypercube image is simply scaled by subtraction of its mean and division by its standard deviation. Mean and maximum image normalization were also applied to the data; for these methods, each image plane in the hypercube was divided by the mean and maximum image, respectively.

The effect of each preprocessing treatment on spectra from segmented regions of the mushroom surface (see Figure 13.3c) is shown in Figure 13.4. In general, the application of spectral and spatial pretreatments to the

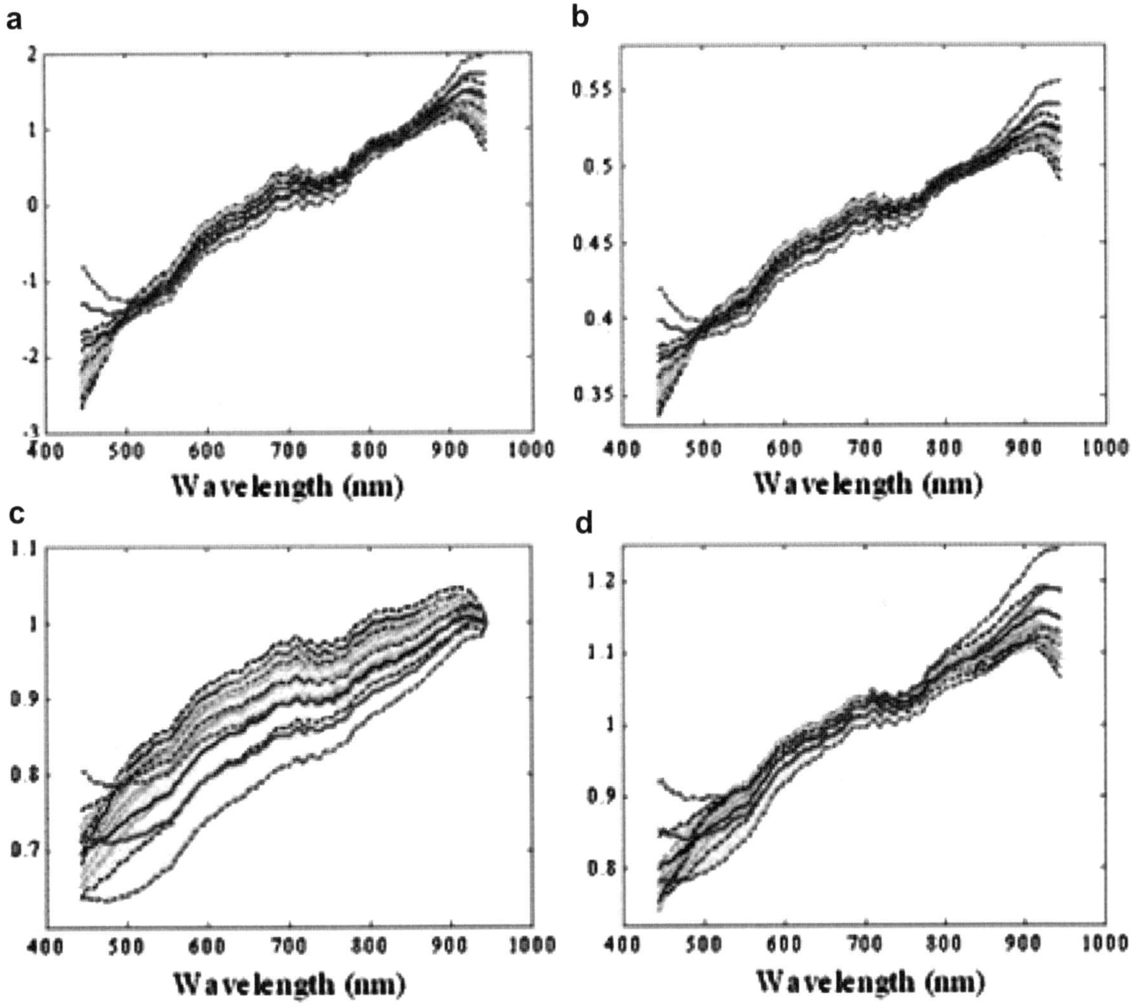

FIGURE 13.4 *Effects of pretreatments on spectra selected from different regions of mushroom surface; solid lines represent mean spectra from each region, dashed lines represent standard deviation spectra from each region (each color represents the corresponding color and region as shown in Figure 13.1(b) and (c)). (a) Spectral pretreatment by MSC; (b) spectral pretreatment by SNV; (c) spatial pretreatment by maximum image normalization; (d) spatial pretreatment by mean image normalization. (Full color version available on http://www.elsevierdirect.com/companions/9780123747532/)*

hyperspectral data decreased the spectral variance resulting from sample morphology. Of the pretreatments studied, SNV and MSC were the most effective for decreasing spectral variance at different regions of the mushroom. Maximum image normalization performed poorest out of those studied and was therefore not included in subsequent analysis. The effect of such

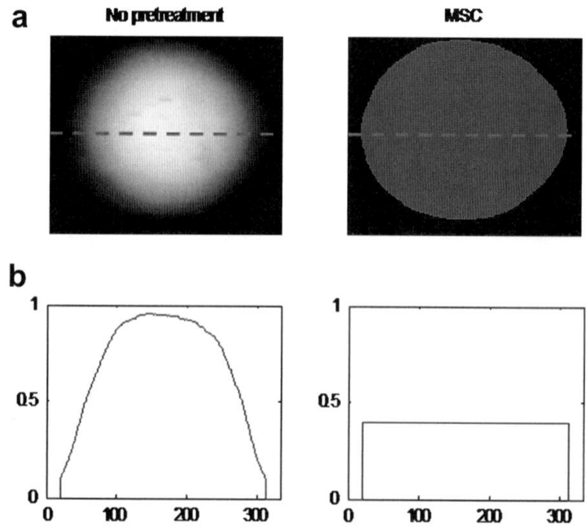

FIGURE 13.5 *Effect of pretreatments on the spatial characteristics of the hyperspectral image: (a) mean intensity image of mushroom; (b) pixel intensity (y - axis) as a function of position (x - axis), where position is indicated by the dashed line in the image. (Full color version available on http://www.elsevierdirect.com/companions/9780123747532/)*

pretreatments on the spatial characteristics of the hyperspectral image may also be examined. As an example, the mean intensity images of a mushroom before and after MSC pretreatment are shown in Figure 13.5(a), from which it can be observed that the effect of mushroom curvature is greatly reduced after MSC pretreatment. Taking a line through the centre of the mean intensity image (Figure 13.5b) further demonstrates the effect of the pretreatment, i.e., the curved intensity profile of the mushroom has now become flat.

13.2.3. Model Building

Hypercubes are data rich. For example, the hyperspectral imaging system employed in this study, which operates in the wavelength range of 400–1 000 nm, with spatial resolution of 580 × 580 pixels, will generate 336 400 spectra in a typical hypercube, each with 121 data points. Numerous model-building strategies for analysis of hyperspectral imaging data may be found in the literature (Gowen *et al.*, 2007). These strategies can be broadly divided into two groups, namely supervised and unsupervised methods. Supervised methods can further be divided into those used for classification and those used for regression. Classification of hyperspectral images aims to identify regions or objects of similar characteristics using the spectral and spatial information contained in the hypercube. Various unsupervised methods, including principal components analysis (PCA), k-nearest neighbours clustering, and fuzzy clustering (Bidhendi *et al.*, 2007), can be applied in either the spectral or spatial domains to achieve classification. These methods are particularly useful in the analysis of samples of unknown composition, facilitating the identification of spectral and spatial similarities within or between images that can further be used for their characterization.

PCA is commonly used as an exploratory tool in hyperspectral imaging, as it represents a computationally fast method for concentrating the spectral variance contained in the >100 image planes of a hyperspectral image into a smaller number (usually <10) of principal component score images. Figure 13.6(a) shows some typical steps involved in performing PCA on a hypercube. In order to apply conventional PCA to a hypercube, it is necessary to "unfold" the three-dimensional hypercube into a two-dimensional matrix in which each row represents the spectrum of one pixel. PCA can be applied to decompose the unfolded hypercube into eigenvectors and eigenvalues. A scores matrix may be obtained by transforming the original data into the directions defined by the eigenvectors. The scores matrix can then be re-folded into a scores cube, such that each plane of the cube represents a principal component, known as a principal component scores image. PCA can also be applied to mean spectra obtained from regions in a hyperspectral image; this is similar to PCA as applied in traditional point spectroscopy.

Supervised classification methods, including partial least squares-discriminant analysis (PLS-DA), neural networks and linear discriminant analysis, require some prior knowledge of the data, as well as the selection of well-defined and representative calibration and training sets for classification optimization. Typical steps in the building of a supervised classification model are shown in Figure 13.6(b). The first step shown is selection of spectra from the hyperspectral imaging data to represent each class of interest. This can be done using just one hyperspectral image, if all classes of interest are present in that image; however, it is preferable to select spectra from a number of hypercubes in order to include in the model potential sources of variability from images taken at different times (e.g. spectral differences arising from changes in the detector response). The categorical variable is a vector of the same length as the spectral data matrix, containing information on the class that each spectrum belongs to. Once a suitable classifier has been trained it can be applied to the entire hypercube and for classification of new hypercubes, resulting in prediction maps, where the class of each pixel can be identified using color mapping.

Hyperspectral image regression enables the prediction of constituent concentration in a sample at the pixel level, thus enabling the spatial distribution or mapping of a particular component in a sample to be

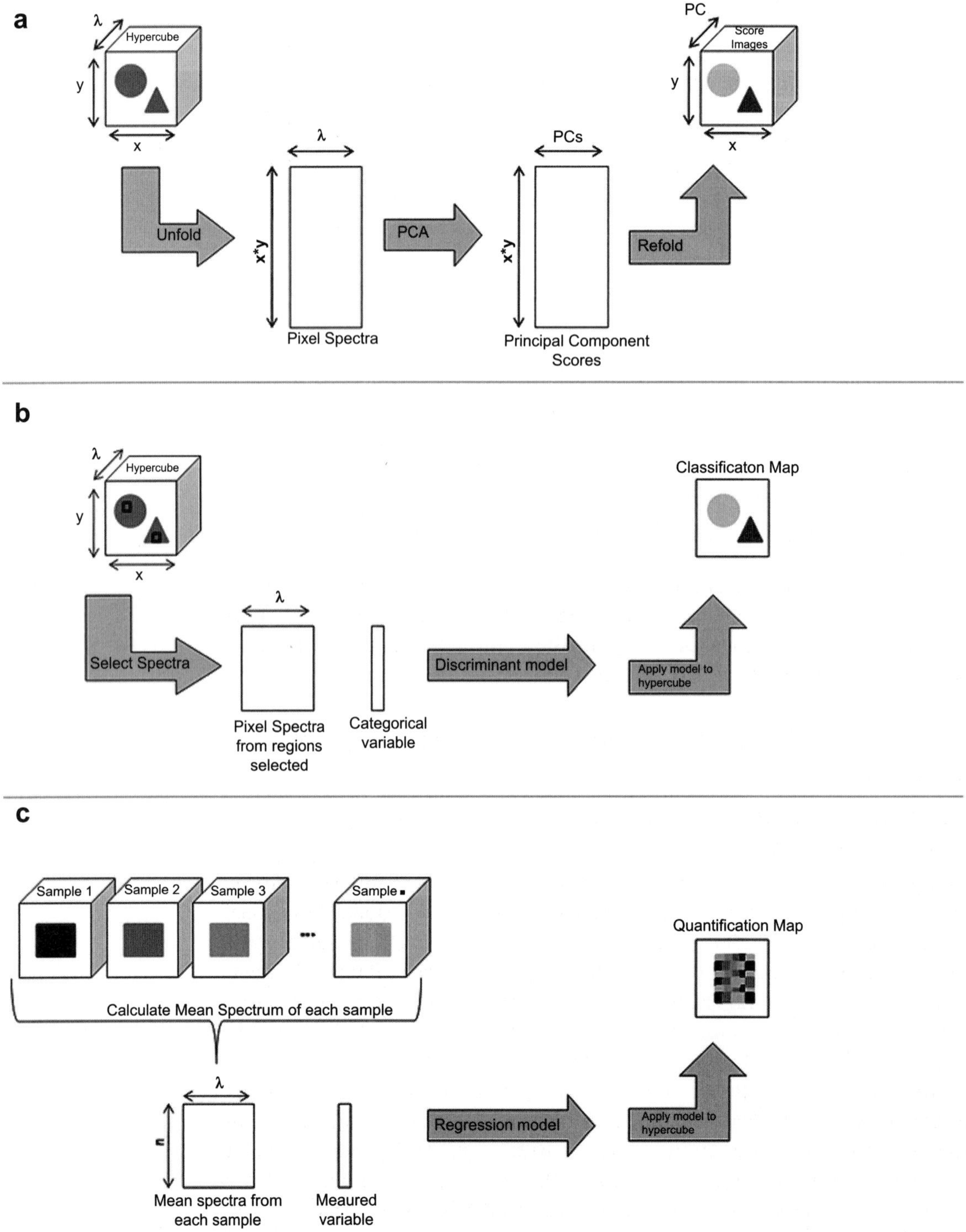

FIGURE 13.6 *Schematics showing typical steps involved in processing of hyperspectral imaging data: (a) PCA; (b) supervised classification; (c) supervised regression. (Full color version available on http://www.elsevierdirect.com/companions/9780123747532/)*

visualized. Many different approaches are available for the development of regression models (e.g. partial least squares regression (PLSR), principal components regression (PCR), stepwise linear regression), all of which require representative calibration sets containing spectra with corresponding measured variables (e.g. fat content, protein content). This poses a problem in hyperspectral imaging: it is practically impossible to measure the precise concentration of components in a sample at the pixel scale and therefore impossible to provide reference values for each pixel spectrum. To overcome this, regression models may be built using mean spectra obtained over the same region of sample (or a representative region) on which the reference value was obtained (Figure 13.6(c)). After model optimization through training and testing, the regression models developed using the mean spectra can be applied to the pixel spectra of the hypercube. This results in a prediction map in which the spatial distribution of the predicted component(s) is easily interpretable.

Selection of the most appropriate modeling strategy is dependent on the final objective of the user; one of the major advantages of HSI in this respect is the sheer volume of data available in each hypercube with which to create calibration, training, and validation sets of data. The following sections present examples of each of the modeling strategies described above as applied to hyperspectral imaging of mushrooms.

13.2.4. Classification Models for Hyperspectral Images of Mushrooms

13.2.4.1. Unsupervised classification: surface damage detection on whole mushrooms

The potential application of HSI for detection of vibration-induced damage on the mushroom surface was investigated (Gowen *et al.*, 2008a). For model development, a set of 100 mushrooms (Group 1) was used: 50 mushrooms that were free from defects were chosen to represent the "undamaged" class, and a further 50 samples were subjected to vibrational damage using a mechanical shaker (Promax 2020, Heidolph Instruments, Schwabach, Germany) set to 400 rpm (revolutions per minute) for 20 min. The "damaged" samples were stored at 21 °C (55% RH) for 24 h prior to imaging to encourage bruise development. A further independent set of 72 mushrooms was tested (Group 2), of which 24 were classified as undamaged, 24 were subjected to damage by shaking at 400 rpm for 20 min, and 24 were subjected to damage by shaking at 200 rpm for 20 min. Representative false-color RGB images (obtained by concatenating hyperspectral images at R = 620 nm, G = 545 nm and B = 450 nm) of the mushrooms under

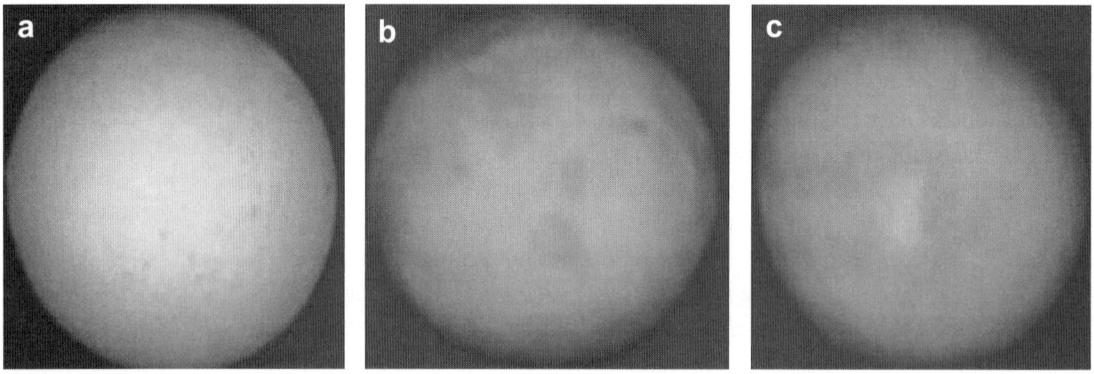

FIGURE 13.7 *False color images obtained by concatenating hyperspectral images at R = 620 nm, G = 545 nm, and B = 450 nm of mushroom: (a) undamaged; (b) 200 rpm shaking damage; (c) 400 rpm shaking damage. (Full color version available on http://www.elsevierdirect.com/companions/9780123747532/)*

investigation in this study are shown in Figure 13.7. Undamaged mushrooms (Figure 13.7a) were generally white in appearance; impact-damaged regions were visibly evident on samples damaged by vibration at 200 rpm (Figure 13.7b), while samples damaged by shaking at 400 rpm (Figure 13.7c) exhibited a more uniform browning of the entire mushroom surface.

Principal components analysis was applied to the hyperspectral image of each mushroom using the steps shown in Figure 13.6(a). The first PC score image (PC1) contained the greatest variance portion of the dataset, which is caused by differences in signal due to curvature on the mushroom surface (Figure 13.8). The second and third PC score images (PC2 and PC3) show contrast between the damaged and undamaged regions on the mushroom, with damaged portions appearing as dark patches on the surface. Noise was dominant from the fourth scores image onwards. Using PCA in this way

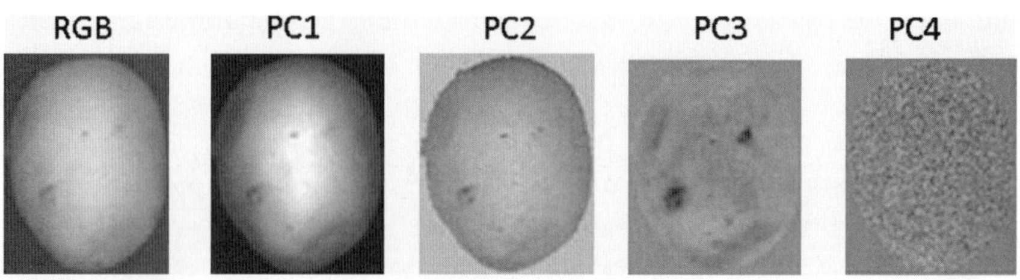

FIGURE 13.8 *False color (obtained by concatenating hyperspectral images at R = 620 nm, G = 545 nm, B = 450 nm) and principal component (PC) images of mushroom. (Full color version available on http://www.elsevierdirect.com/companions/9780123747532/)*

enables reduction of the dimension of the hyperspectral data cube from 101 spectral image planes to just three principal component scores images capturing the greatest variance contained in the data.

An unsupervised classification method could be developed for identification of impact damage on mushrooms by application of PCA to the hypercubes (as described above), followed by analysis of the score image most likely to exhibit differences between sound and damaged tissue. In the present case, the score image that shows greatest contrast between sound and damaged tissue is the third PC image. The main disadvantage of this approach is that applying PCA to each image separately only accounts for the variability contained within the image itself, which includes variability due to size and shape of the sample. A more appealing strategy would be to use spectra from a number of images to build a classifier to separate the spectra from sound and damaged tissue. This can also be achieved using PCA by applying PCA to mean or pixel spectra from each group and examining their distribution in PC scores space. In this example, a dataset comprises of 300 normal spectra and 300 vibration-damaged spectra, which were obtained by interactively selecting spectra from regions of mushroom corresponding to each class (i.e. normal or damaged), from the different images contained in group 1 at different points of elevation on the mushroom surface. These spectra were mean normalized and PCA was applied to the matrix. The score plot of PC1 against PC2 for each spectrum is shown in Figure 13.9, from which it is clear that undamaged and damaged classes are separable along PC1.

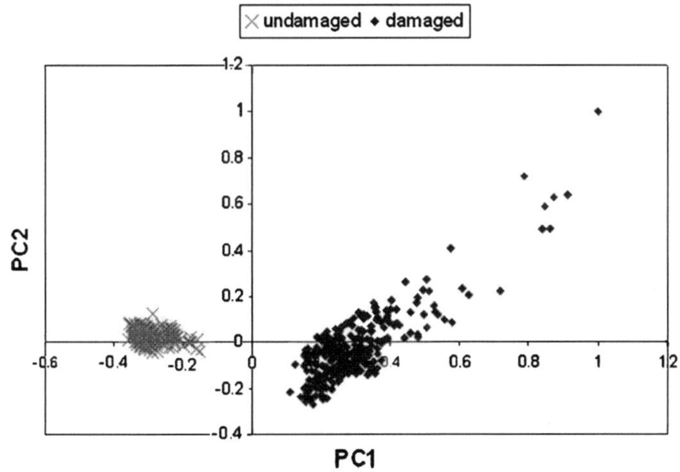

FIGURE 13.9 *PCA scores plot for sample of 600 spectra representing undamaged (n = 300) and vibration damaged (n = 300) mushroom tissue. (Full color version available on http://www.elsevierdirect.com/companions/9780123747532/)*

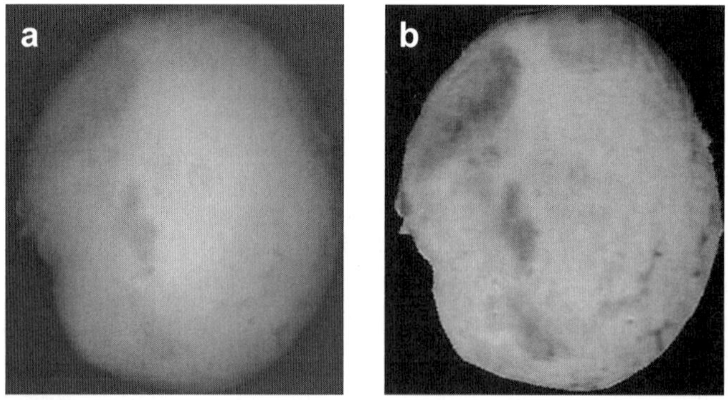

FIGURE 13.10 *Comparison of images of damaged mushroom: (a) RGB image; (b) prediction image obtained after multiplying hypercube by PC 1 loading vector arising from PCA analysis of sample of 600 spectra representing undamaged (n = 300) and vibration-damaged (n = 300) mushroom tissue. (Full color version available on http://www.elsevierdirect.com/companions/9780123747532/)*

Due to the evident separation along PC1, the PC1 eigenvector arising from this analysis represents an operator that can be used to maximize separation between sound and damaged tissue. Multiplying the mean-normalized hyperspectral image of each mushroom sample by this eigenvector results in a 2-D prediction image, in which areas of normal tissue appear brighter than areas of damaged tissue, as shown in Figure 13.10.

13.2.4.2. Supervised classification: PCA-LDA early detection of freezing injury

Mushroom quality is highly dependent on manufacturing processes, transport, and storage conditions (Gormley, 1987). Storage at temperatures below 0 °C causes freezing of intracellular water in mushrooms. When whole mushrooms are frozen, they have a normal appearance just after removal from the freezer; however, as thawing proceeds, water is lost from the mushroom and enzymatic browning occurs. HSI was investigated for identification of mushrooms subjected to freezing before the obvious signs of freeze-damage (i.e. shrinkage and browning) were visibly evident (Gowen *et al.*, 2008c). In order to induce freeze-damage, mushrooms were stored for 24 h in a freezer (Whirlpool, UK) at -30 ± 3 °C. Subsequent to removal from frozen storage the samples were tested after 45 min thawing at 23 ± 2 °C (DD1) and again after a further 24 h after thawing in storage at 4 ± 1 °C (DD2). Undamaged mushrooms were stored at 4 ± 1 °C for the duration of the experiment and tested initially (UD0), after 24 h (UD1) and 48 h (UD2) storage. The experiment was carried out at three different times making three independent sample sets and a total sample size of 144 mushrooms.

Data from the first two time points were grouped together to form a calibration set (sample size of 96 mushrooms) and data from the third time point was used as an independent set (sample size of 48 mushrooms) to test model performance.

For each mushroom, mean reflectance spectra for 10 different regions of interest (each 3 × 3 pixels in size) were obtained from the hyperspectral image around the central top region of the mushroom cap surface. Selecting spectra in this way enabled the construction of a representative calibration set of 2 400 spectra and a test set of 1 200 spectra. Spectra were preprocessed using the SNV transformation to reduce spectral variability (Barnes *et al.*, 1989).

Grayscale images of the mushroom samples investigated are shown in Figure 13.11. Some slight browning on days 1 and 2 is evident on the undamaged samples, due to natural senescence over the storage period. Regarding the frozen samples, no major visible differences can be observed between frozen and frozen–thawed mushrooms on day 1 of storage. Moreover, there is no considerable visible difference between undamaged

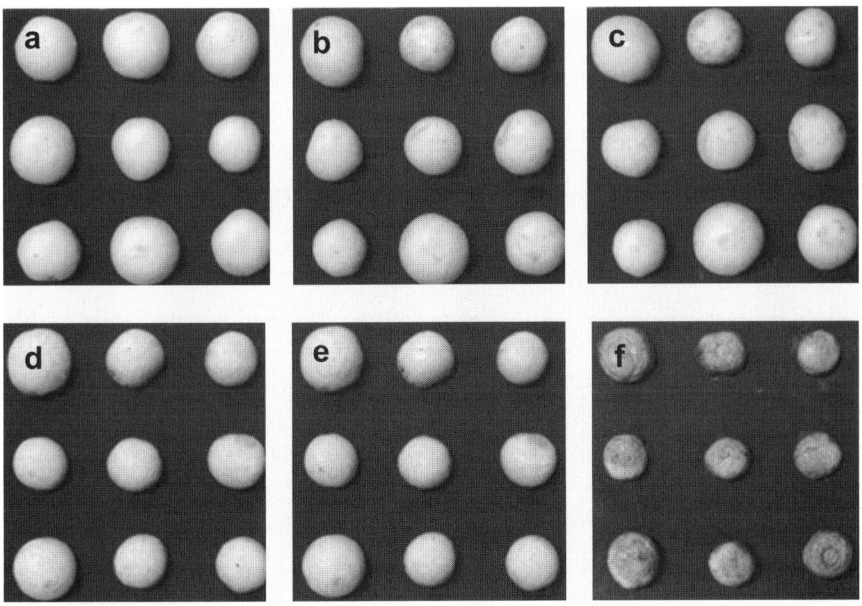

FIGURE 13.11 *Grayscale images of mushrooms under different conditions: (a) undamaged mushrooms at day 0 (UD0) refrigerated at 4°C; (b) undamaged mushrooms at day 1 (UD1) refrigerated at 4°C; (c) undamaged mushrooms at day 2 (UD2) refrigerated at 4°C; (d) frozen mushrooms at day 1 just after removal from freezer at −30°C; (e) frozen mushrooms at day 1 after 45 min thawing at 23°C (DD1), and (f) frozen and thawed mushrooms at day 2 (DD2) after refrigeration at 4°C for 24 h*

mushrooms on days 1 and 2 and the frozen ones at day 1, while frozen–thawed samples at day 2 are shrunken and brown in appearance.

Principal component analysis (PCA) was applied to the calibration set of data to concentrate spectral information into a small number of principal component (PC) scores. The majority of the variance was captured in the first two PC scores, as shown in the eigenvalue plot (Figure 13.12a). The PC1–PC2 score plot for the calibration set is shown in Figure 13.12(b), from which

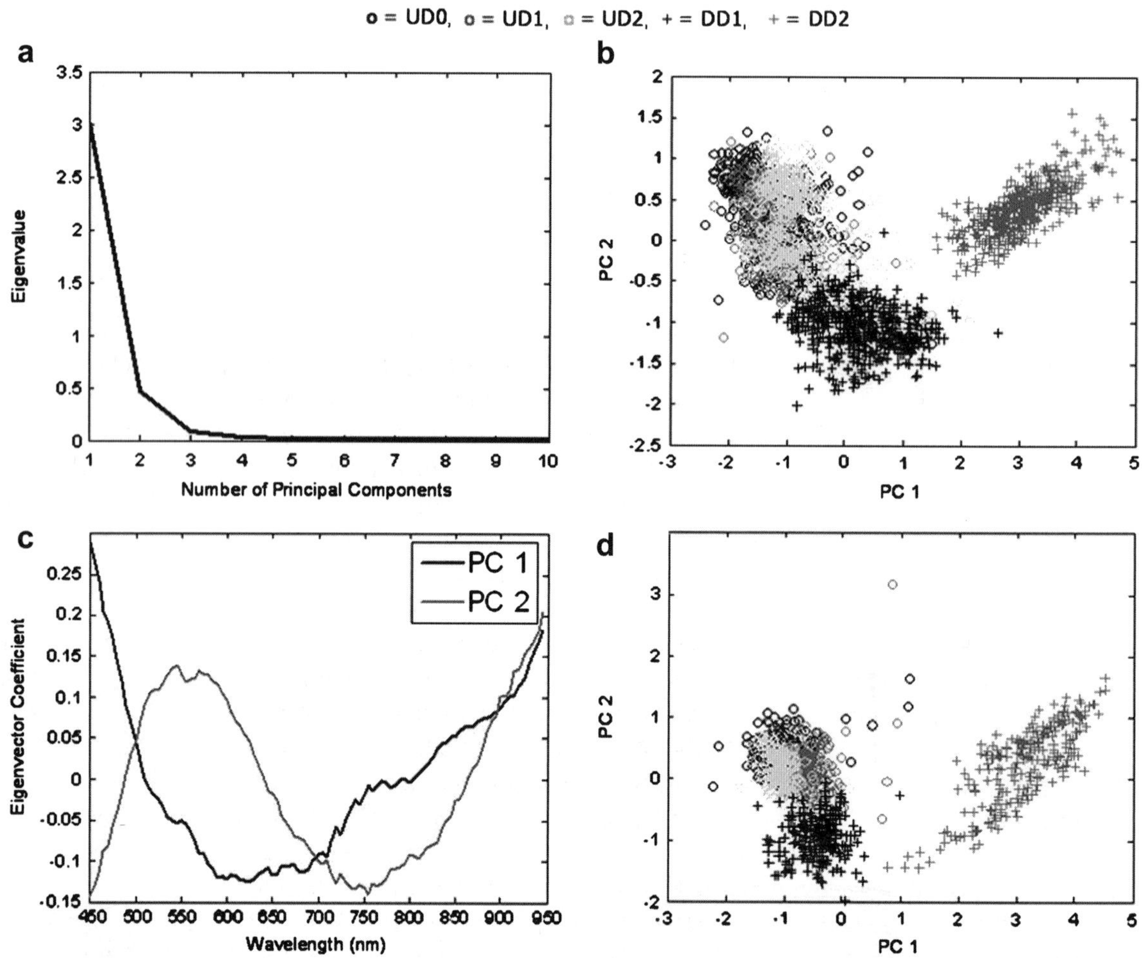

FIGURE 13.12 *Principal components analysis (PCA) of data; (a) eigenvalue as a function of PCs; (b) score plot of PC1 vs. PC2 for calibration set; (c) eigenvector coefficients for PC1 and PC2 of calibration set; (d) score plot of PC1 vs. PC2 for independent test set (scores were obtained by applying eigenvectors in (c) to SNV pretreated test data)*

it can be seen that the undamaged sample spectra (UD0, UD1 and UD2) are overlapped, forming a cluster highly separated from DD2, and largely distinct from DD1. The loadings or eigenvectors (Figure 13.12c) from the PCA transformation can be used to project new data into PC1–PC2 score space. In this way, the SNV preprocessed spectra from the independent test set of data were transformed into the score space defined by the calibration set, and the resultant projected scores are shown in Figure 13.12(d). Again, the undamaged set forms a cluster distinct from the visibly damaged samples (DD2) and the DD1 samples form a cluster which is slightly overlapped with the undamaged cluster.

In order to estimate a boundary to separate the clusters of undamaged and freeze-damaged spectra, LDA was applied. The data from the calibration set were coded with dummy variables as follows: 0 = undamaged (i.e. UD0, UD1, UD2) and 1 = damaged (i.e. DD1, DD2), and LDA was applied to the PC scores (PC1 and PC2) of the calibration set. Prior probability was assigned based on class proportions. Overall, spectra from the calibration set were classified correctly more than 95% of the time. The groups with the lowest correct classification were UD2 (93.1%) and DD1 (92.5%). The LDA model developed on the calibration set was then applied to the projected PC1 and PC2 scores of the spectra from the independent test set. Overall the model performed well for the identification of undamaged sample spectra, but the percentage correct classification for damaged spectra was lower for the test set (87.9%) than for the calibration set (96.25%). In order to test the developed PCA–LDA model performance for classification of hyperspectral images of whole mushrooms, the model was applied to the mean spectrum of each mushroom. Overall, percentage correct classification of mushrooms into their respective classes was high (>95%), and although a relatively high misclassification rate for UD2 samples was obtained (10.4%), all of the DD1 and DD2 mushrooms were correctly classified for the calibration set and greater than 95% of the mushrooms from DD1 and DD2 in the test set were correctly classified.

The developed classification procedure was also applied to entire hyperspectral images to visualize model performance over the surface of the mushroom. The SNV transformation was applied to the unfolded spectra, followed by projection of the data into the directions defined by the PC1 and PC2 (Figure 13.12c). The LDA model was then applied to the PC scores to classify pixels into undamaged (0) or damaged (1) classes. The resultant matrix of predicted class membership for each pixel was "refolded" to form a class prediction map, shown in Figure 13.13 (false-color images of the respective samples are also shown for comparison). Overall, the classification of hyperspectral images was promising: the majority of pixels representing

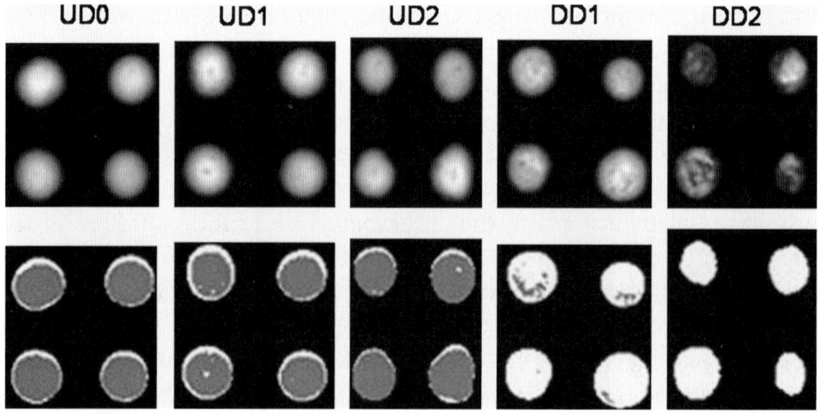

FIGURE 13.13 *Prediction maps for PCA–LDA classification method applied to mushroom hyperspectral data:* (top row) *false color RGB images (obtained by concatenating hyperspectral images at R = 620 nm, G = 545 nm, B = 450 nm);* (bottom row) *prediction maps, where white pixels represent the "damaged" class and gray pixels represent the "undamaged" class*

the undamaged mushrooms were correctly classified; however, edge regions in these images were misclassified as belonging to the damaged class. The prediction maps for the damaged groups, DD1 and DD2, show that the model performed well for identification of freeze-damaged mushrooms, even at early stages of thawing when the effect of freezing was not clearly visible.

13.2.5. Regression Models for Hyperspectral Images of Mushrooms

13.2.5.1. Prediction of quality attributes for sliced mushrooms

Sliced mushrooms are an important sector of the mushroom industry. The recent expansion in demand for them is jointly due to consumers seeking increased convenience and food producers who use them as ingredients (e.g. pizza manufacturers). However, sliced mushrooms are more susceptible to quality deterioration than their whole counterparts. The shelf life of fresh sliced mushrooms is shortened because of the effects of the slicing process, as slicing enables the spread of bacteria over the cut surface and damages the hyphal cells, allowing substrates and enzymes to make contact and form brown pigments. The gills and stipes of sliced mushrooms are more visible than on whole mushrooms and can show spoilage more rapidly than the caps. Additionally, dehydration of slices may cause deformation of the slice shape. Hyperspectral imaging offers a potentially rapid method for non-destructive evaluation of mushroom slice quality (Gowen *et al.*, 2008b).

In this study, approximately 150 second-flush mushrooms with a diameter of 3–5 cm were collected (calibration set) and a further 150 mushrooms were collected one month later (validation set) (Gowen *et al.*, 2008b). Hyperspectral images, color, texture and moisture content of samples were measured on days 0, 1, 2, and 7 of storage at 4 °C and 15 °C. Moisture content of each mushroom slice was measured immediately after HSI experiments using the oven method. Samples were kept in a hot air oven at 110 °C for 48 h and moisture content (*MC*), evaluated by mass difference, was expressed as % w.b. (wet basis). Average color of four randomly selected packages (i.e. 24 slices) for each experimental time/temperature point was calculated. Color measurements were performed using a diffuse CIE standard "D65" illuminant, with an angle of observation of 0° and a measurement area of 25 mm diameter. Color was measured from the middle region of the mushroom cap (the mushroom slice was placed over a black tile during measurement) using a hand-held tristimulus colorimeter (Minolta, Model CR 331, Osaka Japan). Three readings were taken (at the same position each time) per slice and average values were reported. Measurements were recorded in CIE Lab color space, i.e. lightness variable L^* and chromaticity coordinates a^* (redness/greenness) and b^* (yellowness). Only L^* and b^* were used in subsequent modeling, since these were previously identified as important indicators of mushroom slice quality. Texture analysis was carried out on mushroom slices after their color was measured. A texture analyser (Stable Micro Systems, UK) was used for texture analysis of the samples. Each slice was placed on the platform so that the probe would make contact with it at the middle part of the mushroom cap. Texture profile analysis (TPA) was carried out under the following conditions: pre-test speed 2 mm/s; test speed 1 mm/s; post-test speed 5 mm/s; time lag between two compressions 2 s; strain 30% of sample height; data acquisition rate 500 points per second; 6 mm diameter cylindrical stainless steel probe; and load cell 25 kg. TPA hardness (*H*) was used in subsequent analysis.

At each time point, two packages (i.e. 12 slices) at each storage temperature were randomly selected for analysis, making a total of 84 hyperspectral images for each of the calibration and validation sets. Average spectra were extracted from an area of approximately 50 × 50 pixels at the centre of the cap region (corresponding to the region where color and texture measurements were made) of the slice for model building. Principal components regression (PCR) was applied to predict the measured quality indicators (i.e. *MC*, L^*, b^* and *H*) from the extracted mean spectra. The relative prediction deviation (RPD), which is the ratio of the standard deviation to root mean square error of cross-validation (RMSECV) or root mean square error of prediction (RMSEP), was calculated (Table 13.2) to select the best predictive

Table 13.2 Relative prediction deviation PCR predictive models built on full spectrum and a subset of 20 spectra for calibration and test sets of data

Parameter	Model	RPD cal	RPD test	No. LVs
MC	Full λ	1.6	2.8	10
	20 λ	1.8	2	10
L	Full λ	3.4	2	12
	20 λ	3.7	6.5	12
b	Full λ	2.2	1.5	4
	20 λ	2.3	2.3	4
H	Full λ	2.6	1.6	12
	20 λ	2.7	3.1	12

Full λ = full spectrum; 20 λ = subset of 20 spectra; cal = calibration; MC = moisture content, L = L-value, b = b*-value, H = hardness.*

model (Williams, 1987). Rossel *et al.* (2007) stated that RPD values <1.0, between 1.0 and 1.4, between 1.4 and 1.8, between 1.8 and 2, between 2 and 2.5, and greater than 2.5 indicate very poor, poor, fair, good, very good, and excellent model performance, respectively. Based on this classification, regression models performed from poor to excellent for prediction of mushroom quality attributes; when applied to the independent test set RPD ranged from 1.5 (*b*-value) to 6.5 (*L*-value).

A reduced set of 20 wavelengths was obtained for prediction of mushroom quality using exhaustive best subset selection (Development Core Team, 2008). The optimal wavelengths were estimated as 450, 460, 470, 480, 520, 530, 540, 560, 570, 600, 630, 640, 650, 660, 680, 690, 710, 740, 770, and 780 nm. Principal components regression (PCR) was then applied to this reduced set of variables (Table 13.2). When applied to the calibration set of data, PCR on the reduced set of data ($PCR_{reduced}$) performed slightly better than PCR models using the full wavelength range, with RPD ranging from 1.8 (for *MC*) to 3.7 (for L^*). This was also generally the case for the test set of data, with RPD ranging from 2 (for *MC*) to 6.5 (for L^*).

The PCR regression model based on the reduced set of variables was applied to the hypercube data of individual mushroom slices, enabling the generation of virtual prediction images for *MC*, L^*, b^*, and *H*, in which the grayscale intensity value would relate to the values of respective quality parameters at different regions on the sample. For example, in Figure 13.14 the prediction images of *MC* for slices at day 0, day 2 (15 °C) and day 7 (15 °C and 4 °C) are shown. Gills were removed from these images by thresholding, because their spectral characteristics were very different from those of the mushroom cap and were not included in the calibration model.

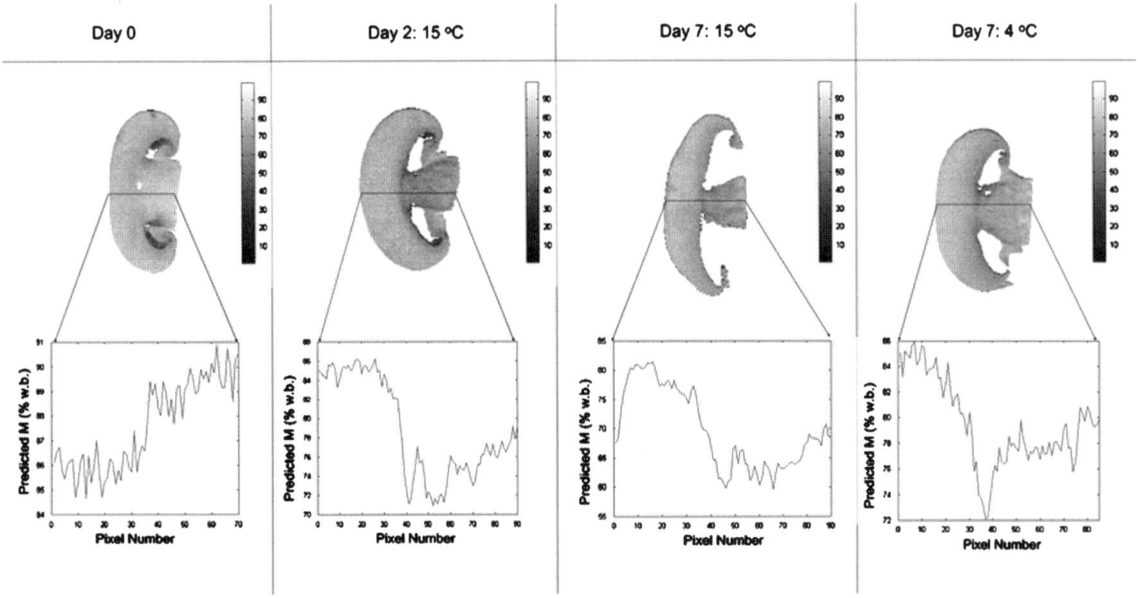

FIGURE 13.14 *Prediction maps (obtained from 10-component PCR calibration model applied to reduced set of wavelengths) for moisture content (M) of sliced mushrooms at day 0, day 2 at 15°C, day 7 at 15°C, and day 7 at 4°C. (Full color version available on http://www.elsevierdirect.com/companions/9780123747532/)*

The average predicted distribution of *MC* in a segment 10-pixels in (vertical) width along the central cap–stipe axis is also shown in Figure 13.14. Visualizing moisture distribution along the surface in this way can offer insight into the mechanisms affecting the deterioration of the slice sample during storage. For example, on day 0, the model predicts a general increase in *MC* from the cap to the stipe region. The predicted distribution of *MC* along the cap–stipe axis for the sample held at 15 °C on day 2 is similar to that of the sample held at 4 °C on day 7, in that the model predicts a much lower amount of water on the stipe region than on the cap. A similar distribution is predicted for the sample held at 15 °C on day 7, but the levels of *MC* are much lower than either the sample held at 4 °C on day 7 or the sample held at 15 °C on day 2. The prediction maps suggest that the majority of moisture is lost through the stipe region of the mushroom.

13.2.5.2. Prediction of quality attributes for whole mushrooms

Moisture content prediction in whole mushrooms

When harvested, whole mushrooms have a moisture content of around 93%; however, they tend to lose moisture during storage, especially at sub-optimal relative humidity (<95% RH) levels (Aguirre *et al.*, 2008). Loss of moisture results in a darkening of the mushroom color and shrinkage of the surface. The typical moisture content *MC* of mushrooms packed in polypropylene (PP) trays and over-wrapped in polyvinyl chloride (PVC) (as is common packaging practice in the mushroom industry) was measured over a duration of one week at ambient conditions (19 °C and relative humidity of 40–60%) and ranged from 93.40 ± 0.62 % w.b. after harvest to 62.72 ± 1.93 % w.b. after one week. The potential of hyperspectral imaging was investigated for prediction of mushroom moisture content within this range. Forty-eight blemish-free second-flush mushrooms, each with a diameter of 3–5 cm, were harvested for the calibration set and a further 48 were harvested a month later for the validation set. Initial mass was noted and mushrooms were dried to four *MC* levels (93.40 ± 0.62 %, 82.76 ± 2.11 %, 73.20 ± 2.60 % and 60.89 ± 4.32 % w b) using a convective air dryer (Gallenkamp Plus II Oven, AGB Scientific, Dublin, Ireland) at 45 ± 1 °C. Samples were removed from the oven at intervals of 0, 30, 60, and 120 min and stored for 30 min in a desiccator prior to weighing and hyperspectral image acquisition. Moisture content of each mushroom was measured using the oven method, i.e., samples were dried in a hot air oven at 110 °C for 48 hours (Roy *et al.*, 1993), and moisture content *MC*, evaluated by mass difference, was expressed as percentage wet basis (% Wb).

Mean spectra were extracted from the hyperspectral image of each mushroom for regression model building using partial least square regression (PLSR). PLSR models were developed to predict *MC* of mushrooms with a four-component PLSR model giving R^2 value = 0.81 and RMSECV = 5.50 for the calibration set and R^2 value = 0.83 and RMSEP = 5.58 for the test set. The RPD values obtained in this study were 2.12 and 2.0 for the calibration and test sets respectively. This compares favourably with the previously reported data on prediction of *MC* in mushrooms using spectra in the 400–1000 wavelength range. Roy *et al.* (1993) reported standard error of difference (SED) of 0.84–0.93 and standard deviation in *MC* of 2.89, giving an RPD of 3.1 for a 10-component PLS model. In order to demonstrate model performance over the surface of the mushroom, *MC* prediction maps of mushrooms dried for different time periods were constructed by applying the 4-component PLSR model to SNV pretreated hyperspectral images (Figure 13.15). The average pixel value of each predicted image, which represents the predicted moisture content for each mushroom image, was calculated and is also shown. Overall, the prediction maps for mushrooms with different *MC* levels show that the model performed well for prediction of mushroom moisture content in the range studied. Using HSI in this way differentiates areas of different moisture content enabling better understanding of dehydration distribution over the mushroom surface.

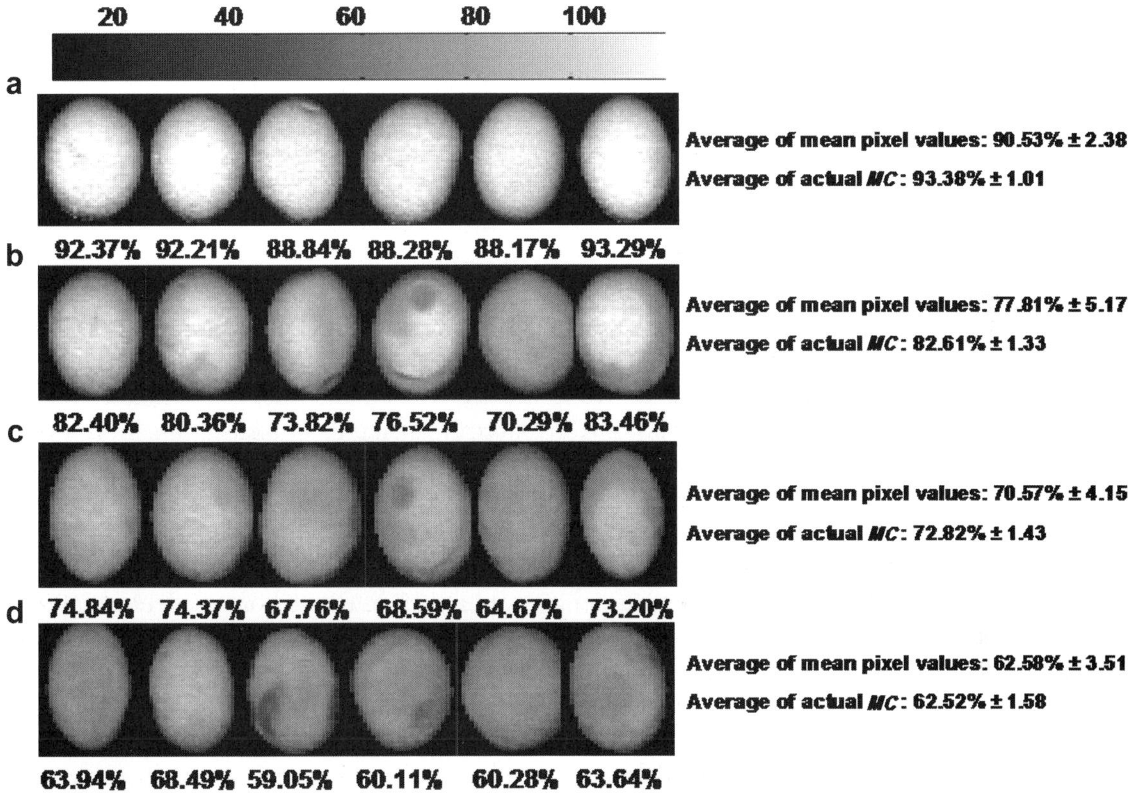

FIGURE 13.15 *Prediction maps for PLSR predictive model applied to mushroom hyperspectral data: (a) fresh mushroom; (b) 30 minutes dried mushrooms; (c) 60 minutes dried mushrooms; (d) 120 minutes dried mushrooms*

Color prediction in whole mushrooms

Color is the most important quality indicator for *Agaricus bisporus* mushrooms, as they are white when fresh, becoming brown and discolored during storage when they reach the end of their shelf life. Conventional mushroom quality grading methods are based on their luminosity or *L*-value (Aguirre *et al.*, 2008; Gormley & O'Sullivan, 1975). Hyperspectral imaging could be used to predict *L*-value for each pixel on the surface of a mushroom, providing valuable information on the distribution of luminosity over the mushroom surface. In order to create groups of varying quality levels, mushrooms were subjected to vibrational damage. This was achieved by shaking 24 mushrooms (placed cap-down) in a plastic mushroom box (3 lb, JF McKenna Ltd, N. Ireland) at 400 rpm for different time periods (30–600 s). The vibration of mushrooms in this manner and the resulting mushroom-to-mushroom impacts induces development of browning on the mushroom surface, and the different damage times included were chosen to artificially generate a sample of mushrooms varying from high to poor quality (*L* ranged from 92 to 63), according to the classification scale shown in Table 13.1.

After vibration, mushrooms were designated into several classes as follows: U = undamaged, D*n* = damaged by vibration for *n* seconds ($n = 30$, 60, 120, 300, 600). For sets 1 and 2, 24 mushrooms were examined per damage level. Hyperspectral images were obtained immediately after impact damage was induced. For each scan, eight mushrooms were placed on a specially designed mushroom holder (incorporating a black paper background) and imaged using the hyperspectral imaging equipment described below. Immediately after hyperspectral imaging, color was measured at the central region of the mushroom cap using a hand-held tristimulus colorimeter (CR-400, Minolta Corp., Japan). Three readings per mushroom were made at different positions on the cap (within a region of approximately 2 cm radius at the centre) and average values recorded. Measurements were taken in Hunter Lab color space, i.e., lightness variable *L* and chromaticity coordinates *a* (redness/greenness) and *b* (yellowness/blueness).

Mean spectra were extracted from each mushroom for model building and SNV was applied. PLSR was applied for prediction of *L*, *a*, and *b*. Models were built on the calibration set using leave-one-out (LOO) cross-validation and then applied to the test set. For prediction of *L*-, *a*- and *b*- value, LOO cross-validation and application of the model to the prediction set indicated that 2-latent variable (LV) regression models were appropriate. Table 13.3 shows the RPD values for each 2-LV model. The RPD values for *b*-value are comparable to those obtained in the experiment for prediction of mushroom slice quality (Table 13.2); however, the prediction of *L*-value is much poorer when compared with the prediction model built for the sliced mushrooms. This could be related to the lower number of latent variables selected in the present case.

The model was then applied to the hypercubes of mushrooms. Figure 13.16 shows the prediction maps for *L*-value resulting from using different numbers of latent variables. The maps show the distribution of *L*-value over the mushroom surface which is not uniform for damaged

Table 13.3 Relative prediction deviation PCR predictive models for calibration and test sets of data

Parameter	RPD cal	RPD test	No. LVs
L	1.7	1.6	2
a	1.5	1.4	2
b	2	1.7	2

L = *L*-value; *a* = *a*-value; *b* = *b*-value (Hunter color coordinates).

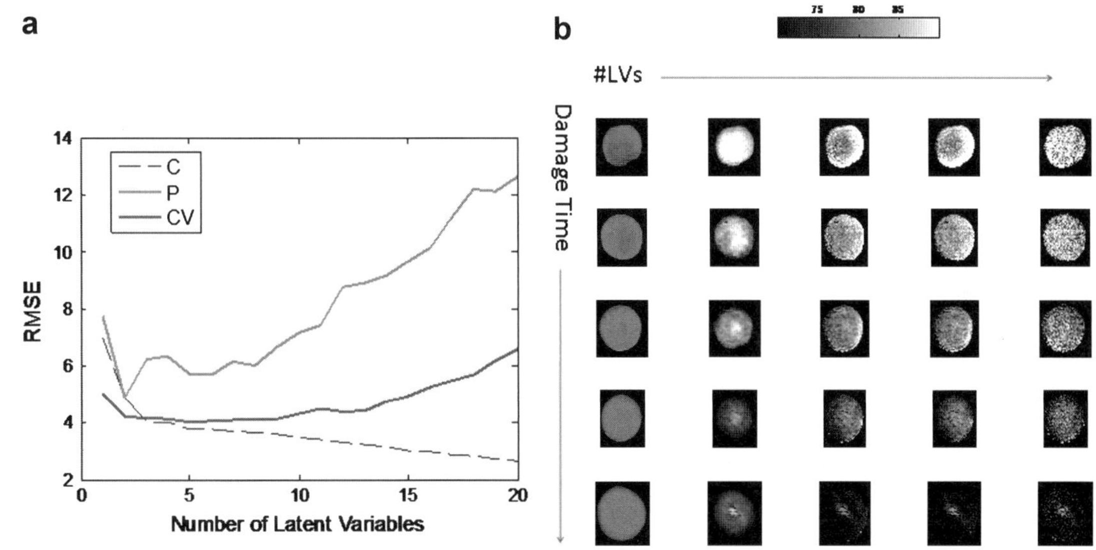

FIGURE 13.16 *Prediction of L-values from hyperspectral images of mushrooms. (a) RMSEC, RMSECV, and RMSEP curves from SNV pre-treated spectra; (b) prediction maps (damage time increases from 0s to 600s along the vertical axis, number of PLS latent variables (LVs) increases from 1 to 5 along the horizontal axis). (Full color version available on http://www.elsevierdirect.com/companions/9780123747532/)*

mushrooms. It can be seen that *L*-value is lower at the mushroom edges for shorter damage times, because the majority of impacts during vibration are at the edges. With the increase in damage time the decrease in *L*-value is more spread out over the surface of the mushroom. The RMSECV and RMSEP curves in Figure 13.16 are reflected in the prediction images, as after 2-LVs the prediction maps are very noisy and outside the range of *L*-values tested (70–90), indicating the unsuitability or overfitting of the models for higher numbers of latent variables. This shows how prediction maps can be used to avoid overfitting in PLS models.

13.3. CONCLUSIONS

Overall, the research shows that HSI is a valuable tool for quality evaluation of mushrooms, with capability for predicting moisture content, color, texture, and identification of surface damage on the mushroom caused by vibration or freeze damage. Different modelling approaches were described and examples for each approach in the evaluation of hyperspectral imaging data of mushrooms were presented. Spectral pretreatments may be applied to decrease variability in hyperspectral images of mushrooms arising from curvature in the mushroom surface. Future work can include the examination of the potential of HSI for quality evaluation of mushrooms in packaging and the capability of this technique for foreign body detection (e.g. presence of casing soil), classification of microbial versus physical damage and prediction of enzyme activity on the mushroom surface. The developed models could be used to identify sub-standard mushroom batches before surface damage is visibly evident, and developed into a tool for non-destructive grading of post-harvest mushroom quality.

NOMENCLATURE

Symbols

a	MSC additive constant
b	MSC multiplicative constant
H	hardness
i	pixel index
I	signal
R	reflectance
S	spectrum
$S_{corrected}$	corrected spectrum
S_{ref}	reference spectrum
W	bright response
λ	wavelength

Abbreviations

CCD	charge-coupled device
D	damaged
DA	discriminant analysis
HSI	hyperspectral imaging
LDA	linear discriminant analysis
LV	latent variable
MC	moisture content
MSC	multiplicative scatter correction
NIR	near-infrared
NIRS	near-infrared spectroscopy
PCA	principal components analysis
PCR	principal components regression
PLSR	partial least squares regression
RGB	red, green, blue
RH	relative humidity
RMSECV	root mean square error of cross-validation

RMSEP	root mean square error of prediction
RPD	relative prediction deviation
SNV	standard normal variate
UD	undamaged
UV	ultraviolet
VIS	visible
Wb	wet basis

ACKNOWLEDGEMENT

Financial support for this reserch from the Irish Department of Agriculture, Fisheris and Food under the FIRM Program is gratefully acknowledged.

REFERENCES

Aguirre, L., Frias, J. M., Barry-Ryan, C., & Grogan, H. (2008). Assessing the effect of product variability on the management of the quality of mushrooms (*Agaricus bisporus*). *Postharvest Biology and Technology, 49*, 247–254.

Aguirre, L., Frias, J. M., Barry-Ryan, C., & Grogan, H. (2009). Modelling browning and brown spotting of mushrooms (*Agaricus bisporus*) stored in controlled environmental conditions using image analysis. *Journal of Food Engineering, 91*, 280–286.

Barnes, R. J., Dhanoa, M. S., & Lister, S. J. (1989). Standard normal variate transformation and de-trending of near-infrared diffuse reflectance spectra. *Applied Spectroscopy, 43*, 772–777.

Bidhendi, S. K., Shirazi, A. S., Fotoohi, N., & Ebadzadeh, M. M. (2007). Material classification of hyperspectral images using unsupervised fuzzy clustering methods. *Third International IEEE Conference on Signal-Image Technologies and Internet-Based Systems, SITIS 2007*, 619–662.

Bord Bia (2009) Factsheet on the Irish Agriculture and Food & Drink Sector (September 2009). Available at http://www.bordbia.ie/industryinfo/agri/pages/default.aspx (accessed September 2009).

Burger, J., & Geladi, P. (2007). Spectral pretreatments of hyperspectral near infrared images: analysis of diffuse reflectance scattering. *Journal of Near Infrared Spectroscopy, 15*, 29–38.

Development Core Team. (2008). *A language and environment for statistical computing*. Vienna: Foundation for Statistical Computing.

Esquerre, C., Gowen, A. A., O'Donnell, C. P., & Downey, G. (2009). Initial studies on the quantitation of bruise damage and freshness in mushrooms using visible-near-infrared spectroscopy. *Journal of Agricultural and Food Chemistry, 57*, 1903–1907.

Gonzalez, F. E., Jimenez, A. S., & Pardo, V. T. (2006). Quality and shelf life of packaged fresh-sliced mushrooms stored at two different temperatures. *Agricultural and Food Science, 5*, 414–422.

Gormley, T. R. (1987). *Handling, packaging and transportation of fresh mushrooms. Proceedings of the 5th National Mushroom Conference*. Dublin, Ireland: Malahide Co.

Gormley, T. R., & O'Sullivan, L. (1975). Use of a simple reflectometer to test mushroom quality. *The Mushroom Journal, 34*, 344–346.

Gowen A. A., & O'Donnell C. P., (2009). Development of algorithms for detection of mechanical injury on white mushrooms (*Agaricus bisporus*) using hyperspectral imaging. In Moon S. Kim, Shu-I Tu, & Kaunglin Chao (Eds.), Sensing for Agriculture and Food Quality and Safety. *Proceedings of SPIE*, 73150G.

Gowen, A. A., O'Donnell, C., Cullen, P. J., Downey, G., & Frias, J. (2007). Hyperspectral imaging—an emerging process analytical tool for food quality and safety control. *Trends in Food Science and Technology, 18*, 590–598.

Gowen, A. A., O'Donnell, C., Taghizadeh, M., Cullen, P. J., & Downey, G. (2008a). Hyperspectral imaging combined with principal component analysis for surface damage detection on white mushrooms (*Agaricus bisporus*). *Journal of Chemometrics, 22*, 259–267.

Gowen, A. A., O'Donnell, C. P., Taghizadeh, M., Gaston, E., O'Gorman, A., *et al.* (2008b). Hyperspectral imaging for the investigation of quality deterioration in sliced mushrooms (*Agaricus bisporus*) during storage. *Sensing and Instrumentation for Food Quality and Safety, 2*(3), 133–143.

Gowen, A. A., Taghizadeh, M., & O'Donnell, C. (2008c). Identification of mushrooms subjected to freeze damage using hyperspectral imaging. *Journal of Food Engineering, 93*, 7–12.

Green, J. M., Grogan, H., Eastwood, D. C., & Burton, K. S. (2008). Investigating genetic and environmental control of brown color development in the cultivated mushroom *Agaricus bisporus* infected with mushroom virus X. *International Society for Mushroom Science, 17*, 41.

Heinemann, P. H., Hughes, R., Morrow, C. T., Sommer, H. J., Beelmam, R. B., & Wuest, P. J. (1994). Grading of mushrooms using a machine vision system. *Transactions of the ASAE, 37*, 1671–1677.

Rossel, R. A. V., Taylor, H. J., & McBratney, A. B. (2007). Multivariate calibration of hyperspectral gamma-ray energy spectra for proximal soil sensing. *European Journal of Soil Science, 58*, 343–353.

Roy, S., Ramaswamy, C. A., Shenk, J. S., Westerhaus, O. M., & Beelman, R. B. (1993). Determination of moisture content of mushrooms by Vis-NIR spectroscopy. *Journal of the Science of Food and Agriculture, 63*, 355–360.

Van de Vooren, J. G., Polder, G., & Van der Heijden, G. W. A. M. (1992). Identification of mushroom cultivars using image analysis. *Transactions of the ASAE, 35*, 347–350.

Vizhanyo, T., & Felfoldi, J. (2000). Enhancing color differences in images of diseased mushrooms. *Computers and Electronics in Agriculture, 26*, 187–198.

Williams, P. C. (1987). Variables affecting near-infrared reflectance spectroscopic analsyis. In P. Williams, & K. Norris (Eds.), *Near-infrared Technology in the Agriculture and Food Industries* (pp. 143–166). St Paul, MN: American Association of Cereal Chemists.

CHAPTER 14

Hyperspectral Imaging for Defect Detection of Pickling Cucumbers

Diwan P. Ariana

Michigan State University, Department of Biosystems and Agricultural Engineering, East Lansing, Michigan, USA

Renfu Lu

USDA ARS Sugarbeet and Bean Research Unit, Michigan State University, East Lansing, Michigan, USA

CONTENTS

14.1. INTRODUCTION

Cucumber (*Cucumis sativus* L.) is believed to have originated on the Indian subcontinent. Cucumbers are members of the cucurbit family and are related to gourds, gherkins, pumpkins, squash, and watermelon. The first horticultural types were selected in the 1700s following introduction into Europe. They were introduced to the Americas by Christopher Columbus, and have been cultivated in the United States for several centuries (Sargent & Maynard, 2009). Cucumbers are an important commercial and garden vegetable. China is the most important cucumber and gherkin producing country, with more than 25 million tons in production. Other important cucumber and gherkin producing countries are Turkey, Iran, Russia, and the United States (FAOSTAT, 2009) (Table 14.1).

There are three basic classes of cucumber marketed in the United States, i.e. field-grown slicers, greenhouse-grown slicers, and processing (pickling) cucumbers. Field-grown slicers (cucumbers for the fresh market) are larger and sweeter, and have a thicker skin than the pickling varieties. The United States produced 920 000 tons of cucumber for all uses in 2007, which are about equally split between the fresh and processing market. Hand (multiple harvests) and machine (once-over) harvesting are practiced in several growing regions of the United States. All fresh-market cucumbers are

Hyperspectral Imaging for Food Quality Analysis and Control

Table 14.1 Cucumber and gherkin production in thousand metric tons of selected top producing countries in the past three years

Country	2005	2006	2007
China	26 558	27 357	28 062
Turkey	1 745	1 800	1 876
Iran	1 721	1 721	1 720
Russia	1 414	1 423	1 410
United States	930	908	920
Others	10 591	10 857	10 623
World	42 958	44 066	44 611

Source: *FAOSTAT (2009)*

harvested by hand, while most pickling cucumbers are harvested by machine (Lucier & Lin, 2000). Although the incidence of cucumber fruit injuries may be higher when harvesting by machine, the acreage planted for mechanically harvested cucumber continues to increase in the United States owing to the scarcity and cost of labor and the continued improvement of harvesting technology.

Cucumbers are prone to damage during fruit enlargements, harvest, transport, and processing, thus steps must be taken in order to minimize losses due to bruising. Severely damaged fruits are visually detected and discarded in cucumber processing plants; however, mechanical injury often causes hidden internal physical damage in the form of carpel separation which can lead to increased bloating in brine-stock cucumbers. Carpel separation is a serious product quality problem, resulting in economic loss for the pickle industry (Wilson & Baker, 1976). Carpel separation in pickling cucumber fruit occurs when the sutures of the three fused carpels form a hollow through part or the entire length of the fruit. As the carpel-suture strength increases, the frequency of carpel separation in green stock would decrease, which in turn would reduce occurrences of carpel balloon bloater formation during fermentation. Bruising triggers numerous biochemical and physiological changes, leading to accelerated aging in harvested fruit already undergoing postharvest senescence (Miller, 1989, Miller *et al.*, 1987). Biochemical methods have been developed to detect and separate bruised fruit in a number of crop species. For example, a catechol test coupled with hydrogen peroxide can be used to detect bruising in whole cucumbers (Hammerschmidt & Marshall, 1991). This test is based on increased peroxidase activity after bruising (Miller & Kelley, 1989). Mechanical injury may result in changes in endogenous levels and rates of biosynthesis of ethylene, indoleacetic acid, zeatin, and elicitors (Miller, 1992), cell wall-degrading enzymes and ethylene production (Miller *et al.*, 1987), and sugar composition of cell walls (Miller, 1989). The aforementioned methods are destructive, not instantaneous, and therefore not suitable for automated grading and sorting in a modern commercial setting.

Researchers have explored various nondestructive methods for detecting mechanical injury in cucumber fruit. Sorting cucumbers by density has been proposed because damaged fruit have internal voids and may have different densities than undamaged fruit (Marshall *et al.*, 1973). However, the rate of misclassification by density was high. Refreshed delayed light emission (RDLE) from chlorophyll was able to consistently distinguish bruised from non-bruised cucumber fruit (Abbott *et al.*, 1991). Although the method is impractical for sorting individual pickling cucumbers owing to the time requirement for dark equilibrium and RDLE measurement, it has the potential to be used as an inspection tool by the processor. Visible/near-infrared light transmission measurement has been studied to evaluate internal quality of pickling cucumbers (Miller *et al.*, 1995). Light transmission increased as the severity of mechanical stress applied to the fruit increased. The technique may be a valuable tool for detecting poor quality cucumbers before processing. Machine vision technology is currently used in many pickling cucumber processing plants, but the technology is designed for inspecting external characteristics, including size, shape, and color.

In recent years, hyperspectral imaging (also called imaging spectroscopy) has been used for quality evaluation and safety inspection of food and agricultural products. Hyperspectral imaging integrates conventional imaging and spectroscopy to obtain both spatial and spectral information from an object. The technique is thus useful for analyzing heterogeneous materials or quantifying properties or characteristics that vary spatially in food items (Park *et al.*, 2006). Reviews on the applications of hyperspectral imaging for food quality and safety evaluation can be found in Gowen *et al.* (2007) and Wang & Paliwal (2007). This chapter presents the application of hyperspectral imaging for defect detection in pickling cucumbers.

14.2. DETECTION OF EXTERNAL BRUISE

The processing quality of cucumbers is a major concern of the pickling industry. Mechanical injury can cause physiological breakdown during postharvest storage and processing. Decreased cucumber quality as evidenced by tissue softening and deterioration has been linked to mechanical injury (Marshall *et al.*, 1972). Miller *et al.* (1987) noted that water-soaked lesions were present in the skin of mechanically stressed cucumbers immediately after treatment, indicating membrane damage at the cellular level. These water-soaked lesions are not very obvious in the visible (VIS) region of the electromagnetic spectrum.

A near-infrared (NIR) hyperspectral imaging system was developed to capture hyperspectral images from pickling cucumbers in the spectral region of 900–1700 nm (Ariana *et al.*, 2006). The system consisted of an imaging spectrograph attached to an InGaAs (indium gallium arsenide) camera with line-light fiber bundles as an illumination source. Two cone-shaped sample holders were used to rotate pickling cucumbers thus scanning the entire surface of the fruits by a line-scan hyperspectral imaging system (Figure 14.1). Hyperspectral images were taken from the pickling cucumbers at 0, 1, 2, 3, and 6 days after they were subjected to dropping or rolling under load which simulated damage caused by mechanical harvesting and handling systems.

Knowledge about bruises and their locations on individual cucumber samples is required to determine the effectiveness of the bruise detection algorithm. The actual individual cucumber class (bruised or normal) was determined by visually comparing relative reflectance NIR images at 1 200 nm before and after mechanical stress was applied. Bruised areas appeared as dark patches in the NIR images (Figure 14.2). Bruised and normal areas had the highest contrast at 1 200 nm. If dark areas appeared on the NIR images only after bruising, the cucumber was designated to be in the bruised class. Rough skin areas, which appeared as dark areas on NIR images both before and after bruising, were not considered as bruised.

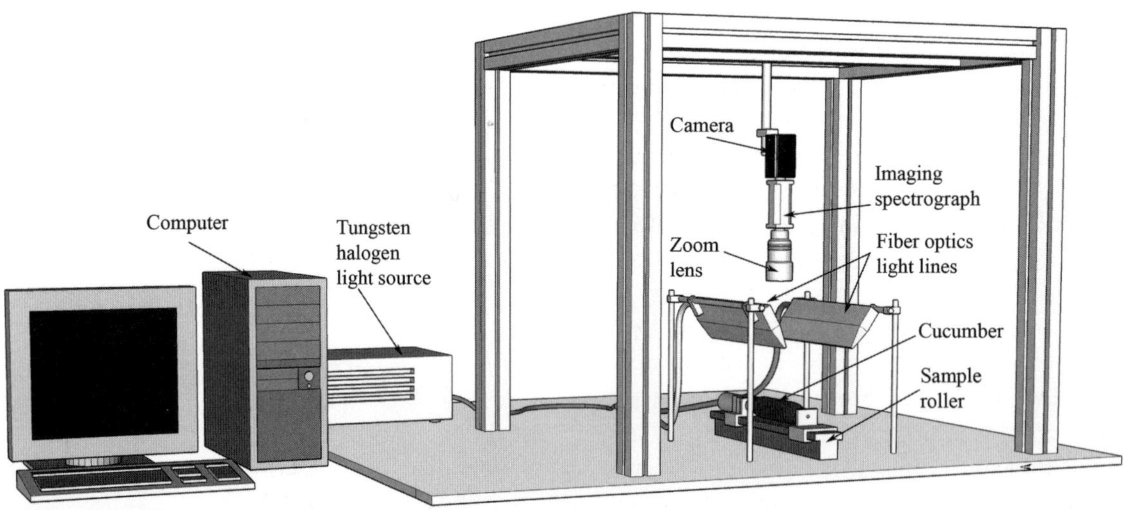

FIGURE 14.1 *Near-infrared hyperspectral imaging system (reproduced from Ariana* et al., *2006. © Elsevier 2006). (Full color version available on http://www.elsevierdirect.com/companions/9780123747532/)*

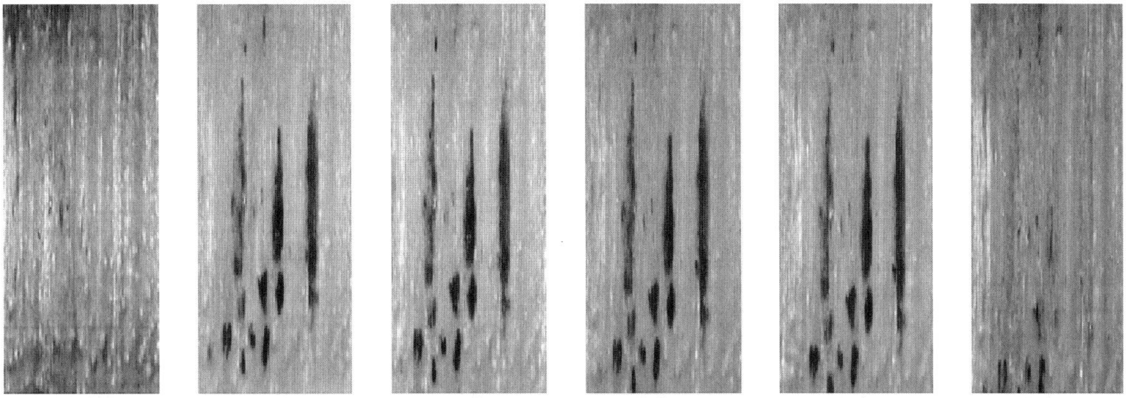

FIGURE 14.2 *Near-infrared spectral images at 1 200 nm showing the changes in bruises over time periods of 0–6 days (reproduced from Ariana* et al.*, 2006. © Elsevier 2006)*

The mean reflectance of bruised tissue was consistently lower than that of the normal tissue over the spectral region of 950–1 650 nm, except in the range of 1 400–1 500 nm where considerable spectral overlapping was observed for the two types of tissue. Water has strong absorption at 1 450 nm (Osborne *et al.*, 1993), which resulted in low reflectance for both normal and bruised tissue (Figure 14.3). The difference in the reflectance between normal and bruised tissue was the greatest in the region between 950 and 1 350 nm. The reflectance of normal tissue was relatively constant over the period of the experiment. On the contrary, the reflectance of bruised tissue increased over time, approaching that of normal tissue (Figure 14.4). This characteristic might be due to the wound healing of the cucumbers in response to mechanical stress (Miller, 1992). Figure 14.2 shows the changes in bruised areas with time from the spectral images at 1 200 nm taken over a period of 6 days. Within 2 hours after bruising (0 day), bruises appeared darker than normal tissue on the image. One day after stress was applied, the relative reflectance of bruised tissue showed the greatest difference from that of normal tissue. This fact is an advantage for machine vision-based sorting because freshly harvested cucumbers are usually sorted within 24 hours after harvesting. The spectral differences between the bruised and normal tissue decreased over time. Some bruise areas were no longer visible on the image after 6 days. Thus, sorting at later days is not desirable.

Principal components of NIR hyperspectral images based on the optimum spectral resolution of 8.8 nm were analyzed in three different spectral regions, 950–1 650 nm, 950–1 350 nm, and 1 150–1 350 nm. Table 14.2 shows the classification accuracies of cucumber samples based on the first principal

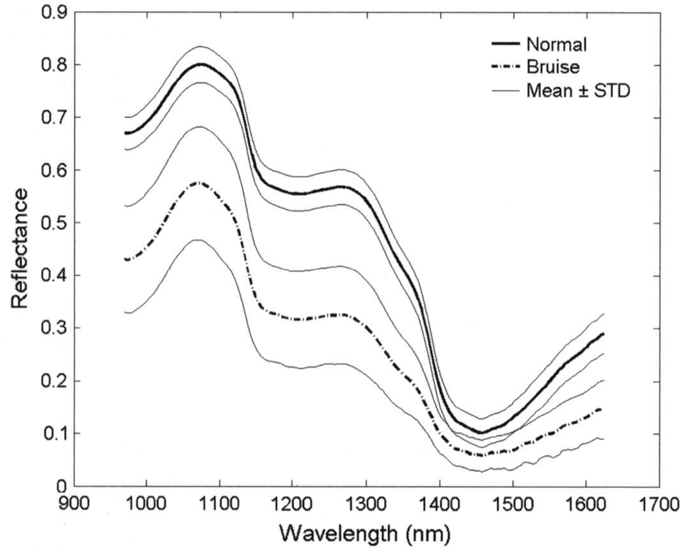

FIGURE 14.3 *Mean relative reflectance spectra of normal and bruised cucumber tissue and their standard deviation (SD) spectra (reproduced from Ariana* et al., *2006. © Elsevier 2006)*

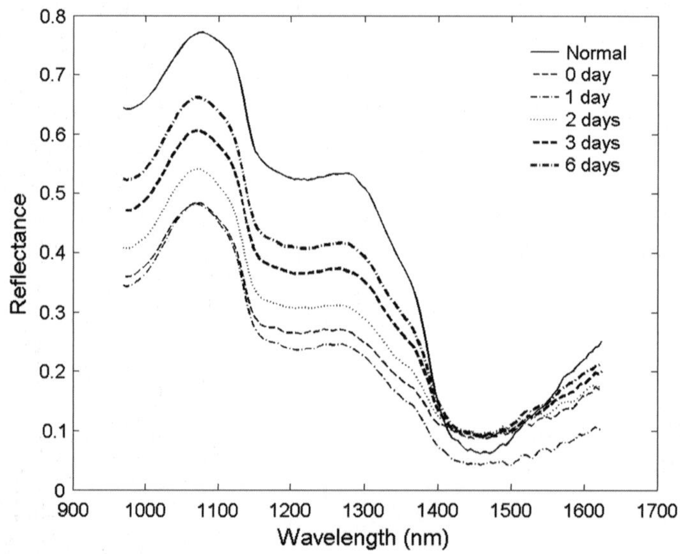

FIGURE 14.4 *Mean relative reflectance changes on bruised tissue of the cucumbers over time periods of 0–6 days after mechanical stress (reproduced from Ariana* et al., *2006. © Elsevier 2006)*

Table 14.2 Classification accuracies (in percentage) based on first principal components of NIR hyperspectral images with spectral resolution of 8.8 nm

	Days after bruising				
Spectral region	**0**	**1**	**2**	**3**	**6**
950–1650 nm	94.6	89.1	89.1	83.6	70.9
950–1350 nm	94.6	92.7	90.9	85.5	74.6
1150–1350 nm	83.6	83.6	85.5	81.8	72.7

Source: *Ariana* et al. *(2006)*

components over the period of 0–6 days. The best classification accuracies were achieved under the spectral region of 950–1 350 nm for all days after mechanical stress. This region represented wavebands where reflectance difference between normal and bruised tissue was the greatest. The classification accuracy within 2 hours of bruising was 94.6% and decreased over time to 74.6% at day 6 after bruising. Decreasing classification accuracy over time was also observed at the other two regions. This pattern was due to the decreased differences in reflectance between bruised and normal tissue.

Although only one component was used in the classification, the computation of the principal component included all spectral images for a spectral region. For real-time applications, it is more desirable to use fewer (two or three) wavelengths in order to accelerate the image acquisition and analysis process. Ratio or difference of two wavelengths followed by image segmentations using a threshold was used in this study. The best two wavelengths for the ratio and difference algorithm were found using correlation analysis of all possible wavelengths. For the ratio of two wavelengths, the best wavelengths are 988 nm and 1 085 nm, as calculated by the following equation:

$$R = \frac{R_{988\text{nm}}}{R_{1085\text{nm}}} \tag{14.1}$$

where $R_{988\text{nm}}$ and $R_{1085\text{nm}}$ are relative reflectance at 988 nm and 1 085 nm, respectively. Wavelengths 1 346 nm and 1 425 nm were found to be the best for difference calculations:

$$D = R_{1346\text{nm}} - R_{1425\text{nm}} \tag{14.2}$$

where $R_{1346\text{nm}}$ and $R_{1425\text{nm}}$ are relative reflectance at 1 346 nm and 1 425 nm, respectively. Classification accuracy based on R and D values

Table 14.3 Classification accuracies (in percentage) based on ratio and difference of two NIR spectral images

	Days after bruising				
Calculation*	0	1	2	3	6
$R = R_{988nm}/R_{1085nm}$	92.7	90.9	89.1	85.5	81.8
$D = R_{1346nm} - R_{1425nm}$	87.3	89.1	89.1	92.7	83.6

*R_{988nm}, R_{1085nm}, R_{1346nm}, *and* R_{1425nm} *are relative reflectance images at their respective wavelengths.*
Source: *Ariana* et al. *(2006)*

using threshold values of 0.79 and 0.16 respectively over a period of 0–6 days are shown in Table 14.3.

The classification accuracy based on the ratio of two wavelengths was slightly better than that based on the difference of two wavelengths for 0 and 1 days after bruising, whereas the difference of two wavelengths was superior for 3 and 6 days. Classification based on the first principal component over the region of 950–1 650 nm (Table 14.2) yielded higher accuracies at 0 and 1 day compared to classification based on *R* or *D* values (Table 14.3). However, its accuracy at 6 days after bruising was much lower than that from band difference or ratio. The general classification performance based on the first principal component was also inferior to that based on the *R* or *D* values. Classification accuracy based on the first principal component seemed more sensitive to the age of tissue bruising. Hence, the method of band ratio or difference is preferable because its classification accuracy was more stable over time.

14.3. DETECTION OF INTERNAL DEFECT

Adverse growing conditions and/or excessive mechanical load during harvest, transport, and postharvest handling are the major causes of internal damage in pickling cucumbers, which often occurs in the form of carpel separation or hollow centers (Miller *et al.*, 1995). Figure 14.5 represents a typical cucumber slice from normal and defective pickling cucumbers. These defective cucumbers would cause a bloating problem during brining if not segregated prior to the brining process. Since the hollow center is largely hidden inside cucumbers, it is difficult to detect with current machine vision systems. Sorting for internal defect is not currently performed on fresh cucumbers but only on whole desalted pickles.

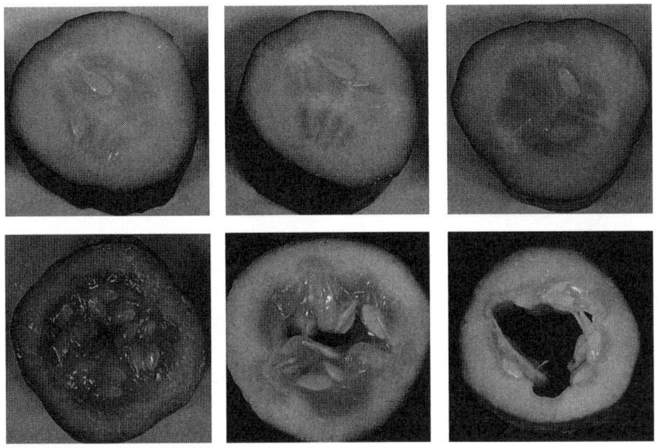

FIGURE 14.5 *Normal and defective cucumber slices (reproduced from Ariana & Lu, 2009. © American Society of Agricultural and Biological Engineers 2009). (Full color version available on http://www.elsevierdirect.com/companions/9780123747532/)*

Defective pickles are separated from normal ones by visual inspection and/or hand touch of pickles moving on conveyor belts. Human sorting and grading of defective cucumbers is not cost-effective and is also prone to error due to speed demand and fatigue. Hence it is desirable that a machine vision system be used for removing defective pickling cucumbers from normal ones prior to brining.

The first study for internal defect detection in pickling cucumbers using hyperspectral imaging was conducted under transmittance mode. Cucumbers were mounted on a rotating stage, illuminated from below, and hyperspectral transmission line scans were captured longitudinally from above using a CCD camera. Three hyperspectral line scans were obtained for each cucumber, separated by 120° (Ariana & Lu, 2008). Examples of the hyperspectral transmittance images from a normal and a defective cucumber are shown in Figure 14.6.

Generally, the defective cucumber had a brighter image (higher pixel intensities) between 700 and 900 nm. The spatial profiles across the 800 nm line showed that transmittance was higher in the defective cucumber (Figure 14.6b) than in the normal cucumber (Figure 14.6a). Further, defective cucumbers generally had a larger variation in transmittance along the scan line than normal cucumbers. Water-soaked lesions and tissue separation could account for increased light transmission in defective cucumbers. The light-scattering abilities of cellular components (e.g., cell walls, starch), which normally diffract or reflect light, might have decreased due to the fluid build-up from the ruptured cells (Miller *et al.*, 1995). When the refractive

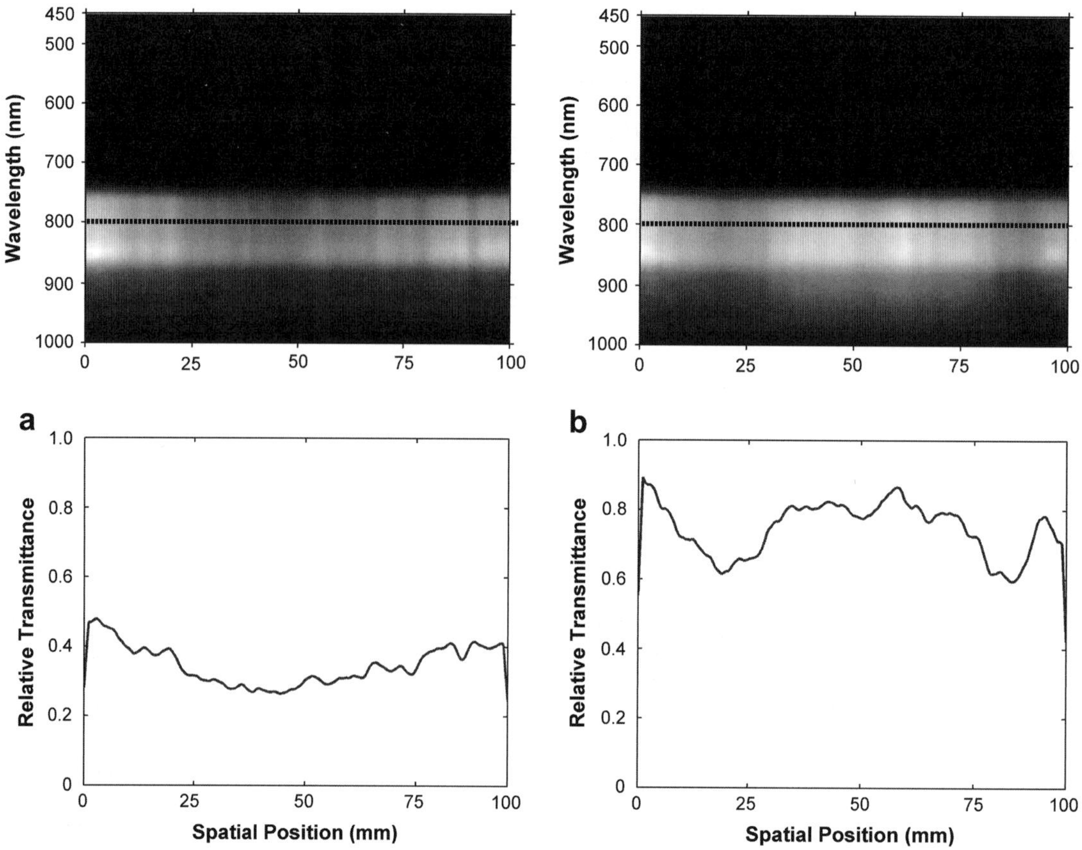

FIGURE 14.6 *Examples of hyperspectral transmittance images of (a) normal and (b) defective cucumber. Pixel intensities along the dotted line at 800 nm are presented in the graph below each image (reproduced from Ariana & Lu, 2008a. © American Society of Agricultural and Biological Engineers 2008).*

index of the cell walls and the infiltration liquid match, reflectance approaches a minimum and transmission a maximum (Vogelmann, 1993). Average classification accuracies of 93.2% and 90.5% were achieved using partial least squares-discriminant analysis (PLS-DA) and Euclidean distance measure respectively. Although resulting in lower accuracy, the Euclidean distance method is preferred because it is simple and requires only normal cucumbers for the model training.

Hyperspectral transmittance imaging is potentially useful for on-line detection of internal defect in pickling cucumbers. Further study to implement hyperspectral imaging for defect detection in an environment close to commercial line situations was conducted. A prototype on-line system using

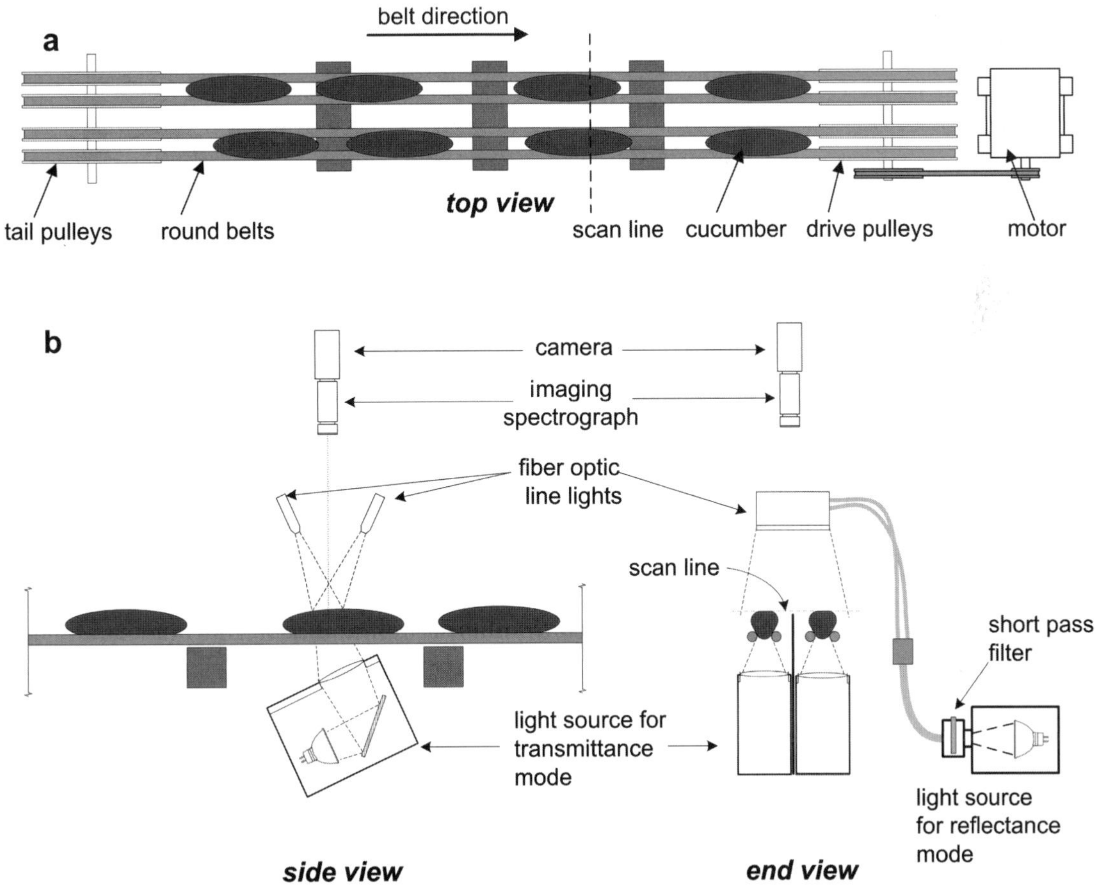

FIGURE 14.7 *Schematic of (a) conveying unit and (b) the illumination and imaging unit locations (reproduced from Ariana & Lu, 2008b. © American Society of Agricultural and Biological Engineers 2008). (Full color version available on http://www.elsevierdirect.com/companions/9780123747532/)*

belt conveyors to carry cucumbers, such as are commonly used in cucumber processing plants, was built (Figure 14.7).

The prototype included three major units: conveying, illumination, and imaging. It was designed to detect hollow center, a common internal defect in pickling cucumbers, as well as to evaluate external quality features, i.e., color and size. In the commercial setting, cucumbers are sized and sorted using multiple lanes of conveyor belts. With this consideration, the prototype was designed to operate in a two-lane mode at a rate of 1–2 pickling cucumbers per second per lane. While this sorting speed is still below that required for commercial application, it would meet the needs of testing the design concept. The two-lane configuration could operate with a single camera and hence simplify the design. The prototype also has a unique feature of simultaneous reflectance and transmittance imaging and continuous measurements of reference spectra. Reflectance imaging in the visible region of 400–740 nm was intended for external quality evaluation such as color, whereas transmittance imaging in the red and near-infrared region of 740–1 000 nm was used for internal defect detection. Separation of the two imaging modes was possible by installing a shortpass filter with the cut-off wavelength at 740 nm in front of the reflectance light source. Spectral calibration references were built in the prototype for correction of each hyperspectral image to minimize the effect of light source fluctuations. In addition, the size of cucumbers was predicted from the hyperspectral images, which could be used for pathlength correction to improve detection accuracy for internal defect.

The camera was set to run continuously with 2 milliseconds exposure time and 8×8 binning, with a conveyor belt speed of 110 mm/s or approximately up to two cucumbers per second. A representative of hypercube data captured by the system at six selected wavelengths from 500 to 1 000 nm along with their corresponding color images is presented in Figure 14.8. Images at 500, 600, and 700 nm were reflectance images that carried mostly color information; meanwhile images at 800, 900, and 1 000 nm

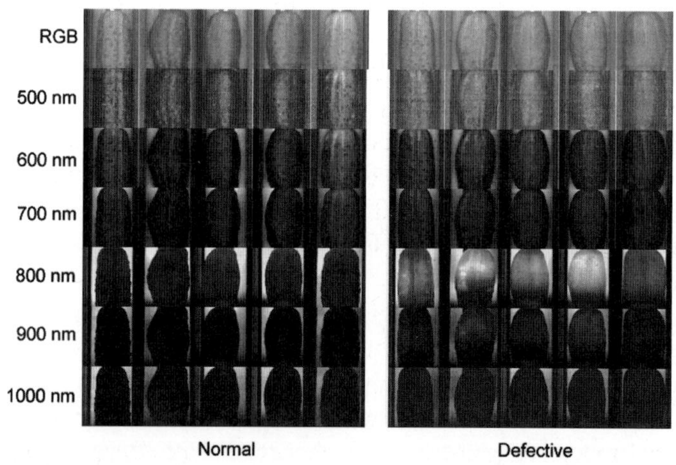

FIGURE 14.8 *Hyperspectral images for the normal (left) and defective (right) groups of pickling cucumbers and the corresponding RGB images (reproduced from Ariana & Lu, 2009. © American Society of Agricultural and Biological Engineers 2009). (Full color version available on http://www.elsevierdirect.com/companions/9780123747532/)*

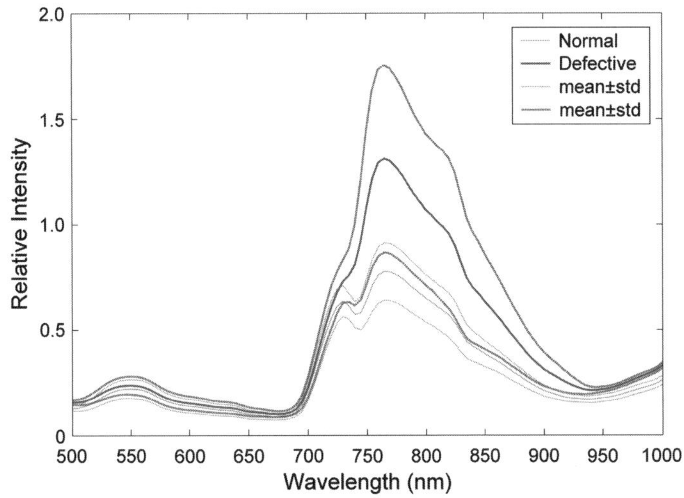

FIGURE 14.9 *Spectra of normal and defective cucumbers under reflectance (500–740 nm) and transmittance (740–1000 nm) mode images (reproduced from Ariana & Lu, 2009. © American Society of Agricultural and Biological Engineers 2009). (Full color version available on http://www.elsevierdirect.com/companions/9780123747532/)*

were transmittance images. The images on each row have been scaled for maximum contrast. For visual observation purposes, the RGB images (top row) were generated from the hyperspectral image cube in the 500–700 nm range by first computing CIE tristimulus values (X, Y, and Z) using the weighted ordinate method, followed by a conversion to RGB values. The intensity of images at 800 nm was higher (appeared brighter) compared to other wavebands. Furthermore, images at 800 and 900 nm appeared brighter in defective cucumbers than in normal cucumbers. In some severely defective cucumbers, bright areas appeared more intense compared to the surrounding pixels in the images at 800 nm (e.g., the second cucumber from the left for the defective group in Figure 14.8). These bright areas did not appear in the visible range.

Typically, the transmittance signal of pickling cucumbers was stronger in the NIR region than in the visible region (Figure 14.9). Both normal and defective spectra exhibited strong absorption at 680 and 950 nm due to chlorophyll and water absorption respectively. Local maxima at 550 nm represented the green color of cucumbers. The overlapping spectra in the visible region of 500–725 nm for normal and defective cucumbers suggested that both classes had similar color. Hence it would be difficult to segregate defective cucumbers from normal ones using hyperspectral reflectance images.

Hyperspectral imaging provides spectral information about a product item in addition to spatial features. Most hyperspectral imaging applications use a pushbroom sensing configuration to build 3-D hyperspectral image cubes, which contain spectral information for each pixel in the 2-D space. This would require a great amount of time for acquiring, processing, and analyzing images, making the technique impractical for on-line sorting and grading applications. In addition, hyperspectral image cubes are high-dimensional data and exhibit a high degree of interband correlation, leading to data redundancy that can cause convergence instability in classification models. Therefore, the use of fewer wavebands is preferable for more stable classification and easier implementation in a multispectral imaging system to meet the speed requirement of a sorting line.

Many techniques are available for selecting wavebands from hyperspectral images. One of them is correlation analysis, a common method to evaluate the relationship between input features and output. The method was proved effective in removing redundant features, and it has been used successfully for defect detection in apples and mandarins (Gomez-Sanchis *et al.*, 2008; Lee *et al.*, 2008). The waveband ratio of 940 and 925 nm resulted in an overall classification accuracy of 85.0%. The best classification accuracy was achieved using the difference of two wavebands at 745 and 850 nm with an overall classification accuracy of 90.8% (Ariana & Lu, 2009).

Representative images for the classification results are shown in Figure 14.10. Coloration was added to the cucumbers and defective areas to

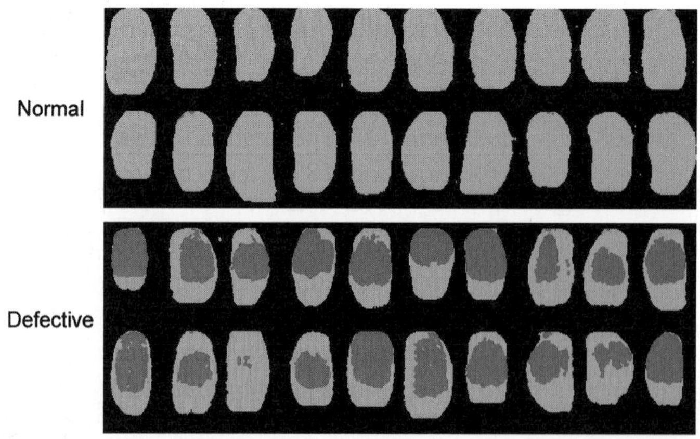

FIGURE 14.10 *Segmented images of normal and defective cucumbers (damage areas are denoted with red color) images (reproduced from Ariana & Lu, 2009. © American Society of Agricultural and Biological Engineers 2009). (Full color version available on http://www.elsevierdirect.com/companions/9780123747532/)*

enhance visual observation. Most of the defective areas in the segmented images appeared large, although smaller areas were also identified, which might have been due to damage in the form of water soak lesions in the mesocarp region near the surface. Severely defective cucumbers would transmit more light because they had a substantial portion of the endocarp missing, which was consequently filled with air.

14.4. CONCLUSIONS

Automated sorting and grading of fruits and vegetable can reduce industry dependence on human inspectors, reduce production cost, and improve product consistency and wholesomeness. Many pickling cucumber processors are currently using machine vision systems in their lines for sorting cucumbers based on size, shape, and color. The systems are not designed for detection of external or internal damage in cucumbers, therefore they are incapable of detecting bruise damage in pickling cucumbers in the form of water soak lesions, carpel separation or hollow center. With the stringent quality control requirements, the presence of external and/or internal defect can lead to rejection and make the processor liable for economic loss.

Studies on the application of hyperspectral imaging for detection of defects in pickling cucumbers show a potential to use the technology in commercial lines. The simultaneous hyperspectral reflectance and transmittance imaging system can simplify the operation and reduce cost by having multiple inspections (size, shape, color, external, and internal bruise) in one station. However, further research is needed to make the technology applicable in the industry. While hyperspectral data are rich in information, processing the hyperspectral data poses several challenges regarding computation speed requirements, information redundancy removal, relevant information identification, and modeling accuracy. Hyperspectral imaging studies are often conducted as a precursor to the design of a multispectral imaging system using 3–4 wavebands for real-time applications.

NOMENCLATURE

Abbreviations

CCD	charge-coupled device
CIE	Commission internationale de l'éclairage (International Commission on Illumination)
InGaAs	indium gallium arsenide

NIR	near infrared
PLS-DA	partial least squares-discriminant analysis
RDLE	refreshed delayed light emission
RGB	red, green, blue

REFERENCES

Abbott, J. A., Miller, A. R., & Campbell, T. A. (1991). Detection of mechanical injury and physiological breakdown of cucumbers using delayed light emission. *Journal of the American Society for Horticultural Science, 116*(1), 52–57.

Ariana, D. P., & Lu, R. (2008a). Detection of internal defect in pickling cucumbers using hyperspectral transmittance imaging. *Transactions of the ASABE, 51*(2), 705–713.

Ariana, D. P., & Lu, R. (2008b). Quality evaluation of pickling cucumbers using hyperspectral reflectance and transmittance imaging. Part I: Development of a prototype. *Sensing and Instrumentation for Food Quality and Safety, 2*(3), 144–151.

Ariana, D. P., & Lu, R. (2009). Wavebands selection for a hyperspectral reflectance and transmittance imaging system for quality evaluation of pickling cucumbers. *ASABE Paper No. 096872*. St Joseph, MI: ASABE.

Ariana, D. P., Lu, R., & Guyer, D. E. (2006). Near-infrared hyperspectral reflectance imaging for detection of bruises on pickling cucumbers. *Computers and Electronics in Agriculture, 53*(1), 60–70.

FAOSTAT (2009). Food and Agriculture Organization of the United Nations. Available online at http://faostat.fao.org/site/567/DesktopDefault.aspx?PageID=567 (accessed April 3, 2009.)

Gomez-Sanchis, J., Camps-Valls, G., Molto, E., Gomez-Chova, L., Aleixos, N., & Blasco, J. (2008). Segmentation of hyperspectral images for the detection of rotten mandarins. *Image analysis and recognition*. Berlin: Springer. 1071–1080.

Gowen, A. A., O'Donnell, C. P., Cullen, P. J., Downey, G., & Frias, J. M. (2007). Hyperspectral imaging—an emerging process analytical tool for food quality and safety control. *Trends in Food Science and Technology, 18*(12), 590–598.

Hammerschmidt, R., & Marshall, D. E. (1991). Potential use of peroxide in external bruise assessment. *Proceedings of the Annual Report to the Pickling Cucumber Industry Committee*.

Lee, K., Kang, S., Delwiche, S. R., Kim, M. S., & Noh, S. (2008). Correlation analysis of hyperspectral imagery for multispectral wavelength selection for detection of defects on apples. *Sensing and Instrumentation for Food Quality and Safety, 2*(2), 90–96.

Lucier, G., & Lin, B. H. (2000). American relish cucumbers. *Agricultural Outlook: Economic Research Service, United States Department of Agriculture*.

Marshall, D. E., Cargill, B. F., & Levin, J. H. (1972). Physical and quality factors of pickling cucumbers as affected by mechanical harvesting. *Transactions of the ASAE, 15*(4), 604–608, 612.

Marshall, D. E., Levin, J. H., & Heldman, D. R. (1973). Density sorting of green stock cucumbers for brine stock quality. *ASAE Paper No. 73-304*. St Joseph, MI: ASAE.

Miller, A. R. (1989). Mechanical stress-induced changes in sugar composition of cell walls from cucumber fruit tissues. *Phytochemistry, 28*(2), 389–392.

Miller, A. R. (1992). Physiology, biochemistry and detection of bruising (mechanical stress) of fruits and vegetables. *Postharvest News and Info, 3*, 53N–58N.

Miller, A. R., & Kelley, T. J. (1989). Mechanical stress stimulates peroxidase activity in cucumber fruit. *HortScience, 24*(4), 650–652.

Miller, A. R., Dalmasso, J. P., & Kretchman, D. W. (1987). Mechanical stress, storage time, and temperature influence cell wall-degrading enzymes, firmness, and ethylene production by cucumbers. *Journal of the American Society for Horticultural Science, 112*(4), 666–671.

Miller, A. R., Kelley, T. J., & White, B. D. (1995). Nondestructive evaluation of pickling cucumbers using visible-infrared light transmission. *Journal of the American Society for Horticultural Science, 120*(6), 1063–1068.

Osborne, B. G., Fearn, T., & Hindle, P. H. (1993). *Practical NIR spectroscopy with applications in food and beverage analysis*. Harlow, UK: Longman Scientific & Technical.

Park, B., Lawrence, K. C., Windham, W. R., & Smith, D. P. (2006). Performance of hyperspectral imaging system for poultry surface fecal contaminant detection. *Journal of Food Engineering, 75*(3), 340–348.

Sargent, S. A., & Maynard, D. N. (2009). Postharvest biology and technology of cucurbits. In J. Janick (Ed.), *Horticultural reviews* (pp. 315–354). Hoboken, NJ: Wiley–Blackwell.

Vogelmann, T. C. (1993). Plant tissue optics. *Annual Review of Plant Physiology and Plant Molecular Biology, 44*, 231–251.

Wang, W., & Paliwal, J. (2007). Near-infrared spectroscopy and imaging in food quality and safety. *Sensing and Instrumentation for Food Quality and Safety, 1*(4), 193–207.

Wilson, J. E., & Baker, L. R. (1976). Inheritance of carpel separation in mature fruits of pickling cucumbers. *Journal of the American Society for Horticultural Science, 101*, 66–69.

CHAPTER 15

Classification of Wheat Kernels Using Near-Infrared Reflectance Hyperspectral Imaging

Digvir S. Jayas, Chandra B. Singh, Jitendra Paliwal
Biosystems Engineering, University of Manitoba, Winnipeg, MB, Canada

CONTENTS

15.1. INTRODUCTION

Wheat is one of the most important staple foods in the world, with an annual global production of 630 million tonnes (FAO, 2006). Wheat is used as raw material in making breads, cakes, cookies, pastries, crackers, and in the manufacturing of pasta products such as macaroni and spaghetti. The processing and quality of these end-products is highly influenced by the class of wheat used as raw material. In trading of wheat, different varieties of the same class are assigned different grades within a class and the market price is decided based on the assigned grade. Grading of wheat is done by taking into consideration chemical and physical properties, growing region, growing season, color, texture, protein content, and hardness/vitreousness. Varietal identification is also important for the plant breeders to differentiate between genotypes. According to growing season, wheat varieties are classified into winter and spring wheat in Canada and the United States. Winter wheat varieties are sown in fall (autumn) and harvested in summer whereas spring wheat varieties are sown in spring and harvested in early fall. Spring wheat has better quality characteristics than the winter variety in terms of grain protein content, grain hardness, milling and flour quality measurements, dough physicochemical properties, and baking characteristics (Maghirang *et al.*, 2006). Wheat hardness is an important parameter in classification of wheat, affecting processing and end-product quality. Hard wheat varieties contain higher protein and are used for making bread. Soft wheat varieties usually have lower protein content and are used for making cakes, cookies, pastries, and crackers (Uri & Beach, 1997). In the United States, wheat is generally classified into three major hardness classes, namely, soft, hard hexaploid, and durum (Maghirang & Dowell, 2003).

Wheat kernels are also grouped into vitreous and non-vitreous kernels. Vitreousness is a measure of wheat quality associated with protein content and is an important classification criterion in the grading of wheat (Wang *et al.*, 2003). Vitreous kernels have a glossy or shiny appearance indicating harder kernels, high protein content, higher semolina yield, superior pasta color, better cooking quality, and thus command a higher price (Xie *et al.*, 2004). Non-vitreous kernels are chalky, opaque, softer, and have lower quality attributes (Wang *et al.*, 2005). Durum wheat is considered the hardest wheat with a very high protein content among all the wheat classes and grouped into a separate class. Durum wheat is specially used in the manufacture of pasta products, and some countries (Italy, France, and Spain) allow only durum wheat in pasta making and inspect for any adulteration by other classes (Cocchi *et al.*, 2006). Wheat is also grouped into commercial classes according to the kernel color as red and white wheat. Red and white wheat have different end uses (Ram *et al.*, 2004) and milling, baking, and taste properties of wheat vary according to its color (Pasikatan & Dowell, 2003). Wheat classes and contrasting color wheat classes are given in Table 15.1. The top three US wheat grades have only 1–3% tolerance for contrasting classes of wheat (Archibald *et al.*, 1998).

Wheat grown in western Canada is officially classified into eight commercial classes: Canada Western Red Spring (CWRS), Canada Western

Table 15.1 Contrasting color wheat classes (USDA, 2004)

Wheat class	Contrasting class
Hard red winter and hard red spring	Durum, hard white, soft white, and unclassed wheat
Durum	Hard red spring, hard red winter, soft red winter, hard white, soft white, and unclassed wheat
Soft red winter	Durum and unclassed wheat
Hard white and soft white	Durum, hard red spring, hard red winter, soft red winter, and unclassed wheat

Red Winter (CWRW), Canada Prairie Spring Red (CPSR), Canada Prairie Spring White (CPSW), Canada Western Soft White Spring (CWSWS), Canada Western Hard White Spring (CWHWS), Canada Western Extra Strong (CWES), and Canada Western Amber Durum (CWAD) wheat (CGC, 2008). The class characteristics of Canadian wheat are given in Table 15.2.

Most of the wheat-exporting/importing countries have regulations enforced by inspecting agencies for the quality inspection of grain. In Canada, the official grain grading system is regulated by the Canadian Grain Commission (CGC) under the Canada Grain Act (1975). For a very long time, the CGC used kernel visual distinguishability (KVD) characteristics for wheat classification and registration of new varieties for commercial production. However, KVD has been removed as a class identification or registration tool from August 1, 2008, due to its constraints in identification of new wheat varieties. In Canada all new wheat varieties must be registered for commercial production. In the United States, the Grain Inspection, Packers and Stockyards Administration (GIPSA) uses visual characteristics to classify wheat into eight standard commercial classes, established under the United States

Table 15.2 Class characteristics of Western Canadian Wheat

Wheat class	Color	Size	Shape	Germ	Brush	Cheeks
Canada Western Red Spring	Translucent red	Small to midsize	Oval to ovate	Round, midsize to large	Varies	
Canada Western Red Winter	Orange to opaque red	Small to midsize	Elliptical	Small, oval to round	Small	Round
Canada Prairie Spring Red	Opaque red to orange	Midsize to large	Ovate to elliptical, incurved base	Midsize to small, oval	Small to midsize	
Canada Prairie Spring White	White	Midsize to large	Ovate to elliptical, incurved base	Midsize, oval	Small to midsize	
Canada Western Soft White Spring	White	Small to midsize	Ovate to oval	Small, oval	Varies	
Canada Western Hard White Spring	White	Small to midsize	Oval to ovate	Round, midsize to large	Varies	
Canada Western Extra Strong	Dark to medium red	Large	Ovate, s-shaped base	Large, wide, typically round	Large, collared ventrally	Round
Canada Western Amber Durum	Amber	Large to midsize	Elliptical	Large, wide oval to rectangular	Varies	Angular

Source: *Canadian Grain Commission. Available online: http://www.grainscanada.gc.ca/wheat-ble/classes/classes-eng.htm.*

Grain Standard Act (USGSA) (Lookhart *et al.*, 1995): hard red winter (HRW), soft red winter (SRW), hard red spring (HRS), soft white (SWH), hard white (HDWH), durum (DU), mixed (XWHT), and unclassified (UNCL).

Australia, which is among the top wheat exporters, also follows a strict protocol on wheat classification. Australian wheat is classified into Australian prime hard, Australian hard, Australian premium white, Australian standard white, Australian premium durum, Australian general purpose, and feed wheat classes based on their protein content that are priced accordingly (Cracknell & Williams, 2004). In France, the Department of Agriculture uses different quality criteria for wheat classification based on Alveograph and bread-making tests and classifies wheat based on end-use products into four classes, namely, high grade bread-making (BPS), regular bread-making (BPC), biscuit wheat (BB), and wheat for other purposes (BU) (Cracknell & Williams, 2004). Other European countries also use similar end-product-based classification criteria.

Though most of the wheat-exporting countries classify wheat into commercial classes considering end-product quality characteristics, methods and accuracy of classification in grading of wheat vary widely. Various methods used for class identification of wheat and their advantages and disadvantages are discussed in the next section.

15.2. CLASSIFICATION METHODS

15.2.1. Visual Identification

Classification of wheat by visually identifying wheat kernels based on color, shape, size, and texture is the most common and simplest method used in the grain handling and trading industry. Canadian Grain Commission inspectors use visual kernel characteristics to assign an official class to procured wheat. In the United States, grain inspectors use morphological (kernel length, kernel width, slope of the back of kernel, germ size, germ angle, brush size, cheek shape, and crease) and surface textural features for official classification of wheat (Lookhart *et al.*, 1995). There are hundreds of varieties of wheat registered in the United States and Canada and sometimes it is difficult even for experienced grain inspectors to assign a correct class and grade to the wheat received at terminal elevators and grain handling facilities. Visual identification requires well-trained personnel and there is subjectivity involved in grade assignment. The grain trading industry is looking for an alternative, low cost, fast, and accurate classification technique. With the removal of KVD in Canada, the need for such a system is very urgent.

15.2.2. Laboratory Methods

15.2.2.1. Phenol test

The phenol test is a simple method used in varietal identification of wheat. Color contrast between wheat varieties is enhanced by soaking the wheat samples in phenol. This method is time-consuming (15 min to 4 h) and subjective as the degree of coloration of the sample is determined manually. Phenol is a toxic compound and extra care should be taken in handling it to avoid skin burns. The phenol test may not be a reliable method for wheat classification due to possible overlap of colors of wheat varieties in different classes.

15.2.2.2. Gel electrophoresis

Electrophoretic identification of wheat by analyzing the wheat-protein is one of the established methods used by the grain processing industry. In electrophoresis analysis, the grain samples are milled and proteins are extracted using specific solvents. A gel material is used and protein is applied to the top of the gel material slab held between parallel plates. An electrolyte buffer is used to conduct an electric current applied to it. Protein molecules migrate on the gel surface according to the electric charge produced by them which is proportional to the size of protein molecules. At the end of each experiment, the gel material is stained to improve visualization and to clearly mark the horizontal lines in the gel representing proteins. Some protein bands are specific to certain wheat varieties and are used for classification and varietal identification. Various electrophoretic methods such as starch gel electrophoresis, polyacrylamide gel electrophoresis analysis (PAGE), acid polyacrylamide gel electrophoresis analysis (A-PAGE), sodium dodecyl sulphate polyacrylamide gel electrophoresis analysis (SDS-PAGE), and gel isoelectric focusing (IEF) are currently being used. Though electrophoresis is reliable, the method has some constraints as it is destructive and time-consuming and cannot be implemented in grain handling facilities for on-line identification and classification of wheat.

15.2.2.3. High performance liquid chromatography (HPLC)

High performance liquid chromatography (HPLC) is a well-established method for protein analysis of cereals and other grains. Protein content to some degree is specific to wheat varieties and classes. This method has been used in quantitative calibrations of proteins in wheat as well as in varietal identification and classification using specific proteins such as albumins, globulins, and glutenins. The reverse-phase high performance liquid chromatography (RP-HPLC) is an advanced form of HPLC. This method is faster and can be used for automatic data collection and computerized analysis of samples. The disadvantages of this method are that it is destructive, expensive, and difficult to implement under field conditions at grain elevators for on-line classification.

15.2.2.4. DNA testing and immunoassay

DNA-based methods have been investigated to identify wheat varieties using polymer chain reaction (PCR) and amplifying it by a wide range of markers such as simple sequence repeat (SSR) and sequence tagged site (STS) markers. DNA of the ground wheat samples is extracted and then markers are derived which have discriminative capability. DNA-based methods are not affected by environmental factors but these methods are destructive and time-consuming.

15.2.3. Non-destructive Methods

15.2.3.1. Near-infrared spectroscopy

Near-infrared (NIR) spectroscopy is a non-destructive and rapid technique which is being used for quality evaluation of many cereals and other grains (Singh *et al.*, 2006). The NIR spectroscopic technique works on the principle that when light strikes an object, the unique chemical composition of the material causes molecules to absorb, reflect or transmit the light. Molecules absorb a part of the incident light in the form of electromagnetic radiation and jump into higher energy levels depending on the wavelength and intensity of the radiation source, and vibrate at unique frequencies (Murray & Williams, 1990). The remainder of the incident light is either reflected or transmitted through the material. The reflected or transmitted light at multiple wavelengths in the NIR region (700–2 500 nm) is recorded by a spectrometer to form the spectra from which qualitative and quantitative information is extracted by chemometric methods. Near-infrared transmittance (NIT) spectroscopy is used for protein content analysis and moisture measurement of wheat in grain handling facilities. The NIR method has also been investigated for hardness measurement (vitreousness) and classification of wheat (Dowell, 2000). However, NIT or NIR reflectance instruments have more than 40 sources of error in them associated with instruments, samples, and operator sources (Williams *et al.*, 1998). This technique has been restricted to lab use only owing to certain drawbacks such as the imprecise nature of the estimates, complex development of robust calibration models, and inconsistency across several individual instruments (Toews *et al.*, 2007).

15.2.3.2. Machine vision

Machine vision is an advanced object recognition technique which is currently being used for quality assessment of many agricultural and food products. Due to advancements in machine vision technology, e.g., high resolution cameras, powerful computers integrated with sophisticated image acquisition and processing software and robust artificial intelligence systems, along with reduced system cost, this technique has emerged as an alternative to human visual inspection with faster, consistent, and greater classification accuracy. Quantitative information from the digital images with high discrimination capability is extracted and given as input to an artificial intelligence system for improved classification. High-speed digital image acquisition cameras operating in different ranges of electromagnetic spectra (X-rays, ultraviolet, visible, NIR, infrared, radiowaves) are available now. The charge-coupled device (CCD) color cameras are low-priced and widely used in machine vision systems. Color images have been used in grain quality analysis to identify different grain types, varieties, classes, impurities, fungal-infected, and insect-damaged kernels. Color images are described by color, textural, and morphological features and are used in the quality assessment of grain. However, color images do not provide information about the chemical composition and its distribution in the kernel. In many wheat varieties and classes these external features are very similar and do not carry any discriminative information for classification and have failed to give satisfactory results when there is high degree of overlap between the classes to be discriminated (Utku, 2000). Therefore, compositional information of the kernels is very desirable to discriminate the wheat classes.

The near-infrared region of the electromagnetic spectrum has absorption bands associated with wheat protein, other kernel compositions, and functionality (Pasikatan & Dowell, 2004). Hyperspectral imaging provides the spectral information in a spatially resolved manner. Spatial information is important for monitoring and visualization of the grain as it can be used to extract the chemical mapping of the sample from the hyperspectral data. NIR hyperspectral imaging is discussed in next section.

15.3. NIR HYPERSPECTRAL IMAGING

A hyperspectral imaging (HSI) system mainly consists of a detector, illumination source, wavelength selection device, image acquisition software, and an integrated computer. There are three types of HSI systems based on sample presentation techniques: point scan, line scan (pushbroom), and

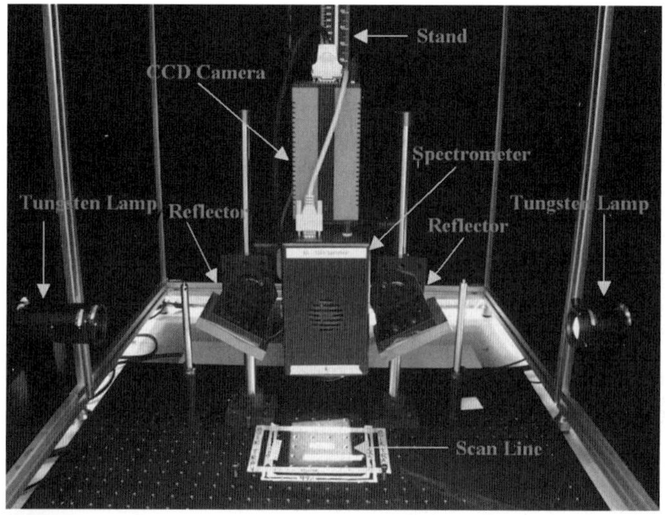

FIGURE 15.1 *Line scan NIR hyperspectral imaging system (pushbroom type) (courtesy: Grain Research Laboratory, Canadian Grain Commission, Winnipeg, MB, Canada). (Full color version available on http://www.elsevierdirect.com/companions/9780123747532/)*

focal plane arrays (FPA) (area scan). The selection of above-mentioned hardware components depends on the choice of imaging system and related application. The line scan and (FPA)-based HSI systems are better suited for food quality inspection (Kim *et al.*, 2001). In line scan imaging full-spectral information for each pixel in one spatial dimension (line) is collected and successive line scans are combined to form a three-dimensional hypercube (Figure 15.1). This system is suitable for the scanning of moving objects. In an FPA-based imaging system (Figure 15.2) the spatial information is collected at each wavelength sequentially to form a 3-D array (hypercube). The first two dimensions of the hypercube represent spatial features (pixels) and the third dimension represents the spectral features (wavelength). This type of imaging system is used mainly to scan images of stationary objects. The main hardware and software components of a HSI system consist of a detector, wavelength filtering device, illumination source, and software to record, transfer, and process the acquired hyperspectral data.

15.3.1. Detectors

Detectors record the sample spectra by selecting the suitable mode of energy wave (reflectance or transmittance). The single-channel NIR detector (point scan) uses lead sulphide (PbS) in the range of 1 100 to 2 500 nm, silicon

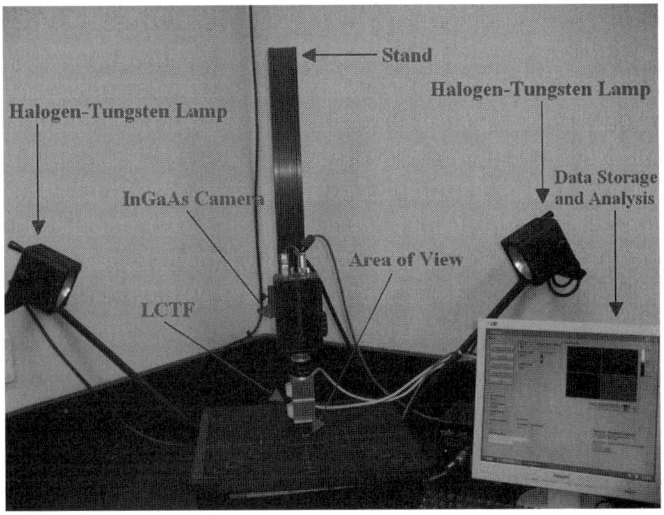

FIGURE 15.2 *Area scan NIR hyperspectral imaging system (focal plane arrays-based system) (courtesy Canadian Wheat Board Centre for Grain Storage Research, Winnipeg, MB, Canada). (Full color version available on http://www.elsevierdirect.com/companions/9780123747532/)*

detectors in the range of 360 to 1 050 nm, and indium gallium arsenide (InGaAs) detectors in the range of 900 to 1 700 nm wavelengths. In line scan imaging, a linear array of detectors (Silicon, InGaAs) is used. In FPA-based imaging, 2-D arrays of detectors, also known as focal plane arrays (FPA), are used. Silicon diode arrays or CCDs are suitable for visible and shortwave NIR regions. Different types of commercial FPAs currently available are: indium antimonide (InSb), platinum silicide (PtSi), indium gallium arsenide (InGaAs), germanium (Ge), mercury cadmium telluride (HgCdTe), and quantum well infrared photodetectors (QWIPs). The InGaAs camera has better sensitivity, wider spectral range, and faster response in the NIR region. High quality chips are produced by changing the thickness of film in $In_xGa_{1-x}As$ sensors, where x and $1-x$ are the concentrations of InAs and GaAs, respectively.

15.3.2. Wavelength Filtering Devices

Wavelength filtering devices such as optical interference filters, grating devices (e.g., prism–grating–prism), and electronically tunable filters (ETF) are used to obtain the light of desired wavebands and remove out-of-band radiation. Acousto–optical tunable filter (AOTF) and liquid crystal tunable filter (LCTF) are two advanced ETFs that have a relatively large optical aperture, high spectral resolution, wide spectral range, and can randomly access tuning wavelengths (Wang & Paliwal, 2007).

15.3.3. Illumination Sources

Tungsten–halogen lamps, quartz–halogen lamps, light emitting diodes (LED), and tunable lasers are used as light illumination sources in NIR instruments. Heated xenon lamps can also be used as sources of illumination in NIR instruments. The application of LED is restricted to only narrow-bands (400–900 nm) of wavelengths. Tungsten–halogen lamps are the most common illumination sources used in NIR hyperspectral imaging due to their durability, stability, and capability to emit light in a broad spectral range (400–2 500 nm).

15.3.4. Integration of Hardware and Software

The image data captured by NIR detectors are digitized and transferred to a computer for storage and analysis. Four standard communication interfaces namely Parallel, FireWire (IEEE 1394), CAMERA Link, and GiGE VISION are used to transfer digital image data between camera and computer. Parallel cameras have a high data transfer rate but they require customized cables due to lack of interface standard. FireWire is a standardized interface but has a lower data transfer speed. CAMERA Link uses a standard channel link chip which is used in the camera and frame grabber for data transmission. An advanced personal computer bus system such as peripheral component interconnect (PCI) express can handle fast data streams transferred by the CAMERA Link cable. GiGE VISION is the latest developed standard interface with a very high data transfer rate but can be used in limited bandwidth. FireWire and GiGE VISION communication interfaces do not require a frame grabber board and are directly connected to the computer. A visual programming platform (e.g., LabVIEW, National Instruments, Austin, Texas) can be used to integrate hardware components, control input parameters, acquire, and store the hyperspectral data. MATLAB can be used as a powerful tool to preprocess, analyze, and classify the hyperspectral data by developing code with the help of several inbuilt functions in image processing, statistics, wavelet, and neural network. Various calibration, preprocessing, data reduction, and classification methods, given in Table 15.3, can be applied to single kernel and bulk analysis of wheat and other grains.

Table 15.3 Analysis of NIR hyperspectral image data for wheat classification

Type of analysis	Calibration and preprocessing	Data reduction	Feature extraction	Classification
Bulk grain analysis	Dark count, normalization, geometric distortion correction, dead pixels removal, image cropping	Averaging, PCA, ICA, FA	Spectral, textural, morphological, wavelet	Statistical classifiers, neural network, GA, fuzzy logic, SVM, ML
Non-touching single kernels	Dark current, normalization, geometric distortion correction, background and dead pixels removal, labeling kernels	Averaging, PCA, ICA, FA	Spectral, textural,* morphological, wavelet*	Statistical classifiers, neural network, GA, fuzzy logic, SVM, ML
Touching single kernels	Dark current, normalization, geometric distortion correction, background and dead pixels removal, separation of touching kernels and labeling	Averaging, PCA, ICA, FA	Spectral, textural,* morphological, wavelet*	Statistical classifiers, neural network, GA, fuzzy logic, SVM, ML

PCA, principal component analysis; ICA, independent component analysis; FA, factorial analysis; GA, genetic algorithm; SVM, support vector machine; ML, maximum likelihood classifier.

**Many textural (e.g., graylevel co-occurrence matrix (GLCM)) and wavelet features require selecting rectangular or square region of interest (ROI) inside the kernel.*

15.4. WHEAT CLASSIFICATION BY NIR HYPERSPECTRAL IMAGING

Archibald *et al.* (1998) developed a shortwave NIR imaging system to classify wheat into color classes. Their system consisted of a CCD monochrome camera, two tungsten–halogen lamps, a LCTF filter (632–1 100 nm), a frame grabber, and an analog board. The reference spectral characteristics of six kernels, three each of hard red spring (HRS) and hard white winter (HDWW) wheat, were first determined by an NIR spectrometer and significant wavelengths were selected to predict the percentage of red color of the sample by multiple linear regression. Then bulk samples of mixtures of HRS and HDWW wheat (50:50) were scanned in a spectral imaging system at 11 selected wavelengths. Principal component analysis (PCA) was used to analyze the data after reshaping into 2-D arrays. Each pixel in the 640×480 size NIR image was considered as a sample and each of the 11 wavelengths as variable, thus resulting into a $307\,200 \times 11$ size 2-D matrix after reshaping the 3-D hyperspectral data.

Principal component scores were mapped into pseudo images and eight PC scores were examined. The score images showed the contrast between red and white wheat; however, the performance was poorer than the spectroscopic method. The imaging approach demonstrated the effect of non-uniform illumination, and saturated and white pixels. In this study, the authors did not develop any supervised classification algorithm for future classification. Image features (morphology, texture, and wavelet) from the bulk images can be extracted and used for training of statistical classifiers (linear, quadratic, and Mahalanobis), support vector machine (SVM), and artificial neural network for future classification.

Gorretta *et al.* (2006) used an NIR hyperspectral imaging system to determine vitreousness of wheat kernels. Durum wheat samples were grouped into three classes: class I (100% vitreous kernels), class II (partial vitreous kernels in which the germ contained 30–60% starchy area), and class III (100% starchy grain). Class II and class III were considered as sub-classes of non-vitreous kernels. Their imaging system included a 1 024 × 1 024 pixel size CCD camera with 16 bit digitizer, linear LCTF filter, and two non-dichroic halogen lamps for illumination. Single durum wheat kernels were scanned at 91 equally distributed wavelengths in the 650–1 100 nm range, i.e., at 5 nm intervals. The mean reflectance spectra of the kernels were obtained after applying erosion to the images to remove kernel contour. Spectral pretreatment methods of standard normal variate (SNV), standard deviation, and logarithmic transform of spectrum followed by second derivative (Savitsky–Golay) were also applied to mean reflectance to improve the classification. The dimensionality of data was further reduced by partial least squares (PLS). Membership degree for each class was calculated by factorial discriminant analysis (FDA). Mahalanobis distance was used to assign a class to an individual sample. Their classification model perfectly discriminated vitreous kernels from non-vitreous kernels and also separated non-vitreous kernels into two subclasses with 94% accuracy. Wavelengths of 910, 990 and 1 030 nm were related to absorption bands of protein, starch, and amino acids, respectively. In this study only spectral analysis in a limited range (650–1 100 nm) was carried out. It is speculated that inclusion of visible imaging and extending the NIR range (up to 2 500 nm) may further improve the classification.

Berman *et al.* (2007) used HSI to classify wheat as sound or stained grains. Stained grains are discolored grains such as black point, field fungi, and pink stain defects caused by pre-harvest weather conditions that affect the commercial value of the wheat. Hyperspectral images were generated by pixel-based (point-based scanning) spectrometer in the 350–2 500 nm range at 1 nm intervals producing 2 151 reflectance values per pixel spectra. The samples were kept in the slots (300 kernels) in a tray in raster arrangement and scanned. In each sub-raster (holding the grain) 60 points were scanned. Thus for each tray 18 000 (300 × 60) points were scanned at 2 151 wavebands to generate a hypercube, which took nearly 10 hours. The dimensionality was reduced by eliminating first 70 bands and co-averaging 10 consecutive bands reducing them to 208 from 2 151. The spectral data were analyzed using 420–2 500 nm (all 208 spectral bands), 420–1 000 nm (58 spectral bands) and 420–700 nm (28 spectral bands) spectral regions. Penalized discriminant analysis (PDA) was used to select pixel classification as sound or as one of the stains, and then grains were classified accordingly. More than 95% classification was achieved and results were similar in the full spectral range as well as reduced spectral ranges. The dimensionality of hyperspectral data in the reduced spectral range (420–1 000) can be further reduced by applying multivariate image analysis (MVI) and only a few selected wavelengths can be used for classification, which will drastically reduce the scanning and classification time.

Shahin & Symons (2008) investigated the potential of NIR hyperspectral imaging to classify wheat into vitreous and non-vitreous kernels, which is one of the most difficult quality parameters to detect by visual inspection. Non-uniform distribution of kernel characteristics (starchy, piebald, and bleached) makes spatial information critical in sample analysis as spectral information can be used for damage detection and spatial information can be used for grain classification. In this study, bulk wheat samples were scanned in the 950–2 450 nm wavelength range. The reflectance spectra obtained by averaging kernel pixels and pretreated with Savitzky–Golay smoothening and second derivative showed clear spectral differences between vitreous, starchy, piebald, and bleached kernels. Their study indicated the enormous potential of hyperspectral imaging to develop supervised classification algorithms for classification of vitreous wheat kernels.

Mahesh *et al.* (2008) investigated the feasibility of NIR hyperspectral imaging to differentiate eight Canadian wheat classes. Their imaging system consisted of a thermoelectrically cooled 640 × 480 spatial resolution InGaAs camera, electronically tunable LCTF filter, lens, two tungsten halogen lamps, data acquisition board, and system control program written in LABVIEW environment. Bulk images of wheat samples were scanned in the 960–1 700 nm wavelength range at 10 nm intervals (total 75 wavelengths) after applying necessary calibrations. An area of 200 × 200 pixels around the central pixel was cropped and pixel intensities at each of the 75 wavelengths were averaged to form a spectrum which was then normalized using a standard 99% reflectance panel. Significant wavelengths from the averaged spectra of the wheat samples were selected by applying PROC STEPDISC (SAS Institute Inc., Cary, NC, USA) using the criteria of partial least square (R^2) and average squared canonical correlation (ASCC). Classification models were developed using statistical discriminant analysis (linear and quadratic) and back propagation neural network (BPNN) classifiers. PROC DISCRIM (SAS Institute Inc., Cary, NC, USA) was used to develop statistical classifiers based on the leave-one-out cross-validation method. Two three-layer neural network architectures (standard BPNN and Wardnet BPNN) were used in developing classifiers. The standard BPNN had 75 inputs, 79 hidden nodes, and eight outputs with linear scaling function in the input layer and logistic activation function in the output layer. The Wardnet structure had three slabs in hidden layers and each slab had 26 nodes with a specific activation function (Gaussian, Gaussian complement, or Tanh). In neural network classification, two sample sets with percentages of 60–30–10 and 70–20–10 for training, testing, and validation, respectively, were used. The BPNN models correctly classified more than 90% wheat samples. A linear discriminant analysis classifier gave the best classification accuracy and correctly classified 94–100% wheat samples (Table 15.4).

In the study of Mahesh *et al.* (2008), only spectral analysis of hyperspectral data was performed for feature extraction. To further improve the classification, Choudhary *et al.* (2008) explored the potential of wavelet texture analysis for the identification of wheat classes using the hyperspectral image data of Mahesh *et al.* (2008). A central area of 256 × 256 pixels in each of 75 image slices of a hypercube was cropped and a Daubechies-4 wavelet transform was applied up to five levels of resolution. Two textural features (energy and entropy) were extracted at each level in horizontal, vertical, and diagonal directions. One additional rotationally invariant feature was obtained by adding these three features at each resolution level resulting in 40 features (8 × 5) per slice and 3 000 features (40 × 75) per hyperspectral

Table 15.4 Classification of Canadian wheat by NIR hyperspectral imaging

	Classification accuracy (%)	
Wheat class	**LDA***	**QDA***
Canada Western Red Spring	>98	>94
Canada Western Red Winter	100	100
Canada Prairie Spring Red	100	100
Canada Prairie Spring White	97.4	>94
Canada Western Soft White Spring	100	>94
Canada Western Hard White Spring	>98	>94
Canada Western Extra Strong	>98	86.0
Canada Western Amber Durum	94.0	>94

**LDA, linear discriminant analysis; QDA, quadratic discriminant analysis.*
Source: *Mahesh* et al., *2008*

image. The size of these data was further reduced by PROC STEPDISC (SAS Institute Inc., Cary, NC, USA) and significant features (top 10–100) were extracted. Then classification models were developed by statistical analysis (linear and quadratic) using PROC DISCRIM (SAS Institute Inc., Cary, NC, USA) and standard BPNN classifiers. In another approach they applied the PCA to normalized hyperspectral data and observed that the first three components retained more than 99% variation. The PC scores images corresponding to the first three PCs were used to extract the same wavelet features resulting in 120 features (40×3). The wavelet features from each of the three score images and in combination were used for the development of statistical classifiers. The top 10–60 features from the combined 120 features were also extracted and used in classification. The linear discriminant classifier discriminated more than 99% of samples using the top 90 features from hyperspectral images (Table 15.5). The wavelet energy features contributed more to classification than the entropy features. Rotational invariant features and features at fine resolution gave better classification accuracy. Wavelet features from score images gave poor classification accuracy. The classification accuracy of BPNN was lower compared to the linear discriminant classifier.

In a recent study, Singh *et al.* (2009) developed supervised classification algorithms to classify artificially sprouted and midge-damaged (naturally sprouted) single wheat kernels using NIR hyperspectral imaging. Sprouting in wheat results in poor bread-making quality due to enzymatic activity of α-amylase and is considered as one of the important grading and pricing factors in all western Canadian wheat classes. Singh *et al.* used the same

Table 15.5 Classification accuracy (%) of Canadian wheat by NIR hyperspectral imaging using wavelet features

	Linear discriminant classifier			BPNN* classifier		
Wheat class	**Top 80 features**	**Top 90 features**	**Top 100 features**	**Top 80 features**	**Top 90 features**	**Top 100 features**
Canada Western Red Spring	98.7	98.7	98.7	50.0	56.7	36.7
Canada Western Red Winter	100	100	100	100	100	100
Canada Prairie Spring Red	100	100	100	100	93.3	100
Canada Prairie Spring White	96.7	97.3	97.3	86.7	93.3	93.3
Canada Western Soft White Spring	98.7	99.0	99.0	100	100	100
Canada Western Hard White Spring	98.0	99.0	98.7	86.7	86.7	83.3
Canada Western Extra Strong	99.7	99.7	100.0	93.3	90.0	96.7
Canada Western Amber Durum	99.0	99.3	99.3	96.7	90	93.3

**BPNN, back propagation neural network.*

Source: *Choudhary* et al., *2008*

imaging system described in Mahesh *et al.* (2008) and imaged five non-touching kernels at a time in the wavelength range of 1 000–1 600 nm at 60 evenly spaced wavebands. Hyperspectral data were analyzed by a programming code developed in MATLAB. Single kernels from five non-touching kernel images were obtained by applying automatic thresholding and labeling the kernels. The dimensionality of the single kernel data was reduced by an MVI (Geladi & Grahn, 1996) program developed in MATLAB. The MVI program reshaped the 3-D single-kernel hyperspectral data into a 2-D array in such a way that all kernel pixels at each wavelength became a column vector at each of the 60 wavelengths thus making wavelength a variable and kernel pixels a sample. Principal component analysis was then applied to the reshaped data set and wavelengths corresponding to the highest factor loadings of the first PC were selected as significant. Image features (maximum, minimum, mean, median, standard deviation, and variance) from the significant wavelengths were extracted and given as input to linear, quadratic, and Mahalanobis discriminant classifiers. The discriminant classifiers classified healthy and damaged wheat kernels with maximum classification accuracies of 98.3% and 100.0%, respectively. The first PC score pseudo color images also showed the clear differences between healthy and sprouted kernels and highlighted the damage caused in the germ area of wheat due to sprouting.

The analysis of hyperspectral images of bulk wheat gave very high classification accuracies; however, grain characteristics obtained from bulk samples do not provide any information about the uniformity of the sample. In bulk analysis, characteristics of individual kernels may be lost (Dowell *et al.*, 2006), which may have a significant effect on the end product. In most of the single kernel HSI studies, manually separated non-touching kernels were imaged and analyzed. Separating the touching kernels in bulk samples (single layer) is a challenging task, and there is a need to develop efficient algorithms for successful separation of touching kernels for implementation of rapid single kernel analysis of wheat and other grains.

15.5. CHALLENGES TO THE HSI TECHNOLOGY

Both pushbroom and FPA-based HSI systems have to deal with optical errors and distortions in the acquired images. Pushbroom systems produce geometric distortions called smile and keystone errors. Smile is the curvature distortion of the horizontal spectral lines in the hyperspectral images and keystone is the transformation of a focal plane rectangle into a trapezoid (Lawrence *et al.*, 2003). The FPA-based imaging system produces chromatic aberrations (CA) in the acquired images if the focus is not changed during tunable wavelength scanning (Wang *et al.*, 2006). Lateral chromatic aberration (LCA) is geometric distortion of images caused by different magnifications at each scanning wavelength. Axial chromatic aberration (ACA) results in blurring of images at specific wavelengths due to defocusing. NIR detectors suffer from non-uniform pixel sensitivity due to detector manufacturing issues, operational conditions, and design limitations (Wang & Paliwal, 2006). Due to this defect, even if uniform light is illuminated across the FPA, the intensities recorded by the detector elements vary (Perry & Derreniak, 1993). Therefore, the images should be corrected for pixel sensitivity before correcting the optical errors. At present, research to tackle these problems is ongoing and once these issues are resolved, HSI technology will find more acceptance in several applications in new areas.

15.6. CONCLUSION

Near-infrared hyperspectral imaging has potential for use in rapid classification of wheat into various commercial classes. The technique can be used to analyze both singulated kernels and bulk samples and simultaneously determine other quality parameters of wheat such as protein content, moisture content, oil content, hardness, dockage, and varietal impurities and to detect sprouted, insect-damaged, and fungal-infected kernels in wheat. Dimensionality of scanned hyperspectral data can be reduced by multivariate analysis. Combined features from spectral and image analysis of hyperspectral data tend to give improved classification. Spectral features can be extracted by chemometric approaches used in NIR spectroscopic analysis. Multivariate image analysis using principal component analysis (PCA) is the most common method to reduce the dimensionality of hyperspectral data. Independent component analysis (ICA) and factorial analysis (FA) have also been explored for data reduction and selection of significant wavelengths. Image features such as morphological, textural, and wavelet features extracted from the hyperspectral images have shown high discriminative power. The PC score images are able to identify damage/defects and provide compositional information of the grain samples. Statistical discriminant classifiers (linear and quadratic) have shown better classification than artificial neural networks. Despite the high potential, the distortion of the images in the HSI systems due to optical errors and non-uniform pixel sensitivity pose a challenge for real-time and precise applications of this technique. Once these problems are solved through ongoing research, the HSI technique will be more acceptable in several fields.

NOMENCLATURE

Abbreviations

ACA	axial chromatic aberration
AOTF	acousto–optical tunable filter
A-PAGE	acid polyacrylamide gel electrophoresis analysis
ASCC	average squared canonical correlation
BB	biscuit wheat
BPC	regular bread-making
BPNN	back propagation neural network
BPS	high grade bread-making
BU	wheat for other purposes
CA	chromatic aberrations
CCD	charge-coupled device
CGC	Canadian Grain Commission
CPSR	Canada Prairie Spring Red
CPSW	Canada Prairie Spring White
CWAD	Canada Western Amber Durum
CWES	Canada Western Extra Strong
CWHWS	Canada Western Hard White Spring
CWRS	Canada Western Red Spring
CWRW	Canada Western Red Winter
CWSWS	Canada Western Soft White Spring
DU	durum
ETF	electronically tunable filters
FA	factorial analysis
FDA	factorial discriminant analysis
FPA	focal plane arrays
GIPSA	Grain Inspection, Packers and Stockyards Administration
HDWH	hard white
HDWW	hard white winter
HPLC	high performance liquid chromatography
HRS	hard red spring
HRW	hard red winter
HSI	hyperspectral imaging
ICA	independent component analysis
IEF	gel isoelectric focusing
KVD	kernel visual distinguishability
LCA	lateral chromatic aberration
LCTF	liquid crystal tunable filter

LED	light emitting diode
MVI	multivariate image analysis
NIR	near-infrared spectroscopy
NIT	near-infrared transmittance
PAGE	polyacrylamide gel electrophoresis analysis
PCA	principal component analysis
PCI	peripheral component interconnect
PCR	polymer chain reaction
PDA	penalized discriminant analysis
PLS	partial least squares
RP-HPLC	reverse-phase high performance liquid chromatography
SDS-PAGE	sodium dodecyl sulphate polyacrylamide gel electrophoresis analysis
SNV	standard normal variate
SRW	soft red winter
SSR	simple sequence repeat
STS	sequence tagged site
SVM	support vector machine
SWH	soft white
UNCL	unclassified
USGSA	United States Grain Standard Act
XWHT	mixed

REFERENCES

Archibald, D. D., Thai, C. N., & Dowell, F. E. (1998). Development of short-wavelength near-infrared spectral imaging for grain color classification. In G. E. Meyer, & J. A. DeShazer (Eds.), Precision Agriculture and Biological Quality. *Proceedings of SPIE, Vol. 3543*, 189–198.

Berman, A., Connor, P. M., Whitbourn, L. B., Coward, D. A., Osborne, B. G., & Southan, M. D. (2007). Classification of sound and stained wheat grains using visible and near infrared hyperspectral image analysis. *Journal of Near Infrared Spectroscopy, 15*, 351–358.

Canada Grain Act. (1975). Canada Grain Regulations. *Canada Gazette, Part II, Vol. 109*(No. 14).

CGC (Canadian Grain Commission) (2008) Western Canadian wheat classes. Available online at http:// www.grainscanada.gc.ca/Quality/Wheat/classes-e.htm (accessed November 30, 2008).

Choudhary, R. L., Mahesh, S., Paliwal, J., & Jayas, D. S. (2008). Identification of wheat classes using wavelet features from near infrared hyperspectral images of bulk samples. *Biosystems Engineering, 102*(2), 115–127.

Cocchi, M., Durante, C., Foca, G., Marchetti, A., Tassi, L., & Ulrici, A. (2006). Durum wheat adulteration detection by NIR spectroscopy multivariate calibration. *Talanta, 68*, 1505–1511.

Cracknell, R. L., & Williams, R. M. (2004). Wheat: grading and segregation. In C. Wrigley, H. Corke, & C. E. Walker (Eds.), *Encyclopaedia of grain science III* (pp. 355–362). London, UK: Elsevier.

Dowell, F. E. (2000). Differentiating vitreous and nonvitreous of durum wheat kernels by using near-infrared spectroscopy. *Cereal Chemistry, 77*(2), 155–158.

Dowell, F. E., Maghirang, E. B., Graybosch, R. A., Baenziger, P. S., Baltensperger, D. D., & Hansen, L. E. (2006). An automated near-infrared system for selecting individual kernels based on specific quality characteristics. *Cereal Chemistry, 83*(5), 537–543.

FAO (Food and Agriculture Organization) (2006). *FAO statistical year book, 2005–2006*. Available online at http://www.fao.org/statistics/yearbook/vol_1_1/pdf/b06.pdf (accessed on November 30, 2008).

Geladi, P., & Grahn, H. (1996). *Multivariate image analysis*. Chichester, UK: John Wiley & Sons.

Gorretta, N., Roger, J. M., Aubert, M., Bellon-Maurel, V., Campan, F., & Roumet, P. (2006). Determining vitreousness of durum wheat kernels using near infrared hyperspectral imaging. *Journal of Near Infrared Spectroscopy, 14*(4), 231–239.

Kim, M. S., Chen, Y. R., & Mehl, P. M. (2001). Hyperspectral reflectance and fluorescence imaging system for food quality and safety. *Transactions of the ASAE, 44*(3), 721–729.

Lawrence, K. C., Park, B., Windham, W. R., & Mao, C. (2003). Calibration of a pushbroom hyperspectral imaging system for agricultural inspection. *Transactions of the ASAE, 46*(2), 513–521.

Lookhart, G. L., Marchylo, B. A., Mellish, V. J., Khan, K., Lowe, D. B., & Seguin, L. (1995). Wheat identification in North America. In C. W. Wrigley (Ed.), *Identification of food-grain varieties* (pp. 201–207). St Paul, MN: American Association of Cereal Chemists Inc.

Maghirang, E. B., & Dowell, F. E. (2003). Hardness measurement of bulk wheat by single-kernel visible and near-infrared reflectance spectroscopy. *Cereal Chemistry, 80*(3), 316–322.

Maghirang, E. B., Lookhart, G. L., Bean, S. R., Pierce, R. O., Xie, F., Caley, M. S., & Dowell, F. E. (2006). Comparison of quality characteristics and breadmaking functionality of hard red winter and hard red spring wheat. *Cereal Chemistry, 83*(5), 520–528.

Mahesh, S., Manickavasagan, A., Jayas, D. S., Paliwal, J., & White, N. D. G. (2008). Feasibility of near-infrared hyperspectral imaging to differentiate Canadian wheat classes. *Biosystems Engineering, 101*, 50–57.

Murray, I., & Williams, P. C. (1990). Chemical principles of near-infrared technology. In P. C. Williams, & K. H. Norris (Eds.), *Near-infrared technology in agriculture and food industries* (pp. 17–34). St Paul, MN: American Association of Cereal Chemists.

Pasikatan, M. C., & Dowell, F. E. (2003). Evaluation of a high-speed color sorter for segregation of red and white wheat. *Applied Engineering in Agriculture, 19*(1), 71–76.

Pasikatan, M. C., & Dowell, F. E. (2004). High-speed NIR segregation of high- and low-protein single wheat seeds. *Cereal Chemistry, 81*(1), 145–150.

Perry, D. L., & Dereniak, E. L. (1993). Linear theory of nonuniformity correction in infrared staring sensors. *Optical Engineering, 32*, 1854–1859.

Ram, M. S., Larry, M., & Dowell, F. E. (2004). Natural fluorescence of red and white wheat kernels. *Cereal Chemistry, 81*(2), 244–248.

Shahin, M. A., & Symons, S. J. (2008). Detection of hard vitreous and starchy kernels in amber durum wheat samples using hyperspectral imaging. *NIR News, 19*(5), 16–18.

Singh, C. B., Jayas, D. S., Paliwal, J., & White, N. D. G. (2009). Detection of sprouted and midge-damaged wheat kernels using near-infrared hyperspectral imaging. *Cereal Chemistry, 86*(3), 256–260.

Singh, C. B., Paliwal, J., Jayas, D. S., & White, N. D. G. (2006). *Near-infrared spectroscopy: applications in the grain industry.* Winnipeg: CSBE, Paper No. CSBE06189.

Toews, M. D., Perez-Mendoza, J., Throne, J. E., Dowell, F. E., Maghirang, E., Arthur, F. H., & Cambell, J. F. (2007). Rapid assessment of insect fragments in flour milled from wheat infested with known densities of immature and adult *Sitophilus oryzae* (Coleoptera: Curculionidae). *Journal of Economic Entomology, 100*, 1704–1723.

Uri, N. D., & Beach, E. D. (1997). A note on quality differences and United States/Canadian trade. *Food Policy, 22*(4), 359–367.

USDA (United States Department of Agriculture) (2004). Wheat. In *Grain Inspection Handbook — Book II Grading Procedures*. Washington, DC: Grain Inspection, Packers & Stockyards Administration (GIPSA), USDA.

Utku, H. (2000). Application of the feature selection method to discriminate digitized wheat varieties. *Journal of Food Engineering, 46*, 211–216.

Wang, N., Dowell, F. E., & Zhang, N. (2003). Determining wheat vitreousness using image processing and a neural network. *Transactions of the ASAE, 46*(4), 1143–1150.

Wang, N., Zhang, N., Dowell, F. E., & Pearson, T. (2005). Determining vitreousness of durum wheat using transmitted and reflected images. *Transactions of the ASAE, 48*(1), 219–222.

Wang, W., Morrison, J., & Paliwal, J. (2006). *Correcting axial chromatic aberration in a fixed focal plane near-infrared imaging system*. Winnipeg: CSBE, Paper No. CSBE06127.

Wang, W., & Paliwal, J. (2006). *Calibration and correction for non-uniform pixel sensitivity in digital NIR imaging*. Winnipeg: CSBE, Paper No. MBSK06108.

Wang, W., & Paliwal, J. (2007). Near-infrared spectroscopy and imaging in food quality and safety. *Sensing and Instrumentation for Food Quality and Safety, 1*(4), 193–207.

Williams, P., Sobering, D., & Antoniszyn, J. (1998). Protein testing methods at the Canadian Grain Commission. In D. B. Fowler, W. E. Geddes, A. M. Johnston, & K. R. Preston (Eds.), *Proceedings of the Wheat Protein Symposium (9–10 March 1998)* (pp. 37–47). Saskatoon, SK: University Extension Press, University of Saskatchewan, Saskatoon, SK.

Xie, F., Pearson, T., Dowell, F. E., & Zhang, N. (2004). Detecting vitreous wheat kernels using reflectance and transmittance image analysis. *Cereal Chemistry, 81*(5), 594–597.

Index